The Soil–Human Health Nexus

Advances in Soil Science

Series Editor:
Rattan Lal

For more information about this series, please visit: https://www.crcpress.com/Advances-in-Soil-Science/book-series/CRCADVSOILSCI

The Soil–Human Health Nexus

Edited by
Rattan Lal, Ph.D.

Distinguished University Professor of Soil Science, and
Director, Carbon Management and Sequestration Center,
The Ohio State University
Columbus, USA

CRC Press is an imprint of the
Taylor & Francis Group, an **informa** business

First edition published 2021
by CRC Press
6000 Broken Sound Parkway NW, Suite 300, Boca Raton, FL 33487-2742

and by CRC Press
2 Park Square, Milton Park, Abingdon, Oxon, OX14 4RN

© 2021 Taylor & Francis Group, LLC

CRC Press is an imprint of Taylor & Francis Group, LLC

Library of Congress Cataloging-in-Publication Data
Names: Lal, R., editor.
Title: The soil-human health-nexus / Rattan Lal.
Other titles: Advances in soil science (Boca Raton, Fla.)
Description: First edition. | Boca Raton, FL : CRC Press, 2020. |
Series: Advances in soil science series | Includes bibliographical references
and index. | Summary: "This volume discusses the relation between soil
health and human health in relation to plant, animal, and human nutrition;
pest and pathogen infection from soil; deficiency of micronutrients;
toxicity of heavy metals; geophagy, and other factors"—Provided by publisher.
Identifiers: LCCN 2020047666 (print) | LCCN 2020047667 (ebook) |
ISBN 9780367422141 (hardback) | ISBN 9780367422134 (paperback) |
ISBN 9780367822736 (ebook)
Subjects: LCSH: Soils—Health aspects. | Environmental health. | Soil ecology.
Classification: LCC S591 .S6943 2020 (print) | LCC S591 (ebook) |
DDC 577.5/7—dc23
LC record available at https://lccn.loc.gov/2020047666
LC ebook record available at https://lccn.loc.gov/2020047667

ISBN: 978-0-367-42214-1 (hbk)
ISBN: 978-0-367-42213-4 (pbk)
ISBN: 978-0-367-82273-6 (ebk)

Typeset in Times
by codeMantra

Contents

Preface

The Soil–Human Health Nexus is based on the premise that the "health of soil, plants, animals, people, and the environment is one and indivisible." This phrase, "health of soil, plants, animals and people is indivisible," proposed by Sir Albert Howard (1920) when he was working in Indore, India, has been expanded to include the term "environment" or "ecosystems" in response to realities of the 21st century. The degradation of soil, an integral component of the environment, affects and is affected by the environment. Further, the importance of a strong "soil–human health nexus" can never be overemphasized. Soil health, dependent on activity and species diversity of soil biota, along with the quality and quantity of soil organic matter content in the root zone, affects human health directly and indirectly. Directly, soil health can affect the intake of heavy metals/contaminants and pathogens. Indirectly, soil health can be a source of medicines (e.g., antibiotics) and micronutrients through the food grown on healthy soil. There are 17 essential micronutrients for human health, and their availability in soil is essential to their uptake and concentration in food grains and other agricultural products. Thus, nutrition-sensitive agriculture is an important strategy to enhance human health. Food is also a potent medicine. As an ancient Vedic proverb states, "When food is right, medicine is of no need; when food is wrong, medicine is of no use".

Environmental quality, an important control of human health, comprises a trinity of environmental parameters: soil, water, and air. The quality of water and air is strongly affected by that of soil. By denaturing and filtration of pollutants, soil purifies the water passing through it. Indeed, soil (as a misnomer called "dirt") is the best media to purify dirty water. Therefore, the sustainable management of soil health is critical to quality and renewability of water and a solution to the problems of anoxia, algal bloom, and other eutrophication issues. Soil properties and processes also affect the quality of air as a source or sink of greenhouse gases (CO_2, CH_4, and N_2O). Soil can also be a major source of dust (wind erosion and dust storms), aerosols, and other pollutants. In contrast, conservation and sustainable management of soil is a major sink of atmospheric CO_2 through sequestration into humus as organic carbon, oxidation of methane (CH_4), and formation of pedogenic or secondary carbonates. There is a strong need to formulate and implement a "soil protection policy" to complement the existing Air Quality Act and Water Quality Act. It is the right time to formulate and enact "Soil Quality Act" and respect the "Rights-of-Soil."

As Barry Commoner stated in his book, "The Closing Circle," there is "no way to throw away," and, certainly, everything released into nature must go somewhere. Indeed, the environmental cost (footprint) of the food produced through agricultural intensification is steep and rising due to the leakage of chemicals (i.e., fertilizers and pesticides) into the environment, and impacting all vital components—air, water, soil, vegetation, and biodiversity. Despite these adverse impacts, the use of agrochemicals is increasing because the demand for the production of victuals (i.e., food, feed, fiber, and fuel) is rising for the growing and increasingly affluent world population.

Therefore, the objective of this 15-chapter volume is to deliberate the technological options that reconcile the need for increasing the global supply of essential victuals with the urgency of restoring and sustaining the environment and decreasing risks to human health. Decline in soil health has a cascading impact on human health through pollution of the environment and degradation of the entire food production chain.

This 15-chapter book explains the linkages between soil health and those of plants, animals, people, and ecosystems. Thus, invited authors, including soil scientists, plant scientists, veterinarians, and practicing physicians, represent diverse disciplines. Therefore, I thank all the authors for their outstanding contributions and for sharing their knowledge and experience with the global soil science community.

Preparation of the manuscript, involving collation and synthesis of the literature and interpretation of the data from context-specific situation, is a time-consuming process that requires dedication and commitment. All authors are dedicated professionals committed to excellence. Source of all material cited has been listed and credited in the text, tables, and or figures, and acknowledged.

Thanks are also due to the editorial staff of Taylor and Francis for their timely help and prompt response to numerous questions and queries from the editor and authors. Appreciations and thanks are also due to the staff of the Carbon Management and Sequestration Center (C-MASC) of The Ohio State University for providing support to the flow of the manuscripts between authors and editors and making valuable contributions in timely submission of the book. In this context, special thanks and appreciation are due to Ms. Maggie Weidner-Willis who formatted the text, checked references, followed the guidelines, compiled list of the contributors, and prepared it for final submission to the publisher. Help received from other staff of C-MASC (i.e., Mr. Kyle Sklenka, Ms. Janelle Watts, and Ms. Gabrielle Collier) is also acknowledged. While it is a major challenge to list all those who made direct and indirect contributions toward completion of this book, it is important to thank contributions of all those who supported the completion of this volume. It is also important to build upon the contributions of all those who study properties, processes, and management of soil in relation to human health over the last century and share the knowledge contained in this volume with others from around the world.

Rattan Lal 20th July 2020
The Ohio State University
Columbus, OH, USA

Editor

Rattan Lal, Ph.D., is a Distinguished University Professor of Soil Science and Director of the Carbon Management and Sequestration Center, The Ohio State University, and an Adjunct Professor of University of Iceland. His current research focus is on climate-resilient agriculture, soil carbon sequestration, eco-intensification, enhancing use efficiency of agroecosystems, and sustainable management of soil resources of the tropics. He has received honorary degrees of Doctor of Science from Punjab Agricultural University (2001), the Norwegian University of Life Sciences, Aas (2005), Alecu Russo Balti State University, Moldova (2010), Technical University of Dresden, Germany (2015), University of Lleida, Spain (2017), Gustavus Adolphus College, Saint Peter, Minnesota, and PUCV, Valparaiso, Chile (2019). He was the President of the World Association of the Soil and Water Conservation (1987–1990), the International Soil Tillage Research Organization (1988–1991), the Soil Science Society of America (2005–2007), and the International Union of Soil Science (2017–2018). He was a member of the Federal Advisory Committee on U.S. National Assessment of Climate Change—NCADAC (2010–2013), member of the SERDP Scientific Advisory Board of the US-DOE (2011–2018), Senior Science Advisor to the Global Soil Forum of Institute for Advanced Sustainability Studies, Potsdam, Germany (2010–2015), member of the Advisory Board of Joint Program Initiative of Agriculture, Food Security and Climate Change (FACCE-JPI) of the European Union (2013–2016), and Chair of the Advisory Board of the Institute for Integrated Management of Material Fluxes and Resources of the United Nation University (UNU-FLORES), Dresden, Germany (2014–2019). Prof. Lal was a lead author of IPCC (1998–2000). He has mentored 115 graduate students, 54 postdoctoral researchers, and hosted 174 visiting scholars. He has authored/co-authored 950 refereed journal articles and has written 20 and edited/co-edited 75 books. For 6 years (2014–2019), Reuter Thomson listed him among the world's most influential scientific minds and having citations of publications among top 1% of scientists in agricultural sciences. He is the recipient of the 2018 GCHERA World Agriculture Prize, 2018 Glinka World Soil Prize, 2019 Japan Prize, 2020 World Food Prize, and 2020 Arrell Global Food Innovation Prize in Research.

Contributors

Balkis Aouadi
Department of Physics and Control
Faculty of Food Science
Szent István University
Budapest, Hungary

Eric C. Brevik
Department of Natural Sciences and
Department of Agriculture and Technical
    Studies
Dickinson State University
Dickinson, North Dakota

David Collier
Brody School of Medicine
Department of Pediatrics
East Carolina University
Greenville, North Carolina

Sudarshan Dutta
African Plant Nutrition Institute
Benguérir, Morocco
and
Université Mohamed VI Polytechnique
Benguérir, Morocco

Steven J. Fonte
Department of Soil and Crop Sciences
Colorado State University
Fort Collins, Colorado

André L.C. Franco
Department of Biology
Colorado State University
Fort Collins, Colorado

Sheng Huang
Institute of Plant Science and Resources
Okayama University
Kurashiki, Japan

Marium Husain
James Comprehensive Cancer Center
The Ohio State University
Columbus, Ohio

Marijana Kapović Solomun
Faculty of Forestry
University of Banja Luka
Republic of Srpska, BIH
Banja Luka

Abul M. Kashem
Department of Soil Science
University of Chittagong
Chittagong, Bangladesh

Kathi J. Kemper
College of Medicine
The Ohio State University
Columbus, Ohio

Ashwin Kotnis
Department of Biochemistry
All India Institute of Medical Sciences
Bhopal, India

Zoltan Kovacs
Department of Physics and Control
Faculty of Food Science
Szent István University
Budapest, Hungary

Jeffrey Lakritz
College of Veterinary Medicine
The Ohio State University
Columbus, Ohio

Rattan Lal
Carbon Management and Sequestration Center
The Ohio State University
Columbus, Ohio

Jian Feng Ma
Institute of Plant Science and Resources
Okayama University
Kurashiki, Japan

Kaushik Majumdar
African Plant Nutrition Institute
Benguérir, Morocco
and
Université Mohamed VI Polytechnique
Benguérir, Morocco

Ivan Malušević
Faculty of Forestry
University of Belgrade
Belgrade, Serbia

Robert L. Mikkelsen
African Plant Nutrition Institute
Benguérir, Morocco
and
Université Mohamed VI Polytechnique
Benguérir, Morocco

Vukašin Milćanović
Faculty of Forestry
University of Belgrade
Belgrade, Serbia

Jelena Muncan
Biomeasurement Technology Laboratory
Graduate School of Agricultural Science
Kobe University
Kobe, Japan

T. Scott Murell
African Plant Nutrition Institute
Benguérir, Morocco
and
Université Mohamed VI Polytechnique
Benguérir, Morocco

Thomas Oberthür
African Plant Nutrition Institute
Benguérir, Morocco
and
Université Mohamed VI Polytechnique
Benguérir, Morocco

Dheeraj Panghaal
Department of Soils
CCS Haryana Agricultural University
Hisar, India

Siniša Polovina
Faculty of Forestry
University of Belgrade
Belgrade, Serbia

Boris Radić
Faculty of Forestry
University of Belgrade
Belgrade, Serbia

Ratko Ristić
Faculty of Forestry
University of Belgrade
Belgrade, Serbia

Ernest Semu
Department of Soil and Geological Sciences
Sokoine University of Agriculture
Morogoro, Tanzania

Darryl D. Siemer
Retired Idaho National Laboratory
Consulting Scientist

Bal Ram Singh
Faculty of Environmental Sciences and Natural
    Resources
Norwegian University of Life Sciences
Ås, Norway

Gavin Sulewski
African Plant Nutrition Institute
Benguérir, Morocco

Neeraj H. Tayal
Division of General Internal Medicine and
    Geriatrics
Department of Internal Medicine
The Ohio State University
Columbus, Ohio

Hamisi Tindwa
Department of Soil and Geological Sciences
Sokoine University of Agriculture
Morogoro, Tanzania

Roumiana Tsenkova
Biomeasurement Technology Laboratory
Graduate School of Agricultural Science
Kobe University
Kobe, Japan

Flora Vitalis
Department of Physics and Control
Faculty of Food Science
Szent István University
Budapest, Hungary

Diana H. Wall
Department of Biology and
School of Global Environmental Sustainability
Colorado State University
Fort Collins, Colorado

Ye Xia
College of Food, Agriculture,
    and Environmental Sciences
The Ohio State University
Columbus, Ohio

Slavko Ždrale
Faculty of Medicine
University of East Sarajevo
Republic of Srpska, BIH

Shamie Zingore
African Plant Nutrition Institute
Benguérir, Morocco
and
Université Mohamed VI Polytechnique
Benguérir, Morocco

1 The Soil–Human Health Environment Trinity

Rattan Lal
The Ohio State University

CONTENTS

1.1 INTRODUCTION

The prime soil resources are finite, unequally distributed globally, and fragile upon land misuse and soil mismanagement. Humanity, despite all of its scientific advances and modern discoveries in every aspect of the terrestrial and extra-terrestrial processes, is even more fragile than the soil on which it depends. This truism has been documented by the rapid spread of the COVID-19 pandemic which paralyzed 8 billion people over a short span of 10 months from December 2019 to September 2020, and there is no end in sight. As of September 20, 30 million people had tested positive and ~ 1 million had perished. The mighty COVID-19 does not differentiate between rich or poor nations, developed or developing countries, scientifically advanced or less progressive societies, and even among those with or without the so-called "weapons of mass destruction" (WMDs). As a matter of fact, COVID-19 is the WMD that humans have yet to get their grip on. Surprisingly, humanity also has a short-lived memory. Not only has it forgotten the demise of once thriving civilizations (i.e., Mesopotamia, Indus, Maya, Aztec), but the transient nature of humanity's memory is vividly demonstrated by complacency not of the relatively distant plague, smallpox, and HIV but also more recent Ebola, Zika, and other pandemics of modern era.

The societal lockdown enforced by COVID-19 has not only documented the vulnerability of the economy of all rich and poor nations alike but also how rapidly the atmospheric chemistry responds to the industrial shutdown. By the end of March 2020, within 3 months of onslaught of COVID-19 on humanity, the atmospheric concentration of NO_2 declined over several cities in Europe (Abnett 2020; Holcombe and O'Key 2020). The concentration of NO_2 levels is a contributing factor to coronavirus fatalities (Ogen 2020). Furthermore, seasonal changes in the atmospheric concentration of CO_2, as shown by the Keeling curve, may also be affected by the long-term shut down. It is estimated that the global fossil fuel use would have to decline by 10% for a full year to show up in carbon dioxide concentrations of only about 0.5 parts per million (Monroe 2020). Across the whole year, Betts et al. (2020) estimated that CO_2 levels will rise by 2.48 parts per million (ppm),

1

which is 0.32 ppm smaller than if there had been no lockdown. This decrease is equivalent to 11% of the expected rise (Betts et al. 2020). Disappointingly, carbon dioxide recorded at the Mauna Loa Observatory in Hawaii reached 417 parts per million (ppm) in May 2020, higher than the record of 414.8 ppm set last year (Reuters 2020).

Pollution levels declined across cities in India because of the lockdown related to COVID-19. Mahato, Pal, and Ghosh (2020) reported that concentrations of PM_{10} and $PM_{2.5}$ were reduced by >50% as compared with the prelockdown phase. In comparison with 2019 for the same period, the reduction of PM_{10} and $PM_{2.5}$ is 60% and 39%, respectively. Among other pollutants, NO_2 has been decreased by −53% and CO by −30% during the lockdown phase (Mahato, Pal, and Ghosh 2020). Goswami (2020) outlined four environmental changes in India due to the COVID-19 lockdown: (i) the air quality index in Delhi dropped from 90 in the worst-case scenarios to below 20 in May because 11 million registered cars were taken off the roads, and factories and constructions were stopped, and the $PM_{2.5}$ concentration declined from 71% to 26%; (ii) South Asian river dolphins, listed among endangered species, were spotted at different locations in Kolkata Ghats; (iii) thousands of flamingos were seen in Mumbai; and (iv) the water of the river Ganges was of a drinkable quality (Goswami 2020). Cities of Europe also reported cleaner air during the lockdown (Abnett 2020).

Therefore, prudential management along with a planned reduction in the use of fossil fuel and its substitution by noncarbon fuel sources may bring about the much-needed decline in the increase of the atmospheric concentration of greenhouse gases (GHGs). This is an important lesson that must be fallowed upon by the policymakers after the COVID-19 pandemic. Such a rapid response indicates a possibility that a strong commitment to recarbonization of the terrestrial biosphere (i.e., soil, vegetation, and wetlands) at a global scale (Lal et al. 2018) may also make a measurable impact on the atmospheric chemistry within a foreseeable future. Admittedly, the tragic pandemic of COVID-19 cannot be dubbed having a silver lining, but the fact that atmospheric chemistry can be altered rapidly provides strong motivation toward adoption of negative emission technologies (NET).

Improvement of soil health, through conversion of degraded and agriculturally marginal lands to restorative ecosystems (i.e., by afforestation and set-aside or land retirement programs) and adoption of recommended management practices (RMPs), is a pertinent example of NET. These strategies would restore soil organic matter (SOM) content because of the progressive development of the positive soil/ecosystem carbon (C) budget on a decadal scale. The positive soil C budget would restore soil health and strengthen the provisioning of essential ecosystem services (ESs) for an effective functioning of nature via nutrient cycling, increase in activity and species diversity of biota, water purification and renewability, and above all, sequestration of atmospheric CO_2 into SOM and in the above and below-ground biomass. Therefore, the objective of this chapter is to deliberate the soil–human health–environment trinity, and how restoration of the soil health can simultaneously improve the resilience of humanity through improvement of the environment and functioning of nature. The specific objective of this article is to deliberate the impact of soil degradation caused by SOC depletion on the nutritional quality of crops and on human health.

1.2   SOIL ORGANIC MATTER IN RELATION TO THE HEALTH OF SOIL AND THE ENVIRONMENT

Soil components have an important impact on human health (Nieder, Benbi, and Reichl 2018). An important among these components is soil organic carbon (SOC) content. In conjunction with afforestation of degraded and agriculturally marginal lands, enhancing and sustainable management of SOC stock on depleted agricultural soils is critical to mandatory reduction in the atmospheric concentration of carbon dioxide (CO_2) and other GHGs (i.e., nitrous oxide or N_2O and methane or CH_4). The SOC stock and its management is a critical climate variable (IPCC 2019). Whereas measuring SOC stock requires access to facilities and understanding of the processes, there is also a large variation in SOC concentration and stock at soil scape or landscape levels because of the large variations in control factors both vertically and horizontally even at short distances.

Yet, credible assessments of SOC stocks are needed to understand the impact on a range of pedological processes whose degradation may lead to reversal of ESs (Figure 1.1) which are essential to human and nature. Assessment of SOC stocks and their temporal changes are also needed to evaluate impacts on atmospheric concentrations of GHGs in relation to anthropogenic climate change. A severe decline in SOC stock, below the critical threshold for the specific land use and soil type, can create numerous disservices (Figure 1.2) with adverse impacts on human well-being and nature

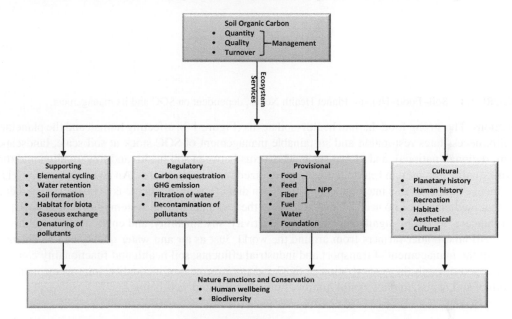

FIGURE 1.1 SOC and ESs (NPP, net primary production; GHG, greenhouse gases).

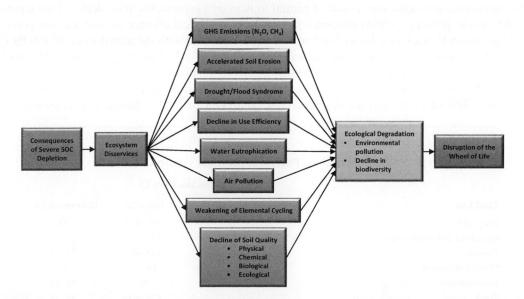

FIGURE 1.2 Ecosystem disservices created by severe depletion of SOC stock leading to the disruption of the Wheel of Life (Birth—Growth—Reproduction—Death—Birth).

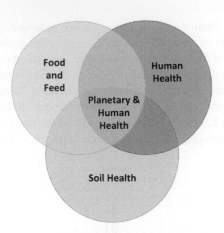

FIGURE 1.3 Soil–Food–Human–Planet Health Nexus, dependent on SOC and its management.

functions. The strong food–human health–soil nexus (Figure 1.3), affecting human and the planetary health, necessitates restoration and sustainable management of SOC stock at soil scale, landscape, farm, regional, national, and global scale. The nexus between soil health and food quality, and that of diet quality and human health, has been recognized in ancient cultures. An Ayurvedic proverb is a pertinent example of this interconnectivity: "When diet is wrong, medicine is of no use; when diet is correct, medicine is of no need." Contextualizing the soil and farming system effects of SOC depletion indicates its strong significance to the productivity, sustainability, and environmental quality of the small land-holder farmers from around the world. Just as air and water quality can be changed through the management of transport and industrial effluents, soil health and functionality can also be changed through the sequestration of SOC and improvement of its quality. Improvements in the health of soil, water, and air would improve human health and well-being.

1.3 HISTORIC DEPLETION OF THE TERRESTRIAL C STOCK

Anthropogenic activities, conversion of natural to managed ecosystems, have depleted the terrestrial C stock. Erb et al. (2018) estimated that in the hypothetical absence of land use, the potential vegetation C stock may be as much as 916 Pg C compared with the actual stock of 450 Pg C (Table 1.1). Similarly, compared with the actual SOC stock of 1,477.2 Pg, the prehistoric SOC stock to 1-M depth was estimated by Buringh (1984) at 2,014.1 Pg (Table 1.2) with the SOC loss of 537 Pg C or 27% of the global SOC stock. The depletion of SOC stock is exacerbated by soil degradation, and ~33% of world soils are degraded (FAO and ITPS 2015). The human impact on soil and

TABLE 1.1

Land Use Type and the Actual vs. Potential Biomass Carbon Stocks

	Biomass Stock (Pg C)		
Land Use	Actual	Potential	Difference (%)
Cropland	10	139–141	93
Grassland and grazing land	119–121	374–379	69–70
Forests	297–368	443–460	22–33
Unused nonforest	16–17	16–17	0
Infrastructure	1	12	92–93
Total	407–476	875–906	48–54

Source: Adapted from Erb et al. (2018).

TABLE 1.2

Prehistoric and the Present Soil Organic Carbon Stock to 1-m Depth

Soil Order	SOC Stock (Pg C) Prehistoric	Present	Percent Lost
Alfisols	388.9	254.8	34.5
Aridsols	33.8	33.0	2.4
Entisols	198.7	144.6	27.2
Histosols	41.5	41.5	0
Inceptisols	279.4	206.6	26.1
Mollisols	165.6	156.9	5.3
Oxisols	255.5	173.0	32.3
Spodosols	62.9	50.7	19.4
Ultisols	165.2	112.9	31.7
Vertisols	40.6	32.2	45.3
Mountain Soils	382.0	281.0	26.4
Total	2,014.1	1,477.2	26.7

Source: Adapted from Buringh (1984).

environment is not new (Balfour 1949) and has been known for millennia (Lal 2013). What is new is the magnitude of the impact of a large population with insatiable demands on soil and environment quality on one hand and on the pace at which the change is occurring is on the other. These are the characteristics of the Anthropocene (Crutzen 2002; Crutzen and Stoermer 2000). The magnitude of change (i.e., extent and severity of degradation, multiple GHG emission, water depletion and eutrophication, and biodiversity loss) is alarming, and the term "big" is of the past. Degradation of soil and the environment has a strong impact on human health and well-being.

1.4 SOIL ORGANIC MATTER DEPLETION AND SOIL/ ENVIRONMENT DEGRADATION

There exists a strong relationship between SOM content and soil quality (Johnston 1986; Mahajan et al. 2019; Page, Dang, and Dalal 2020). SOM originates from the transfer of atmospheric CO_2 into plant biomass and the latter's decomposition by microbial processes into organo-mineral complexes. Because the input of chemical fertilizers alone cannot restore soil quality, adopting the strategy of integrated nutrient management (INM) is essential. The latter involves restoring and sustaining SOM content, maintaining nutrient supply, and improving soil physical properties and processes by following the Law of Return (Howard 1943). The ever-increasing inputs of fertilizers, irrigation, and tillage required to achieving the desired crop yield in soils depleted of their SOM stocks is indicative of the need for a paradigm shift in the management of agroecosystems (Ball, Hargreaves, and Watson 2018). But for the paradigm shift, the rate and magnitude of SOM depletion may increase with the current and projected increase in global temperature (Semenchuk et al. 2019).

1.5 THE SOIL–HUMAN HEALTH NEXUS

Soil health has both direct and indirect effects on human health (Figure 1.4, Oliver and Gregory 2015). Directly, soil can affect human health as a source of infectious pests and pathogens, medicines and antibodies, and by producing healthy and nutritious foods. Indirectly, degraded soil can affect human health by producing nutrient-deficient food of low yield and aggravate both under and malnutrition. Degraded soil, through low yield, can reduce the farm income and aggravate poverty

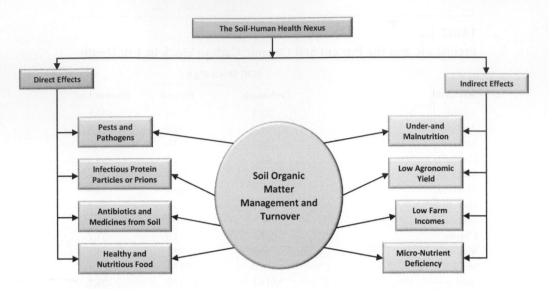

FIGURE 1.4 Direct and indirect effects of soil health on human health through SOM management and turnover.

(Figure 1.4). Because of the strong implications of soil quality to human health (Abrahams 2002), there is a growing concern that soil degradation may cause human malnutrition (Lal 2009; Long 1999; Mayer 1997; Ball, Hargreaves, and Watson 2018; Jack 1998; Jones et al. 2013). Long (1999) observed that the vitamin and mineral content of US and the British food may be decreasing. Contamination of soil with organic pollutants (Cachada et al. 2012) and with heavy metals such as lead can also adversely impact human health (Filippelli and Laidlaw 2010). Deficiency of micronutrients in the degraded soils is an important factor impacting human health (Lal 2009; Abrahams 2002; Oliver and Gregory 2015; Jones et al. 2013). Thus, adoption of sustainable management practices is critical to simultaneously advancing both food and nutritional security (Abrahams 2002; Bindraban et al. 2020; Ball, Hargreaves, and Watson 2018; Watson et al. 2002). There is a strong need to develop nutrition-sensitive agriculture and identify appropriate cropping systems in developing countries for improving human nutrition without degrading soil resources (Amede, Stroud, and Aune 2004).

1.6 INTERCONNECTED VICIOUS CYCLES PREVALENT IN DEGRADED SOILS OF AGROECOSYSTEMS

Despite a strong progress in the amount of total grain and animal-based products since the onset of Green Revolution circa 1960s, there are about 820 million people prone to undernourishment (FAO et al. 2018) and about 1.9 billion affected by malnourishment (WHO 2019). Incidences of both under and malnourishment are aggravated by increase in the extent and severity of soil degradation. There are a series of interconnected degradation processes, the so-called vicious circles that feed one another, which are set-in-motion by land misuse and soil mismanagement (Lal 2020). Oliver and Gregory (2015) and Benton (2019) described scenario analyses that affect human health and food security through soil degradation. Such mutually enhancing circles started by land misuse and soil mismanagement are the basic causes of food-insecurity, malnourishment and of the global warming. The latter is aggravated by the emission of GHGs from degraded soils of agroecosystems and increased by inappropriate use of inputs (i.e., fertilizers, pesticides, tillage, and irrigation) to mitigate the adverse effects of soil degradation. The food-insecurity and global warming circles in turn set-in-motion by others which adversely affect human health, exacerbate soil degradation, and aggravate the effects of pandemics (Lal 2020).

There is a strong link between soil degradation and human health by direct and indirect effects (Figure 1.4), especially those related to the deficiency of micronutrients and protein. A new infectious proteinaceous particle called "proline" has been defined by Prusiner (1982, 1998) and explained by Oliver and Gregory (2015). Deficiency of some and toxicity of other trace elements (heavy metals) can also have a strong adverse effect on human health. Important among heavy metals with adverse impacts on human health are lead (Pb), cadmium (Cd), arsenic (As), and mercury (Hg). Micronutrients whose deficiency affects human health include iron (Fe), iodine (I), cobalt (Co), chromium (Cr), copper (Cu), manganese (Mg), molybdenum (Mn), nickel (Ni), selenium (Se), and Zinc (Zn) (Lal 2009). The last among the series of interconnected circles is the hydrological cycle characterized by the increase in intensity and frequency of the drought/flood syndrome. The desperateness of the growing population leads to encroachment of the remaining forces and natural ecosystems but also to civil strife and unrest and flux of the so-called "soil-refugees" or "climate refugees" (Lal 2020). It is important to identify the entry point(s) to break these self-feeding and self-perpetuating vicious circles. Such interdependent and mutually reinforcing degradation processes have made food systems dependent on the intensive use of external inputs (e.g., fertilizer, pesticides, and irrigation) with adverse effects on the environment and eventually on human health (Benton 2019). Furthermore, these circles have introduced a degree of fragility into the food systems which is vulnerable to numerous factors. Increase in agricultural produce and the cheaper availability of food have aggravated malnourishment and obesity on one hand and soil and environmental degradation is on the other (Benton and Bailey 2019). Therefore, the proposed paradigm shift is to focus on the overall production and efficiency in terms of the number of people that can be fed healthily and sustainably while reducing the inputs (Benton and Bailey 2019).

1.7 CHOICE OF ENTRY POINTS FOR BREAKING THE VICIOUS CIRCLES

The interconnected problems of food insecurity, malnourishment, and soil degradation are widespread in Sub-Saharan Africa (Kim et al. 2019), South Asia (Singh 2015), in the Andean region (Vanek, Jones, and Drinkwater 2016), and elsewhere in the developing world. The problem of degrading soil fertility, poor crop nutrition, and climate change is akin to opening of Pandora's box in developing countries where recurring drought, low soil fertility, and increasing population are posing a daunting challenge (St. Clair and Lynch 2010). Soil degradation being the principal reason of malnourishment (Lal 2009), zero tolerance on soil degradation (Nguyen, Zapata, and Dercon 2010; Lal, Safriel, and Boer 2012; Cowie et al. 2018) is the prudent strategy and the first choice of an entry point to break the vicious circle. Among other important entry points include the following: (i) diversification of the food production systems (Dwivedi et al. 2017) involving judicious integration of diverse crops with trees and livestock to produce healthy diets and reduce diet-related illnesses, (ii) restoration of soil health for improving plant nutrition (El-Ramady et al. 2014) with a view to consider soil and human health as a whole, (iii) integration of plant nutrition research with plant genetics and molecular biology (Cakmak 2002; Clark and Duncan 1991), (iv) use of soil amendments such as biochar (Martos et al. 2020) and phytoremediation techniques (Radwan, Al-Awadhi, and El-Nemr 2000) to enhance crop safety as food and mitigate pollution by agrochemicals, (v) adoption of soil-less culture (Villagra et al. 2012) and to promote urban farming based on the recycling of urban waste (Lal and Stewart 2017), (vi) reduction of air pollution (SO_2, NO_2, O_3) and its effects on shoot–root interactions (Rennenberg, Herschbach, and Polle 1996) and of the minimization of environmental pollution (i.e., acid deposition, excess N deposition, and soil solution) on soil and forest nutrition (Johnson and Ball 1990), and (vii) adoption of innovative techniques of balanced input of macro- and micronutrients in the developed and developing countries to ensure the production of healthy, nutritious, and safe food (Loneragan 1997). Above all, there is an urgent need of safeguarding human and planetary health through critical and an objective review of the global fertilizer sector (Bindraban et al. 2020). Innovative technologies, as entry points to break the vicious circles, are needed to address the interconnectivity of food, soil, and environmental trinity (Figure 1.3).

1.8   ENHANCING HUMAN NUTRITION BY SOIL MANAGEMENT

Soil degradation, low soil quality, inadequate nutrient management, and lack of recycling are among the major concerns of human malnutrition. Yang, Chen, and Feng (2007) observed that more than 374 million people in China suffer from goiter disease, caused by deficiency of iodine (I). These constraints must be alleviated through the adoption of RMPs based on the strategies of INM and restoration of SOM content and stocks. A total of 14 nutrients are essential for plant growth (Lal 2009; Grusak, Broadley, and White 2016), yet most cropland soils, especially those in developing countries, have severe growth-limiting constraints of the nutrient imbalance (deficiency or toxicity). Thus, there is a need for the adoption of innovative options including that of biotechnology to enhance human nutrition (Kishore and Shewmaker 1999). The strategy is to advance human nutrition and alleviate the deficiency of micronutrients along with protein and macronutrients in developing countries without degrading soil resources (Amede, Stroud, and Aune 2004). Micronutrient uptake in food crops can also be increased by the foliar application of Zn and Ca fertilizers and that of Fe and Zn in grains and roots/tuber by plant breeding and genetic engineering (Frossard et al. 2000). In China, Yang, Chen, and Feng (2007) observed that 40% of the total land area is deficient in Fe and Zn, and they proposed an ecologically sound strategy of sustainable flow of micronutrients in soil-plant systems to improve human micronutrient nutrition. Enhancing soil health by improving SOM content is the long-term strategy to ensure an adequate mineral composition in plant produce. Root-induced changes in the rhizosphere can also impact nutrient availability and uptake. Management-induced changes in rhizosphere pH may be 2 units higher or lower than the bulk pH with a strong impact on nutrient uptake. Marschner et al. (1986) observed that low-molecular weight root exudates may also enhance the mobilization of nutrients, such as mobilization of phosphorus in the rhizosphere in white lupin (*Lupinus albus*). Thus, mixed cropping of wheat and white lupin can enhance the uptake of P by wheat (Marschner et al. 1986).

The outline in Figure 1.5 depicts the strategy of "One Health" concept through a sustainable management of soil health. The 3-step process involves (i) integration of crops with trees and livestock to restore SOM content and stock; (ii) restoration of soil, plant health, and animal health over a decadal scale, and improvement in nutritional quality of food leading to improvement in health of plants, animals, and people, which eventually translates into restoration and sustainable management of the overall environment. The major driver of the process depicted in Figure 1.5 is the buildup of SOM content and stock that strengthens ESs while also reconciling the need for advancing food and nutritional quality with the necessity of improving the environment.

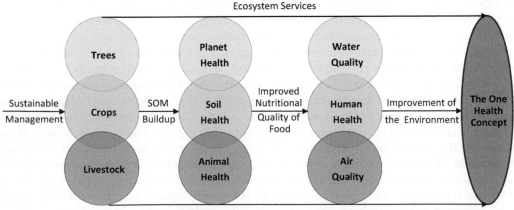

FIGURE 1.5   Integration of crops with trees and livestock and sustainable management of soil is the basis of the "One Health" concept: the health of soils, plants, animals, people, and the environment is one and indivisible.

1.9 RESEARCH AND DEVELOPMENT PRIORITIES

Soil health is being impacted by industrialization and disposal of effluents on soil including the use of wastewater in irrigation (Becerra-Castro et al. 2015). Because of the strong interconnectivity, the future research on the soil management and restoration must focus on the "One Health" initiative (Ohno and Hettiarachchi 2018). The strategy is to enhance environmental quality through multidisciplinary research that links, plant, animal, human, and environmental health through a sustainable management of soil health. Thus, there is a need for a systematic review of the way by which food is produced, processed, distributed, cooked, consumed, and the waste disposed. The entire food chain has been based on producing more and more with utter disregard to the quality of the food, health of the people, and status of the environment. It is important to produce more of a quality food while simultaneously restoring and sustaining the environment: health of soil, water, air, and biodiversity. The fragility of the current food system must be addressed. In addition, the research and education must be reoriented toward the "One Health" concept. Future agriculture must be nutrition-sensitive. Rather than the gross production, quality of the produce must be judged on the basis of nutrients, protein, and vitamins harvested. Furthermore, the produce harvested must have restorative and positive impacts on the environment.

1.10 CONCLUSIONS

The importance of a strong link between soil health and human healthy can never be overemphasized. Nature, properties, and processes of soil affect human health through their effects on plant growth and nutritional quality. Soil properties also affect human health through inhalation, ingestion, and dermal absorption of pollutants and pathogens. Indirect effects of soil on human health are through alterations in the properties and processes of atmosphere, hydrosphere, and the biosphere. Degradation of soil health by overexploitation, extractive farming practices, and soil mismanagement can adversely impact the environment and human health. Soil pollution, with industrial effluents and reuse of wastewater in irrigation, can adversely impact soil and human health. Thus, restoration and sustainable management of SOM content, a principal determinant of soil health, is important to improving quality and quantity of food and feed while improving the environment. Being the foundation of human and environmental health, a sustainable management of soil is also critical to well-being of the present and future generations and of the health of the planet. In pursuant of the "One Health" concept, multidisciplinary research must be pursed to understand interdependence between human, animal, and environmental health. In addition to research and education, it is also pertinent to translate science into action through direct dialogue with the policymakers.

REFERENCES

Abnett, K. 2020. Coronavirus Lockdowns Give Europe's Cities Cleaner Air. *Reuters*. https://www.reuters.com/article/us-health-coronavirus-air-pollution/coronavirus-lockdowns-give-europes-cities-cleaner-air-idUSKBN21G0XA.

Abrahams, P.W. 2002. Soils: Their Implications to Human Health. *Science of the Total Environment* 291, no. 1–3 (May 27): 1–32.

Amede, T., A. Stroud, and J. Aune. 2004. Advancing Human Nutrition without Degrading Land Resources through Modeling Cropping Systems in the Ethiopian Highlands. *Food and Nutrition Bulletin* 25, no. 4: 344–353.

Balfour, E. 1949. *The Living Soil : Evidence of the Importance to Human Health of Soil Vitality, with Special Reference to National Planning*. Rev. ed. London, UK: Faber and Faber Ltd.

Ball, B.C., P.R. Hargreaves, and C.A. Watson. 2018. A Framework of Connections between Soil and People Can Help Improve Sustainability of the Food System and Soil Functions. *Ambio* 47, no. 3 (April 1): 269–283. doi:10.1007/s13280-017-0965-z.

Becerra-Castro, C., A.R. Lopes, I. Vaz-Moreira, E.F. Silva, C.M. Manaia, and O.C. Nunes. 2015. Wastewater Reuse in Irrigation: A Microbiological Perspective on Implications in Soil Fertility and Human and Environmental Health. *Environment International* 75: 117–135. http://www.sciencedirect.com/science/article/pii/S0160412014003237.

Benton, T.G. 2019. Using Scenario Analyses to Address the Future of Food. *EFSA Journal* 17, no. S1 (July 1): e170703.

Benton, T.G., and R. Bailey. 2019. The Paradox of Productivity: Agricultural Productivity Promotes Food System Inefficiency. *Global Sustainability* 2: e6. https://www.cambridge.org/core/article/paradox-of-productivity-agricultural-productivity-promotes-food-system-inefficiency/4D5924AF2AD829EC1719F52B73529CE4.

Betts, R., C. Jones, Y. Jin, R. Keeling, J. Kennedy, J. Knight, and A. Scaife. 2020. Analysis: What Impact Will the Coronavirus Pandemic Have On Atmospheric CO2? *Carbon Brief*. https://www.carbonbrief.org/analysis-what-impact-will-the-coronavirus-pandemic-have-on-atmospheric-co2.

Bindraban, P.S., C.O. Dimkpa, J.C. White, F.A. Franklin, A. Melse-Boonstra, N. Koele, R. Pandey, et al. 2020. Safeguarding Human and Planetary Health Demands a Fertilizer Sector Transformation. *Plants, People, Planet*: ppp3.10098. https://onlinelibrary.wiley.com/doi/abs/10.1002/ppp3.10098.

Buringh, P. 1984. Organic Carbon in Soils of the World. In *The Role of Terrestrial Vegetation in the Global Carbon Cycle: Measurement by Remote Sensing*, 23, 91–109. New York: Wiley.

Cachada, A., P. Pato, T. Rocha-Santos, E.F. da Silva, and A.C. Duarte. 2012. Levels, Sources and Potential Human Health Risks of Organic Pollutants in Urban Soils. *Science of The Total Environment* 430: 184–192. http://www.sciencedirect.com/science/article/pii/S004896971200647X.

Cakmak, I. 2002. Plant Nutrition Research: Priorities to Meet Human Needs for Food in Sustainable Ways. *Plant and Soil* 247: 3–24. Springer.

St. Clair, S.B., and J.P. Lynch. 2010. The Opening of Pandora's Box: Climate Change Impacts on Soil Fertility and Crop Nutrition in Developing Countries. *Plant and Soil* 335, no. 1: 101–115.

Clark, R.B., and R.R. Duncan. 1991. Improvement of Plant Mineral Nutrition through Breeding. *Field Crops Research* 27, no. 3 (October 1): 219–240.

Cowie, A.L., B.J. Orr, V.M. Castillo Sanchez, P. Chasek, N.D. Crossman, A. Erlewein, G. Louwagie, et al. 2018. Land in Balance: The Scientific Conceptual Framework for Land Degradation Neutrality. *Environmental Science & Policy* 79: 25–35. http://www.sciencedirect.com/science/article/pii/S1462901117308146.

Crutzen, P.J. 2002. Geology of Mankind. *Nature* 415: 23. Nature Publishing Group.

Crutzen, P.J., and E.F. Stoermer. 2000. The Anthropocene, Global Change. International Geosphere–Biosphere Programme (IGBP).

Dwivedi, S.L., E.T. Lammerts van Bueren, S. Ceccarelli, S. Grando, H.D. Upadhyaya, and R. Ortiz. 2017. Diversifying Food Systems in the Pursuit of Sustainable Food Production and Healthy Diets. *Trends in Plant Science* 22: 842–856. Elsevier Ltd.

El-Ramady, H.R., T.A. Alshaal, M. Amer, É. Domokos-Szabolcsy, N. Elhawat, J. Prokisch, and M. Fári. 2014. Soil Quality and Plant Nutrition. In *Sustainable Agriculture Reviews*, vol. 14, 345–447. Cham, Switzerland: Springer.

Erb, K.H., T. Kastner, C. Plutzar, A.L.S. Bais, N. Carvalhais, T. Fetzel, S. Gingrich, et al. 2018. Unexpectedly Large Impact of Forest Management and Grazing on Global Vegetation Biomass. *Nature* 553, no. 7686 (January 4): 73–76.

FAO, IFAD, UNICEF, WFP, and WHO. 2018. *The State of Food Security and Nutrition in the World 2018: Building Climate Resilience for Food Security and Nutrition*. Rome, Italy: FAO.

FAO, and ITPS. 2015. *Status of the World's Soil Resources (SWSR) – Main Report*. Rome, Italy: Food and Agriculture Organization of the United Nations and Intergovernmental Technical Panel on Soils.

Filippelli, G.M., and M.A. Laidlaw. 2010. The Elephant in the Playground: Confronting Lead-Contaminated Soils as an Important Source of Lead Burdens to Urban Populations. *Perspectives in Biology and Medicine* 53, no. 1: 31–45.

Frossard, E., M. Bucher, F. Mächler, A. Mozafar, and R. Hurrell. 2000. Potential for Increasing the Content and Bioavailability of Fe, Zn and Ca in Plants for Human Nutrition. *Journal of the Science of Food and Agriculture* 80, no. 7: 861–879. https://onlinelibrary.wiley.com/doi/abs/10.1002/(SICI)1097-0010(20000515)80:7%3C861::AID-JSFA601%3E3.0.CO;2-P?casa_token=y0SxDWWkrMAAAAAA:UrrJSPaXfaElvNz_ISGKLoN_dlow9qPl_Aw0qJwm4pQ1uFoLshgN2rV4ThKSSXkLcvGS9jwZ24QGLHg.

Goswami, K. 2020. Covid-19: 4 Unbelievable Environmental Changes Seen in India since Lockdown. *India Today*. https://www.indiatoday.in/education-today/gk-current-affairs/story/covid-19-4-vital-environmental-changes-evidenced-in-india-since-lockdown-1673726-2020-05-02.

Grusak, M.A., M.R. Broadley, and P.J. White. 2016. Plant Macro- and Micronutrient Minerals. In *ELS*, 1–6. John Wiley & Sons, Ltd. doi: 10.1002/9780470015902.a0001306.pub2

Holcombe, M., and S. O'Key. 2020. Satellite Images Show Less Pollution over the US as Coronavirus Shuts down Public Places. *The Mercury News*. https://www.mercurynews.com/2020/03/23/satellite-images-show-less-pollution-over-the-us-as-coronavirus-shuts-down-public-places/.

Howard, A. 1943. *An Agricultural Testament.* New York: Oxford University Press.

IPCC. 2019. *Special Report on Climate Change and Land.* Ed. P.R. Shukla, J. Skea, E. Calvo Buendia, V. Masson-Delmotte, H.-O. Pörtner, D.C. Roberts, P. Zhai, et al. Geneva, Switzerland: Intergovernmental Panel on Climate Change. https://www.ipcc.ch/srccl/.

Jack, A. 1998. *Nutrition under Siege.* Vol. 1. Becket, MA: One Peaceful World.

Johnson, D.W., and J.T. Ball. 1990. Environmental Pollution and Impacts on Soils and Forests Nutrition in North America. *Water, Air, & Soil Pollution* 54, no. 1 (March): 3–20.

Johnston, A.E. 1986. Soil Organic Matter, Effects on Soils and Crops. *Soil Use and Management* 2, no. 3: 97–105.

Jones, D.L., P. Cross, P.J.A. Withers, T.H. DeLuca, D.A. Robinson, R.S. Quilliam, I.M. Harris, D.R. Chadwick, and G. Edwards-Jones. 2013. REVIEW: Nutrient Stripping: The Global Disparity between Food Security and Soil Nutrient Stocks. *Journal of Applied Ecology* 50, no. 4 (August 1): 851–862. doi:10.1111/1365-2664.12089.

Kim, J., N.M. Mason, S. Snapp, and F. Wu. 2019. Does Sustainable Intensification of Maize Production Enhance Child Nutrition? Evidence from Rural Tanzania. *Agricultural Economics* 50, no. 6 (November 29): 723–734. https://onlinelibrary.wiley.com/doi/abs/10.1111/agec.12520.

Kishore, G.M., and C. Shewmaker. 1999. Biotechnology: Enhancing Human Nutrition in Developing and Developed Worlds. *Proceedings of the National Academy of Sciences of the United States of America* 96, no. 11: 5968–5972. https://www.pnas.org/content/96/11/5968.short.

Lal, R. 2009. Soil Degradation as a Reason for Inadequate Human Nutrition. *Food Security* 1, no. 1 (February): 45–57.

Lal, R. 2013. Soil and Sanskriti. *Journal of Indian Society of Soil Science* no. 61: 267–274.

Lal, R. 2020. Soil Science Beyond COVID-19. *Journal of Soil and Water Conservation* 75, no. 4: 1–3.

Lal, R., U. Safriel, and B. Boer. 2012. Zero Net Land Degradation. *UNCCD Position Paper for Rio+20.* Bonn, Germany.

Lal, R., P. Smith, H.F. Jungkunst, W.J. Mitsch, J. Lehmann, P.K.R. Nair, A.B. McBratney, et al. 2018. The Carbon Sequestration Potential of Terrestrial Ecosystems. *Journal of Soil and Water Conservation* 73, no. 6: A145–A152.

Lal, R., and B.A. Stewart (Eds.). 2017. *Urban Soils.* CRC Press. doi: 10.1201/9781315154251

Loneragan, J.F. 1997. Plant Nutrition in the 20th and Perspectives for the 21st Century. Plant and Soil 196: 163–174. doi: 10.1023/A:1004208621263

Long, C. 1999. Is Chemical Farming Making Our Food Less Nutritious. *Organic Gardening* 46(6): 12.

Mahajan, N.C., K. Mrunalini, K.S. Krishna Prasad, R.K. Naresh, and L. Sirisha. 2019. Soil Quality Indicators, Building Soil Organic Matter and Microbial Derived Inputs to Soil Organic Matter under Conservation Agriculture Ecosystem: A Review. *International Journal of Current Microbiology and Applied Sciences* 8, no. 2: 1859–1879.

Mahato, S., S. Pal, and K.G. Ghosh. 2020. Effect of Lockdown amid COVID-19 Pandemic on Air Quality of the Megacity Delhi, India. *Science of The Total Environment* 730: 139086. http://www.sciencedirect.com/science/article/pii/S0048969720326036.

Marschner, H., V. Römheld, W.J. Horst, and P. Martin. 1986. Root-induced Changes in the Rhizosphere: Importance for the Mineral Nutrition of Plants. *Zeitschrift Für Pflanzenernährung Und Bodenkunde* 149, no. 4: 441–456.

Martos, S., S. Mattana, A. Ribas, E. Albanell, and X. Domene. 2020. Biochar Application as a Win-Win Strategy to Mitigate Soil Nitrate Pollution without Compromising Crop Yields: A Case Study in a Mediterranean Calcareous Soil. *Journal of Soils and Sediments* 20, no. 1 (January 1): 220–233.

Mayer, A.M. 1997. Historical Changes in the Mineral Content of Fruits and Vegetables. *British Food Journal* 99(6): 207–211. doi: 10.1108/00070709710181540.

Monroe, R. 2020. What Does It Take for the Coronavirus (or Other Major Economic Events) to Affect Global Carbon Dioxide Readings? *Scripps Institution of Oceanography.* https://scripps.ucsd.edu/programs/keelingcurve/2020/03/11/what-does-it-take-for-the-coronavirus-or-other-major-economic-events-to-affect-global-carbon-dioxide-readings/.

Nguyen, M.-L., F. Zapata, and G. Dercon. 2010. "Zero-Tolerance" on Land Degradation for Sustainable Intensification of Agricultural Production. In: Zdruli P., M. Pagliai, S. Kapur, A. Faz Cano (eds.), *Land Degradation and Desertification: Assessment, Mitigation and Remediation*, 37–47. Dordrecht: Springer Netherlands.

Nieder, R., D.K. Benbi, and F.-X. Reichl. 2018. *Soil Components and Human Health.* Heidelberg: Springer Netherlands.

Ogen, Y. 2020. Assessing Nitrogen Dioxide (NO_2) Levels as a Contributing Factor to Coronavirus (COVID-19) Fatality. *Science of the Total Environment* 726 (April): 138605.

Ohno, T., and G.M. Hettiarachchi. 2018. Soil Chemistry and the One Health Initiative: Introduction to the Special Section. *Journal of Environmental Quality* 47, no. 6 (November): 1305–1309. https://www.ncbi. nlm.nih.gov/pubmed/30512058.

Oliver, M.A., and P.J. Gregory. 2015. Soil, Food Security and Human Health: A Review. *European Journal of Soil Science* 66, no. 2 (March 1): 257–276. http://doi.wiley.com/10.1111/ejss.12216.

Page, K.L., Y.P. Dang, and R.C. Dalal. 2020. The Ability of Conservation Agriculture to Conserve Soil Organic Carbon and the Subsequent Impact on Soil Physical, Chemical, and Biological Properties and Yield. *Frontiers in Sustainable Food Systems*. https://www.frontiersin.org/article/10.3389/fsufs.2020.00031.

Prusiner, S.B. 1982. Novel Proteinaceous Infectious Particles Cause Scrapie. *Science* 216, no. 4542: 136–144.

Prusiner, S.B. 1998. Prions. *Proceedings of the National Academy of Sciences of the United States of America* 95: 13363–13383. National Academy of Sciences.

Radwan, S.S., H. Al-Awadhi, and I.M. El-Nemr. 2000. Cropping as a Phytoremediation Practice for Oily Desert Soil with Reference to Crop Safety as Food. *International Journal of Phytoremediation* 2, no. 4: 383–396.

Rennenberg, H., C. Herschbach, and A. Polle. 1996. Consequences of Air Pollution on Shoot-Root Interactions. *Journal of Plant Physiology* 148, no. 3–4 (January 1): 296–301.

Reuters. 2020. Global CO2 Measurement Hits Record High in May Despite Pandemic Induced Lockdown. *News 18*. https://www.news18.com/news/world/global-co2-measurement-hits-record-high-in-may-despite-pandemic-induced-lockdown-2653289.html.

Semenchuk, P.R., E.J. Krab, M. Hedenström, C.A. Phillips, F.J. Ancin-Murguzur, and E.J. Cooper. 2019. Soil Organic Carbon Depletion and Degradation in Surface Soil after Long-Term Non-Growing Season Warming in High Arctic Svalbard. *Science of The Total Environment* 646: 158–167.

Singh, G. 2015. Agriculture Diversification for Food, Nutrition, Livelihood and Environmental Security: Challenges and Opportunities. *Indian Journal of Agronomy* 60, no. 2: 172–184. http://www.indianjournals.com/ijor.aspx?target=ijor:ija&volume=60&issue=2&article=001.

Vanek, S.J., A.D. Jones, and L.E. Drinkwater. 2016. Coupling of Soil Regeneration, Food Security, and Nutrition Outcomes in Andean Subsistence Agroecosystems. *Food Security* 8, no. 4 (August 1): 727–742.

Villagra, E.L., M.G. Minervini, E.Z. Brandán, and R.R. Fernández. 2012. Effects of Mineral Nutrition and Biofertilization on Lettuce Production under Conventional and Soilless Culture. *Acta Horticulturae* 947, no. 947 (May 1): 395–400. https://www.actahort.org/books/947/947_51.htm.

Watson, C.A., D. Atkinson, P. Gosling, L.R. Jackson, and F.W. Rayns. 2002. Managing Soil Fertility in Organic Farming Systems. *Soil Use and Management* 18, no. s1 (September 1): 239–247. doi:10.1111/j.1475-2743.2002. tb00265.x.

WHO. 2019. Malnutrition Is a World Health Crisis. *World Health Organization: Nutrition*. https://www.who. int/nutrition/topics/world-food-day-2019-malnutrition-world-health-crisis/en/.

Yang, X.E., W.R. Chen, and Y. Feng. 2007. Improving Human Micronutrient Nutrition through Biofortification in the Soil-Plant System: China as a Case Study. *Environmental Geochemistry and Health* 29, no. 5: 413–428.

2 Health of Soil, Plants, Animals, and People

Kathi J. Kemper, Ye Xia, Jeffrey Lakritz, and Rattan Lal
The Ohio State University

CONTENTS

2.1 OVERVIEW

This chapter is based on the concept of "One Health," a global, collaborative, multidisciplinary effort to attain optimal health for plants, animals, people, and the environment. "One Health" recognizes the fundamental interconnectedness of ecosystems, plants, animals, and humans (Figure 2.1). The health or toxicity of one affects all.

1. Healthy soil is a vital, living ecosystem that sustains productive populations of soil organisms. Healthy soil contains nutrients that sustain all land-based plant, animal, and human life; it is also the source of some of our most important antibiotics. On the other hand, soil that has been compacted, eroded, contaminated (e.g., with toxic heavy metals), or which

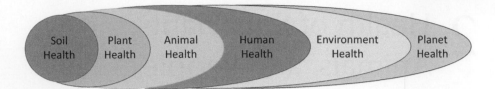

FIGURE 2.1 Relationships between ecosystem, plant, animal, and human health. The "One Health" concept: the continuum between the health of the soil, planet, and its ecosphere (e.g., hydrosphere, atmosphere, biosphere, and pedosphere).

has been subjected to pesticides or imbalances of healthy and pathogenic organisms (e.g., helminths, protozoa, bacteria, fungi, and viruses), reduces the viability of plants, animals, and humans. Optimal agricultural and forestry practices can also promote soil's role as a carbon sink, offsetting anthropogenic increases in atmospheric carbon dioxide.

2. Healthy plants are critical for ecosystem and for the security of food for animal and human beings. Unhealthy plants cannot provide optimal nutrients for livestock or humans; plants contaminated with heavy metals (e.g., arsenic) or pathogenic organisms (e.g., *E. coli* or *Salmonella*) can be toxic for humans and livestock.

3. Healthy animals promote healthy soil, contribute nutrients to healthy plants, and promote optimal human nutrition. Livestock consume significant amounts of grain and produce methane, a potent greenhouse gas; animal waste can also contaminate food crops with pathogenic organisms leading to human illnesses.

4. The human population has grown from 2.5 billion in 1950 to over 7 billion today, and is expected to exceed 9 billion by 2050, accelerating expanding needs for healthy crops and livestock. At the same time, the amount of land available to grow food has been contracted by expanding cities, rising sea levels, desertification, and salinization as the climate crisis escalates; furthermore, rising CO_2 levels promote rapid carbohydrate accumulation in plants without concomitant increases in protein and essential minerals, leading to a relative decrease in the nutritional value of crops and animals that feed on those crops. Human behaviors such as mining, agriculture, and industry can result in toxic dumping that adversely affects the health of soil, plants, and animals and reduces biodiversity. Human health is inextricably linked to the health of the soil as well as the diverse plants and animals with whom we share this planet, and humans can act to protect this ecosystem or destroy it.

2.2   ONE HEALTH

"One Health" (Figure 2.1) has been defined as "the collaborative effort of multiple disciplines—working locally, nationally, and globally—to attain optimal health for people, animals, plants, and our environment" (King et al. 2008). "One Health" has a rich history. In the Bible, Moses (Anon. 2002) recommended that the people observe as they entered Canaan,

> See what the land is like and whether the people who live there are strong or weak, few or many. What kind of land do they live in? Is it good or bad? How is the soil? Is it fertile or poor? Are there trees on it or not? Do your best to bring back some fruit of the land.

> *(Numbers 13:18–20)*

For example, in the 19th century, prominent human biomedical researchers such as the microbiologist, Robert Koch, and immunologist, Paul Ehrlich, worked in the fields of both animal and human medicine (Shomaker, Green, and Yandow 2013). The father of cell biology, Rudolf Virchow, also recognized the relationship between human and animal health and passed it along to his student, Sir William Osler, who brought the concept to North America. Sir Albert Howard, President of the National Academy of Sciences in India, said in the early 20th century, "Health of soil, plants, animal, and people is one and indivisible" (Howard 1943, 1945). The concept of "One Health" has continued into the 21st century by both veterinarians and physicians who recognized that many animal illnesses have much to teach physicians about human health.

The US Centers for Disease Control and Prevention (CDC) has a "One Health" office (https://www.cdc.gov/onehealth/index), recognizing approximately six out of every ten infectious diseases in humans are spread from animals. For example, nearly 75% of infectious diseases emerging in the past 30 years that affect humans originated in animals. These include HIV/AIDS (from West African primates), global influenza H1N1 pandemic (from swine), SARS (from Chinese bats and palm civets), and Covid-19 (Coronavirus Disease 2019). In addition, the "One Health" Commission (https://www.onehealthcommission.org) is rooted into the understanding of the *interdependence* of human and natural systems, and addressing the loss of biodiversity, pollution, climate change, ecosystem function, and the social determinants of health, including social justice. The "One Health" Commission is a partnership of multiple organizations including the American Medical Association, the American Public Health Association, the Association of American Medical Colleges, the American Veterinary Medical Association, and the Association of American Veterinary Medical Colleges. As awareness of the impact of climate change on human health has grown, the concept of "One Health" has expanded to include not only animal and human health but also the health of ecosystems.

The relationships between ecosystems, plant, animal, and human health are complex and interconnected (Figure 2.1). The driving force of all ecosystems is plant-based photosynthesis, which uses the sun's energy to convert atmospheric CO_2 to energy-rich compounds, which can then transfer solar energy via carbon-based compounds to other organisms within the ecosystem (Reicosky and Janzen 2018). Although the flow of carbon-based energy drives ecosystem function, the reserve of stored soil carbon provides resilience to the system, serving as a buffer of energy and nutrients and sustaining the system during stresses and shortfalls (Reicosky and Janzen 2018). The ecosystem (of which soil is a vital component) affects human health both directly and indirectly. For example, waterborne illnesses kill an estimated 2.5 million people globally annually. While some of these illnesses come directly from contaminated water, some come indirectly from contaminated soil, which contaminates water, thereby affecting food plants and livestock.

2.3 SOIL HEALTH

2.3.1 Soil and Life

Soil is the essence of all terrestrial life. Charles Kellogg (1938) stated that "essentially all life depends upon the soil—There can be no life without soil and no soil without life; they have evolved together." The rhizosphere, root-soil interphase at the nanoscale, is the only site in the universe which possesses the divine powers of resurrecting death into life. In "A Sand County Almanac," Wendell Berry (1977) reiterated that "The soil is the great connector of our lives, the source and destination of it all." Aldo Leopold (1949) opined that "Land, then, is not merely soil; it is a fountain of energy flowing through a circuit of soils, plants, and animals." John Muir (1838–1914) emphasized the importance of interconnectivity by stating that "When we try to pick out anything by itself, we find it hitched to everything else in the Universe" (Muir 1911). It is this interconnectivity, indicating that everything is connected to everything else (Commoner 1971), which is depicted in Figure 2.1 from the scale of a pedon to that of the planet.

2.3.2 Soil Health

Terms "soil quality" and "soil health," often used interchangeably, are indeed different (Lal 2016). Soil quality denotes soil functions or what it does (Doran and Werner 1990; Doran and Zeiss 2000; Doran 1990). In comparison, soil health is defined as the "capacity of soil to function as a vital living system to sustain biological productivity, maintain environment quality, and promote plant, animal, and human health (Doran, Sarrantonio, and Liebig 1996; Doran and Zeiss 2000). In essence, soil health is the biological component of soil quality (Figures 2.2 and 2.3).

Soil quality can be quantified by measurements of soil properties and processes. In comparison, soil health is still a qualitative aspect, and efforts need to be made to quantify its attributes and

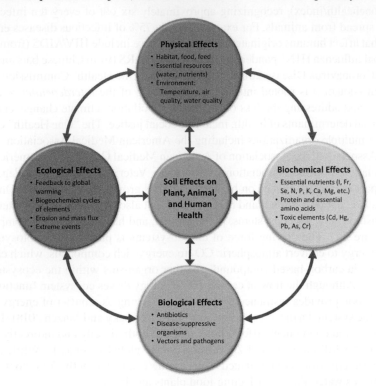

FIGURE 2.2 Soil effects on plant, animal, and human health (Kemper, Lakritz, and Lal 2017).

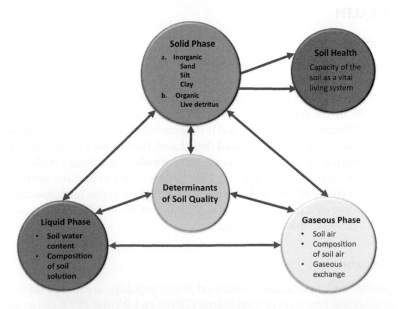

FIGURE 2.3 Relationship between soil quality and soil health.

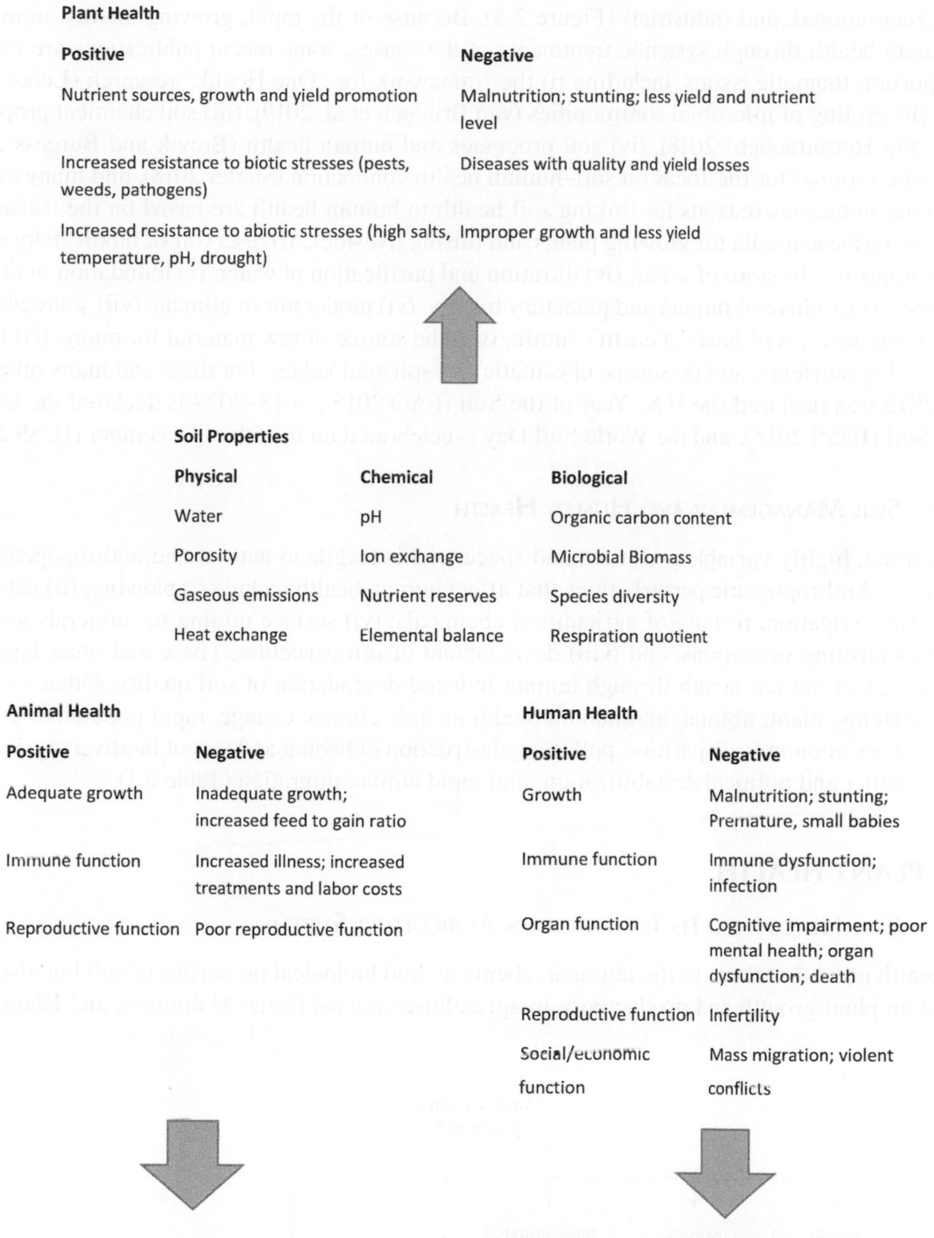

Plant Health

Positive	Negative
Nutrient sources, growth and yield promotion	Malnutrition; stunting; less yield and nutrient level
Increased resistance to biotic stresses (pests, weeds, pathogens)	Diseases with quality and yield losses
Increased resistance to abiotic stresses (high salts, temperature, pH, drought)	Improper growth and less yield

Soil Properties

Physical	Chemical	Biological
Water	pH	Organic carbon content
Porosity	Ion exchange	Microbial Biomass
Gaseous emissions	Nutrient reserves	Species diversity
Heat exchange	Elemental balance	Respiration quotient

Animal Health

Positive	Negative
Adequate growth	Inadequate growth; increased feed to gain ratio
Immune function	Increased illness; increased treatments and labor costs
Reproductive function	Poor reproductive function

Human Health

Positive	Negative
Growth	Malnutrition; stunting; Premature, small babies
Immune function	Immune dysfunction; infection
Organ function	Cognitive impairment; poor mental health; organ dysfunction; death
Reproductive function	Infertility
Social/economic function	Mass migration; violent conflicts

FIGURE 2.4 Positive and negative effects of soil properties on plant, animal, and human health (Kemper, Lakritz, and Lal 2019).

indicators. However, being a biological component, it is highly dynamic and strongly influenced by natural and anthropogenic perturbations. Soil health has important positive and negative effects on the health of plants, animals, and human health (Figure 2.4).

2.3.3 SOIL HEALTH AND HUMAN HEALTH

There is a strong interest in managing human health by a holistic approach of managing the environment in which humans live. Thus, there is a growing interest in linking human health to that of soil health. Soil health can be managed through specific land uses (e.g., arable, pastoral, silvicultural, urban, recreational, and industrial) (Figure 2.5). Because of the rapid, growing interest in managing human health through systemic treatments of the causes, some recent publications are focused on important thematic issues, including (i) the framework for "One Health" research (Lebov et al. 2017), (ii) cycling of microbial communities (van Bruggen et al. 2019), (iii) soil chemical properties (Ohno and Hettiarachchi 2018), (iv) soil processes and human health (Brevik and Burgess 2014), and (v) the rational for the focus on soil–human health connection (Shafer 2018), and many others.

Among numerous reasons for linking soil health to human health are based on the truism that soil is the (i) basic media for growing plants and raising livestock, ii) reservoir of biodiversity and of the germplasm, (iii) store of water, (iv) filtration and purification of water, (v) foundation of all civil structures, (v) archive of human and planetary history, (vi) moderator of climate, (vii) source/sink of greenhouse gases, (vii) basis of earth's future, (vii) the source of raw material for many, (ix) media for recycling nutrients, and (x) source of esthetic and spiritual values. For these and many other reasons, 2015 was declared the U.N. Year of the Soil (FAO 2015), 2015–2024 is declared the Decade of the Soil (IUSS 2015), and the World Soil Day is celebrated on the 5th of December (IUSS 2002).

2.3.4 Soil Management and Human Health

Soil is finite, highly variable over time and space, and is fragile to natural and anthropogenic perturbations. Anthropogenic perturbations that affect human health include (i) plowing, (ii) vehicular traffic, (iii) irrigation, (iv) use of agricultural chemicals, (vi) surface mining for minerals and topsoil, (vii) farming operations, and (viii) development of infrastructure. These and other land uses can also affect human health through human-induced degradation of soil quality. Other stressors on ecosystems, plant, animal, and human health include climate change, rapid population growth, urbanization, economic disparities, pollution, destruction of habitat and loss of biodiversity, international conflict and political destabilization, and rapid human migration (Table 2.1).

2.4 PLANT HEALTH

2.4.1 Soil Health and Its Indicators in Agriculture System

Soil health not only relates to the physical, chemical, and biological properties of soil but also to its impact on plant growth and productivity in agriculture systems (Aziz, Mahmood, and Islam 2013;

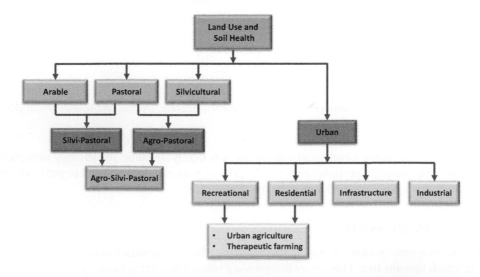

FIGURE 2.5 Strategies of managing soil health through specific functions or land uses.

TABLE 2.1

Stressors on Ecosystems, Plant, Animal, and Human Health

1. Climate change, warming, ocean acidification and sea-level rise, severe storms, flooding; fires
2. Rapid human population growth from over 7 billion to an anticipated 9.5 billion by 2050
3. Urbanization
4. Economic disparities
5. Pollution
6. Habitat destruction and loss of biodiversity
7. International conflict, terrorism, political destabilization, revolution
8. Rapid human migration (e.g., refugees from climate, conflict, and famine) and international travel allowing rapid spread of diseases

Bal et al. 2013; Yucel et al. 2015). Soil health is strongly influenced by the concentration and composition of the soil organic carbon (SOC) and nitrogen (Lal 2007).

Improved soil health helps to enhance the resilience of croplands to extreme weather events and biotic stress, and directly impacts the local jobs and the economic stability of rural communities (Clay et al. 2015). In conventional systems, however, soil health is often a lesser priority amid others, such as higher yields, protection of crops from pests and diseases, and most importantly net income. Commonly, the accepted indicators of soil health include a combination of microbial biomass, biodiversity and efficiency, aggregate stability, drainage and water retention, nutrient cycling, active and passive pools of soil organic matter (SOM), soil enzyme activities, etc. (Romig, Garlynd, and Harris 1997; Islam and Weil 2000; Aziz, Mahmood, and Islam 2013). Of all the indicators, SOM is a composite indicator of healthy soil and contributes to food security and ecosystem resilience (Zebarth et al. 1999; Islam and Weil 2000; Aziz et al. 2015; Lal 2017). The balance among the natural processes of primary production, decomposition, and transformation contributes to the SOM formation (Wilson 1991). SOC, through its impact on soil health and crop productivity, has a high societal value (Lal 2014).

Agricultural management practices known to improve the SOM and overall soil health include but are not limited to the no-till, crop rotation, cover crops, soil amendments, and beneficial microbes, such as the plant growth promotion bacteria and arbuscular mycorrhizal fungi (AMF) (González et al. 2003; Slepetiene 2001). Soil organic amendments have been used as the source of organic matters and essential nutrients (Celik, Ortas, and Kilic 2004; Tejada et al. 2008). Soils with high and labile organic matter may provide a better environment for enhanced microbial diversity and efficiency. In addition, soil with high organic matter contents has a greater abundance of water-stable aggregates with higher exchange capacity for plant nutrients. Complex organic nutrients need to be broken down by diverse microbes for plants to utilize them efficiently (Tejada et al. 2008). A higher SOM content also improves the water-infiltration and water-holding capacity, providing plant resilience to drought and other stresses to stabilize long-term farm economics (Celik, Ortas, and Kilic 2004; Lal 2007, 2017).

2.4.2 Plant Pathogens and Their Caused Destructive Diseases on Plants

There are many microbial pathogens that can infect different plant species in the greenhouse and field conditions, which can cause huge yield and economic losses. Of all the problems related to the pathogen-caused plant diseases, those caused by the soilborne pathogens can be the most frustrating ones (Table 2.2). In the fields, the destructive soilborne pathogens, such as *Rhizoctonia* spp, *Sclerotinia* spp., *Pythium* spp., *Fusarium* spp., and *Verticillium* spp., could cause infections on plants separately or together (Gao et al. 2014). These pathogens can survive in the soil even when their usual host crops are no longer present. For example, *Rhizoctonia solani* is a soilborne pathogen and can cause severe disease pressure and huge yield losses on tomato plants and many other plant species (Meyer et al. 2006). The related destructive plant disease symptoms caused by these soil-borne pathogens include but are not limited to the damping-off of seedlings, aerial blights, root rots, and stem cankers on the plants. Currently, many plant species are short of resistant varieties to confer resistance to most of the soilborne pathogens; therefore, the chemical control by the fungicide/bactericide application is the main approach (Stevens et al. 2003). However, the long residue of these chemicals can lead to the harmful effects on human health and environment. To reduce chemical applications, the combined use of fungicides/bactericides with proper agricultural practices could potentially enhance the efficacy of the related disease controls (Hüberli et al. 2015). Many studies have shown that plant diseases caused by the soilborne pathogens, such as the *R. solani*, could also

TABLE 2.2
Soilborne Infectious Diseases of Plants, Animals, and Humans

Category	Examples in Plants	Examples in Animals	Examples in Humans
Helminths	Phylum Nematoda, nematodes, such as root-knot nematode (*Meloidogyne* species); cyst nematode (*Heterodera* species); Sting Nematode; needle nematodes (*Longidorus africanus*); stubby root nematodes (*Paratrichodorus* species)	Nematodes—Hookworm (*Bunostomum phlebotomum*); Trichostrongylid nematodes (*Haemonchus contortus, Trichostrongylus* spp., *Ostertagia ostertagia, Cooperia* spp.), Trichuriasis, *Ascaris suum, Dictyocaulus viviparous, Protostrongylus* spp., *Muellerius* spp.	Ascariasis (roundworms), Echinococcosis, Hookworm, Strongyloidosis (threadworms), Trichuriasis (whipworms), Trichinellosis (Trichinosis)
Protozoa	Naked amebae, testate amebae, flagellates, ciliates, microsporidia, and sporozoans	*Eimeria* spp., *Entamoeba histolytica*, Cryptosporidiosis, Giardiasis	*Entamoeba histolytica* (Amebiasis), Cryptosporidiosis, Cylcosporiasis, Giardiasis, Toxoplasmosis
Fungi (oomycetes)	*Microdochium, Sclerotinia, Fusarium, Rhizoctonia, Verticillium, Monosporascus, Aphanomyces, Bremia, Phytophthora, Pythium, Macrophomina, Plasmodiophora* and *Spongospora*	Endophytes (*Neotyphodium coenophialum*), *Trichphyton verrucosum*, Histoplasmosis, Blastomycosis, Cryptococcosis, Coccidioidomycosis, Aspergillosis, Mucormycosis	Aspergillosis, Blastomycosis, Coccidioidomycosis (Valley fever), Histoplasmosis, Sporotrichosis, Mucormycosis, Nocardia
Bacteria	*Rhizobium radiobacter* (*Agrobacterium tumefaciens*), *Erwinia, Rhizomonas, Streptomyces. Pseudomonas* and *Xanthomonas* for short time in soil.	*Listeria monocytogenes, Salmonella* spp., *E. coli, Campylobacter* spp., *Yersinia entercolitica, Enterococcus* spp., *M. avium,* spp. *pseudotuberculosis,* Clostridial intoxications (*Clostridium chauvoeii,* Cl. novyi Type D, Cl. Septicum, Cl. Sordelli, Cl. Perfringens Type C, Cl. Perfringens Type D, Cl. Tetani, Cl. Botulinum, *Bacillus anthracis,* Leptospirosis, Dermatophilosis, *Actinobacillus ligneresii, Actinomyces bovis, Brucella abortus*	Actinomycosis, Anthrax, *Bacillus cereus, Campylobacter, Clostridium botulinum* (Botulism), *C. perfringens* (gas gangrene), *Clostridium tetanus, Coxiella burnetti* (Q fever), *Escherichia coli* (Enterohemorrhagic, Enterotoxigenic, Verotoxigenic, and Enteropathogenic types), Legionella, Leptospirosis, *Listeria monocytogenes,* Lyme disease, Melioidosis, Pseudomonas, Salmonellosis (including typhoid fever and non-typhoid salmonella infections), Shigellosis, Tularemia, *Yersinia pestis* (plague)

(*Continued*)

TABLE 2.2 (*Continued*)

Soilborne Infectious Diseases of Plants, Animals, and Humans

Category	Examples in Plants	Examples in Animals	Examples in Humans
Viruses	Lettuce necrotic stunt virus (LNSV); Beet necrotic yellow vein virus (BNYVV; genus *Benyvirus*), Potatomop-topvirus (PMTV; genus *Pomovirus*); *Furovirus* (type species soilborne wheat mosaic virus); *Bymovirus* (type species Barley yellow mosaic virus); Peanut clump virus (PCV; genus *Pecluvirus*); infection of lettuce big-vein associated virus (LBVaV; genus *Varicosavirus*); Mirafiori lettuce big-vein virus (MiLBVV; genus *Ophiovirus*); Viruses of the genera *Tombusvirus* (cucumber *necrosis virus*; CNV), *Carmovirus* (i.e., melon necrotic spot virus, MNSV); Grapevine Fan leaf virus (GFLV, genus *Nepovirus*); Tobacco rattle virus (TRV, genus *Tobravirus*).	Coronavirus, *Vesicular stomatitis, Rotavirus,* Contagious ecthyma virus, Goat pox, Peste des petitis ruminants,	Enteroviruses (poliovirus, coxsackieviruses, and echoviruses), *Hantavirus* (hantavirus pulmonary syndrome or hemorrhagic fever with renal syndrome)

be suppressed by either adding SOM for general suppression, or/and specific microbial antagonists by beneficial microbes, such as the bacteria *Bacillus* species (Weller et al. 2002; Stone, Scheuerell, and Darby 2004) and AMF (Abdel-Fattah et al. 2011).

2.4.3 Agricultural Practices Affect Plant and Soil Health

More and more studies have investigated the comparisons between conventional and sustainable farming systems, such as the organic farming system. These studies reveal that sustainable agricultural practices can influence soil composition and functions, which could be associated with higher biological efficiency and diversity of the soil-associated microbial communities (Monokrousos et al. 2006; Esperschütz et al. 2007; Araujo et al. 2009; R. Li et al. 2012). To conserve system biodiversity, including the soil microbes, sustainable practices excluding the pesticide use could reduce their impacts on the nontarget organisms (Kuepper and Gegner 2004). Our previous studies examined the culturable endophytic bacterial and fungal communities from four economically important crops, including corn, tomato, pepper, and melon, grown under low-input organic or conventional management practices. Our results showed that the plant culturable endophytic bacterial and fungal communities were significantly impacted by different agricultural production practices (conventional vs. organic farming system) (Xia et al. 2015, 2019).

The agricultural practices, such as crop rotation with cover crops and green manures, conservation tillage, and microbial products usage contribute to building an active, diverse, disease-suppressive soil associated with certain microbial communities, which have been shown to effectively reduce most of the major soilborne diseases and enhance plant tolerance to abiotic/biotic stress and yield within their respective agro-ecosystems (Marzano 2012).

2.4.4 PLANT- AND SOIL-ASSOCIATED MICROBIOMES FOR PLANT GROWTH AND HEALTH

The plant-associated microbiomes are the naturally occurring complex community of microbes, which together with the soil microbiomes, and play critical roles in plant growth, resilience, and yield (Berendsen, Pieterse, and Bakker 2012). The whole microbial community (not a single microbe) functions together to help plants grow and develop and, meanwhile, protects the plants from diverse biotic and abiotic stresses. Among them, a group of fungi, known as AMF, develop strong relationships with their associated plant hosts. The fossils revealed that AMF existed for more than 460 million years even before plants first appeared in lands. AMF could infect and colonize the plant roots and form specialized structures within plant tissues. Approximately, 80% of the land plants form symbiotic relationships with AMF (Parniske 2008). In this relationship, the fungi benefit from carbon resources provided by the plants, and in most cases, AMF enhance plant health through different mechanisms. For instance, AMF can contribute to the soil aggregate formation (Rillig 2004), nutrient cycling in soil (Lehmann et al. 2014), and have been primarily recognized for their ability to scavenge phosphorous (P) and enhance plants' P uptake for their growth and development (Parniske 2008). Agricultural practices, such as cover crops, promote the maintenance and diversity of AMF in soil (Säle et al. 2015; Bowles et al. 2017). AMF have been reported to inhibit diverse plant pathogens, such as soilborne fungal pathogen *Rhizoctonia solani* and *Phytophthora parasitica* (Abdel-Fattah et al. 2011).

Plant disease suppression by beneficial microbes, such as AMF and plant growth-promoting rhizobacteria (PGPR), could be through the competition for habitat and nutrient and/or antagonism by producing antibiotic compounds (Haas and Défago 2005). Plant disease suppression could also be achieved by the induction of systemic-induced host resistance (ISR). In ISR, nonpathogenic microorganisms, often below ground dwellers, can prime the plants to resist against diverse diseases or pests. This phenomenon, also known as priming of defenses, has been shown to occur upon the AMF colonization (Pozo and Azcón-Aguilar 2007; Bennett, Bever, and Deane Bowers 2009) as well as in response to the other fungi and bacteria (Kloepper, Ryu, and Zhang 2004; Shoresh and Harman 2008). For instance, AMF priming has been shown to alleviate the effects of soilborne pathogens, leaf pathogens, and insect damage; some of these responses could be mediated through the changes in both salicylic acid and jasmonic acid-dependent pathways (Pozo and Azcón-Aguilar 2007). Priming is also important for plant tolerance to drought and ozone stresses (Beckers and Conrath 2007).

By harnessing the soil- and plant-associated microbiomes, the approaches to benefit plants for their growth and health become a sustainable strategy for plant disease control; considerable research has been carried out in the related fields in recent years with quickly developing next-generation sequencing and bioinformatics techniques (Lundberg et al. 2012, 2013). For instance, researchers found that the key bacterial taxa in disease-suppressive soil such as Proteobacteria (*Pseudomonadaceae*, *Burkholderiaceae*, *Xanthomonadales*) and Firmicutes (*Lactobacillaceae*) as well as gene-nonribosomal peptide synthetases were highly associated with the disease suppression activity to fungal root pathogen—*Rhizoctonia solani* that caused severe disease in sugar beets (Mendes et al. 2011; Mendes, Garbeva, and Raaijmakers 2013). Another study showed that the Huanglongbing disease of citrus was associated with the shifts in the microbiome of citrus rhizosphere (Trivedi et al. 2012). Some researchers also reported that the disease-resistant cultivars of tomato recruited more diverse microbes than susceptible cultivars by the culturable approach (Upreti and Thomas 2015). All these studies indicated that plants could manipulate the microbial consortia from soil for the protection against the infections upon pathogens and benefit their health and growth.

In 2012, two research groups from US and Germany investigated the *Arabidopsis*-associated root endophytic microbiome, and both of them found that there was a very similar core group of microbes in (endophytic part) and around (rhizosphere part) the roots of *Arabidopsis* plants grown in different soils, which meant that *Arabidopsis* plant might need to recruit some specific microbes to provide a particular function. They also showed that plant genotype might play important roles in selecting these associated microbes (Lundberg et al. 2012; Bulgarelli et al., 2012).

There are multiscale interactions among plants, microbes, soil, and their interactive environment. Characterization of the soil- and plant-associated microbiome has been developed rapidly in recent years (Beckers et al. 2016; Lundberg et al. 2012, 2013). The plant–microbiome–soil associations have been documented for almost a century (Perotti 1926), but many aspects remain poorly understood. For example, the genetic and molecular mechanisms on how the host plants recruit mutual microbes but not pathogenic microbes and protect themselves from the pathogenic microbes in the environment are not well known (Lundberg et al. 2012, 2013; Tkacz and Poole 2015; Jansson and Hofmockel 2018). Currently, the environmental factors and the management practices that affect the plant- and soil-associated microbial community and the mechanisms in which plant–microbiome-soil associations occur in agro-ecosystems for plant production need more in-depth studies. Nonetheless, some progress has been made in recent years, and new studies are needed to build upon recent accomplishments (Rosenblueth and Martínez-Romero 2006; Reinhold-Hurek and Hurek 2011; Bowles et al. 2017).

2.5 ANIMAL HEALTH

2.5.1 LIVESTOCK PRODUCTION-SOIL AND PLANT HEALTH

Livestock are crucial to the food security of hundreds of millions of humans. Livestock production provides economic growth and, through income and employment, brings food security and improved health and resiliency to humans through the consumption of high protein, high-energy foods, containing micronutrient profiles that closely align with those needed for growth, development, and overall health of humans (Ussiri and Lal 2018). Based upon the predictions of human population increases of approximately 25% by the year 2050 (to 9 billion), livestock production will need to increase dramatically to provide enough high-quality protein for the anticipated growth of human population (Alexandratos and Bruinsma 2012).

Livestock production provides benefits to humans through expanding the availability of high-quality protein foods (meat, milk, cheese, and eggs), energy, and essential micronutrients (Ussiri and Lal 2018). Ruminant production centers on feeding plant-based feeds (grass, forage, crop residues and by-products, and cereal grains) to ruminants. Livestock convert low-quality, less-palatable feed to high-quality, protein-rich, palatable, and nutrient-dense foods for humans.

The global livestock industries also produce manure and urine useful as fertilizer for soil to replenish important microbes, organic matter, and nutrients required in the efficient production of healthy plants and plant products. Together, livestock production provides high-quality nutrition and food security, income, and fertilizer to produce products more efficiently, resulting in decreased prices and improved access of human population.

2.5.2 EFFECTS OF SOIL ON ANIMAL HEALTH

"Soil health" is a critical aspect of the earth's biosphere and, by association, agricultural productivity and ecosystem sustainability (Doran and Zeiss 2000). Healthy soil has been defined as "the capacity of soil to function as a vital living system to *sustain biological productivity*, maintain environmental quality and promote plant, animal and human health" (Larkin 2015; Doran, Sarrantonio, and Liebig 1996; Doran and Zeiss 2000; van Bruggen and Semenov 2000). Soil health's prominent role in the efficient production of healthy animals centers around the maintenance of soil organic material, moisture, air, *nutrients* (*organic carbon*, *micro-*, and *macrominerals*), healthy roots to support soil and large and diverse population of organisms (bacteria, *actinomycetes*, fungi, algae, protozoa, nematodes, *Annelids*, mites, insects, and soil-dwelling animals) (Ussiri and Lal 2018). Together, these organisms decompose organic materials, facilitate nutrient cycling, and mitigate heavy metal contamination (Shrestha, Bellitürk, and Görres 2019). These features allow soil to collect and filter water and provide space for roots and, in some cases, outcompete microbial pathogens. Factors involved in the soil stress include physical (excess cycling of rain and drought, soil compaction), chemical (altered soil osmolality, pH, salinity, and soil contamination with heavy metals, pesticides, and hydrocarbons), and biological (nutrient imbalances, pathogenic organisms, or organisms with a competitive advantages) (Doran and Zeiss 2000). Soil resilience declines as a result of the alterations in concurrent physical, chemical, biochemical factors or may result from a single stressor. Soil health decline has been associated with tillage and compaction of the soil by farm equipment or large ruminants. Agricultural productivity has led to modifications of the environment, climate change, and air and water quality declines, and resulted in critical changes in soil health. With the myriad changes in ecosystems, biodiversity losses are significant.

The critical nature of soil health with a human population that is expanding centers on the provision of food. Approximately 90% of the caloric intakes of humans are derived from soil (Brevik 2013). Healthy soils are the basis of food production for animals and humans (Franzluebbers and Sharpley 2017). Healthy soil increases the production of forage, grains, and crop-associated byproducts essential for an efficient animal production of food (meat, milk, eggs and other products) and fiber. In addition, livestock are also fed byproducts derived from human food, fiber, and fuel production, providing valuable feed sources that are not consumed by humans (Capper, Berger, and Brashears 2013). Some estimates suggest that approximately 80% of livestock feed is derived from products that are not consumed by humans (Mottet et al. 2017). In this analysis, Mottet et al. (2017) found that 46% of 6 billion tons of dry matter consumed by livestock (in 2010) comprise grass and leaves. Nineteen percent were from crop residues, 5% of byproducts, and 8% of fodder crops. Only 5% and 13% of 6 billion tons of dry matter were from oilseed cakes and grains, respectively (Mottet et al. 2017). A significant proportion of global livestock production utilizes extensive means to feed cattle (grazing systems). These grasslands are generally associated with soil that is not sufficient for conversion to cropland (Teague 2018).

Livestock productivity is intimately linked to the essential nutrients accumulated within the forage, cereal grains, and oilseeds from nutrient-sufficient soil. Animal forages and grains produced from healthy soil contain most of the macro- and trace minerals required for human health (Suttle 2010; Haes et al. 2012). Consumption of animal-based foods by humans requires the growth of healthy forage, cereal, or oil-seed crops to meet the animal and human mineral requirements. To raise healthy, productive, and nutrient-sufficient animals for human food, soil health and soil trace- and macromineral sufficiency is important.

Minerals accumulated by plants vary greatly based upon plant genetics (varieties), the presence of weeds, soil microorganism content and diversity, environmental differences (solar radiation, heat, cold, water), and geologic types of soil present (Franzluebbers and Sharpley 2017; Cox and Amador 2018; Pérez-Gutiérrez and Kumar 2019). The efficiency at which soil minerals are absorbed by plants is critical to human and animal health. Soil mineral content and bioavailability are influenced by soil type, soil pH, electrical conductivity, crop type, SOC, management strategy, and environment (temperature, humidity, rainfall, irrigation) (Sigua, Chase, and Albano 2014; Adviento-Borbe et al. 2006). Some minerals (cobalt, nickel, manganese) accumulate in grass growing on acidic soils. However, alkaline soils favor the absorption of other nutrients such as molybdenum (Suttle 2010). In addition, management factors such as fertilization, augmenting SOM (cover crops), irrigation, and other factors such as nutrient competition for absorption play important roles in the availability of nutrients in plants and therefore within the animals that consume them (Suttle 2010). Soils deficient in copper are associated with copper deficiency in animals (Table 2.3). Soils with elevated zinc, molybdenum, iron, or sulfur can also lead to copper deficiency in livestock. In a recent examination of beef cattle with reproductive failure, liver biopsies demonstrated tissue copper concentrations

TABLE 2.3

Soil Minerals and Effects of Their Deficiencies on Plant, Animal, and Human Health

Deficient Soil Mineral	Plant Disease and Physiological Deficiency	Animal Disease	Human Disease
Boron	Plant size can be significantly reduced. Root tips often become swollen and discolored. Leaves eventually become brittle and may be curled with yellow spotting	Seizure-like activity in goats dosed with 2 g of borate fertilizer; nephritis in other species	Osteoporosis, poor memory and concentration; weak muscles; aging skin
Calcium	Young leaves can be affected before older leaves become distorted or dead. Bud development can be inhibited, and root tips may die. The fruit development can be affected.	Milk fever/parturient paresis (hypocalcemia),	Rickets, osteomalacia
Chromium	The seed germination, plant growth and yield can be decreased. The functional enzymatic activities can be inactivated. The photosynthesis can be impaired and nutrient and oxidative activities can be imbalanced.	Glucose tolerance factor (Cr 3+) (deficiency); 20 mg L^{-1} at 48 h—100% mortality (Carp) Toxic level 100-fold>nutrient requirements Cr 6+—toxicity suspected in cattle exposed to oil-field wastes	Hypertension, high cholesterol, impaired glucose tolerance
Copper	Stunted growth. Leaves can become limp, curl, or drop. Seed stalks also become limp and bend over.	Copper toxicity: sheep most susceptible species Acute toxic dose (sheep) 9–20 mg kg^{-1} BW; acute toxic dose (Cattle 200 mg kg^{-1}) depends on feed levels of Mo, S, Fe, Zn Chronic toxic dose (sheep)—long-term accumulation; Liver histological lesions @ 350 mg Cu kg^{-1} liver; Hepatic necrosis @ 1,000 mg Cu Kg^{-1} liver Hemolysis, hemoglobinuria, hemoglobinuria	Anemia, neuropathies; arrhythmias; osteoporosis
Iron	Interveinal chlorosis on the youngest leaves.	Iron deficiency anemia Iron toxicity—reactive oxidation (liver, heart, pancreas) 200 mg/piglet—33% survival to 21 days 16 mg kg^{-1} ferrous fumarate—foals (3/5 foals died in 7 days) Salers cattle—Hemosiderosis	Anemia, fatigue, weakened immune function, inattention, prematurity
Iodine	It can affect plant vigor and fruit quality and development.	Hypothyroidism, goiter, weak, inability to regulate temperature	Hypothyroidism; goiter; cretinism
Magnesium	low growth and leaves turn pale yellow, sometimes just on the outer edges. New leave may be yellow with dark spots.	Deficiency — "tetany" Toxicosis—reduced muscle tone, recumbence, bradycardia, CNS depression, death High acute doses—diarrhea	Constipation; irritability; anxiety; fatigue; weakness; loss of appetite; nausea; muscle spasms; tremors; apathy; ataxia; hypertension

(Continued)

TABLE 2.3 (*Continued*)
Soil Minerals and Effects of Their Deficiencies on Plant, Animal, and Human Health

Deficient Soil Mineral	Plant Disease and Physiological Deficiency	Animal Disease	Human Disease
Manganese	Growth slows. Younger leaves turn pale yellow, often starting between veins. May develop dark or dead spots. Leaves, shoots, and fruit diminished in size. Failure to bloom	Limited toxicity in large animals. Iron deficiency; serum Fe, may occur with high intakes of Mn. Low dietary iron >anemia (young animals more sensitive)	Impaired carbohydrate metabolism; poor wound healing
Molybdenum	Generally, it can cause "whiptail"-like elongated, misshapen central leaves.	Deficiency—Copper toxicosis or loss of critical enzymatic reactions (aldehyde oxidase, sulfite oxidase, xanthine oxidase); Toxicosis—Copper deficiency	Intellectual disability; seizures; opisthotonus; renal failure; coma
Sulfur	New growth turns pale yellow, older growth stays green. Stunts growth.	Sulfur—amino acid metabolism. Toxicosis—H_2S, isothiocyanates (thyroid function), hemolytic anemia, cerebro-cortical necrosis	Weak connective tissue; impaired detoxification; fatigue; impaired glucose metabolism
Selenium	Plants will be sensitive to UV and oxidative stress. Plants may also show increased senescence and decreased plant growth.	White Muscle Disease; impaired growth and reproductive capacity; retained placenta; impaired immune function	Muscle weakness; myalgia; heart failure; cardiac arrhythmia; Keshan disease; hypothyroidism; increased susceptibility to infection; reproductive difficulties; increased risk of cancer; Kashim-Beck (bone and joint) disease
Zinc	Yellowing between veins of new growth. Terminal (end) leaves may form a rosette.	Deficiency—Carbonic anhydrase; protein, carbohydrate, nucleic acid, lipid metabolism. Toxicity—Excess Zn—Cu deficiency;	Acrodermatitis enteropathica; diarrhea; respiratory infections; growth retardation; delayed sexual development; alopecia; anorexia; skin rashes; increased susceptibility to infections
Phosphorus (P)	Leaves could turn the reddish-purple color. Leaves could become almost black. Fruit or seed production could be reduced.	P deficiency—bright alert downers (hypophosphatemia)—muscle weakness, hemolysis. Toxicity—urolithiasis; nutritional bone diseases	
Potassium (K)	Leaves may become scorched around the edges and/or look wilted. Interveinal chlorosis could be developed.	Acid base balance regulation	
Nitrogen (N)	Plants grow slowly. Older leaves, generally at the bottom of the plant, will turn yellow. The remaining foliage is often light green. Stems may also turn yellow and may become spindly.	Hyperkalemic periodic paralysis in horses	

Source: Adapted from Kemper, Lakritz, and Lal (2019). Animal disease information obtained from National Research Council (2005).

below the sufficient range (9.66 ppm; reference: 40–650 ppm) and elevated liver molybdenum (7.42 ppm; reference: 1.80–4.70 ppm). The ratio of liver Cu:Mo was approximately 1.3 (reference: 3:1–4:1). Liver levels of zinc and iron were within the range and considered adequate in these animals (Corah and Dargatz 1996).

While zinc is a common soil mineral, the bioavailability of Zn to the crop can be limited to less than that required for optimal plant growth of specific crops (Haes et al. 2012). This can occur with inadequate release of mineral from the subsoil horizon, limited acquisition from mineral complexes, or depletion of soil by farming without replenishment (soils treated only with nitrogen–phosphate–potassium fertilizers). In general, the concentration of minerals in the plants reflects those of the soil they were produced from. By extension, mineral deficiencies in animals reflect the concentrations accumulated in plants and, therefore, provided by the soil (Suttle 2010). Specific deficiencies of essential minerals result in the reduced growth of crops and lower crop production at harvest (Suttle 2010). This limits the commodities necessary for growth, health, and reproductive functions of the animals. Trace mineral deficiencies in soils and plants have fostered the supplementation of animal feeds with relatively high concentrations of zinc to maintain productivity.

Trace minerals are critical for growth, health, and productivity because they function in a wide array of important biological processes in plants, animals, and humans (Table 2.3). Some minerals such as calcium, magnesium, zinc, sulfur, silica and phosphorous play critical roles in the musculoskeletal system. Sodium, chloride, potassium, calcium, and magnesium regulate body fluid volumes, acid base balance, and function of excitable tissues (Suttle 2010). Many minerals are involved in critical enzyme catalytic functions. For example, over 300 unique enzymes contain zinc atoms that play roles in protein, nucleic acid, carbohydrate, and lipid metabolism (King and Cousins 2006; Chasapis et al. 2012). As such, trace mineral deficiency in animals often results in poor growth, reduced dry matter intake and feed- to-gain ratios (López-Alonso 2012). If untreated, animals suffer a variety of maladies such as perinatal morbidity and mortality, pneumonia, or reduced growth (calves, lambs, kids), retained placenta and mastitis in dairy cows, reduced hatchability in quail, and decreased resistance to gastrointestinal (GI) parasites in chickens (Suttle 2010). The myopathies in young ruminants, wool growth and color in sheep and lambs, weak or nonviable calves should suggest a potential relationship with trace mineral deficiencies, or imbalances along with other common disorders (Suttle 2010). One mineral deficiency may often be associated with the excess of others.

With most mineral deficiencies, the clinical presentation may resemble those associated with a variety of maladies (nonspecific, poor growth due to limited milk production of dam, immunodeficiency) (Suttle 2010). Other deficiencies can be strongly suggestive of mineral deficiency. Iodine, not an essential nutrient for most plants, is generally supplemented to the diet. In some regions, elevated soil or water iodine can increase the plant acquisition of iodine and result in uptake of iodine by grazing this forage by the dam. Low thyroid hormone levels (T3, T4) resulting in low viability of newborn young stock are observed in some locations associated with low soil iodine (low dietary iodine), plant goitrogens consumed during pregnancy, and selenium deficiency (Figure 2.6a) (Suttle 2010). In one group of Boer goats, noniodized salt was provided to pregnant goats. When the kids were born, most were too weak to stand and nurse. Some were born dead and others died within days of birth. Affected kids were born with enlarged, symmetrical masses in the cranial cervical region (Figure 2.6a). Autopsy of the dead kids confirmed that these masses were the thyroids. (Image provided by Dr. Angie Sherman.)

Selenium is also not an absolute requirement for the growth of crops. However, Se is an absolute requirement for health, growth, and productivity of humans and animals (Figure 2.6b). In some regions, soil selenium deficiency commonly results in nutritional myopathies (skeletal and cardiac) in animals and cardiovascular, endocrine, musculoskeletal, neurodegenerative diseases, and infertility in humans (Shreenath and Dooley 2020). Selenium can also be accumulated from seleniferous soils by some plants (rice, corn, soybeans) (Song et al. 2018). Grazing plants produced on seleniferous soils can also result in Se toxicity (Raisbeck 2000).

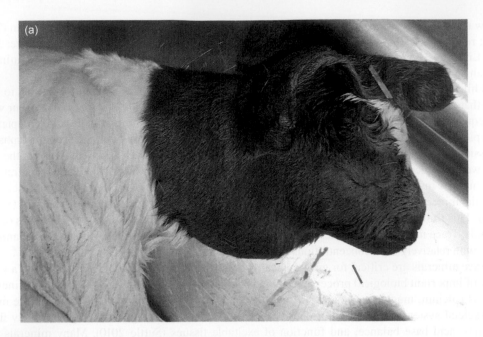

FIGURE 2.6A One of several goat kids born to dams consuming an iodine-deficient diet during gestation. The kids were either born dead or weak and did not survive in the winter weather in the Great Lakes region of Ohio (enlarged thyroids, elevated serum creatine kinase, and serum globulins were determined in this group of kids, suggesting low dietary I and Se during pregnancy, with potential adequate passive transfer of maternal antibodies). The arrow indicates the location of the enlarged thyroids.

FIGURE 2.6B A six-month-old Charolais heifer, presented down, with a body weight of 117 kg (reference: 245–272 kg at this age). Serum analysis of AST (355 IU L^{-1}; reference 54–177 IU L^{-1}) and creatine phospho-kinase (3,621 IU L^{-1}; reference 101–466 IU L^{-1}) levels were supportive of chronic deficiency of selenium, confirmed by the serum mineral analyses (9 ng mL^{-1}; reference: 65–100 ng mL^{-1}). In addition, serum Zn (0.65 µg mL^{-1}; reference 0.9–2.0 µg mL^{-1}) and copper (0.5 µg mL^{-1}; reference 0.6–0.8 µg mL^{-1}) were below lower reference intervals seen at this laboratory.

Low soil Se and inadequate supplementation (Great Lakes region, Northwest areas of the USA) commonly result in severe myopathies in animals. These may be manifested as sudden death (cardiac form) or poor growth associated with chronic muscle inflammation and fibrosis. In the fall of 2018, one local production unit (Ohio) housed cows on pastures where they eventually calved outdoors in a cold and wet winter. The cattle were fed round bale forage of late maturity. Trace mineral supplements were not available during gestation, and the calves did not receive the common mineral supplements required for this region immediately after birth. Approximately half of the cows died or aborted their calves and 50% of the calves born, died, or developed severe Se-deficiency myopathy (Figure 2.6b).

Farming practices that do not focus on soil replenishment or fertilizing soil with a simple nitrogen–phosphorous–potassium routine can also result in deficiencies of many of the essential nutrients (Haes et al. 2012). Addition of phosphorus to the soil can increase forage and grain production by uptake of phosphorus. However, these products may have lower nutritional (mineral) value to animals. Excess phosphorous can antagonize the absorption of other minerals compounding the problem. Addition of excess phosphorus is associated with lower plant-available zinc, iron, and copper (Suttle 2010).

Soil fertilization with manure from livestock can be beneficial since most ruminant diets are supplemented with levels of mineral "to prevent deficiency" based upon the current feed supplementation guidelines (Suttle 2010). Excess trace- and macrominerals are excreted in the urine and feces. Like Zn, other essential nutrients (boron, copper, iron, manganese, and molybdenum) play critical roles in plant growth. However, limited mining reserves suggest that utilization of manures could limit the use of these scarce commodities by resupplying the soil (Haes et al. 2012; Ahsin et al. 2019). More efficient recycling of liquid or solid manures or perhaps expansion of the use of anaerobic fermentation to capture GHG from this material can reduce the footprint of GHG from this sector. Returning manures to the soil it came from should also support the availability of trace- and macrominerals from these soils.

2.5.3 EFFECTS OF PLANTS ON ANIMAL HEALTH

Similar to soil, the fermentation compartment of the ruminant forestomach contains a large and diverse populations of microorganisms that play key roles in the ability of the ruminant to acquire energy and protein from the plant material they consume. Ruminants ingest forage, grains, and byproducts which are metabolized to sugars and converted anaerobically to volatile fatty acids to produce energy for the animal (Capper, Berger, and Brashears 2013). Rumen flora also utilizes plant nitrogen to produce bacterial proteins that are converted into amino acids and absorbed by the ruminant intestine. The energy and protein produced by these symbionts support growth and reproductive functions of these animals (Capper, Berger, and Brashears 2013). The rumen flora also has requirements for trace minerals provided by the animals' consumption of feedstuffs containing these critical elements (Takashima, Shimada, and Speece 2011; He et al. 2014).

Anaerobic microbial fermentation also results in the production of carbon dioxide (CO_2), methane (CH_4), ammonia (NH_3), and hydrogen sulfide (H_2S) gases that are eructated periodically into the atmosphere (15–20 times h^{-1} on dry hay) (Rosenberger et al. 1979). Livestock production systems are currently under intense scrutiny for the production of approximately 14.5% of the anthropogenic greenhouse gas emissions worldwide (Gerber et al. 2013).

The proportions of major GHG produced by the livestock sector include CO_2 emissions (27%), CH_4 emissions (44%), and N_2O emissions (29%) (Gerber et al. 2013). Beef cattle produce approximately two-thirds of this sector's emissions, whereas swine, poultry, buffaloes, and small ruminants (sheep and goats) contribute approximately 8.5%. These emissions include production, processing, and transport of feed. Land use changes (deforestation for croplands) account for approximately 10% of the agricultural sectors GHG emissions (Gerber et al. 2013).

2.5.4 EFFECT OF ANIMALS ON SOIL AND PLANT HEALTH

The global land use footprint of cattle production (growing feedstuffs, feeding cattle) approximates 20%–25% of dry land (Steinfeld et al. 2006). The global water use in beef cattle production approximates 10% of the global agricultural production use of this essential commodity (Mekonnen and Hoekstra 2012).

Grazing livestock on grasslands in an environmentally conscious manner can be successful. However, poorly managed soils are not productive or sustainable due to erosion, compaction, limited water, and nutrient retention and are more prone to plant diseases (Larkin 2015). Depleted soil is less resilient and leads to reduced productivity. Reduced soil fertility may progress to desertification under poorly managed systems associated with soil compaction, addition of excess nutrients, and altered nutrient cycling. Soil compaction reduces the soil absorption of water that can increase soil erosion in rainfall. Compacted soils have limited porosity, so water infiltration is also limited, soil oxygen is limited, and the microbial communities cannot function properly. Grazing animals affect soils differently depending upon their size and plant preferences (Cox and Amador 2018; Wang et al. 2018).

Organic fertilizer (manures, bio-solids) applications are of value as they also promote soil ecosystems (bacterial, algal, fungal, protozoan, nematode, Annelid, mite, insects, plant diversity, and root growth) and support trace- and macromineral content. Organic fertilizer adds structure and porosity to soil and promotes retention of soil water. When healthy, soil retains moisture, filters water, prevents erosion, traps more organic material, and provides nutrients to soil for plants and the organisms within. Healthy and diverse soil organism populations reduce the impact of plant and soil pathogens that can negatively affect the growth of forage and grains needed to feed humans and animals. Healthy soil also reduces the impact of water damage associated with run-off and buffering the impact of fertilization or use of pesticides and other products on water quality. By way of the effects of soil organisms and soil organic material on the production of nitrogen- and carbon-containing gases, healthy soil also impacts the atmosphere.

Fertilization of soil with manure, urine, and bio-solids returns macro- and trace minerals to the soil. Assuming supplementation of minerals is provided to animals in sufficient amounts for the growth, reproductive function, and replacement of lost minerals, the manure should provide minerals lost to harvest. However, fertilization with mineral-laden manures, slurries, or treated waste may increase the mineral footprint of the soil and water leading to environmental concerns. Oversupplementation with minerals may have direct effects on animals. Excess copper supplementation of sheep can cause hepatocellular necrosis and hemolytic anemia. Magnesium and phosphorus may result in the formation of urinary calculi in male ruminants (Suttle 2010). Conversely, copper deficiency may occur on copper-deficient soils, no supplementation of animals, or when the concentration of mineral competitors is present in soil, forage, or grains. Iron, sulfur, zinc, and molybdenum compete with copper for absorption. High dietary calcium may reduce the absorption of zinc.

In 1996, a NAHMS survey of forage mineral content (Corah and Dargatz 1996), *Forage Analysis from Cow/Calf Herds in 18 States* (https://naldc.nal.usda.gov/download/33018/PDF), demonstrated deficient forage levels of copper (14%), manganese (5%), zinc (63%), selenium (64%), with 10% of the samples having very high levels of iron and molybdenum. This survey suggests that minerals essential for plant growth and animal productivity are limited in some areas. Further, high levels of iron (>400 ppm) and molybdenum (>3 ppm) appear to be of concern in the antagonism of copper assimilation.

If the forage is grown on biosolid-supplemented pasture, some species may not tolerate the levels of soil minerals. For example, swine are relatively insensitive to the ingestion of excess copper, whereas sheep are not. It would not be prudent to allow sheep to graze pasture with excess soil copper (unless a competitor mineral was present in the soil or provided as a supplement). In pig slurry-supplemented areas of northwestern Spain, 2.2% of the calves grazing these pastures had liver copper concentrations in excess of 100 ppm (López Alonso et al. 2000). Calves born to copper-supplemented cows (late gestation), who were given 12.5 g copper oxide wire particles within 2 days of birth, died from liver failure and hemolytic anemia at 6–8 weeks of age. Liver copper levels in excess of 250 mg kg^{-1} were identified after autopsy examination of the animals. A similar study with cow-only supplementation demonstrated 435 mg kg^{-1} (wet weight) in livers of the calves that died. Pregnant cows grazing pig slurry-fertilized pastures should be monitored for excess copper exposure because some soils arise from rock with higher copper content, and when soil is supplemented with additional minerals, this may lead to costly losses of animals (López Alonso et al. 2000). Excess copper may be problematic because calves accumulate copper in utero, using manures containing excess mineral may require soil or animal mineral analyses to prevent livestock losses.

Calves grazing pasture fertilized with swine slurries have been observed to accumulate salt, heavy metals, and xenobiotics (Table 2.4). Some slurries also contain very high concentrations of salt, which if applied regularly may result in lower growth of forages, altered solubility of metal ions, and promote greater compaction (Moral et al. 2008). Other plants may accumulate heavy metals such as arsenic, cadmium, nickel, and lead from soils (Table 2.4) (Reeves et al. 2017). Lead poisoning occurs in grazing cattle through the consumption of forage grown on Pb-contaminated soil, disposal of Pb-acid batteries, and lead-based paint from old barns or gasoline and lubricants (Bischoff et al. 2012). While animal illness and/or death may be significant, a more concerning problem is the half-life of lead in the animals without clinical signs. Bischoff et al. (2012) demonstrated that the elimination half-life of Pb in contaminated animals was 135 ± 125 days (range 3–577 days). Potential human exposure occurring due to the consumption of animal tissues is of great concern. Recent outbreaks in our region of the USA included grass hay contaminated with lead-containing lubricants and animals housed on reclaimed Pb mine ground. In these situations, concern for human exposure to Pb may also be relevant.

Other exposures of concern include veterinary medicines (antibiotics, endectocides, anticoccidial), hormones (implants, injectables), euthanasia products, and other medications (Boxall 2010). Many of these medications when administered to animals may be excreted in the urine and feces or may be washed-off animal hides after topical application. Contamination of soil and water may have significant effects on the soil microbiome and long-term impacts on soil health. Antimicrobial agents used to sustain animal health may influence the soil resistome, and the hormones excreted can influence the development and survival of microorganisms and invertebrates when exposed.

TABLE 2.4
Soil Toxicants Effects on Plants, Animals, and Human Health

Contaminant	Effect on Plants	Effect on Animals	Effect on Humans
Heavy Metals			
Arsenic	Reduction of seed germination, height of the shoots, length of the roots, and plant biomass. Decreased production of chlorophyll and protein content of exposed plants.	GI exposure: Sudden death, abdominal pain, vomiting, staggering, weakness, shock, diarrhea (dark and bloody) collapse and death; Dermal exposure—blisters	Damage to GI track, skin, heart, liver, nerves, brain. Miscarriage, pre-term birth; cancer; death
Asbestos	Reduction in seed germination, seedling height; reduced leaf area and dry matter production, and plant yield; decrease in leaf fresh weight. The developed symptoms of wilting and chlorosis.	Mesothelioma has been observed; however, no reports associate with asbestos.	Pulmonary asbestosis; lung cancer

(Continued)

TABLE 2.4 (*Continued*)

Soil Toxicants Effects on Plants, Animals, and Human Health

Contaminant	Effect on Plants	Effect on Animals	Effect on Humans
		Heavy Metals	
Cadmium	Reduction in seed germination; decrease in plant nutrient content; reduced shoot and root length.	Cattle accumulate Cd readily when consuming contaminated pasture; kidney disease/lesions (Equine), Osteochondrosis, osteomalacia (Equine),	Bone, liver, and kidney damage; cancer; itai-itai disease
Chromium, Hexavalent	Reduced seed germination, shoot and root growth; Decrease in plant nutrient acquisition Decreased yield of plants.	Acute toxicity associated with death, severe congestion and inflammatory lesions of GI tract, renal and hepatic damage Chronic toxicity associated with gastroenteritis and dermatitis	Cancer
Lead	Reduction in seed germination; suppressed root and shoot growth; reduced plant biomass; decrease in plant protein content; decreased number of leaves and leaf area; reduced plant height and biomass.	Readily absorbed from the GI tract especially in mineral deficient or calcium deficient animals; accumulates widely (bone and muscle); anorexia, anemia, CNS disease (depression, weakness, ataxia, blindness, seizures, death	Neurodevelopmental delay; lower intelligence; bone damage; high blood pressure; kidney disease
Mercury	Decrease in seed germination and plant height; reduction in flowering number; increased leaf chlorosis; reduced shoot growth and development; yield reduction.	Anorexia, diarrhea, skin/eye lesions, ataxia, incoordination, renal damage/failure, convulsions, death	Brain and gastrointestinal track damage; lower IQ; poor coordination Liver, heart, kidney damage Teratogenic
Endocrine-Disrupting Compounds and Persistent Organic Pollutants (Dioxin, PCB, BPA, PBDEs, DDT, phthalates etc.)	Can lead to all the symptoms and defects mentioned above.	Early signs - ↓ milk production, appetite, loss of weight, lacrimation, urination, resorption of pregnancies, hematomas (10% of animals), increased growth of hoof horn, alopecia, thickened skin, dystocia, calf loss	Early thelarche, disrupted male reproductive development, thyroid dysfunction, neurodevelopmental delay and dysfunction, obesity, immune system damage, skin damage, liver damage, cancer
Pesticides and *Herbicides*	Can lead to all the symptoms and defects mentioned above.	OP: Acute toxicity—SLUD signs, lung edema, cyanosis, twitching muscles, tremors, ataxia, stiff gait, convulsions, death; Dinitrophenol—methemoglobinemia, intravascular hemolysis	Non-Hodgkin lymphoma, Parkinson's disease, cognitive and psychomotor dysfunction, neurodegenerative and neurodevelopmental effects, leukemia, birth defects, pediatric brain tumors, cancer of the kidney, prostate, and pancreas

Source: Modified and adapted from Science Communication Unit (2013) and Gupta (2007).

2.6 HUMAN HEALTH

World population increased from 2.5 billion in 1950 to 7.4 billion in 2016. As the human population will grow to nearly 10 billion by mid-century, increasing demands will be made upon soil to enable agriculture to feed additional humans at the same time soils have become degraded and will increasingly be lost as cities expand into former farmland, sea-level rises, desertification expands, and run-off expands with intensifying storms secondary to climate change (UN 2019). Nutritious, high-quality food depends on healthy soil because over 97% of global food needs come from land and less than 3% from oceans, based on caloric intake. Despite tremendous gains in decreasing widespread hunger over the past 50 years, still as of 2016, over 800 million people are undernourished and billions more suffer from "hidden hunger," specific micronutrient deficiencies such as anemia (insufficient iron), weak immune function (inadequate vitamin A), and hypothyroidism (insufficient iodine) (Amoroso 2016).

2.6.1 HUMAN ACTIVITY IMPACTS SOIL HEALTH

Human activity affects soil health both directly and indirectly (Table 2.5). Negative direct effects on soil come from human activities such as mining, dumping industrial pollutants, sewage, agricultural practices (such as extensive tillage, monoculture, and overuse of pesticides and herbicides), and building practices that lead to toxicity, decreased microbial diversity, compaction, nutrient depletion, and erosion. Indirect effects arise from inadvertently transporting invasive species (Buhk and Jungkunst 2018), forestry practices that increase runoff and erosion, emissions from burning fossil fuels and releasing methane and other potent greenhouse gasses, growing livestock (replacing forests and perennial prairies with cropland to feed cattle and other livestock), and climate change. Climate change increases global temperatures which dry the soil, leading to desertification and wildfires and making it susceptible to blowing away or running off; climate change also increases sea levels which can flood low lying areas or contribute to low-level soil salinization; climate change-related storms accelerate runoff and erosion of topsoil.

TABLE 2.5
Human Activities, Animal Activities, and Their Impact on Soil Health

Human Activities	Animal Activities	Soil Health
Sewage sludge containing microbes, antibiotics, medications, illicit drugs	Antibiotic use as growth promoters	Microbial contamination, selective pressure for antimicrobial resistance, microbiome alterations
Mining, smelting	Mineral supplementation in diet	Heavy metal contamination
Deforestation to clear for cropland	Heavy grazing	Erosion; loss of CO_2 sequestration
Deforestation to clear for grazing	Heavy grazing	Erosion; loss of CO_2 and methane sequestration
Laundry and other waste releasing microplastic fibers		Microplastic contamination
Trash management		Release of toxins and endocrine disrupting chemicals
Agricultural activities: Excess nitrogen fertilization	Rumen gas production of CO_2 and CH_4	Reduced microbial diversity
Increasing greenhouse gas emissions; climate change		Desertification, salinization, loss of topsoil from heavy winds and rain in intense storms and flooding

On the other hand, prudent human activities can restore degraded and depleted soil, build SOC, sequester atmospheric carbon in the ground, and promote biodiversity. These practices include sustainable forestry and conservation agriculture. Sustainable conservation agriculture means meeting the twin goals of increased productivity and reduced environmental damage (Reicosky and Janzen 2018). The first facet of conservation agriculture is preserving soil cover with either live crops or crop residues (mulches) to prevent erosion, reduce the loss of soil moisture, enrich the SOC content, and prevent the growth of weeds (Reicosky and Janzen 2018). The second facet of conservation agriculture is plant diversity, with a minimum of three different crops including at least one nitrogen-fixing legume. The third facet of conservation agriculture is reducing or eliminating tillage to minimize soil disturbance (Reicosky and Janzen 2018).

2.6.2 SOIL IMPACT PHYSICALLY

Soil health impacts human health physically and structurally, biochemically through plant- and animal-based nutrition, and biologically through infectious diseases and as a source of antimicrobial medications. Physically, soil forms the solid ground on which we live, work, play, and move. Soil can be crafted into bricks to build dwellings. As soil breathes, it participates in the exchange of gases such as carbon dioxide, oxygen, and methane. Healthy soil acts as a physical water filter and container, and it sequesters carbon, potentially offsetting human emissions of carbon dioxide (Lal 2004, 2016).

Soil plays a key role in air and water quality, and both soil and plants can moderate the climate. For example, air quality is adversely affected by wind erosion (dust), and as human activity increases global temperatures, there will be more desertification, erosion, and dust storms, carrying soil, potential pathogens, and pollen, increasing the risk of asthma and other respiratory diseases. Soil can also release methane (CH_4), nitrous oxide (N_2O), hydrogen sulfide (H_2S), and other gases, depending on the soil temperature, moisture, and number of anaerobic bacteria; these gasses can contribute to climate change and air pollution, increasing the risk of systemic inflammation.

2.6.3 SOIL IMPACT ON HUMAN NUTRITION

Humans require macronutrients such as carbohydrates, protein, and fat as well as micronutrients. Macrominerals found in soil include carbon, sodium, potassium, calcium, magnesium, nitrogen, sulfur, phosphorous, and chloride.[1] For example, to build proteins, humans and animals require nitrogen. To transform energy in basic biochemical pathways involving ADP and ATP, humans require phosphorous. In addition, humans depend on healthy soil for essential micronutrients such as iron, zinc, copper, manganese, iodine, boron, selenium, molybdenum, lithium, and vanadium (Table 2.3). The formation of healthy bones requires calcium, magnesium, iron, copper, and manganese. Healthy cardiac function requires adequate potassium, magnesium, copper, and selenium. A proper balance of sodium and potassium is needed for proper cellular function and cell signaling. Among agricultural soils globally, 49% are deficient in zinc, 31% in boron, and 15% are deficient in molybdenum (Franzluebbers and Sharpley 2017). Chronic protein-calorie malnutrition from inadequate macronutrients can arise from poor soils, crops, and livestock. Protein-calorie malnutrition leads to stunted growth, frailty, weakened immune function, impaired intellectual development, apathy or anger, depression, fatigue, decreased capacity for work and learning, and, in the most severe cases, death (Batool et al. 2015). Chronic food insecurity from poor soils and insufficient harvests, exacerbated by climate change, can lead to famine, mass migration, and violent conflicts (Butler 2018). Currently, this is already occurring in central Asian and sub-Saharan Africa. Poor-quality soil leads to lower crop yields, food insecurity, famine, migration, and/or violent conflicts (Reuveny 2007). For example, global grain shortages due to drought in China and flooding in Canada, combined with regional deforestation, overgrazing, soil erosion, water pollution, drought, and desertification,

---

[1] Note that although hydrogen and oxygen are also essential, they can be obtained from water and air.

amplified by climate change and population growth, led to widespread food shortages and the 2011 Arab Spring that devolved into a violent civil war in Syria, which has led to a massive migration of Syrian refugees to Europe (Fischetti 2015). Similarly, dry soil led to poor crop production and food shortages in South Sudan, Yemen, Somalia, and parts of Nigeria which have contributed to widespread violence and massive migration. Thus, maintaining and restoring healthy soil has become a matter of global and national security.

Globally, over 800 million people (mostly in Southeast Asia and sub-Saharan Africa) are at risk of protein-calorie malnutrition or food insecurity (Pérez-Escamilla 2017). In addition, billions more suffer from deficiencies of one or more micronutrients such as iodine, iron, selenium, or zinc. Healthy soil contains an optimal balance of macro- and micronutrients to support the health of plants, animals, and humans. For example, soils deficient in iodine lead to low iodine intakes and are associated with hypothyroidism, cretinism, and goiter. Iron deficiency leads to fatigue, inattention, irritability, and anemia. Zinc deficiency increases the risk of diarrhea, respiratory infections, anorexia, stunted growth, and skin rashes (Table 2.3) (Golden 1982).

Due to the degradation of soil quality over the past 50 years of modern agricultural methods, fruits and vegetables grown today contain significantly lower levels of essential vitamins and minerals than those grown 50 years ago. According to the USDA, in a study of 13 nutrients in 43 different vegetable crops, varieties grown today contain lower amounts of calcium, phosphorous, iron, riboflavin, and ascorbic acid (Davis et al. 2004). Similarly, a comparison of the mineral content of 20 fruits and 20 vegetables grown in the UK showed several marked reductions from the 1930s to the 1980s; these reductions were particularly significant for calcium, magnesium, copper, and sodium in vegetables and magnesium, iron, copper, and potassium in fruit (Mayer 1997). Soil can contain toxic chemicals, heavy metals, and pathogens that adversely affect human health (Table 2.4). Humans can be exposed directly by ingesting (geophagia or pica) or inhaling contaminated soil (dust), through dermal (skin, intact or abraded, cut or punctured) absorption. Exposure can also occur indirectly by water runoff from contaminated soil, by plants grown on soil that take up toxins, or by crops that become contaminated through contact with water or soil (Table 2.6). Because humans are generally at the top of the food chain, we can also be exposed to soil pathogens by ingesting contaminated meat (Table 2.6).

2.6.4 TOXIC COMPOUNDS IN SOIL IMPACT HUMAN HEALTH

Soil may contain toxic compounds naturally or as a result of contamination (Table 2.4). Naturally occurring toxic compounds include radon, which is linked to an increased risk of lung cancer (Khan, Gomes, and Krewski 2019). Naturally occurring arsenic (As) can leach into the ground water and then be taken up into foods grown with that water, such as rice, which has affected people in several parts of Asia including Bangladesh, China, Cambodia, Pakistan, and India (Murphy et al. 2018; Nawab et al. 2018; Biswas et al. 2019; Wang et al. 2019). Toxic heavy metals that contaminate soil can include cadmium, chromium, copper, lead, mercury, nickel, uranium, and zinc (Zhuang et al. 2009; Trujillo-González et al. 2016); high levels of these toxic metals are found in and near hazardous waste sites, exposing millions, particularly those with low incomes and few housing options, to potentially toxic metals, increasing their risks of birth defects, neurodevelopmental problems, and cancer (Ericson et al. 2013). For example, 90% of cadmium exposure in nonsmokers comes from food grown in cadmium-containing soil; the bioavailability of cadmium to crops is influenced by soil pH, organic matter, the type of crop, and the presence of other elements. Similarly, lead exposure in children occurs when infants and toddlers ingest dust-containing lead from paint or leaded gasoline; elevated lead levels cause persistent problems in neurodevelopment, lowering long-term intelligence, and increasing the risk of hypertension and kidney disease (Landrigan et al. 2002). Contaminated soil from extractive industries, such as mining, deposits from manufacturing waste and construction, the use of sewage sludge as fertilizer, and air pollution from burning coal which settles on land and water, adversely affects the health of tens of millions of people (Ericson et al. 2013).

TABLE 2.6

Examples of Human Illnesses Transmitted by Contaminated Plants or Animals

Human Illness	Plants Transmitting	Animal Transmitting	Infectious Agent
Listeriosis	Cantaloupes (CDC 2011)		Listeria
Food poisoning	Bean sprouts (Buchholz et al. 2011)		E. coli O104:H4
Food poisoning	Alfalfa sprouts (Van Beneden et al. 1999)	Cheese (Altekruse et al. 1998)	Salmonella
	Cucumbers (Bardsley et al. 2019)	Eggs (Whiley and Ross 2015)	
	Pistachio nuts (Harris et al. 2016)	Poultry (Antunes et al. 2016)	
	Tomato (Lim, Lee, and Heu 2014)	Seafood (Kumar, Datta, and Lalitha 2015)	
	Nut butter (Maki 2009)	Pet turtles (Gambino-Shirley et al. 2018)	
Mad cow disease; scrapie	Wheat (Pritzkow et al. 2015)	Cattle (BSE) (Greenlee and Heather West Greenlee 2015)	Prions
		Sheep and goats (scrapie) (Greenlee and Heather West Greenlee 2015)	
Liver disease and cancer	Contaminated grains/beans/ peanuts/seeds(Li et al. 2009)	Contaminated milk, poultry, and meat (Fouad et al. 2019; Milićević et al. 2019)	Aflatoxin-producing fungi

Endocrine-disrupting compounds such as polychlorinated biphenyls (PCBs), bisphenol A (BPA), polybrominated biphenyls, dioxin, phthalates, and persistent pesticides such as DDT are present in soil and can contribute to early thelarche, thyroid dysfunction, delayed or disrupted male sexual development, thyroid dysfunction, obesity, and neurodevelopmental problems, particularly in poor, minority, and vulnerable populations such as pregnant women and young children (Gore et al. 2015; Ruiz et al. 2018). Overall, soil contamination from a variety of industrial and agricultural practices adversely affects human health (Table 2.4).

2.6.5 SOIL PATHOGENS

Healthy soil supports beneficial organisms such as earthworms, fungi, and health-promoting microbes, and it limits pathogens. On the other hand, soil can host pathogenic microbes that cause tetanus (*Clostridium tetani*), botulism (*Clostridium botulinum*), listeriosis (*Listeria monocytogenes*), nocardiosis (*Nocardia* species), and anthrax (*Bacillus anthracis*) (Table 2.2) (Wells and Wilkins 1996). Some soil microbes pathogenic to humans, such as the Gram-negative bacteria, *Pseudomonas*, and *Enterobacteriaceae*, protect plants against soil pathogens, promote plant growth, or assist in decomposing decaying matter (Berg, Eberl, and Hartmann 2005; Berg et al. 2005). Pathogenic organisms that live in soil range from macroscopic helminths (such as hookworms) to microscopic (fungi, bacteria, protozoa, and viruses) (Jeffery and van der Putten 2011). For example, pathogenic soil fungi include *Aspergillus*, which can cause sinusitis, allergic bronchopulmonary aspergillosis, pneumonia, or sepsis (usually in immunocompromised patients or those with genetic disorders like cystic fibrosis); *Blastomyces*, *Histoplasma*, and *Coccidiodes* species, which can cause pneumonia; *Mucormycetes*, which can cause serious infections in immunocompromised people; and *Cryptococcus*, which can cause meningitis in those already infected with human immunodeficiency virus (HIV). Pathogenic soil protozoa include *Entamoeba histolytica, Toxoplasma gondii, Giardia*, and *Cryptosporidium parvum*.

Many cases of food poisoning arise from contamination of irrigation water with runoff from fields that contain manure from livestock (Table 2.6). This is an example of indirect exposure because the bacteria (such as *E. coli* or *Salmonella*) are transmitted from manure to soil and to water to plants. Human sewage can also contaminate soil with Gram-negative enteric pathogens such as *E. coli*, *Salmonella* species, and *Shigella* species; again, when this soil washes into water used to irrigate crops, humans may be exposed, perpetuating infections.

The survival of both beneficial (e.g., earthworms) and noxious soilborne organisms is affected by soil pH, the amount and quality of organic matter, moisture content, soil particle size, sunlight exposure, and temperature (Hackenberger et al. 2018). For example, hantavirus normally lives in rodents (which rarely transmit it directly to humans), but it can be deposited onto soil through rodent urine and feces; wind can carry Hantavirus on dry dust particles, which are then inhaled by humans, causing hantavirus cardiopulmonary syndrome which has a 30%–40% mortality rate (Hjelle and Torres-Pérez 2010).

Exposure to pesticides generally reduces the number of neutral or beneficial organisms and facilitates the overgrowth of pathogenic organisms in soil, whereas the addition of biochar can increase the soil's microbial diversity and plant health (Lekberg et al. 2017; Meng et al. 2019).

Over 80% of the antibiotics used in the US are sold for us in farm animals, most of it to promote growth and weight gain rather than to treat infections. Veterinary antibiotics excreted by livestock and/or humans that enter waterways and are deposited on land may also place selective pressure on antimicrobial communities, promoting the growth and spread of multiple drug-resistant organisms in the soil, which can then be transferred to crops and livestock, which are then consumed by humans, exposing downstream victims to antibiotic-resistant organisms (Marshall and Levy 2011; Ben et al. 2019).

2.6.6 SOIL AS SOURCE OF MODERN ANTIBIOTICS AND OTHER BENEFITS

Overall, the majority of soil microbes are nonpathogenic and may serve as a food source for plants (Paungfoo-Lonhienne et al. 2010). Soil organisms are also essential for decomposing organic residues from dead plants and animals. In addition, soil hosts a diverse microbial environment that has served as a source of modern antibiotics (Table 2.7). For example, one genus of bacteria, *Streptomyces*, from the *Actinomycetes* family, is responsible for over half of the clinically useful antibiotics of natural origin used in human medical care, including antitumor drugs (such as adriamycin, bleomycin, and anthracyclines), and both antibacterial agents (such as erythromycin, neomycin, gentamycin, and vancomycin) and antifungal agents (such as amphotericin B and nystatin) (Bhatti, Haq, and Bhat 2017). Furthermore, some strains of Actinomycetes can be used to degrade toxic pollutants in bioremediation.

TABLE 2.7

Soil-Based Microbes and Their Antibiotic Derivatives (Aminov 2017)

Soil-Based Microbe	Examples of Antimicrobial Compounds
Acremonium chrysogenum	Cephalosporins
Actinomycetes family of bacteria, including *Streptomyces*	Actinomysin, Amphotericin B, Chloramphenicol, Cycloserine, Daptomycin, Erythromycin, Imipenem, Kanamycin, Lincomycin, Neomycin, Nystatin, Rifamycin, Spectinomycin, Streptomycin, Tetracycline, Vancomycin, Virginiamycin
Aspergillus family	Fumagillin
Bacillus subtilis	Bacitracin
Micromonospora echinospora	Gentamicin
Paenibacillus polymyxa	Colistin (Polymyxin B)
Penicillium	Penicillin
Saccharopolyspora erythraea	Erythromycin
Streptoalloteichus tenebrarius	Tobramycin

As understanding grows about the characteristics and importance of the human microbiome, interest has increased in the relationship between the human microbiome and the soil microbiome. The soil microbiome can be directly transferred to humans by ingesting soil (pica or geophagia), topically (mud baths, outdoor exposure), and by inhalation; it can also be indirectly transferred to humans through the plants and animals we consume. Many questions remain about the relationship between soil and human microbiota, but it is clear that the diversity of soil microbiota is impacted by the pesticide use. For example, a single application of the fungicide tetraconazole to the soil in an apple orchard resulted in a significant decrease in the microbial activity 28 days later (Sułowicz and Piotrowska-Seget 2016). The impact of pesticides on the soil microbiota also depends on factors such as soil type (sandy, loam, or clay), moisture, and pH (Merlin et al. 2016).

2.7 CONCLUSIONS

"One Health" implies the interconnectedness of the ecosystem, plant, animal, and human health. Healthy soil is essential to plant, animal, and human health. Therefore, sustainable soil management is essential to promote human health and advance the United Nations Sustainable Development Goals. Healthy plants depend on healthy soil, containing the optimal balance of macro- and micro-nutrients and water, free of contamination with heavy metals or toxic chemicals. Healthy animals, including livestock, depend on healthy plants for nutrition. Human well-being depends on healthy soil to help sequester carbon, provide a foundation for housing, to filter water, as an indirect source of nutrients (from plants and animals) and as an important source of antibiotics. In turn, human activity can support or harm the health of soil, plants, and animals on which human life depends.

REFERENCES

Abdel-Fattah, G.M., S.A. El-Haddad, E.E. Hafez, and Y.M. Rashad. 2011. Induction of Defense Responses in Common Bean Plants by Arbuscular Mycorrhizal Fungi. *Microbiological Research* 166, no. 4 (May): 268–281.

Adviento-Borbe, M., J.W. Doran, R. Drijber, and A. Dobermann. 2006. Soil Electrical Conductivity and Water Content Affect Nitrous Oxide and Carbon Dioxide Emissions in Intensively Managed Soils. *Journal of Environmental Quality* 35, no. 6 (November 1): 1999–2010.

Ahsin, M., S. Hussain, Z. Rengel, and M. Amir. 2019. Zinc Status and Its Requirement by Rural Adults Consuming Wheat from Control or Zinc-Treated Fields. *Environmental Geochemistry and Health* 42, no. 7 (November 6): 1877–1892.

Alexandratos, N., and J. Bruinsma. 2012. *World Agriculture Towards 2030/2050: The 2012 Revision*. ESA Working Paper. Rome, Italy. http://www.fao.org/3/a-ap106e.pdf.

Altekruse, S.F., B.B. Timbo, J.C. Mowbray, N.H. Bean, and M.E. Potter. 1998. Cheese-Associated Outbreaks of Human Illness in the United States, 1973 to 1992: Sanitary Manufacturing Practices Protect Consumers. *Journal of Food Protection* 61, no. 10: 1405–1407.

Aminov, R. 2017. History of Antimicrobial Drug Discovery: Major Classes and Health Impact. *Biochemical Pharmacology* 133: 4–19. Elsevier Inc.

Amoroso, L. 2016. The Second International Conference on Nutrition: Implications for Hidden Hunger. *World Review of Nutrition and Dietetics* 115: 142–152. https://www.karger.com/Article/FullText/442100.

Anon. 2002. Numbers 12:18–20. In *The New American Bible*. Washington, DC: United States Conference of Catholic Bishops.

Antunes, P., J. Mourão, J. Campos, and L. Peixe. 2016. Salmonellosis: The Role of Poultry Meat. *Clinical Microbiology and Infection* 22, no. 2 (February 1): 110–121.

Araujo, A., L. Leite, V. Santos, and R.F.V Carneiro. 2009. Soil Microbial Activity in Conventional and Organic Agricultural Systems. *Sustainability* 1 (June 1): 268–276.

Aziz, I., N. Bangash, T. Mahmood, and K.R. Islam. 2015. Impact of No-till and Conventional Tillage Practices on Soil Chemical Properties. *Pakistan Journal of Botany* 47, no. 1 (February 1): 297–303.

Aziz, I., T. Mahmood, and K.R. Islam. 2013. Effect of Long Term No-till and Conventional Tillage Practices on Soil Quality. *Soil and Tillage Research* 131: 28–35. http://www.sciencedirect.com/science/article/pii/S0167198713000597.

Bal, H.B., L. Nayak, S. Das, and T.K. Adhya. 2013. Isolation of ACC Deaminase Producing PGPR from Rice Rhizosphere and Evaluating Their Plant Growth Promoting Activity under Salt Stress. *Plant and Soil* 366, no. 1: 93–105. doi:10.1007/s11104-012-1402-5.

Bardsley, C.A., L.N. Truitt, R.C. Pfuntner, M.D. Danyluk, S.L. Rideout, and L.K. Strawn. 2019. Growth and Survival of Listeria Monocytogenes and Salmonella on Whole and Sliced Cucumbers. *Journal of Food Protection* 82: 301–309. International Association for Food Protection.

Batool, R., M.S. Butt, M.T. Sultan, F. Saeed, and R. Naz. 2015. Protein-Energy Malnutrition: A Risk Factor for Various Ailments. *Critical Reviews in Food Science and Nutrition*. http://www.tandfonline.com/doi/abs/10.1080/10408398.2011.651543.

Beckers, B., M. Op De Beeck, S. Thijs, S. Truyens, N. Weyens, W. Boerjan, and J. Vangronsveld. 2016. Performance of 16s RDNA Primer Pairs in the Study of Rhizosphere and Endosphere Bacterial Microbiomes in Metabarcoding Studies. *Frontiers in Microbiology* 7 (May 13): 650. https://pubmed.ncbi.nlm.nih.gov/27242686.

Beckers, G.J.M., and U. Conrath. 2007. Priming for Stress Resistance: From the Lab to the Field. *Current Opinion in Plant Biology* 10, no. 4: 425–431. http://www.sciencedirect.com/science/article/pii/S1369526607000726.

Ben, Y., C. Fu, M. Hu, L. Liu, M.H. Wong, and C. Zheng. 2019. Human Health Risk Assessment of Antibiotic Resistance Associated with Antibiotic Residues in the Environment: A Review. *Environmental Research* 169: 483–493. Academic Press Inc.

Van Beneden, C.A., W.E. Keene, R.A. Strang, D.H. Werker, A.S. King, B. Mahon, K. Hedberg, et al. 1999. Multinational Outbreak of Salmonella Enterica Serotype Newport Infections Due to Contaminated Alfalfa Sprouts. *Journal of the American Medical Association* 281, no. 2: 158–162. https://jamanetwork.com/journals/jama/article-abstract/188356.

Bennett, A.E., J.D. Bever, and M. Deane Bowers. 2009. Arbuscular Mycorrhizal Fungal Species Suppress Inducible Plant Responses and Alter Defensive Strategies Following Herbivory. *Oecologia* 160, no. 4: 771–779. doi:10.1007/s00442-009-1338-5.

Berendsen, R.L., C.M.J. Pieterse, and P.A.H.M. Bakker. 2012. The Rhizosphere Microbiome and Plant Health. *Trends in Plant Science* 17, no. 8: 478–486. http://www.sciencedirect.com/science/article/pii/S1360138512000799.

Berg, G., L. Eberl, and A. Hartmann. 2005. The Rhizosphere as a Reservoir for Opportunistic Human Pathogenic Bacteria. *Environmental Microbiology*. http://doi.wiley.com/10.1111/j.1462-2920.2005.00891.x.

Berg, G., A. Krechel, M. Ditz, R.A. Sikora, A. Ulrich, and J. Hallmann. 2005. Endophytic and Ectophytic Potato-Associated Bacterial Communities Differ in Structure and Antagonistic Function against Plant Pathogenic Fungi. *FEMS Microbiology Ecology* 51, no. 2 (January): 215–229. https://academic.oup.com/femsec/article-lookup/doi/10.1016/j.femsec.2004.08.006.

Berry, W. 1977. *The Unsettling of America: Culture & Agriculture*. Sierra Club Books. https://books.google.com/books?id=DohgAAAAIAAJ.

Bhatti, A.A., S. Haq, and R.A. Bhat. 2017. Actinomycetes Benefaction Role in Soil and Plant Health. *Microbial Pathogenesis* 111: 458–467. Academic Press.

Bischoff, K., B. Thompson, H.N. Erb, W.P. Higgins, J.G. Ebel, and J.R. Hillebrandt. 2012. Declines in Blood Lead Concentrations in Clinically Affected and Unaffected Cattle Accidentally Exposed to Lead. *Journal of Veterinary Diagnostic Investigation: Official Publication of the American Association of Veterinary Laboratory Diagnosticians, Inc* 24, no. 1 (January): 182–187. https://www.ncbi.nlm.nih.gov/pubmed/22362951.

Biswas, A., S. Swain, N.R. Chowdhury, M. Joardar, A. Das, M. Mukherjee, and T. Roychowdhury. 2019. Arsenic Contamination in Kolkata Metropolitan City: Perspective of Transportation of Agricultural Products from Arsenic-Endemic Areas. *Environmental Science and Pollution Research* 26, no. 22 (August 1): 22929–22944.

Bowles, T.M., L.E. Jackson, M. Loeher, and T.R. Cavagnaro. 2017. Ecological Intensification and Arbuscular Mycorrhizas: A Meta-Analysis of Tillage and Cover Crop Effects. *Journal of Applied Ecology* 54, no. 6 (December 1): 1785–1793. doi:10.1111/1365-2664.12815.

Boxall, A.B.A. 2010. Veterinary Medicines and the Environment. In *Comparative and Veterinary Pharmacology*, eds. F. Cunningham, J. Elliott, and P. Lees, 291–314. Berlin, Heidelberg: Springer Berlin Heidelberg. doi:10.1007/978-3-642-10324-7_12.

Brevik, E.C. 2013. Soils and Human Health: An Overview. In *Soils and Human Health*, eds. E.C. Brevik and L. Burgess, 29–56. Boca Raton, FL: CRC Press.

Brevik, E.C., and L.C. Burgess. 2014. The Influence of Soils on Human Health. *Nature Education Knowledge* 5, no. 12: 1. https://www.nature.com/scitable/knowledge/library/the-influence-of-soils-on-human-health-127878980/.

van Bruggen, A.H.C., E.M. Goss, A. Havelaar, A.D. van Diepeningen, M.R. Finckh, and J.G. Morris. 2019. One Health - Cycling of Diverse Microbial Communities as a Connecting Force for Soil, Plant, Animal, Human and Ecosystem Health. *Science of The Total Environment* 664: 927–937. http://www.sciencedirect.com/science/article/pii/S0048969719305728.

van Bruggen, A.H.C., and A.M. Semenov. 2000. In Search of Biological Indicators for Soil Health and Disease Suppression. *Applied Soil Ecology* 15, no. 1: 13–24. http://www.sciencedirect.com/science/article/pii/S0929139300000688.

Buchholz, U., H. Bernard, D. Werber, M.M. Böhmer, C. Remschmidt, H. Wilking, Y. Deleré, et al. 2011. German Outbreak of Escherichia Coli O104:H4 Associated with Sprouts. *New England Journal of Medicine* 365, no. 19 (November 10): 1763–1770. http://www.nejm.org/doi/abs/10.1056/NEJMoa1106482.

Buhk, C., and H.F. Jungkunst. 2018. Effects of Plant Invasions on the Soil Carbon Storage in the Light of Climate Change. In *Soil and Climate*, eds. R. Lal and B.A. Stewart, 283–300. Boca Raton, FL: CRC Press.

Bulgarelli, D., M. Rott, K. Schlaeppi, et al., 2012. Revealing structure and assembly cues for Arabidopsis root-inhabiting bacterial microbiota. *Nature* 488: 91–95.

Butler, C.D. 2018. Climate Change, Health and Existential Risks to Civilization: A Comprehensive Review (1989–2013). *International Journal of Environmental Research and Public Health*. http://www.mdpi.com/1660-4601/15/10/2266.

Capper, J.L., L. Berger, and M.M. Brashears. 2013. Animal Feed vs. Human Food: Challenges and Opportunities in Sustaining Animal Agriculture Toward 2050. *Council for Agricultural Science and Technology* 53, no. 53 (January 1): 1–16. http://www.cast-science.org/download.cfm?PublicationID=278268&File=1e30d1 11d2654524a7967353314f1529765aTR.

CDC. 2011. Centers for Disease Control and Prevention. Multistate Outbreak of Listeriosis Linked to Whole Cantaloupes from Jensen Farms, Colorado (FINAL UPDATE). https://www.cdc.gov/listeria/outbreaks/cantaloupes-jensen-farms/index.html.

Celik, I., I. Ortas, and S. Kilic. 2004. Effects of Compost, Mycorrhiza, Manure and Fertilizer on Some Physical Properties of a Chromoxerert Soil. *Soil and Tillage Research* 78, no. 1: 59–67. http://www.sciencedirect.com/science/article/pii/S0167198704000492.

Chasapis, C.T., C.A. Spiliopoulou, A.C. Loutsidou, and M.E. Stefanidou. 2012. Zinc and Human Health: An Update. *Archives of Toxicology* 86, no. 4 (April): 521–534.

Clay, D.E., G. Reicks, C.G. Carlson, J. Moriles-Miller, J.J. Stone, and S.A. Clay. 2015. Tillage and Corn Residue Harvesting Impact Surface and Subsurface Carbon Sequestration. *Journal of Environmental Quality* 44, no. 3 (May 1): 803–809. doi:10.2134/jeq2014.07.0322.

Commoner, B. 1971. *The Closing Circle: Nature, Man, and Technology*. Bantam Books. Knopf. https://books.google.com/books?id=lpYwAAAAMAAJ.

Corah, L.R., and D. Dargatz. 1996. *Forage Analyses from Cow / Calf Herds in 18 States. Health (San Francisco)*. Beef CHAPA Cow/Calf Health & Productivity Audit. Fort Collins. https://naldc.nal.usda.gov/download/33018/PDF.

Cox, A.H., and J.A. Amador. 2018. How Grazing Affects Soil Quality of Soils Formed in the Glaciated Northeastern United States. *Environmental Monitoring and Assessment* 190, no. 3 (February 21): 159. https://www.ncbi.nlm.nih.gov/pubmed/29468318.

Davis, D.R., M.D. Epp, H.D. Riordan, and D.R. Davis. 2004. Changes in USDA Food Composition Data for 43 Garden Crops, 1950 to 1999. *Journal of the American College of Nutrition* 23, no. 6 (December): 669–682. http://www.tandfonline.com/doi/abs/10.1080/07315724.2004.10719409.

Doran, J.W. 1990. Microbial Acitvity and N Transformation in Sustainable Systems. In *Proceedings of the Conference Extending Sustainable Systems*, 109–116. St. Cloud, MN: Minnesota Extension Service and Farm Management. https://www.statpearls.com/as/nutrition/28855/

Doran, J.W., M. Sarrantonio, and M.A. Liebig. 1996. Soil Health and Sustainability. In *Advances in Agronomy*, ed. D. Sparks, Vol. 56, 1–54. Academic Press. http://www.sciencedirect.com/science/article/pii/S0065211308601789.

Doran, J.W., and M. Werner. 1990. Management and Soil Biology. In *Sustainable Agriculture in Temperate Zones*, eds. C.A. Francis, C.B. Flora, and L.D. King, 205–230. New York: Jhon Wiley and Sons.

Doran, J.W., and M.R. Zeiss. 2000. Soil Health and Sustainability: Managing the Biotic Component of Soil Quality. *Applied Soil Ecology* 15, no. 1 (August): 3–11. http://www.sciencedirect.com/science/article/pii/S0929139300000676.

Ericson, B., J. Caravanos, K. Chatham-Stephens, P. Landrigan, and R. Fuller. 2013. Approaches to Systematic Assessment of Environmental Exposures Posed at Hazardous Waste Sites in the Developing World: The Toxic Sites Identification Program. *Environmental Monitoring and Assessment* 185, no. 2 (February): 1755–1766.

Esperschütz, J., A. Gattinger, P. Mäder, M. Schloter, and A. Fliessbach. 2007. Response of Soil Microbial Biomass and Community Structures to Conventional and Organic Farming Systems under Identical Crop Rotations. *FEMS Microbiology Ecology* 61, no. 1 (July): 26–37.

FAO. 2015. 2015: International Year of Soils. http://www.fao.org/soils-2015/en/.

Fischetti, M. 2015. Climate Change Hastened Syria's Civil War. *Scientific American*. https://www.scientificamerican.com/article/climate-change-hastened-the-syrian-war/.

Fouad, A.M., D. Ruan, H.A.K. El Senousey, W. Chen, S. Jiang, and C. Zheng. 2019. Harmful Effects and Control Strategies of Aflatoxin B1 Produced by Aspergillus Flavus and Aspergillus Parasiticus Strains on Poultry: Review. *Toxins*. https://www.mdpi.com/2072-6651/11/3/176.

Franzluebbers, A.J., and A.N. Sharpley. 2017a. Macronutrients in Soils and Plants, and Their Impacts on Animal and Human Health. In *The Nexus of Soils, Plants, Animals and Human Health*, eds. B.R. Singh, M.J. McLaughlin, and E.C. Brevik, 58–63. Stuttgart: Catena-Schweizerbart.

Gambino-Shirley, K., L. Stevenson, J. Concepción-Acevedo, E. Trees, D. Wagner, L. Whitlock, J. Roberts, et al. 2018. Flea Market Finds and Global Exports: Four Multistate Outbreaks of Human Salmonella Infections Linked to Small Turtles, United States–2015. *Zoonoses and Public Health* 65, no. 5 (August): 560–568. http://doi.wiley.com/10.1111/zph.12466.

Gao, X., M. Wu, R. Xu, X. Wang, R. Pan, H.-J. Kim, and H. Liao. 2014. Root Interactions in a Maize/Soybean Intercropping System Control Soybean Soil-Borne Disease, Red Crown Rot. *PLoS One* 9, no. 5: e95031.

Gerber, P.J., H. Steinfeld, B. Henderson, A. Mottet, C. Opio, J. Dijkman, A. Falcucci, and G. Tempio. 2013. *Tackling Climate Change through Livestock – A Global Assessment of Emissions and Mitigation Opportunities*. Rome, Italy: Food and Agricultural Organization of the United Nations (FAO). http://www.fao.org/3/a-i3437e.pdf.

Golden, M.H.N. 1982. Trace Elements in Human Nutrition. *Human Nutrition. Clinical Nutrition* 36, no. 3: 185–202. https://pascal-francis.inist.fr/vibad/index.php?action=getRecordDetail&idt=PASCAL82X0333666.

González, M.G., M.E. Conti, R.M. Palma, and N.M. Arrigo. 2003. Dynamics of Humic Fractions and Microbial Activity under No-Tillage or Reduced Tillage, as Compared with Native Pasture (Pampa Argentina). *Biology and Fertility of Soils* 39, no. 2: 135–138. doi:10.1007/s00374-003-0691-5.

Gore, A.C., V.A. Chappell, S.E. Fenton, J.A. Flaws, A. Nadal, G.S. Prins, J. Toppari, and R.T. Zoeller. 2015. Executive Summary to EDC-2: The Endocrine Society's Second Scientific Statement on Endocrine-Disrupting Chemicals. *Endocrine Reviews* 36, no. 6 (December): 593–602. https://www.ncbi.nlm.nih.gov/pubmed/26414233.

Greenlee, J.J., and M. Heather West Greenlee. 2015. The Transmissible Spongiform Encephalopathies of Livestock. *ILAR Journal* 56, no. 1 (May 19): 7–25.

Gupta, R.C., ed. 2007. *Veterinary Toxicology*. Oxford: Academic Press. http://www.sciencedirect.com/science/article/pii/B9780123704672501930.

Haas, D., and G. Défago. 2005. Biological Control of Soil-Borne Pathogens by Fluorescent Pseudomonads. *Nature Reviews Microbiology* 3, no. 4: 307–319. doi:10.1038/nrmicro1129.

Hackenberger, D.K., G. Palijan, Ž. Lončarić, O. Jovanović Glavaš, and B.K. Hackenberger. 2018. Influence of Soil Temperature and Moisture on Biochemical Biomarkers in Earthworm and Microbial Activity after Exposure to Propiconazole and Chlorantraniliprole. *Ecotoxicology and Environmental Safety* 148 (February 1): 480–489.

De Haes, H.A.U., R.L. Voortman, D.W. Bussink, C.W. Rougoor, and W.J. van der Weijden. 2012. *Scarcity of Micronutrients in Soil, Feed, Food and Mineral Reserves*. Culemborg: IATP (Institute for Agriculture & Trade Policy). https://www.iatp.org/documents/scarcity-of-micronutrients-in-soil-feed-food-and-mineral-reserves.

Harris, L.J., V. Lieberman, R.P. Mashiana, E. Atwill, M. Yang, J.C. Chandler, B. Bisha, and T. Jones. 2016. Prevalence and Amounts of Salmonella Found on Raw California Inshell Pistachios. *Journal of Food Protection* 79, no. 8 (August 1): 1304–1315.

He, Y., Z. Chen, X. Liu, C. Wang, and W. Lu. 2014. Influence of Trace Elements Mixture on Bacterial Diversity and Fermentation Characteristics of Liquid Diet Fermented with Probiotics under Air-Tight Condition. *PLoS One* 9, no. 12 (December 8): e114218. doi:10.1371/journal.pone.0114218.

Hjelle, B., and F. Torres-Pérez. 2010. Hantaviruses in the Americas and Their Role as Emerging Pathogens. *Viruses*. http://www.mdpi.com/1999-4915/2/12/2559.

Howard, A. 1943. *An Agricultural Testament*. New York: Oxford University Press.

Howard, A. 1945. *The Soil and Health: A Study of Organic Agriculture*. Lexington, KY: University Press of Kentucky.

Hüberli, D., M. Connor, S. Miyan, W. McLeod, R. Battaglia, L. Forsyth, B. Parkin, T. Klein, and M.A. Clarke M, Robertson G, Correll R, Desbiolles J, Bogacki P. 2015. Rhizoctonia Solant AG8: New

Breakthroughs in Control and Management. *GRDC Crop Research Updates*. Perth, WA. https://grdc. com.au/resources-and-publications/grdc-update-papers/tab-content/grdc-update-papers/2015/02/ rhizoctonia-solani-ag8-new-breakthroughs-in-control-and-management.

Islam, K.R., and R.R. Weil. 2000. Soil Quality Indicator Properties in Mid-Atlantic Soils as Influenced by Conservation Management. *Journal of Soil and Water Conservation* 55, no. 1 (January 1): 69–78. http:// www.jswconline.org/content/55/1/69.abstract.

IUSS. 2002. World Soil Day. https://www.iuss.org/meetings-events/world-soil-day/.

IUSS. 2015. IUSS Proclaims the International Decade of Soils 2015–2024. *Vienna Soil Declaration*. Vienna. http://www-naweb.iaea.org/nafa/swmn/Vienna-Soil-Declaration-Dec6–2015.pdf.

Jansson, J.K., and K.S. Hofmockel. 2018. The Soil Microbiome–from Metagenomics to Metaphenomics. *Current Opinion in Microbiology* 43: 162–168. http://www.sciencedirect.com/science/article/pii/ S1369527417302205.

Jeffery, S., and W.H. van der Putten. 2011. *Soil Borne Diseases of Humans*. http://www.jrc.ec.europa.eu/.

Kellogg, C.E. 1938. Soil and Society. In *Soils and Men: USDA Yearbook of Agriculture*, eds. H. Knight, C.E. Kellogg, C.P. Barnes, M.A. McCall, B.W. Allin, and A.L. Patrick. Washington, DC: United States Government Printing Office. https://naldc.nal.usda.gov/download/IND50000140/PDF.

Kemper, K.J., J. Lakritz, and R. Lal. 2017. The Soil-Animal-Human Health Nexus. In *The Nexus of Soils, Plants, Animals, and Human Health*, eds. E. Singh, B.R. McLaughlin, and M.J. Brevik, 16–20. Stuttgart: GeoEcology Essays.

Kemper, K.J., J. Lakritz, and R. Lal. 2019. Soil and Human Health in a Changing Climate. In *Soil and Climate*, eds. R. Lal and B.A. Stewart, 403–418. Boca Raton, FL: CRC Press.

Khan, S.M., J. Gomes, and D.R. Krewski. 2019. Radon Interventions around the Globe: A Systematic Review. *Heliyon*. https://www.sciencedirect.com/science/article/pii/S2405844019336813.

King, J., and R. Cousins. 2006. Zinc. In *Modern Nutrition in Health and Disease*, eds. M. Shils, M. Shike, A. Ross, B. Caballero, and R. Cousins, 271–285. 10th ed. Baltimore, MD: Lippincott Williams & Wilkins.

King, L.J., L.R. Anderson, C.G. Blackmore, M.J. Blackwell, E.A. Lautner, L.C. Marcus, T.E. Meyer, et al. 2008. Executive Summary of the AVMA One Health Initiative Task Force Report. *Journal of the American Veterinary Medical Association* 233, no. 2 (July 15): 259–261.

Kloepper, J.W., C.-M. Ryu, and S. Zhang. 2004. Induced Systemic Resistance and Promotion of Plant Growth by Bacillus Spp. *Phytopathology* 94, no. 11 (November): 1259–1266.

Kuepper, G., and L. Gegner. 2004. Organic Crop Production Overview: Fundamentals of Sustainable Agriculture. *ATTRA National Sustainable Agriculture Information Service*. Washington, DC: U.S. Department of Agriculture. http://www.attra.ncat.org/attra-pub/PDF/organiccrop.pdf.

Kumar, R., T.K. Datta, and K.V. Lalitha. 2015. Salmonella Grows Vigorously on Seafood and Expresses Its Virulence and Stress Genes at Different Temperature Exposure. *BMC Microbiology* 15, no. 1 (December 1): 254.

Lal, R. 2004. Soil Carbon Sequestration Impacts on Global Climate Change and Food Security. *Science* 304, no. 5677 (June): 1623–1627.

Lal, R. 2007. Carbon Management in Agricultural Soils. *Mitigation and Adaptation Strategies for Global Change* 12, no. 2: 303–322. doi:10.1007/s11027-006-9036-7.

Lal, R. 2014. Societal Value of Soil Carbon. *Journal of Soil and Water Conservation* 69, no. 6: 186A–192A. www.swcs.org.

Lal, R. 2016. Soil Health and Carbon Management. *Food and Energy Security* 5, no. 4 (November 1): 212–222. http://doi.wiley.com/10.1002/fes3.96.

Lal, R. 2017. Urban Agriculture and Food Security. In *Soils within Cities*, eds. M. Levin, K. Kim, J. Morel, W. Burghardt, P. Charzyński, and R. Shaw, 177–180. Stuttgart, Germany: Catena Soil Sciences.

Landrigan, P.J., C.B. Schechter, J.M. Lipton, M.C. Fahs, and J. Schwartz. 2002. Environmental Pollutants and Disease in American Children: Estimates of Morbidity, Mortality, and Costs for Lead Poisoning, Asthma, Cancer, and Developmental Disabilities. *Environmental Health Perspectives* 110, no. 7 (July): 721–728. https://www.ncbi.nlm.nih.gov/pubmed/12117650.

Larkin, R.P. 2015. Soil Health Paradigms and Implications for Disease Management. *Annual Review of Phytopathology* 53: 199–221. https://www.ncbi.nlm.nih.gov/pubmed/26002292.

Lebov, J., K. Grieger, D. Womack, D. Zaccaro, N. Whitehead, B. Kowalcyk, and P.D.M. MacDonald. 2017. A Framework for One Health Research. *One Health* 3: 44–50. http://www.sciencedirect.com/science/ article/pii/S2352771416300696.

Lehmann, A., S.D. Veresoglou, E.F. Leifheit, and M.C. Rillig. 2014. Arbuscular Mycorrhizal Influence on Zinc Nutrition in Crop Plants – A Meta-Analysis. *Soil Biology and Biochemistry* 69: 123–131. http://www. sciencedirect.com/science/article/pii/S0038071713004045.

Lekberg, Y., V. Wagner, A. Rummel, M. McLeod, and P.W. Ramsey. 2017. Strong Indirect Herbicide Effects on Mycorrhizal Associations through Plant Community Shifts and Secondary Invasions. *Ecological Applications* 27, no. 8 (December 1): 2359–2368.

Leopold, A. 1949. *A Sand County Almanac: And Sketches Here and There.* New York: Oxford University Press.

Li, F.-Q., Y.-W. Li, Y.-R. Wang, and X.-Y. Luo. 2009. Natural Occurrence of Aflatoxins in Chinese Peanut Butter and Sesame Paste. *Journal of Agricultural and Food Chemistry* 57, no. 9 (May 13): 3519–3524. https://www.ncbi.nlm.nih.gov/pubmed/19338351.

Li, R., E. Khafipour, D.O. Krause, M.H. Entz, T.R. de Kievit, and W.G.D. Fernando. 2012. Pyrosequencing Reveals the Influence of Organic and Conventional Farming Systems on Bacterial Communities. *PloS One* 7, no. 12: e51897.

Lim, J.-A., D.H. Lee, and S. Heu. 2014. The Interaction of Human Enteric Pathogens with Plants. *The Plant Pathology Journal* 30, no. 2 (June): 109–116. https://pubmed.ncbi.nlm.nih.gov/25288993.

López-Alonso, M. 2012. Trace Minerals and Livestock: Not Too Much Not Too Little. Ed. Ø Bergh, P Butaye, M H Kogut, and S Whisnant. *ISRN Veterinary Science* 2012: 704825. doi:10.5402/2012/704825.

López Alonso, M., J.L. Benedito, M. Miranda, C. Castillo, J. Hernández, and R.F. Shore. 2000. The Effect of Pig Farming on Copper and Zinc Accumulation in Cattle in Galicia (North-Western Spain). *Veterinary Journal (London, England: 1997)* 160, no. 3 (November): 259–266. https://www.ncbi.nlm.nih.gov/pubmed/11061963.

Lundberg, D.S., S.L. Lebeis, S.H. Paredes, S. Yourstone, J. Gehring, S. Malfatti, J. Tremblay, et al. 2012. Defining the Core Arabidopsis Thaliana Root Microbiome. *Nature* 488, no. 7409: 86–90. doi:10.1038/nature11237.

Lundberg, D.S., S. Yourstone, P. Mieczkowski, C.D. Jones, and J.L. Dangl. 2013. Practical Innovations for High-Throughput Amplicon Sequencing. *Nature Methods* 10, no. 10: 999–1002. doi:10.1038/nmeth.2634.

Maki, D.G. 2009. Coming to Grips with Foodborne Infection – Peanut Butter, Peppers, and Nationwide Salmonella Outbreaks. *New England Journal of Medicine* 360, no. 10 (March 5): 949–953.

Marshall, B.M., and S.B. Levy. 2011. Food Animals and Antimicrobials: Impacts on Human Health. *Clinical Microbiology Reviews.* http://cmr.asm.org/.

Marzano, S.-Y.L. 2012. Assessment of Disease Suppression in Organic Transitional Cropping Systems. PhD Dissertation. University of Illinois at Urbana-Champaign.

Mayer, A.M. 1997. Historical Changes in the Mineral Content of Fruits and Vegetables. *British Food Journal* 99, no. 6: 207–211.

Mekonnen, M.M., and A.Y. Hoekstra. 2012. A Global Assessment of the Water Footprint of Farm Animal Products. *Ecosystems* 15, no. 3 (April 24): 401–415. doi:10.1007/s10021-011-9517-8.

Mendes, R., P. Garbeva, and J.M. Raaijmakers. 2013. The Rhizosphere Microbiome: Significance of Plant Beneficial, Plant Pathogenic, and Human Pathogenic Microorganisms. *FEMS Microbiology Reviews* 37, no. 5 (September): 634–663.

Mendes, R., M. Kruijt, I. de Bruijn, E. Dekkers, M. van der Voort, J.H.M. Schneider, Y.M. Piceno, et al. 2011. Deciphering the Rhizosphere Microbiome for Disease-Suppressive Bacteria. *Science* 332, no. 6033 (May 27): 1097–1100. https://www.ncbi.nlm.nih.gov/pubmed/21551032.

Meng, L., T. Sun, M. Li, M. Saleem, Q. Zhang, and C. Wang. 2019. Soil-Applied Biochar Increases Microbial Diversity and Wheat Plant Performance under Herbicide Fomesafen Stress. *Ecotoxicology and Environmental Safety* 171: 75–83. https://www.sciencedirect.com/science/article/pii/S0147651318313642.

Merlin, C., M. Devers, J. Béguet, B. Boggio, N. Rouard, and F. Martin-Laurent. 2016. Evaluation of the Ecotoxicological Impact of the Organochlorine Chlordecone on Soil Microbial Community Structure, Abundance, and Function. *Environmental Science and Pollution Research* 23, no. 5 (March 1): 4185–4198.

Meyer, M.C., C.J. Bueno, N.L. de Souza, and J.T. Yorinori. 2006. Effect of Doses of Fungicides and Plant Resistance Activators on the Control of Rhizoctonia Foliar Blight of Soybean, and on Rhizoctonia Solani AG1–IA in Vitro Development. *Crop Protection* 25, no. 8: 848–854. http://www.sciencedirect.com/science/article/pii/S0261219405003157.

Milićević, D., R. Petronijević, Z. Petrović, J. Djinović-Stojanović, J. Jovanović, T. Baltić, and S. Janković. 2019. Impact of Climate Change on Aflatoxin M1 Contamination of Raw Milk with Special Focus on Climate Conditions in Serbia. *Journal of the Science of Food and Agriculture* 99, no. 11 (August 30): 5202–5210.

Monokrousos, N., E.M. Papatheodorou, J.D. Diamantopoulos, and G.P. Stamou. 2006. Soil Quality Variables in Organically and Conventionally Cultivated Field Sites. *Soil Biology and Biochemistry* 38, no. 6: 1282–1289. http://www.sciencedirect.com/science/article/pii/S0038071705003421.

Moral, R., M.D. Perez-Murcia, A. Perez-Espinosa, J. Moreno-Caselles, C. Paredes, and B. Rufete. 2008. Salinity, Organic Content, Micronutrients and Heavy Metals in Pig Slurries from South-Eastern Spain. *Waste Management* 28, no. 2: 367–371. http://www.sciencedirect.com/science/article/pii/S0956053X07000384.

Mottet, A., C. de Haan, A. Falcucci, G. Tempio, C. Opio, and P. Gerber. 2017. Livestock: On Our Plates or Eating at Our Table? A New Analysis of the Feed/Food Debate. *Global Food Security* 14: 1–8. http://www.sciencedirect.com/science/article/pii/S2211912416300013.

Muir, J. 1911. *My First Summer in the Sierra*. San Francisco, CA: Houghton Mifflin Company.

Murphy, T., K. Phan, E. Yumvihoze, K. Irvine, K. Wilson, D. Lean, B. Ty, A. Poulain, B. Laird, and L.H.M. Chan. 2018. Groundwater Irrigation and Arsenic Speciation in Rice in Cambodia. *Journal of Health and Pollution* 8, no. 19 (September 1): 180911.

National Research Council. 2005. *Mineral Tolerance of Animals: Second Revised Edition, 2005*. Washington, DC: The National Academies Press. https://www.nap.edu/catalog/11309/mineral-tolerance-of-animals-second-revised-edition-2005.

Nawab, J., S. Farooqi, W. Xiaoping, S. Khan, and A. Khan. 2018. Levels, Dietary Intake, and Health Risk of Potentially Toxic Metals in Vegetables, Fruits, and Cereal Crops in Pakistan. *Environmental Science and Pollution Research* 25, no. 6 (February 1): 5558–5571.

Ohno, T., and G.M. Hettiarachchi. 2018. Soil Chemistry and the One Health Initiative: Introduction to the Special Section. *Journal of Environmental Quality* 47, no. 6 (November): 1305–1309. https://www.ncbi.nlm.nih.gov/pubmed/30512058.

Parniske, M. 2008. Arbuscular Mycorrhiza: The Mother of Plant Root Endosymbioses. *Nature Reviews Microbiology* 6, no. 10: 763–775. doi:10.1038/nrmicro1987.

Paungfoo-Lonhienne, C., D. Rentsch, S. Robatzek, R.I. Webb, E. Sagulenko, T. Näsholm, S. Schmidt, and T.G.A. Lonhienne. 2010. Turning the Table: Plants Consume Microbes as a Source of Nutrients. *PLoS One* 5, no. 7. https://www.ncbi.nlm.nih.gov/pmc/articles/PMC2912860/.

Pérez-Escamilla, R. 2017. Food Security and the 2015–2030 Sustainable Development Goals: From Human to Planetary Health. *Current Developments in Nutrition* 1, no. 7: e000513. https://academic.oup.com/cdn/article-abstract/1/7/e000513/4259862.

Pérez-Gutiérrez, J.D., and S. Kumar. 2019. Simulating the Influence of Integrated Crop-Livestock Systems on Water Yield at Watershed Scale. *Journal of Environmental Management* 239: 385–394. http://www.sciencedirect.com/science/article/pii/S0301479719303718.

Perotti, R. 1926. On the Limits of Biological Inquiry on Soil Science. *Proceeding of International Society of Soil Science* 2: 146–161.

Pozo, M.J., and C. Azcón-Aguilar. 2007. Unraveling Mycorrhiza-Induced Resistance. *Current Opinion in Plant Biology* 10, no. 4: 393–398. http://www.sciencedirect.com/science/article/pii/S1369526607000702.

Pritzkow, S., R. Morales, F. Moda, U. Khan, G.C. Telling, E. Hoover, and C. Soto. 2015. Grass Plants Bind, Retain, Uptake, and Transport Infectious Prions. *Cell Reports* 11, no. 8 (May 26): 1168–1175. https://pubmed.ncbi.nlm.nih.gov/25981035.

Raisbeck, M.F. 2000. Selenosis. *Veterinary Clinics of North America: Food Animal Practice* 16, no. 3: 465–480. http://www.sciencedirect.com/science/article/pii/S0749072015300815.

Reeves, R.D., A.J.M. Baker, T. Jaffré, P.D. Erskine, G. Echevarria, and A. van der Ent. 2017. A Global Database for Plants That Hyperaccumulate Metal and Metalloid Trace Elements. *New Phytologist* 218: 407–411.

Reicosky, D., and H. Janzen. 2018. Conservation Agriculture: Maintaining Land Productivity and Health by Managing Carbon Flows. In *Soil and Climate*, eds. R. Lal and B.A. Stewart, 32. Boca Raton, FL: CRC Press. https://www.taylorfrancis.com/books/e/9780429487262/chapters/10.1201/b21225-4.

Reinhold-Hurek, B., and T. Hurek. 2011. Living inside Plants: Bacterial Endophytes. *Current Opinion in Plant Biology* 14, no. 4: 435–443. http://www.sciencedirect.com/science/article/pii/S1369526611000549.

Reuveny, R. 2007. Climate Change-Induced Migration and Violent Conflict. *Political Geography* 26, no. 6: 656–673. https://www.sciencedirect.com/science/article/pii/S0962629807000601.

Rillig, M.C. 2004. Arbuscular Mycorrhizae and Terrestrial Ecosystem Processes. *Ecology Letters* 7, no. 8 (August 1): 740–754. doi:10.1111/j.1461-0248.2004.00620.x.

Romig, D.E., M.J. Garlynd, and R.F. Harris. 1997. Farmer-Based Assessment of Soil Quality: A Soil Health Scorecard. *Methods for Assessing Soil Quality*. SSSA Special Publications. doi:10.2136/sssaspecpub49.c3.

Rosenberger, G., G. Dirksen, H.D. Grunder, E. Grunert, D. Krause, and M. Stober. 1979. *Clinical Examination of Cattle*. Parey. https://books.google.com/books?id=U_0OAQAAMAAJ.

Rosenblueth, M., and E. Martínez-Romero. 2006. Bacterial Endophytes and Their Interactions with Hosts. *Molecular Plant-Microbe Interactions : MPMI* 19, no. 8 (August): 827–837.

Ruiz, D., M. Becerra, J.S. Jagai, K. Ard, and R.M. Sargis. 2018. Disparities in Environmental Exposures to Endocrine-Disrupting Chemicals and Diabetes Risk in Vulnerable Populations. *Diabetes Care* 41, no. 1: 193–205. http://care.diabetesjournals.org/lookup/suppl/doi:10.2337/dc16-2765/-/DC1.

Säle, V., P. Aguilera, E. Laczko, P. Mäder, A. Berner, U. Zihlmann, M.G.A. van der Heijden, and F. Oehl. 2015. Impact of Conservation Tillage and Organic Farming on the Diversity of Arbuscular Mycorrhizal Fungi. *Soil Biology and Biochemistry* 84: 38–52. http://www.sciencedirect.com/science/article/pii/S0038071715000504.

Science Communication Unit. 2013. Science for Environment Policy In-Depth Report: Soil Contamination: Impacts on Human Health. Report Produced for the European Commission DG Environment. Bristol, UK. https://ec.europa.eu/environment/integration/research/newsalert/pdf/IR5_en.pdf.

Shafer, S. 2018. Opinion: Why It's Important to Connect Soil Health and Human Health Science. *AgriPulse*. https://www.agri-pulse.com/articles/11482-opinion-why-its-important-to-connect-soil-health-and-human-health-science

Shomaker, T.S., E.M. Green, and S.M. Yandow. 2013. Perspective: One Health: A Compelling Convergence. *Academic Medicine*. https://journals.lww.com/academicmedicine/fulltext/2013/01000/Perspective___One_Health___A_Compelling.19.aspx.

Shoresh, M., and G.E. Harman. 2008. The Molecular Basis of Shoot Responses of Maize Seedlings to Trichoderma Harzianum T22 Inoculation of the Root: A Proteomic Approach. *Plant Physiology* 147, no. 4 (August): 2147–2163. https://pubmed.ncbi.nlm.nih.gov/18562766.

Shreenath, A.P., and J. Dooley. 2020. Selenium Deficiency. In *StatPearls [Internet]*. Treasure Island, FL: StatPearls Publishing.

Shrestha, P., K. Bellitürk, and J.H. Görres. 2019. Phytoremediation of Heavy Metal-Contaminated Soil by Switchgrass: A Comparative Study Utilizing Different Composts and Coir Fiber on Pollution Remediation, Plant Productivity, and Nutrient Leaching. *International Journal of Environmental Research and Public Health* 16, no. 7 (April 9): 1261. https://www.ncbi.nlm.nih.gov/pubmed/30970575.

Sigua, G.C., C.C. Chase, and J. Albano. 2014. Soil-Extractable Phosphorus and Phosphorus Saturation Threshold in Beef Cattle Pastures as Affected by Grazing Management and Forage Type. *Environmental Science and Pollution Research* 21, no. 3: 1691–1700. doi:10.1007/s11356-013-2050-x.

Slepetiene, A. 2001. Soil Organic Matter Characteristics as a Result of Management Systems and Geochemical Factors. In *Biogeochemical Processes and Cycling of Elements in the Environment*, eds. J. Weber, E. Jamroz, J. Drozd, and A. Karczewska, 301–302. Wroclaw, Poland: Polish Society of Humic Substances.

Song, T., X. Su, J. He, Y. Liang, T. Zhou, and C. Liu. 2018. Selenium (Se) Uptake and Dynamic Changes of Se Content in Soil-Plant Systems. *Environmental Science and Pollution Research International* 25, no. 34 (December): 34343–34350. https://www.ncbi.nlm.nih.gov/pubmed/30298355.

Steinfeld, H., P. Gerber, T.D. Wassenaar, FAO, V. Castel, M. Rosales, M.R. M, and C. de Haan. 2006. *Livestock's Long Shadow: Environmental Issues and Options*. Rome, Italy: Food and Agriculture Organization of the United Nations.

Stevens, C., V.A. Khan, R. Rodriguez-Kabana, L.D. Ploper, P.A. Backman, D.J. Collins, J.E. Brown, M.A. Wilson, and E.C.K. Igwegbe. 2003. Integration of Soil Solarization with Chemical, Biological and Cultural Control for the Management of Soilborne Diseases of Vegetables. *Plant and Soil* 253, no. 2: 493–506. doi:10.1023/A:1024895131775.

Stone, A.G., S.J. Scheuerell, and H.M. Darby. 2004. Chapter 5. Suppression of Soilborne Diseases in Field Agricultural Systems: Organic Matter Management, Cover Cropping, and Other Cultural Practices. In *Soil Organic Matter in Sustainable Agriculture*, eds. R.R. Weil and F. Magdoff., 131–177. Boca Raton, FL: CRC Press.

Sułowicz, S., and Z. Piotrowska-Seget. 2016. Response of Microbial Communities from an Apple Orchard and Grassland Soils to the First-Time Application of the Fungicide Tetraconazole. *Ecotoxicology and Environmental Safety* 124: 193–201. https://www.sciencedirect.com/science/article/pii/S0147651315301408.

Suttle, N.F. 2010. *Mineral Nutrition of Livestock*. Ed. S. Hulbert. Cabi Series. Cambridge, MA: CABI. https://books.google.com/books?id=SRcEZVPbVRQC.

Takashima, M., K. Shimada, and R.E. Speece. 2011. Minimum Requirements for Trace Metals (Iron, Nickel, Cobalt, and Zinc) in Thermophilic and Mesophilic Methane Fermentation from Glucose. *Water Environment Research: A Research Publication of the Water Environment Federation* 83, no. 4 (April): 339–346. https://www.ncbi.nlm.nih.gov/pubmed/21553589.

Teague, W.R. 2018. Forages and Pastures Symposium: Cover Crops in Livestock Production: Whole-System Approach: Managing Grazing to Restore Soil Health and Farm Livelihoods. *Journal of Animal Science* 96, no. 4 (April 14): 1519–1530. https://www.ncbi.nlm.nih.gov/pubmed/29401363.

Tejada, M., J.L. Gonzalez, A.M. García-Martínez, and J. Parrado. 2008. Effects of Different Green Manures on Soil Biological Properties and Maize Yield. *Bioresource Technology* 99, no. 6: 1758–1767. http://www.sciencedirect.com/science/article/pii/S096085240700301X.

Tkacz, A., and P. Poole. 2015. Role of Root Microbiota in Plant Productivity. *Journal of Experimental Botany* 66, no. 8 (April): 2167–2175.

Trivedi, P., Z. He, J.D. Van Nostrand, G. Albrigo, J. Zhou, and N. Wang. 2012. Huanglongbing Alters the Structure and Functional Diversity of Microbial Communities Associated with Citrus Rhizosphere. *The ISME Journal* 6, no. 2 (February): 363–383.

Trujillo-González, J.M., M.A. Torres-Mora, S. Keesstra, E.C. Brevik, and R. Jiménez-Ballesta. 2016. Heavy Metal Accumulation Related to Population Density in Road Dust Samples Taken from Urban Sites under Different Land Uses. *Science of the Total Environment* 553: 636–642. https://www.sciencedirect.com/science/article/pii/S0048969716303126.

UN. 2019. *World Population Prospects 2019: Data Booklet (ST/ESA/SER.A/424)*. Rome, Italy: United Nations, Department of Economic and Social Affairs, Population Division. https://population.un.org/wpp/Publications/Files/WPP2019_DataBooklet.pdf.

Upreti, R., and P. Thomas. 2015. Root-Associated Bacterial Endophytes from Ralstonia Solanacearum Resistant and Susceptible Tomato Cultivars and Their Pathogen Antagonistic Effects. *Frontiers in Microbiology* 6 (April 14): 255. https://pubmed.ncbi.nlm.nih.gov/25926818.

Ussiri, D.A.N., and R. Lal. 2018. The Soil-Livestock-Climate Nexus. In *Soil and Climate*, eds. R. Lal and B.A. Stewart, 44. Boca Raton, FL: CRC Press.

Wang, L., S. Gao, X. Yin, X. Yu, and L. Luan. 2019. Arsenic Accumulation, Distribution and Source Analysis of Rice in a Typical Growing Area in North China. *Ecotoxicology and Environmental Safety* 167: 429–434. https://www.sciencedirect.com/science/article/pii/S0147651318310133.

Wang, Z., X. Yuan, D. Wang, Y. Zhang, Z. Zhong, Q. Guo, and C. Feng. 2018. Large Herbivores Influence Plant Litter Decomposition by Altering Soil Properties and Plant Quality in a Meadow Steppe. *Scientific Reports* 8, no. 1: 9089. doi:10.1038/s41598-018-26835-1.

Weller, D.M., J.M. Raaijmakers, B.B.M. Gardener, and L.S. Thomashow. 2002. Microbial Populations Responsible for Specific Soil Suppressiveness to Plant Pathogens. *Annual Review of Phytopathology* 40: 309–348.

Wells, C.L., and T.D. Wilkins. 1996. *Clostridia: Sporeforming Anaerobic Bacilli. Medical Microbiology*. https://www.ncbi.nlm.nih.gov/books/NBK8219/.

Whiley, H., and K. Ross. 2015. Salmonella and Eggs: From Production to Plate. *International Journal of Environmental Research and Public Health* 12, no. 3: 2543–2556. www.mdpi.com/journal/ijerph.

Wilson, W.S. 1991. *Advances in Soil Organic Matter Research the Impact on Agriculture and the Environment*. Ed. W.S. Wilson, T.R.G. Gray, D.J. Greenslade, R.M. Harrison, and M.H.B. Hayes. Oxford: Woodhead Publishing. http://www.sciencedirect.com/science/article/pii/B9781855738133500020.

Xia, Y., S. DeBolt, J. Dreyer, D. Scott, and M.A. Williams. 2015. Characterization of Culturable Bacterial Endophytes and Their Capacity to Promote Plant Growth from Plants Grown Using Organic or Conventional Practices. *Frontiers in Plant Science* 6: 490.

Xia, Y., M.R. Sahib, A. Amna, S.O. Opiyo, Z. Zhao, and Y.G. Gao. 2019. Culturable Endophytic Fungal Communities Associated with Plants in Organic and Conventional Farming Systems and Their Effects on Plant Growth. *Scientific Reports* 9, no. 1: 1669. doi:10.1038/s41598-018-38230-x.

Yucel, D., C. Yucel, E.L. Aksakal, K. Barik, M. Khosa, I. Aziz, and K.R. Islam. 2015. Impacts of Biosolids Application on Soil Quality under Alternate Year No-till Corn–Soybean Rotation. *Water, Air, & Soil Pollution* 226, no. 6: 168. doi:10.1007/s11270-015-2430-6.

Zebarth, B.J., G.H. Neilsen, E. Hogue, and D. Neilsen. 1999. Influence of Organic Waste Amendments on Selected Soil Physical and Chemical Properties. *Canadian Journal of Soil Science* 79, no. 3: 501–504.

Zhuang, P., M.B. McBride, H. Xia, N. Li, and Z. Li. 2009. Health Risk from Heavy Metals via Consumption of Food Crops in the Vicinity of Dabaoshan Mine, South China. *The Science of the Total Environment* 407, no. 5 (February 15): 1551–1561. https://www.ncbi.nlm.nih.gov/pubmed/19068266.

3 Transport of Mineral Elements from Soil and Human Health

Sheng Huang and Jian Feng Ma
Okayama University

CONTENTS

3.1 INTRODUCTION

Soil can affect human health in both beneficial and toxic ways in terms of mineral nutrition. The beneficial way is to provide mineral nutrients required for human health. There are 23 mineral elements for human heath, which are directly or indirectly taken from soil through soil–crop–human continuum. On the other hand, the toxic way is to supply toxic mineral elements to the human diets. Due to rapid urbanization and industrialization, many soils for the crop production are contaminated by toxic elements such as cadmium (Cd) and arsenic (As) (Harvey et al., 2002; Zhao et al., 2010; Clemens and Ma, 2016). The crops produced on these contaminated soils accumulate high-toxic elements, which has become a global concern of food safety.

Mineral elements in soil affect human health through crops. There are many steps for the transport of mineral elements from soil to the edible parts for human consumption. Mineral elements in soil are first taken up by the plant roots and then translocated to the shoots and finally delivered to the edible parts for human consumption (Huang et al., 2020b). All these processes are mediated by various different transporters (Sasaki et al., 2016; Mitani-Ueno et al., 2018). Therefore, to link soil and human health, it is very important to understand and control the transport system of mineral elements in crops. In this chapter, recent progresses made in rice are described on the essential mineral elements; iron (Fe), zinc (Zn), and toxic elements including Cd and As. Rice is a staple food for half of world's population, therefore, it is a major source of both essential and toxic elements for humans.

3.2 TRANSPORT OF ESSENTIAL MINERAL ELEMENTS IN RICE

Dietary deficiency of Fe and Zn has been estimated as the 16th and 40th leading risk factors, respectively, underlying global burden of disease (GBD, 2016). It is estimated that nearly 2 billion people are suffering from micronutrients such as Fe and Zn (Graham et al., 2012). Deficiency of Fe and Zn causes various diseases. For example, Fe deficiency causes anemia, while Zn deficiency results in dysgeusia, which happens not only in developing countries but also in developed countries (White and Broadley, 2009). This so-called "hidden hunger" is especially a serious health problem for people subsisting on cereal-based diets such as rice (Kennedy et al., 2003; Von Grebmer et al., 2014; Nakandalage and Seneweera, 2018). Rice is a major source of energy and protein for the rice-eating population but cannot provide sufficient amount of Fe and Zn for these people due to low concentration and low bioavailability (WHO, 2002; White and Broadley, 2005; Huang et al., 2020b). Brown rice contains 12 mg Fe kg^{-1} and 34 mg Zn kg^{-1} (average of 140 rice cultivars from different countries grown under the same soil) (Huang et al., 2020b). The polished rice contains even less Fe and Zn because most micronutrients are accumulated in the aleurone layers (White and Broadley, 2009), and 90% of Fe and 40% of Zn are lost during polishing. Furthermore, rice grain contains high phytic acid as a major storage form of P, which can strongly chelate with Fe and Zn, inhibiting their uptake by the guts and, therefore, decreasing their bioavailability (Raboy et al., 2009).

3.2.1 Transport of Fe in Rice

Iron is relatively abundant in many cultivated soils, but its solubility is very low, especially in calcareous and alkaline soils. However, under flooded condition, the Fe solubility greatly increases due to soil reduction (Wang et al., 2020). The major Fe form is oxidized insoluble Fe^{3+} compounds in upland soil and ferrous iron (Fe^{2+}) in paddy soil (Wang et al., 2020). Rice takes up both Fe^{2+} and Fe^{3+} depending on the growth condition. Uptake of Fe^{2+} is proposed to be mediated by OsIRT1 and OsIRT2, two members of the ZIP transporter family (Figure 3.1) (Bughio et al., 2002; Ishimaru et al., 2006; Wang et al., 2020). By contrast, to acquire the insoluble Fe in upland soil, rice roots secrete 2'-deoxymugineic acid (DMA), a phytosiderophore, through TOM1 transporter (Nozoye et al., 2011). DMA is able to chelate Fe^{3+}, and the resulting Fe(III)-DMA complex is taken up by the Yellow Stripe1-Like 15 (OsYSL15) transporter (Figure 3.1) (Inoue et al., 2009; Lee et al., 2009).

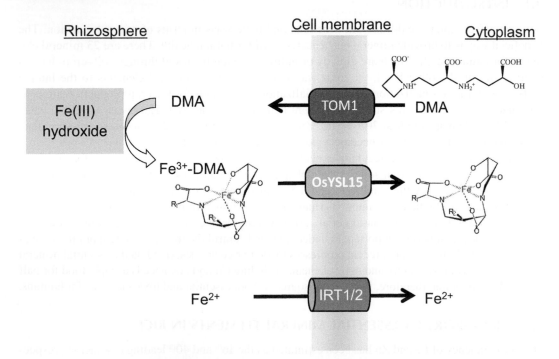

FIGURE 3.1   Fe uptake system in rice. Rice roots can take up Fe in the form of Fe^{2+} by IRT1/2 transporter, or Fe^{3+}-DMA complex by OsYSL15 depending on soil conditions. DMA, 2'-deoxymugineci acid, is secreted from cytoplasm to rhizosphere by TOM1. For details, refer to the text.

Since rice roots have two Casparian strips at both the exodermis and endodermis (Enstone et al., 2002) and a highly developed aerenchyma in mature roots in which almost all of the cortex cells between the exodermis and endodermis are destroyed. Therefore, to transport Fe across Casparian strip and aerenchyma toward the stele for subsequent translocation to the shoots, an efflux transporter for Fe is required, but this transporter has not been identified so far.

After uptake, a portion of Fe is sequestered into the vacuoles by Vacuolar Iron Transporter 1 and 2 (OsVIT1 and OsVIT2) (Zhang et al., 2012), while the remaining portion is loaded into the xylem by unknown transporters for Fe. However, FRD3-like protein 1 (OsFRDL1), a citrate efflux transporter is required for the efficient root-to-shoot translocation of Fe in rice (Yokosho et al., 2009). OsFRDL1 localized at the root pericycle is responsible for releasing citrate to the xylem, which chelates Fe(III) for translocation to the shoots.

Distribution of Fe to developing tissues including grain occurs in the nodes. Rice nodes have highly developed vascular systems, mainly consisting of enlarged vascular bundles (EVBs) and diffuse vascular bundles (DVBs) (Yamaji et al., 2014, 2017). EVBs come from the two lower nodes and are connected to the leaf attached to the node, while DVBs start at the node and are connected to the upper two nodes or panicle (Yamaji and Ma, 2014). Therefore, an intervascular transfer of mineral elements from EVBs to DVBs is required for their preferential distribution to developing tissues and panicles. However, transporters required for intervascular transfer of Fe have not been identified, but OsFRDL1 is also required for the distribution of Fe to panicles by releasing citrate to solubilize apoplastic Fe deposited in the parenchyma cells in node (Yokosho et al., 2016). Recently, OsVMT, a transporter for DMA, was found to be involved in the distribution of Fe in the nodes (Che et al., 2019). It is localized to the tonoplast in the node cells and responsible for the sequestration of DMA into the vacuoles. In addition, two members belonging to the rice Yellow Strip 1-Like (YSL) family, OsYSL2 and OsYSL9, are involved in Fe loading and distribution in developing rice grains, respectively (Ishimaru et al., 2010; Senoura et al., 2017).

Many attempts have been made to increase Fe concentration in the polished rice grain by manipulation of Fe transporters. For example, introduction of *OsYSL2* (encoding a Fe-NA and Mn-NA transporter) with *HvNAS1* (barley (*Hordeum vulgare*) NA synthase 1 gene) and *SferH2* (soybean (*Glycine max*) Ferritin gene) resulted in 4.4-fold Fe increases in the polished grain of transgenic lines grown in the field (Masuda et al., 2012). Knockout of *OsVMT* also resulted in a significant increase in Fe concentration in the polished rice grain (Che et al., 2019). This is because knockout of this gene decreased sequestration of DMA into the vacuoles, resulting in increased solubilization of Fe deposited in the node (Che et al., 2019).

3.2.2 TRANSPORT OF ZN IN RICE

Compared with Fe, the content of Zn in soil is very low. Zn in soil solution is present in the form of Zn^{2+}. Zn^{2+} is taken up by OsZIP9 as well as OsZIP1 (Figure 3.2) (Ramesh et al., 2003; Huang et al., 2020a); two members of ZIP family. OsZIP9 is localized at the exodermis and endodermis of the rice mature roots and is responsible for primary Zn uptake from soil (Figure 3.2) (Huang et al., 2020a). Its expression was induced by Zn deficiency. Knockout of this gene significantly decreases Zn uptake under Zn-limited condition in nutrient solution and soil (Huang et al., 2020a). However, the exact role of OsZIP1 in Zn uptake is not clear.

Sequestration of Zn to the root vacuoles is mediated by heavy metal P-type ATPase 3 (OsHMA3) (Ueno et al., 2010; Cai et al., 2019), while its root-to-shoot translocation is mediated by OsHMA2 localized at the root pericycle cells (Figure 3.2) (Yamaji et al., 2013). Distribution of Zn to the grain is mediated by two transporters; OsZIP3 and OsHMA2, which are localized at the nodes (Yamaji et al., 2013; Sasaki et al., 2015). OsZIP3 is localized at the xylem transfer cells in EVBs and is involved in unloading of Zn from the xylem of EVBs (Sasaki et al., 2015), while OsHMA2 is localized at the phloem region of both EVBs and DVBs and is responsible for loading Zn to the

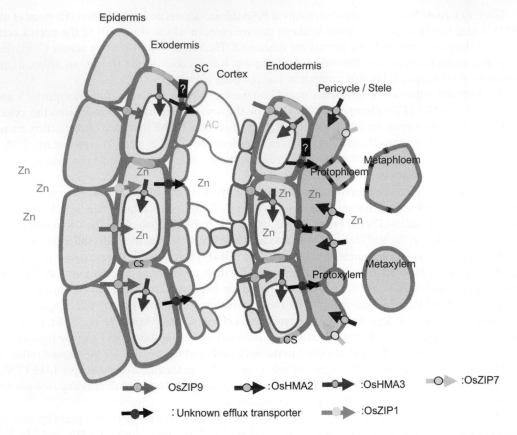

FIGURE 3.2 Schematic diagram of zinc (Zn) transport in rice roots. Zn is first taken up by OsZIP9/OsZIP1 localized at the exodermis from soil solution to exodermal cells and further taken up by OsZIP9 from the apoplastic solution in aerenchyma to endodermal cells. Part of Zn in the exodermal and endodermal cells is sequestered by OsHMA3 localized at the tonoplast into the vacuoles, and the remaining Zn is released by an unknown transporter. The root-to-shoot translocation is mediated by OsHMA2 and probably by OsZIP7 too localized at the pericycle cells. SC, schelanchyma cell; CS, Casparian strip; AC, aerenchyma.

phloem of DVBs and EVBs (Yamaji et al., 2013). In addition, OsZIP7 was also implicated in Zn xylem loading and distribution (Tan et al., 2019). Knockout of *OsVMT* also resulted in a significant increase in the Zn concentration in the polished rice grain (Che et al., 2019).

3.3 TRANSPORT OF TOXIC ELEMENTS

Although plants do not require Cd and As for their growth, these minerals will also be transported from soil to the edible parts, which threaten human health through the food chain. Excessive intake of Cd causes kidney failure, osteoporosis, and osteomalacia (Bertin and Averbeck, 2006; Nawrot et al., 2006), while chronic exposure to As causes numerous diseases, such as cancers, diabetes, cardiovascular disease, and developmental disorders (Abdul et al., 2015). Rice is a major dietary source of Cd and As, accounting for about 50% of total intake (Zhao et al., 2010; Clemens and Ma, 2016). The threshold value in rice grain has been set to 0.4 mg kg^{-1} for Cd and 0.2 mg kg^{-1} for As by the Codex Alimentarius Commission/World Health Organization with responsibility for the safety of food and human health (Codex Alimentarius Commission of Food and Agriculture Organization, 2007); however, rice produced on contaminated soil often exceeds these limits, representing a global concern of food safety. Therefore, it is important to reduce the transfer of these toxic elements from soil to grains in rice.

3.3.1 Transport of As in Rice

A number of As species may be present in soil depending on soil conditions, including arsenate [As(V)], arsenite [As(III)], and methylated As species such as monomethylarsonic acid [MMA(V)] and dimethylarsinic acid [DMA(V)]. Under aerobic conditions, arsenate (As(V)) is the dominant As species, which is taken up by phosphate transporters (Figure 3.3) (Abedin et al., 2002; Wu et al., 2011). Among 13 phosphate transporters (OsPht1;1-Pht1;13) localized on the plasma membrane in rice (*Oryza sativa*.), Pht1;1, Pht1;4, and Pht1;8 have been reported to be involved in As(V) uptake (Cao et al., 2017; Kamiya et al., 2013; Wang et al., 2016; Wu et al., 2011; Ye et al., 2017). However, the contribution of phosphate transport pathway to As accumulation in grain is small under flooded conditions (Wu et al., 2011). After uptake of As(V), it is readily reduced to As(III) by As(V) reductase HAC1;1, HAC1;2, and HAC4 (Figure 3.3) (Shi et al., 2016; Xu et al., 2017). A part of reduced As(III) is immediately effluxed outside the cell by Lsi1 and other unknown transporters (Zhao et al., 2010).

In the anaerobic paddy field, As is mainly present in the form of arsenite [As(III)], which is taken up by two Si transporters, namely, Lsi1 and Lsi2 (Ma et al., 2008) (Figure 3.3). Lsi1 and Lsi2 were initially identified as influx and efflux transporters for Si, respectively (Ma et al., 2006, 2007), but also show transport activity for As(III) (Ma et al., 2008). They are polarly localized at the exodermis and endodermis of rice roots and show high expression (Ma et al., 2006, 2007). This is one of the reasons for higher As accumulation in rice compared with other cereal crops such as barley and wheat. Knockout of *Lsi2* decreased 40%–50% As in the grain compared with wild-type rice (Ma et al., 2008). MMA and DMA can also be transported into the roots by Lsi1 but not by Lsi2 (Figure 3.3) (Li et al., 2009). Part of As(III) forms complexion with thiol compounds, like phytochelatins (PCs), which is subsequently sequestered into vacuoles by OsABCC1 (Song et al., 2014) (Figure 3.3).

At reproductive stage, the allocation of As to the grain is mediated by at least two transporters; Lsi2 and OsABCC1 (Figure 3.4) (Song et al., 2014; Chen et al., 2015). Knockout of *OsABCC1* resulted in 13 times As accumulation in brown rice (Song et al., 2014). Targeted overexpression of *OsABCC1* in

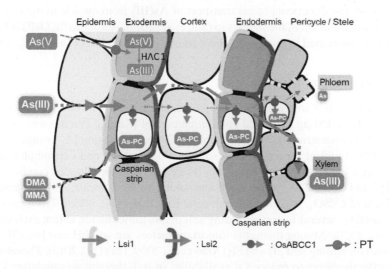

FIGURE 3.3 Transport pathway of arsenic in rice. As(V) is taken up by phosphate transporters, followed by reduction by HAC1 from As(V) to As(III). As(III) is taken up by two Si transporters; Lsi1 and Lsi2, which are polarly localized at the distal and proximal side of both exodermis and endodermis of the roots. Sequestration of As(III)-PC into the vacuoles is mediated by an ABC transporter, OsABCC1. (Modified from Clemens and Ma (2016).)

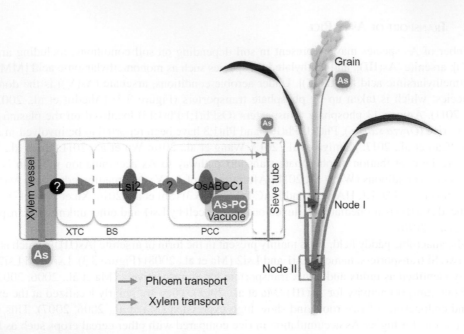

FIGURE 3.4 Transporters involved in the distribution of arsenic in the rice node. An efflux transporter Lsi2 localized at the bundle sheath (BS) is responsible for the intervascular transfer of As(III), while OsABCC1 localized at the tonoplast of phloem companion cell (PCC) is responsible for the sequestration of As into the vacuoles. XTC, xylem transfer cell. (Modified from Clemens and Ma (2016).)

the root cortical and internode phloem cells with other genes showed great effect in decreasing As accumulation in rice grain (Deng et al., 2018). Recently, it was reported that overexpression of two NIP genes, *OsNIP1;1* and *OsNIP3;3*, also resulted in decreased As accumulation in the rice grain (Sun et al., 2018). OsNIP1;1 and OsNIP3;3 also showed permeability to As(III) in *Xenopus* oocyte (Ma et al., 2008; Sun et al., 2018). Strong constitutive expression of *OsNIP1;1* and *OsNIP3;3* on all cell sides may disrupt the directional radial transport of As(III) from outside to the stele, leading to decreased As(III) concentration in the xylem sap (Sun et al., 2018). Additionally, OsPTR7 (OsNPF8.1) expressed in roots is implicated in the root-to-shoot translocation of DMA (Tang et al., 2017).

3.3.2 Transport of Cd in Rice

Cd concentration in the soil solution lies between 0.2 and 6 µg L^{-1} in non-Cd-contaminated soil and could be as high as 400 µg L^{-1} in Cd-contaminated soil (Kabata-Pendias and Gondek, 1978). However, soluble Cd concentration in soil solution rarely exceeds 1 µmol L^{-1} range (Smolders and Mertens, 2013). The chemical forms of Cd in soil are CdS, CdSO$_4$, and Cd complex with organic chelates. These forms change with soil conditions. For example, under flooded condition, Cd is precipitated with HS and forms CdS, which has a low solubility. By contrast, under upland soil, Cd is present in the form of CdSO$_4$, which has a high solubility (Ma et al., 2020). The availability of Cd in soil is determined by several factors including soil pH, organic matter, cation exchange capacity, and Eh (Ma et al., 2020). Among them, the important factors are soil pH and Eh; Cd solubility in soil decreases with increasing soil pH and Eh (Arao et al., 2009; Hu et al., 2015). Therefore, flooding and liming are effective ways to reduce Cd availability in soil, thereby accumulating Cd in grains (Ma et al., 2020).

Cd in soil solution is taken up by OsNramp5, an influx transporter for Cd (Sasaki et al., 2012). It is polarly localized to the distal side of both the exodermis and endodermis of roots (Figure 3.5) (Sasaki et al., 2012). Knockout of *OsNramp5* resulted in a significant reduction of Cd in the grain

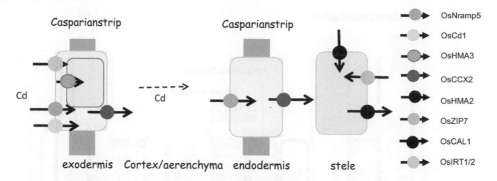

FIGURE 3.5 Transporters involved in cadmium (Cd) uptake in rice roots. Cd is mainly taken up by OsNramp5 polarly localized at the distal side of exodermis and endodermis. OsIRT1/2, OsCd1, and OsCCX2 are also implicated in Cd uptake although their exact cellular localization is unknown. Vacuolar sequestration of Cd is mediated by OsHMA3, and the root-to-shoot translocation of Cd is mediated by OsHMA2, OsZIP7, and OsCAL1. (Modified from Ma et al. (2020).)

(Sasaki et al., 2012; Tang et al., 2017; Ishikawa et al., 2012; Liu et al., 2019; Yang et al., 2019;). For example, a rice mutant with the loss of function of *OsNramp5* generated by ion-beam irradiation showed nearly undetectable Cd in the grains when grown in Cd-contaminated paddy field (Ishikawa et al., 2012). Knockout of *OsNramp5* by CRISPR/Cas9 also resulted in 90% reduction in grain Cd in the hybrid rice (Tang et al., 2017). However, knockout of these genes often showed negative effect on plant growth and grain yield (Sasaki et al., 2012; Yang et al., 2019). This is because OsNramp5 is originally used for transporting Mn (Sasaki et al., 2012); therefore, knockout of this gene causes Mn deficiency. In addition to OsNramp5, other transporters such as OsIRT1, OsIRT2, OsNramp1, and OsCd1 are potentially implicated in Cd uptake (Nakanishi et al., 2006; Takahashi et al., 2011; Yan et al., 2019). However, the expression of *OsIRT1*, *OsIRT2*, and *OsNramp1* was induced by Fe deficiency; this raises a question on whether these transporters contribute to Cd uptake in paddy soil with rich Fe^{2+} (Wang et al., 2020). In addition, a putative cation/Ca exchanger OsCCX2 may function as an efflux transporter of Cd and cooperatively work with OsNramp5 (Hao et al., 2018).

After uptake, part of Cd is sequestered into vacuoles in the roots by tonoplast-localized transporter, OsHMA3 (Figure 3.5) (Ueno et al., 2010; Miyadate et al., 2011). The expression of native *OsHMA3* is quite low in rice roots (Ueno et al., 2010); therefore, overexpression of this gene resulted in a significant reduction of Cd accumulation in the grains without yield penalty (Ueno et al., 2010; Shao et al., 2018). Overexpression of functional *OsHMA3* increases vacuolar sequestration of Cd in the root cells, resulting in low root-to-shoot translocation of Cd (Ueno et al., 2010; Sasaki et al., 2014). Although OsHMA3 is also a transporter for Zn, the concentration of Zn in the shoots was not affected in the overexpression lines because five genes related to Zn transport were upregulated in the overexpression line to maintain Zn homeostasis (Sasaki et al., 2014).

The root-to-shoot translocation of Cd is medicated by OsHMA2 (Satoh-Nagasawa et al., 2012; Takahashi et al., 2012; Yamaji et al., 2013), a plasma membrane-localized transporter for Cd. Knockout of this gene has resulted in decreased Cd accumulation in the grain but also caused yield loss because OsHMA2 is also required for Zn translocation (Yamaji et al., 2013). In addition, OsZIP7 and OsCAL1 are also implicated in the root-to-shoot translocation of Cd (Luo et al., 2018; Tan et al., 2019). Knockout of *OsZIP7* and *OsCAL1* decreased the root-to-shoot translocation of Cd, but knockout of *OsCAL1* did not affect Cd accumulation in the grain (Luo et al., 2018).

Distribution of Cd to the grains is mediated by OsHMA2, OsLCT1 (*Oryza sativa* low-affinity cation transporter1), OsZIP7, and OsCCX2 (Figure 3.6) (Uraguchi et al., 2011; Yamaji et al., 2013; Hao et al., 2018; Tan et al., 2019). In the rice node, OsHMA2 is localized at the phloem of EVBs and DVBs (Yamaji et al., 2013), which is responsible for reloading Cd from the intervening parenchyma tissues

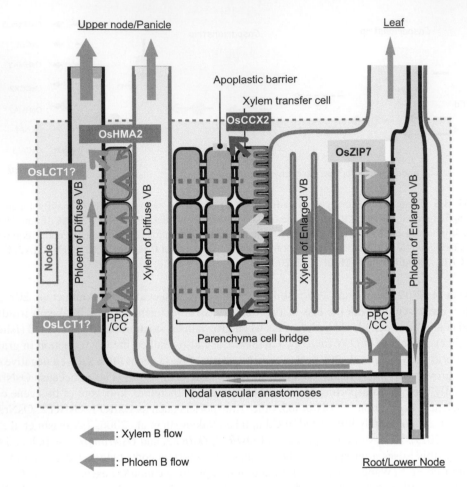

FIGURE 3.6 Transporters involved in the distribution of cadmium to rice grain. Influx transporters OsZIP7, OsHMA2, OsLCT1, and efflux transporter OsCCX2 are implicated in the intervascular transfer of Cd from EVB to DVB. For details, refer to the text. (Modified from Ma et al. (2020).)

into the phloem of EVBs and DVBs (Figure 3.6). OsZIP7 seems to play a similar role in the inter-vascular transfer of Cd (Tan et al., 2019), and OsLCT1 seems to be also involved in the intervascular transfer of Cd through efflux of Cd from the phloem within the nodes (Uraguchi et al., 2011). On the other hand, OsCCX2 is also highly expressed in the xylem region of the EVBs in the nodes. It may function in releasing Cd from these cells (Hao et al., 2018). Knockout or knockdown of *OsCCX2* and *OsLCT1* also decreased the Cd concentration in the rice grain (Uraguchi et al., 2011; Hao et al., 2018).

3.4 FUTURE PERSPECTIVE

Soil provides mineral elements essential for human health but also is a source of toxic elements when soil is contaminated. Therefore, it is necessary to understand the processes for the transport of these mineral elements from soil to edible parts (e.g., grain in rice). The transport of mineral elements requires various different transporters. Although great progress in understanding the transport system in different plants, especially in Arabidopsis and rice, have been made during last decade, and a number of transporters for mineral elements have been identified, most transporters remain to be identified in the future using different approaches. There is a great variation in the accumulation of mineral elements including essential and toxic elements in rice grains (Huang et al., 2020b). These variations will be useful for hunting responsible genes and breeding cultivars for human health.

ACKNOWLEDGEMENTS

Some work presented in this chapter was supported by Grant-in-Aid for Specially Promoted Research (JSPS KAKENHI Grant Number 16H06296 to J.F.M.).

REFERENCES

Abdul, K. S. M., Jayasinghe, S. S., Chandana, E. P., Jayasumana, C., and De Silva, P. M. C. 2015. Arsenic and human health effects: a review. *Environ Toxicol Pharmacol.* 40:828–846.

Abedin, M. J., Feldmann, J., and Meharg, A. A. 2002. Uptake kinetics of arsenic species in rice plants. *Plant Physiol.* 128:1120–1128.

Arao, T., Kawasaki, A., Baba, K., Mori, S., and Matsumoto, S. 2009. Effects of water management on cadmium and arsenic accumulation and dimethylarsinic acid concentrations in Japanese rice. *Environ Sci Technol.* 43:9361–9367.

Bertin, G., and Averbeck, D. 2006. Cadmium: cellular effects, modifications of biomolecules, modulation of DNA repair and genotoxic consequences (a review). *Biochimie.* 88:1549–1559.

Bughio, N., Yamaguchi, H., Nishizawa, N. K., Nakanishi, H., and Mori, S. 2002. Cloning an iron-regulated metal transporter from rice. *J Exp Bot.* 53:1677–1682.

Cai, H., Huang, S., Che, J., Yamaji, N., and Ma, J. F. 2019. The tonoplast–localized transporter OsHMA3 plays an important role in maintaining Zn homeostasis in rice. *J Exp Bot.* 70:2717–2725.

Cao, Y., Sun, D., Ai, H., Mei, H., Liu, X., Sun, S., Xu, G., Liu, Y., Chen, Y., and Ma, L. Q. 2017. Knocking out *OsPT4* gene decreases arsenate uptake by rice plants and inorganic arsenic accumulation in rice grains. *Environ Sci Tech.* 51:12131–12138.

Che, J., Yokosho, K., Yamaji, N., and Ma, J. F. 2019. A vacuolar phytosiderophore transporter alters iron and zinc accumulation in polished rice grains. *Plant Physiol.* 181:276–288.

Chen, Y., Moore, K. L., Miller, A. J., McGrath, S. P., Ma, J. F., and Zhao, F. J. 2015. The role of nodes in arsenic storage and distribution in rice. *J Exp Bot.* 66:3717–3724.

Clemens, S., and Ma, J. F. 2016. Toxic heavy metal and metalloid accumulation in crop plants and foods. *Annu Rev Plant Biol.* 67:489–512.

Codex Alimentarius Commission, Joint FAO/WHO Food Standards Programme, and World Health Organization. 2007. *Codex Alimentarius Commission: Procedural Manual.* Food and Agriculture Organization.

Deng, F., Yamaji, N., Ma, J. F., Lee, S. K., Jeon, J. S., Martinoia, E., Lee, Y. K., and Song, W. Y. 2018. Engineering rice with lower grain arsenic. *Plant Biotechnol J.* 16:1691–1699.

Enstone, D. E., Peterson, C. A., and Ma, F. 2002. Root endodermis and exodermis: structure, function, and responses to the environment. *J Plant Growth Regul.* 21:335–351.

GBD. 2016. Mortality and causes of death collaborators. Global, regional, and national life expectancy, all-cause mortality, and cause-specific mortality for 249 causes of death, 1980–2015: a systematic analysis for the Global Burden of Disease Study 2015. *Lancet.* 388:1459–1544.

Graham, R. D., Knez, M., and Welch, R. M. 2012. How much nutritional iron deficiency in humans globally is due to an underlying zinc deficiency? Adv Agron. 115: 1–40.

Hao, X., Zeng, M., Wang, J., Zeng, Z., Dai, J., Xie, Z., Yang, Y., Tian, L., Chen, L., and Li, D. 2018. A node-expressed transporter OsCCX2 is involved in grain cadmium accumulation of rice. *Front Plant Sci.* 9:476.

Harvey, C. F., Swartz, C. H., Badruzzaman, A. B. M., Keon–Blute, N., Yu, W., Ali, M. A., Jay, J., Beckie, R., Niedan, V., Brabander, D., and Oates, P. M. 2002. Arsenic mobility and groundwater extraction in Bangladesh. *Science.* 298:1602–1606.

Hu, P. J., Ouyang, Y. N., Wu, L. H., Shen, L. B., Luo, Y. M., and Christie, P. 2015. Effects of water management on arsenic and cadmium speciation and accumulation in an upland rice cultivar. *J Environ Sci.* 27:225–231.

Huang, S., Sasaki, A., Yamaji, N., Mitani-Ueno, N., and Ma, J. F. 2020a. The ZIP transporter family member OsZIP9 contributes to root Zn uptake in rice under Zn-limited conditions. *Plant Physiol.* 183:1224–1234.

Huang, S., Wang, P. T., Yamaji, N., and Ma, J. F. 2020b. Plant nutrition for human nutrition: hints from rice research and future perspective. *Mol Plant.* 13:825–835.

Inoue, H., Kobayashi, T., Nozoye, T., Takahashi, M., Kakei, Y., Suzuki, K., Nakazono, M., Nakanishi, H., Mori, S., and Nishizawa, N. K. 2009. Rice OsYSL15 is an iron-regulated iron (III)-deoxymugineic acid transporter expressed in the roots and is essential for iron uptake in early growth of the seedlings. *J Biol Chem.* 284:3470–3479.

Ishikawa, S., Ishimaru, Y., Igura, M., Kuramata, M., Abe, T., Senoura, T., Hase, Y., Arao, T., Nishizawa, N. K., and Nakanishi, H. 2012. Ion-beam irradiation, gene identification, and marker-assisted breeding in the development of low-cadmium rice. *Proc Natl Acad Sci USA*, 109:19166–19171.

Ishimaru, Y., Masuda, H., Bashir, K., Inoue, H., Tsukamoto, T., Takahashi, M., Nikanishi, H., Aoki, N., Hirose, T., Ohsugi, R., and Nishizawa, N. K. 2010. Rice metal-nicotianamine transporter, OsYSL2, is required for the long-distance transport of iron and manganese. *Plant J.* 62:379–390.

Ishimaru, Y., Suzuki, M., Tsukamoto, T., Suzuki, K., Nakazono, M., Kobayashi, T., Wada, Y., Watanabe, S., Matsuhashi, S., Takahashi, M., Nikanishi, H., Mori, S., and Nishizawa, N. K. 2006. Rice plants take up iron as an Fe^{3+}-phytosiderophore and as Fe^{2+}. *Plant J.* 45:335–346.

Kabata-Pendias, A., and B. Gondek. 1978. Bioavailability of heavy metals in the vicinity of a copper smelter. *Trace Subst Environ Health*. 12:523

Kamiya, T., Islam, R., Duan, G., Uraguchi, S., and Fujiwara, T. 2013. Phosphate deficiency signaling pathway is a target of arsenate and phosphate transporter OsPT1 is involved in As accumulation in shoots of rice. *Soil Sci Plant Nutri*. 59:580–590.

Kennedy, G., Nantel, G., and Shetty, P. 2003. The scourge of "hidden hunger": global dimensions of micronutrient deficiencies. *Food Nutr Agric*. 32:8–16.

Lee, S., Chiecko, J. C., Kim, S. A., Walker, E. L., Lee, Y., Guerinot, M. L., and An, G. 2009. Disruption of *OsYSL15* leads to iron inefficiency in rice plants. *Plant Physiol*. 150:786–800.

Li, R. Y., Ago, Y., Liu, W. J., Mitani, N., Feldmann, J., McGrath, S. P., Ma, J. F., and Zhao, F. J. 2009. The rice aquaporin Lsi1 mediates uptake of methylated arsenic species. *Plant Physiol*. 150:2071–2080.

Liu, S. M., Jiang, J., Liu, Y., Meng, J., X, S. L., Tan, Y. Y., Li, Y. F., Shu, Q. Y., and Huang, J. Z. 2019. Characterization and evaluation of OsLCT1 and OsNramp5 mutants generated through CRISPR/Cas9-mediated mutagenesis for breeding low Cd rice. *Rice Sci*. 26:88–97.

Luo, J. S., Huang, J., Zeng, D. L., Peng, J. S., Zhang, G. B., Ma, H. L., Guan, Y., Yi, H. Y., Fu, Y. L., Han, B., Lin, H. X., Qian, Q., and Gong, J. M. 2018. A defensin-like protein drives cadmium efflux and allocation in rice. *Nature Commun*. 9:1–9.

Ma, J. F., Shen, R. F., and Shao, J. F. 2020. Transport of cadmium from soil to grain in cereal crops. *Pedosphere*. doi:10.1016/S1002-0160(20)60015-7.

Ma, J. F., Tamai, K., Yamaji, N., Mitani, N., Konishi, S., Katsuhara, M., Ishiguro, M., Murata, Y. and Yano, M. 2006. A silicon transporter in rice. *Nature*. 440:688–691.

Ma, J. F., Yamaji, N., Mitani, N., Tamai, K., Konishi, S., Fujiwara, T., Katsuhara, M., and Yano, M. 2007. An efflux transporter of silicon in rice. *Nature*. 448:209–212.

Ma, J. F., Yamaji, N., Mitani, N., Xu, X. Y., Su, Y. H., McGrath, S. P., and Zhao, F. J. 2008. Transporters of arsenite in rice and their role in arsenic accumulation in rice grain. *Proc Natl Acad Sci USA*. 105:9931–9935.

Masuda, H., Ishimaru, Y., Aung, M. S., Kobayashi, T., Kakei, Y., Takahashi, M., Higuchi, K., Nakanishi, H., and Nishizawa, N. K. 2012. Iron biofortification in rice by the introduction of multiple genes involved in iron nutrition. *Sci Rep*. 2:543–549.

Mitani-Ueno, N., Yamaji, N., and Ma, J. F. 2018. Transport system of mineral elements in rice. In Sasaki, T., and Ashikari, M., (eds.), *Rice Genomics, Genetics and Breeding* . Singapore: Springer, pp. 223–240.

Miyadate, H., Adachi, S., Hiraizumi, A., Tezuka, K., Nakazawa, N., Kawamoto, T., Katou, K., Kodama, I., Sakurai, K., Takahashi, H., and Satoh-Nagasawa, N. 2011. OsHMA3, a P1B-type of ATPase affects root-to-shoot cadmium translocation in rice by mediating efflux into vacuoles. *New Phytol*. 189:190–199.

Nakandalage, N., and Seneweera, S. 2018. Micronutrients use efficiency of crop-plants under changing climate. In Hossain, M. A., Kamiya, T., Burritt, D. J., Tran L.-S. P, Fujiwara, T. (eds.), *Plant Micronutrient Use Efficiency* (. Cambridge: Academic Press, pp. 209–224.

Nakanishi, H., Ogawa, I., Ishimaru, Y., Mori, S., and Nishizawa, N. K. 2006. Iron deficiency enhances cadmium uptake and translocation mediated by the Fe^{2+} transporters OsIRT1 and OsIRT2 in rice. *Soil Sci Plant Nutri*. 52:464–469.

Nawrot, T., Plusquin, M., Hogervorst, J., Roels, H. A., Celis, H., Thijs, L., Vangronsveld, J., Van Hecke, E., and Staessen, J. A. 2006. Environmental exposure to cadmium and risk of cancer: a prospective population–based study. *Lancet Oncol*. 7:119–126.

Nozoye, T., Nagasaka, S., Kobayashi, T., Takahashi, M., Sato, Y., Sato, Y., Uozumi, N., Nakanishi, H., and Nishizawa, N. K. 2011. Phytosiderophore efflux transporters are crucial for iron acquisition in graminaceous plants. *J Biol Chem*. 286:5446–5454.

Raboy, V. 2009. Approaches and challenges to *engineering* seed phytate and total phosphorus. *Plant Sci*. 177:281–296.

Ramesh, S. A., Shin, R., Eide, D. J., and Schachtman, D. P. 2003. Differential metal selectivity and gene expression of two zinc transporters from rice. *Plant Physiol*. 133:126–134.

Sasaki, A., Yamaji, N., and Ma, J. F. 2014. Overexpression of *OsHMA3* enhances Cd tolerance and expression of Zn transporter genes in rice. *J Exp Bot.* 65:6013–6021.

Sasaki, A., Yamaji, N., and Ma, J. F. 2016. Transporters involved in mineral nutrient uptake in rice. *J Exp Bot.* 67:3645–3653.

Sasaki, A., Yamaji, N., Mitani-Ueno, N., Kashino, M., and Ma, J. F. 2015. A node-localized transporter OsZIP3 is responsible for the preferential distribution of Zn to developing tissues in rice. *Plant J.* 84:374–384.

Sasaki, A., Yamaji, N., Yokosho, K., and Ma, J. F. 2012. Nramp5 is a major transporter responsible for manganese and cadmium uptake in rice. *Plant Cell.* 24:2155–2167.

Satoh-Nagasawa, N., Mori, M., Nakazawa, N., Kawamoto, T., Nagato, Y., Sakurai, K., Takahashi, H., Watanabe, A., and Akagi, H. 2012. Mutations in rice (*Oryza sativa*) heavy metal ATPase 2 (OsHMA2) restrict the translocation of zinc and cadmium. *Plant Cell Physiol.* 53:213–224.

Senoura, T., Sakashita, E., Kobayashi, T., Takahashi, M., Aung, M. S., Masuda, H., Nakanishi, H., and Nishizawa, N. K. 2017. The iron-chelate transporter OsYSL9 plays a role in iron distribution in developing rice grains. *Plant Mol Biol.* 95:375–387.

Shao, J. F., Xia, J., Yamaji, N., Shen, R. F., and Ma, J. F. 2018. Effective reduction of cadmium accumulation in rice grain by expressing *OsHMA3* under the control of the *OsHMA2* promoter. *J Exp Bot.* 69:2743–2752.

Shi, S., Wang, T., Chen, Z., Tang, Z., Wu, Z., Salt, D. E., Chao, D. Y., and Zhao, F. J. 2016. OsHAC1;1 and OsHAC1;2 function as arsenate reductases and regulate arsenic accumulation. *Plant Physiol.* 172:1708–1719.

Smolders, E., and Mertens, J. 2013. Chapter10. Cadmium. In Alloway, B. J. (ed.), *Heavy Metals in Soils: Trace Metals and Metalloids in Soils and Their Bioavailability.* Dordrecht: Springer Science+Business Media, pp 283–311.

Song, W. Y., Yamaki, T., Yamaji, N., Ko, D., Jung, K. H., Fujii-Kashino, M., An, G., Martinoia, E., Lee, Y., and Ma, J. F. 2014. A rice ABC transporter, OsABCC1, reduces arsenic accumulation in the grain. *Proc Natl Acad Sci USA.* 111:15699–15704.

Sun, S. K., Chen, Y., Che, J., Konishi, N., Tang, Z., Miller, A. J., Ma, J. F., and Zhao, F. J. 2018. Decreasing arsenic accumulation in rice by overexpressing *OsNIP1;1* and *OsNIP3;3* through disrupting arsenite radial transport in roots. *New Phytol.* 219:641–653.

Takahashi, R., Ishimaru, Y., Senoura, T., Shimo, H., Ishikawa, S., Arao, T., Nakanishi, H., and Nishizawa, N. K. 2011. The OsNRAMP1 iron transporter is involved in Cd accumulation in rice. *J Exp Bot.* 62:4843–4850.

Takahashi, R., Ishimaru, Y., Shimo, H., Ogo, Y., Senoura, T., Nishizawa, N. K., and Nakanishi, H. 2012. The OsHMA2 transporter is involved in root-to-shoot translocation of Zn and Cd in rice. *Plant Cell Environ.* 35:1948–1957.

Tan, L., Zhu, Y., Fan, T., Peng, C., Wang, J., Sun, L., and Chen, C. 2019. OsZIP7 functions in xylem loading in roots and inter–vascular transfer in nodes to deliver Zn/Cd to grain in rice. *Biochem Biophys Res Commun.* 512:112–118.

Tang, L., Mao, B., Li, Y., Lv, Q., Zhang, L., Chen, C., He, H., Wang, W., Zeng, X., Shao, Y., and Pan, Y. 2017. Knockout of *OsNramp5* using the CRISPR/Cas9 system produces low Cd-accumulating indica rice without compromising yield. *Sci Rep.* 7:1–12.

Ueno, D., Yamaji, N., Kono, I., Huang, C. F., Ando, T., Yano, M., and Ma, J. F. 2010. Gene limiting cadmium accumulation in rice. *Proc Natl Acad Sci USA.* 107:16500–16505.

Uraguchi, S., Kamiya, T., Sakamoto, T., Kasai, K., Sato, Y., Nagamura, Y., Yoshida, A., Kyozuka, J., Ishikawa, S., and Fujiwara, T. 2011. Low-affinity cation transporter (OsLCT1) regulates cadmium transport into rice grains. *Proc Natl Acad Sci USA.* 108:20959–20964.

Von Grebmer, K., Saltzman, A., Birol, E., Wiesman, D., Prasai, N., Yin, S., Yohannes, Y., Menon, P., Thompson, J., and Sonntag, A. 2014. *Synopsis: 2014 Global Hunger Index: The Challenge of Hidden Hunger.* Bonn, Washington, D.C., and Dublin: Intl Food Policy Res Inst.

Wang, P., Yamaji, N., Inoue, K., Mochida, K., and Ma, J. F. 2020. Plastic transport systems of rice for mineral elements in response to diverse soil environmental changes. *New Phytol.* 226:156–169.

Wang, P., Zhang, W., Mao, C., Xu, G., and Zhao, F. J. 2016. The role of OsPT8 in arsenate uptake and varietal difference in arsenate tolerance in rice. *J Exp Bot.* 67:6051–6059.

White, P. J., and Broadley, M. R. 2005. Biofortifying crops with essential mineral elements. *Trends Plant Sci.* 10:586–593.

White, P. J., and Broadley, M. R. 2009. Biofortification of crops with seven mineral elements often lacking in human diets-iron, zinc, copper, calcium, magnesium, selenium and iodine. *New Phytol.* 182:49–84.

WHO. 2002. *The World Health Report 2002-Reducing Risks, Promoting Healthy Life.* Geneva, Switzerland: World Health Organization (WHO).

Wu, Z., Ren, H., McGrath, S. P., Wu, P., and Zhao, F. J. 2011. Investigating the contribution of the phosphate transport pathway to arsenic accumulation in rice. *Plant Physiol*. 157:498–508.

Yamaji, N., and Ma, J. F. 2014. The node, a hub for mineral nutrient distribution in graminaceous plants. *Trends Plant Sci*. 19:556–563.

Yamaji, N., and Ma, J. F. 2017. Node-controlled allocation of mineral elements in Poaceae. *Curr Opin Plant Biol*. 39:18–24.

Yamaji, N., Xia, J., Mitani-Ueno, N., Yokosho, K., and Ma, J. F. 2013. Preferential delivery of zinc to developing tissues in rice is mediated by P-type heavy metal ATPase OsHMA2. *Plant Physiol*. 162:927–939.

Yan, H., Xu, W., Xie, J., Gao, Y., Wu, L., Sun, L., Feng, L., Chen, X., Zhang, T., Dai, C., and Li, T. 2019. Variation of a major facilitator superfamily gene contributes to differential cadmium accumulation between rice subspecies. *Nature Commun*. 10:2562.

Yang, C. H., Zhang, Y., and Huang, C. F. 2019. Reduction in cadmium accumulation in japonica rice grains by CRISPR/Cas9-mediated editing of *OsNRAMP5*. *J Integr Agr*. 18:688–697.

Ye, Y., Li, P., Xu, T., Zeng, L., Cheng, D., Yang, M., Luo, J., and Lian, X. 2017. OsPT4 contributes to arsenate uptake and transport in rice. *Front Plant Sci*. 8:2197.

Yokosho, K., Yamaji, N., and Ma, J. F. 2016. *OsFRDL1* expressed in nodes is required for distribution of iron to grains in rice. *J Exp Bot*. 67:5485–5494.

Yokosho, K., Yamaji, N., Ueno, D., Mitani, N., and Ma, J. F. 2009. OsFRDL1 is a citrate transporter required for efficient translocation of iron in rice. *Plant Physiol*. 149:297–305.

Zhang, L. Zhang, L., Hu, B., Li, W., Che, R., Deng, K., Li, H., Yu, F., Ling, H., Li, Y., and Chu, C. 2014. OsPT2, a phosphate transporter, is involved in the active uptake of selenite in rice. *New Phytol*. 201:1183–1191.

Zhang, Y., Xu, Y. H., Yi, H. Y., and Gong, J. M. 2012. Vacuolar membrane transporters OsVIT1 and OsVIT2 modulate iron translocation between flag leaves and seeds in rice. *Plant J*. 72:400–410.

Zhao, F. J., McGrath, S. P. and Meharg, A. A. 2010. Arsenic as a food chain contaminant: mechanisms of plant uptake and metabolism and mitigation strategies. *Annu Rev Plant Biol*. 61:535–559.

4 Soils and Human Health
Communication between Soil Scientists and Health Care Providers

David Collier
East Carolina University

Eric C. Brevik
Dickinson State University

CONTENTS

4.1 INTRODUCTION

4.1.1 BRIEF OVERVIEW OF SOILS AND HUMAN HEALTH CONNECTIONS

It has been appreciated since antiquity that certain properties of soils have negative effects on human health (Brevik and Sauer, 2015; Farkhutdinov et al., 2020). Public health and healthcare practitioners (HCPs) recognize etiological connections between soils and human heath, but these connections are usually focused on the negative aspects of what soils "do to" human health rather than positive aspects of what soils "do for" human health. For example, HCPs are broadly aware that human disease can be caused by exposure to soilborne toxins including asbestos, arsenic, lead, cadmium, and other heavy metals (Oliver, 1997) or to soilborne infectious agents such as viruses, bacteria, fungi, and parasites (Brevik, 2009b). Likewise, HCPs also appreciate that diseases such as hypothyroidism/multinodular goiter, anemia, or Keshan/Kashin Beck's disease are associated with soil deficiencies in iodine, iron, and selenium, respectively (Lazarus, 2015; Rayman, 2012).

The study of soils and soil health, defined as "the continued capacity of soil to function as a vital living ecosystem that sustains plants, animals and humans" (USDA-NRCS, 2019), by soil scientists, geologists, and ecologists ("soil science practitioners" [SSPs]) increasingly employs a holistic framework that is rooted in ecological theory allowing a highly nuanced and complex understanding of what soils "do for" human health. This includes a myriad of critical ecosystem services that healthy soils provide beyond serving as the basis for most food production. These services include detoxification, carbon sequestration, water and nutrient retention, and maintenance of biodiversity (Brevik et al., 2018; NASEM, 2017). Of particular salience are ecosystem services that mitigate global climate change. This is because more frequent, severe, and prolonged heat events, fires, deterioration of air quality, sea level rise, flooding, and expansion of vector-borne diseases and increases in food- and weather-related infections caused by global climate change (USGCRP, 2016) are thought to pose the greatest threat to global health this century.

It is now widely recognized that human health, as defined by the World Health Organization as "a state of complete physical, mental and social well-being, and not merely the absence of disease and infirmity" (WHO, 1946), is influenced to a great extent by so-called social determinants of health (SDOH) (McGinnis et al., 2002; USDHHS, 2019; WHO, 2008). SDOH are social and physical factors that impact the conditions in which people are born, grow, live, work, and age. Governments, healthcare entities, and HCPs now recognize the necessity of addressing SDOH in order to deliver effective care and promote optimal health (Alley at al., 2016; McGinnis et al., 2002; WHO, 2008; Brevik et al., 2020).

Soils and soil health impact employment, food insecurity, and poverty; influence access to foods that support healthy eating patterns; influence environmental conditions; and impact air and water quality (NASEM, 2017; WHO, 2008) and, therefore, must be considered a key SDOH. Yet soils and soil health are currently underrecognized as a SDOH in the medical field. We posit that this is due in large part to ineffective communication across relevant disciplines. Here, we examine opportunities for elevating communication between soil scientists and HCPs.

4.1.1.1 Writings by Soil Scientists and Geologists

A comprehensive review of work published by soil scientists addressing connections between soil and human health is beyond the scope of this chapter. However, some examples of thoughtful, integrative reviews and perspectives are included here. First, the work of Steffan et al. (2018) and Pepper (2013) provide excellent overviews of multiple connections between soils and human health, while Oliver and Gregory (2015) principally focus on how soils and human health are linked through food security. Brevik et al. (2019) highlights the role of soils in human health through the provision of shelter, clothing, and fuel. Keith et al. (2016) provides a framework for connecting ecosystem services provided by soils to "One Health"—, an initiative aiming to integrate disparate disciplines concerned with human, animal, and ecosystem health. The work of Hanyu et al. (2014) suggests that, compared to control behaviors, the deliberate observation (touching, seeing, and smelling) of forest soils has salutary effects on the autonomic nervous system and mood in humans. In their editorial, Kemper and Lal (2017) recognize three domains of soil health: chemical/nutritional, physical, and biological, and they then briefly discuss how these three domains of soil health may influence human health. They posit that there may be a broad spectrum in the level of understanding of soil health–human health connections among clinicians that depend on provider demographics. Finally, Wall et al. (2015) provides a conceptual framework for how land use and management practices are linked to human health through their impacts on soil biodiversity. To meet the United Nations Sustainable Development Goals (i.e., to address SDOH globally), they recommend inclusion of biodiversity experts as well as human, plant, and animal health experts in integrated research programs that ultimately will improve human health.

Several publications from the field of geology, or more specifically medical geology, which describe the core concepts of medical geology as a re-emerging field and/or highlight key examples of how human health relates to geologic materials, are also highlighted here. Medical geology has been broadly defined as "the study of the interaction between the environment and health" (Boulos and Le Blond, 2016; Buck et al., 2016). A more precise definition specifies that the environment component refers specifically to geologic materials or processes and, strictly speaking, includes only natural geological material (rocks, ores, minerals, water, etc.) and processes (weathering, erosion, volcanism, etc.). However, given the extent of human activity, it is also useful to include anthropogenic factors (building materials, alloys, byproducts of mining and ore processing, etc.) in the field of study (Boulos and Le Blond, 2016; Buck et al., 2016; Farkhutdinov et al., 2020). A prime example of a geological material whose effects on human health are predominately due to anthropogenic activity is the lingering problem of lead exposure, particularly in urban and postindustrialized settings (Filippelli et al., 2018). Of note, health can be affected by either the presence of, and/or the absence of, a particular geologic factor as evidenced by the fact that inadequate fluoride intake predisposes to development of dental caries while excessive intake is linked to dental and skeletal fluorosis and cancer (Buck et al., 2016).

Clay is a natural geologic material that is thought to be unique in its ability to impact health via all routes of exposure: inhalation, ingestion, and dermal contact (Finkelman, 2019). Like many geogenic materials, clay can have either harmful or salutary effects (Finkelman, 2019; Williams, 2017; Williams and Hillier, 2014). Of particular interest are the remarkable broad antibacterial, even bactericidal, properties of certain clays that were found to effectively treat severe Buruli ulcer caused by the acid-fast bacterium *Mycobacterium ulcerans* (Williams, 2017). The nascent understanding of what imparts antibacterial properties to circa 5% of clays worldwide has resulted from microbiological, physical, geochemical, mineralogical, chemical, and imaging studies as well as historical empirical clinical observations (Williams, 2017). The promise of this research is that by understanding and mimicking the natural geochemical processes that impart antibacterial properties on these clays, we will be able to develop novel treatments for the emerging problem of antibiotic-resistant bacteria.

As evidenced by the study of clay's antibacterial properties, medical geology exists at the intersection of multiple disciplines. Typically, these are thought to include geoscience, epidemiology, toxicology, and medicine (and informatics!). Hence, a key to the success of this interdisciplinary approach is the willingness "to speak a language that is translational across the sciences to encourage an integrated understanding of the problem" (Boulos and Le Bond, 2016).

It is important to note that most of the soil science and medical geology works cited to this point were published in journals not read by HCPs. Physicians, even PhD microbiologists, are unlikely to be aware of articles published in journals such as *Clay Minerals* and other similar outlets. While these are respected journals in the world of soil science and geology, the material published in these journals is almost in a sense buried from the perspective of an HCP. This is despite the fact that much of the clay minerals' work reviewed here was supported by NIH, an agency dedicated to funding health-oriented research. This highlights the communications gap that currently exists between SSPs and HCPs.

4.1.1.2 Writings by Human Health Professionals

To estimate the extent to which the roles of soils and soil health in human health are recognized within the clinical medical literature, we performed an Ovid search of Medline/Epub using the key word "soil" and journal title for seven key medical journals. The journals included the top three general medical journals: *The New England Journal of Medicine* (NEJM), *The Lancet*, and the *Journal of the American Medical Association* (JAMA), as well as *Annals of Internal Medicine*, *American Family Physician*, *Obstetrics and Gynecology*, and *Pediatrics*. The later four are the official journals of the American College of Physicians (Internal Medicine), American Academy of Family Physicians, American College of Obstetricians and Gynecologists, and the American Academy of Pediatrics, respectively.

Remarkably, no articles with "soil" as a keyword were found in *American Family Physician*. *Obstetrics and Gynecology* had two, one published in 1968 and one in 1969, and the journal *Pediatrics* also contained two which were both published in 1980. *Annals of Internal Medicine* had four, all published in 1997 and all related to Legionnaire's disease. NEJM had nine; the first published in 1951 and the most recent in 2012, while JAMA had 13; the first published in 1962 and the most recent in 2008. Finally, *The Lancet* had 30 with the most recent published in 2019. While neither a comprehensive nor particularly scientific search of the entire medical literature, the fact that only 58 articles, and very few recent ones, among the tens of thousands of manuscripts published in these general medical journals could be identified using "soil" as a keyword suggests an opportunity for better communication of soil health concepts in the medical literature. Lastly, we recognize that there are relevant manuscripts published by HCPs in other journals such as the excellent clinical review published in the *Journal of the American Board of Family Medicine* (Baumgardner, 2012). Baumgardner effectively incorporates soil health and soil microbial ecology into his review of soil-related bacterial and fungal infections, highlighting the importance of obtaining accurate travel and soil exposure histories for making timely and accurate diagnoses.

4.1.2 Soil Health (As Defined) Represents a Key Life Sustaining "Factor" Just as Clean Air and Clean and Abundant Water

The WHO acknowledges that climate change poses the most significant health risks globally this century. The deleterious effects of global climate change on human health are also recognized by the Medical Society Consortium on Climate and Health, an organization of 18 societies representing in excess of 450,000 physicians and more than half of the physicians in the US. Hence, the work of SSPs documenting how ecosystem services provided by healthy soils help to mitigate global climate change should be of interest to HCPs. Likewise, other ecosystem services including detoxification, water and nutrient retention, maintenance of biodiversity, and suppression of pathogens that together help provide clean air, water, food, and a healthy environment can be readily framed as SDOH. The public health sector has long recognized the necessity of addressing SDOH in order to promote optimal human health, and there is increasing demand and support for HCPs to directly address key SDOH in the clinical setting. Hence, an effective strategy for SSPs to better engage HCPs is to communicate soil heath concepts in terms of SDOH.

4.2 SOILS KNOWLEDGE USEFUL TO THE PRACTITIONER

Soils impact human health through their physical, chemical, and biological properties. Therefore, it is important that HCPs understand some basic aspects of each of these areas to understand how and why soils influence human health. These three property categories are all important in determining soil health as well, and links between soil health and human health have been demonstrated (Kemper and Lal, 2017; Pepper, 2013; Wrench, 1939). Finally, as important parts of cycles such as the carbon, nitrogen, and hydrologic cycles, these properties are important to human health. Therefore, it is critical that SSPs communicate the importance of these properties to their medical colleagues. A brief summary of soil properties and how they impact human health is given below; more complete discussions of soil properties from the perspective of soils and human health are given in Brevik (2013) and Pereg et al. (2021).

4.2.1 Basic Soil Physical and Chemical Properties

Soil physical and chemical properties have long been studied by soil scientists, and the fundamental tests and procedures for determining them are well established (Brevik, 2009a). Major physical properties of soil include texture (sand, silt, and clay content), structure (the arrangement of textural units into regular shapes in the soil), porosity (the amount open space in the soil), infiltration rate, depth to water table, bulk density, penetration resistance, water holding capacity (how much water the soil can store for plant use between rain events), and color. Each of these properties is important to the health of a given soil. For example, texture is an important control on a soil's cation exchange capacity (CEC), which determines how many cation nutrients may be available for plant growth. Bulk density is an important control on how much space roots have to grow and soil organisms have to move through the soil, and penetration resistance is a measure of how difficult it is for roots or larger organisms to push through the soil. All of these properties are important to crop growth and potential contaminant movement through the soil, and thus to human health. Color is important because it tells about the chemical conditions (aerobic versus anaerobic, or fluctuations between these conditions) the soil formed under, what types of organisms are likely able to survive in the soil, and what types of secondary minerals are present. Infiltration rate and depth to water table are important properties that determine the suitability of a given soil for the installation of a septic system, which again can be an important component of human health.

Important soil chemical properties include pH, soil organic matter (SOM), major and minor nutrients contents, CEC, electrical conductivity, and the potential presence of heavy metals or other elements that may serve as toxins at high enough bioavailability levels. Soil pH controls the availability of soil nutrients and potential toxins, thereby influencing human health through the soil's ability to supply nutritious foods and the potential for the accumulation of toxic levels of certain elements in those foods. It is important to realize that the bioavailability of an element is more important than its absolute concentration in the soil when determining whether or not an element will pose human health risks (see Soil and Human Health Case Study—Itai Itai Disease); therefore, various indices have been developed to estimate risk levels (Aitta et al., 2019; Brevik et al., 2020). It is also important to realize that the difference between deficient and toxic can be very small with some trace elements (Steffan et al., 2018), so when attempting to correct trace element deficiencies, it is important not to over-apply the nutrient. SOM is the base of the food web in the soil ecosystem and releases plant and microbial nutrients as it decomposes; therefore, it is extremely important that soils have adequate levels of SOM. Electrical conductivity is a measure of the salt content of a soil; EC levels that are too high tend to restrict plant growth and other aspects of the soil ecosystem. Ways that soil physical and chemical properties influence human health are summarized in Table 4.1.

TABLE 4.1

Examples of Ways Selected Soil Physical and Chemical Properties May Affect Human Health

Soil Property	Positive/ Negative Effects	Effects on Human Health	References
Texture	Positive	Compensation for nutrient deficiencies and/or detoxification within the digestive system through geophagy. Various soil clays in particular have served as sources of medications. Clays offer CEC to a soil that is important for holding available plant nutrients.	Oliver and Gregory (2015), Henry and Cring (2013), Mbila (2013), and Brady and Weil (2016)
	Negative	Binding of nutrients by soil particles in the digestive system leading to deficiencies, exposure to pathogens and poisons through geophagy. Various soil textures can introduce challenges in crop production (e.g., sandy soils have low water-holding capacity and low natural fertility), threatening food security.	Henry and Cring (2013) and Brady and Weil (2016)

(Continued)

TABLE 4.1 (*Continued*)

Examples of Ways Selected Soil Physical and Chemical Properties May Affect Human Health

Soil Property	Positive/ Negative Effects	Effects on Human Health	References
Structure, porosity, and infiltration rate (three interrelated properties)	Positive	Soil structure is important for porosity, infiltration rate, and water-holding capacity, which are in turn important in crop production and the maintenance of a healthy soil ecosystem. Soil pores (porosity) are where air and water enter soil, smaller pores (micropores) are where water is stored in the soil, and pores take place in the physical filtration of water that passes through the soil, helping prevent contamination of groundwater resources. Pores are also important for root penetration and the movement of soil organisms, all of which are important to a healthy soil ecosystem and good crop yields that support food security. Infiltration rate is important in determining how much of a given rainfall enters the soil, therefore having the opportunity to be taken up by crops or stored for future crop use.	Brevik (2009a), Brady and Weil (2016), and Helmke and Losco (2013)
	Negative	Rapid infiltration rates may introduce contaminants into groundwater systems before the natural filtering processes in soil have time to clean the water. Infiltration rates that are too high or too low cause problems for septic systems, which use the natural filtering properties of soil to clean human waste and prevent disease.	Brady and Weil (2016) and Helmke and Losco (2013)
pH	Positive	Maintaining an appropriate soil pH level makes plant nutrients more available. This allows production of increased quality and quantity of crops, aiding in food security and human health.	Cockx and Simonne (2003) and Havlin et al. (2013)
	Negative	pH levels that are too low make the majority of heavy metals plant available, threatening crop quality and human health.	Brevik (2009b) and Morgan (2013)
Soil organic matter	Positive	Influences many essential soil processes, such as formation of structure, release of plant nutrients, and serves as a food source for soil organisms forming the base of the soil food web.	Brevik (2009a) and Wolf and Snyder (2003)
	Negative	Adding large amounts of organic matter to a soil can increase soil carbon dioxide levels, prevent good seed/soil contact and hamper germination, or introduce allelopathic effects to the soil that can restrict crop growth	Brevik (2009a) and Wolf and Snyder (2003)
Cation exchange capacity	Positive	Holds cation nutrients on soil particle surfaces that are then available for crop access, enhancing crop production and improving food security; important indicator of soil health/quality.	Havlin et al. (2013) and Khaledian et al. (2017)
	Negative	Binding of nutrients in the digestive system to cation exchange sites on consumed soil particles can lead to nutrient deficiencies through geophagy.	Henry and Cring (2013)

(Continued)

TABLE 4.1 (*Continued*)
Examples of Ways Selected Soil Physical and Chemical Properties May Affect Human Health

Soil Property	Positive/ Negative Effects	Effects on Human Health	References
Salinity	Positive	Increased salinity may increase crop tolerance to selected elements such as boron; this may prevent boron toxicity in some cases. Moderate salinity may also increase crop yield if high levels of soil boron are present (many caveats accompany these relationships).	Smith et al. (2013)
	Negative	Increased salinity usually leads to decreased crop yields, reducing food security. Salinity can also cause nutrient imbalances in crops grown in saline soils; these imbalances can be passed up the food web.	De Pascale and Barbieri (1995) and Butcher et al. (2018)
Heavy metals	Positive	Some heavy metals (e.g., Co, Cu, Ni, Sn, Zn) are trace elements required for human nutrition.	Brevik (2009b)
	Negative	Many heavy metals have no known human health benefits, and most cause health problems at high levels.	Brevik (2009b) and Morgan (2013)

Soil and Human Health Case Study: Itai Itai Disease

Itai itai disease, which was documented in the Toyama Prefecture of Japan, is a classic case of trace element toxicity and the influence of soil and/or other environmental variables on the final health impacts of the trace element. Citizens of Toyama Prefecture began to experience severe health problems, including weak brittle bones, pain in the legs and spine, coughing, anemia, and kidney failure after consuming rice that had been irrigated with cadmium-contaminated river water. However, residents of the village of Shipham, in England, lived in an area with average total Cd content in their soils that were about 30 times higher than those documented to cause problems in Japan but did not suffer any of the negative health effects noted in Japan. Several hypotheses have been offered to account for this. One is that Cd bioavailability in the soil is influenced by soil aeration status, with the anaerobic soils of the rice paddies making Cd more available than in aerobic soils such as those around Shipham. The pH range (7.0–7.8) and nutrient competition in the Shipham soils have also been offered as reasons the people of Shipham have not suffered the same ill-effects as their Japanese counterparts. Because of differences in soil properties, the soil solution Cd levels in England and Japan were quite similar even though the English soils had much higher total Cd levels on average. The more highly varied diet of the English as opposed to the Japanese is another possible reason for the observed discrepancy. The Japanese itai-itai victims' diet was relatively Fe- and Zn-deficient. Cadmium retention is increased by 15 times in those with a Fe- and Zn-deficient diet compared to individuals who have adequate Fe and Zn intake. This case study demonstrates that simply knowing the content of an element, such as Cd, in a soil is not enough to reach conclusions regarding resulting health problems. Other environmental factors, including the physical and chemical properties of the soil, must also be considered.

Case study based on Brevik (2013) and Morgan (2013).

4.2.2   Basic Soil Biological Properties

It has been estimated that approximately 25% of the species diversity on Earth is in the soil (Bultman et al., 2013), and diversity is very important to the health of a soil (Ferris and Tuomisto, 2015). The tests for determining soil biological properties are less well-established than those for soil physical and chemical properties (Pereg et al., 2021), but soil biology is one of the focus areas for soil science research today. While this statement is true independent of any human health connections, soil biology is very important to human health. A healthy soil ecosystem increases crop yields and plant nutrient uptake and reduces the leaching of nutrients such as nitrogen (Bender and van der Heijden, 2015), all of which is beneficial to human health. Healthy soils also cycle carbon and nitrogen through biogeochemical reactions and are therefore important parts of these natural cycles. Soil biology can represent a threat to human health in some instances (Table 4.2; Brevik and Burgess, 2013a), but there are also documented benefits from exposure to soil organisms or from the actions of soil organisms. Burrowing organisms such as earthworms and ants are important in conditioning soil physical properties and in soil nutrient cycling, which can help crop yield (Ernakovich et al., 2016). Large percentages of the medications available today have their origins in soil (Mbila, 2013). Mouse models indicate that exposure to *Mycobacterium vaccae*, a common soil saprophyte, is involved in immune system activation and specific serotonergic pathways that influence emotional and behavioral response to stress (Lowry et al., 2007; Smith et al., 2019). Administration of a heat-killed *M. vaccae* preparation to humans has resulted in improved chemotherapy response, suggesting a potential role for soilborne microbial products in immunotherapy (O'Brien et al., 2000). Some scientists now state that we are too removed from environmental biodiversity in our modern world and that this separation is leading to problems with allergic and chronic inflammatory disorders (Haahtela, 2019). It also seems increasingly apparent that there are links between the soil microbiome and the human microbiome (Blum et al., 2019). However, there is still much we do not know about soil ecology and the influence of soil organisms on human health, so additional research in these areas is badly needed.

4.2.3   Soil Health

The various soil physical, chemical, and biological properties come together to determine the health of any given soil. Soil health is a very difficult thing to quantify, because it depends in part of the use that one wants to make of a given soil (Brevik, 2009a). The perfect pH to grow blueberries is different than the perfect pH to grow corn. Some crops like sandy soils, others prefer a high clay content. The perfect soil conditions to grow a rice crop would be horrible conditions to install a septic filter field in. All of the above examples involve items that are important to human health (crop production and proper waste disposal), yet the ideal healthy soil is going to be different for these different uses. That being said, there are soil properties that take us outside the realm of healthy soil. Severe compaction and high penetration resistance, loss of SOM and soil structure, extreme pH levels (either high or low), nutrient deficiencies or oversupply, accumulations of organic chemicals or trace elements to toxic levels, and a soil ecosystem characterized by low numbers and diversity of organisms are all things that indicate an unhealthy soil. So, while it is hard to quantitatively determine when a soil is healthy, there are definite indicators of a lack of health. Fortunately, appropriate soil management is able to correct most, if not all, of the properties that make a soil unhealthy (Brevik, 2009a; Njira and Nabwami, 2013; Turmel et al., 2015). Actions taken to improve soil health are also generally expected to improve human health, as they will increase the quantity and/or quality of crops grown, enhance the soil's ability to act as a natural filter, increase soil ecological diversity, etc.

4.2.4   Soil and Climate Connections

Climate change is widely expected to be one of the largest human health challenges in the coming century, with negative impacts on issues such as food and water security, distributions of infectious

TABLE 4.2
Examples of Soil Organisms That Can Have Negative Impacts on Human Health

Organism	Common Name	Point of Infection/ Gateway	Disease Name	Associated Problems	Soil Residency	Incidence	Geographic Distribution
Necator americanus	Hookworm	Intestine	Ancylostomiasis	Chronic anemia, diarrhea, cramps	Periodic	1.2 billion/year (includes *A duodenale*)	Central and South America, southern Asia, Australia, Pacific Islands
Strongyloides stercoralis	Roundworm	Small intestine	Strongyloidiasis	Abdominal pain, diarrhea, rash	Permanent	100 million/year	Tropical, subtropical, and temperate regions
Taenia saginata	Beef tapeworm	Small intestine		Vitamin deficiency, abdominal pain, weakness, change in appetite, weight loss	Transient	50 million/year	Worldwide
Giardia lamblia	Giardia	Large intestine	Giardiasis	Diarrhea, abdominal cramps, bloating, fatigue, weight loss	Transient-incidental	500,000	Worldwide
Toxoplasma gondii		Eye, brain, heart, skeletal muscles	Toxoplasmosis	Flu-like symptoms, birth defects, hepatitis, pneumonia, blindness neurological disorders, myocarditis	Transient	60 million in the USA are probably carriers	Worldwide, most common in warm climate, low altitudes
Coccidioides		Respiratory, sometimes trauma to skin	Coccidioicomycosis (Valley fever)	Flu-like symptoms, fever, cough, rash, headache, muscle aches	Permanent	15/100,000	Southwestern United States, northern Mexico, microfoci in Central and South America
Trichophyton sp., Microsporum sp., Epidermophyton sp.		Skin contact	Tinea corporis (Ringworm)	Itchy, red circular rash with healthy-looking skin in the middle		Common	Worldwide

(Continued)

TABLE 4.2 (*Continued*)
Examples of Soil Organisms That Can Have Negative Impacts on Human Health

Organism	Common Name	Point of Infection/ Gateway	Disease Name	Associated Problems	Soil Residency	Incidence	Geographic Distribution
Bacillus anthracis		Respiration, skin trauma, ingestion	Anthrax	Ulcer w/black necrotic center, severe breathing problems, shock, nausea, fever, vomiting, diarrhea, abdominal pain	Periodic	Rare	South and Central America, southern and eastern Europe, Asia, Africa, the Caribbean, the Middle East
Clostridium tetani		Skin trauma	Tetanus (Lockjaw)	Painful muscle tightening	Permanent	500,000 annual deaths	Worldwide
Clostridium botulinum		Ingestion	Botulism	Vision problems, slurred speech, dry mouth, difficulty swallowing, muscle weakness	Permanent	Rare	Worldwide
Clostridium sp. (other than 2 above)		Skin trauma	Gas gangrene	Air under skin, blisters, fever, drainage, pain	Permanent	Common before antibiotics	Worldwide
Rickettsia sp.		Tick bite	Rocky Mountain spotted fever, other tick fevers and spotted fevers	Fever, nausea, vomiting, severe headache, pain, rash	Periodic	250–1,200 cases/year in the United States	Worldwide

Source: Table modified from Brevik (2009b). More extensive information is provided in Brevik (2009b) and Pereg et al. (2021).

diseases, increased heat waves and violent storms, and rising sea levels that will create hundreds of millions of environmental refuges (Barrett et al., 2015). Soil is an integral part of the carbon, nitrogen, and hydrologic cycles, and as such, is also an important part of the climate change picture (Lal, 2007; Brevik, 2012). Soils have the ability to serve as carbon sinks, removing carbon dioxide from the atmosphere, or as carbon sources, increasing atmospheric carbon dioxide levels, depending on human management of the soil resource. Activities such as tillage and removal of crop residues from fields lead to losses in soil organic carbon (SOC), while additions of organic matter through manure additions, returning crop residues to the soil, and use of cover crops increase SOC. Investigations of techniques like no-till have led to some mixed results, but no-till does seem to increase SOC under at least some environmental conditions (Carr et al., 2015; Abbas et al., 2020). In addition, no-till creates other desirable soil properties such as an enhanced soil ecosystem, improved soil structure, reduced erosion, and increased resilience (Lal, 2013).

The big picture for this discussion is that soils have a major impact on global biogeochemical cycles that in turn affect global climate change. Those effects can be either positive or negative, depending on human management and how that management influences soil physical, chemical, and biological properties. A warming climate due to increased atmosphere greenhouse gas concentrations is widely expected to increase human health concerns. Therefore, appropriate human–soil interactions are essential to mitigate climate change and its expected negative human health impacts. Actions that build soil health are also generally positive from the soil and climate change perspective.

4.3 WAYS TO FACILITATE COMMUNICATION

4.3.1 PUBLICATIONS IN JOINTLY READ JOURNALS

One way to facilitate communication between soil scientists and healthcare providers is to publish in outlets that might be read by both. Google Scholar gives lists of the top 20 journals in several academic fields, including soil science and health and medical sciences. It is important to note that the health and medical sciences category is broad in itself, and several subfields of the medical sciences are also tracked by Google Scholar, but this analysis focused on the broader category. Investigation of these lists shows no overlap between the two category's top 20 journals (Table 4.3). However, there are at least a small number of examples of soil scientists publishing in leading

TABLE 4.3

Top 20 Soil Science, Environmental Science, and Health and Medical Sciences Journals According to Google Scholar as of April 2020

Ranking	Soil Science Journals	Environmental Science Journals	Health and Medical Sciences Journals
1	Soil Biology and Biochemistry	Nature Climate Change	The New England Journal of Medicine
2	Geoderma	Environmental Science & Technology	The Lancet
3	Catena	Science of The Total Environment	Cell
4	Plant and Soil	Water Research	Proceedings of the National Academy of Sciences
5	Soil and Tillage Research	Bioresource Technology	JAMA
6	Applied Soil Ecology	Global Change Biology	Journal of Clinical Oncology
7	Biology and Fertility of Soils	Journal of Hazardous Materials	The Lancet Oncology
8	Journal of Environmental Quality	Environment International	PLoS ONE
9	Journal of Soils and Sediments	Chemosphere	Nature Genetics

(Continued)

TABLE 4.3 (*Continued*)

Top 20 Soil Science, Environmental Science, and Health and Medical Sciences Journals According to Google Scholar as of April 2020

Ranking	Soil Science Journals	Environmental Science Journals	Health and Medical Sciences Journals
10	European Journal of Soil Science	Environmental Pollution	Nature Medicine
11	Agronomy Journal	Journal of Environmental Management	Journal of the American College of Cardiology
12	Soil Science Society of America Journal	Marine Pollution Bulletin	Circulation
13	Pedosphere	Desalination	European Heart Journal
14	Journal of Plant Nutrition and Soil Science	Waste Management	Blood
15	European Journal of Soil Biology	Environmental Science and Pollution Research	Gastroenterology
16	Archives of Agronomy and Soil Science	Atmospheric Environment	Neuron
17	SOIL	Ecotoxicology and Environmental Safety	Nature Neuroscience
18	Journal of Soil Science and Plant Nutrition	Environmental Modelling & Software	BMJ
19	Nutrient Cycling in Agroecosystems	Resources, Conservation and Recycling	Cochrane Database of Systematic Reviews
20	Vadose Zone Journal	Journal of Environmental Chemical Engineering	Science Translational Medicine

medical journals such as *The Lancet* (Sanchez and Swaminathan, 2005). Another possibility would be publications in leading journals from a field that straddles the intersection between soil science and human health, such as environmental science. While there is again no overlap between the environmental science, soil science, and health and medical sciences journals, there are several journals on the environmental sciences list that SSPs commonly publish in and read (Table 4.3). However, readership of these journals by HCPs is much less common.

Every field has their specialty journals, but there are some broad leading scientific journals that are read by researchers across the scientific spectrum, including SSPs and HCPs. The top 25 overall journals, according to Google Scholar, are shown in Table 4.4. Several of these, including *Nature*, *Science*, *Nature Communications*, *Proceedings of the National Academy of Sciences*, and *PLoS One* publish papers ranging from soil science to medical science. While these journals do offer the opportunity to bridge the communications gap between SSPs and HCPs through their broad publication and readership, many of these are also very selective which makes publication in them difficult. Many papers that might be meaningful and useful in terms of cross-disciplinary communication would likely not make it through the review and acceptance process in these highly competitive journals.

A third approach to crossing the disciplinary communications divide would be to encourage both fields to read journals that cater to the publication of soil and human health information. One way to identify journals that publish soil and human health information is to use the Journal/Author Name Estimator (JANE; http://jane.biosemantics.org/), maintained by the Biosemantics Group and funded by the Netherlands Bioinformatics Center (NBIC). Eight example searches in JANE using keywords focused on various soil and human health issues are shown in Table 4.5. For the most part, the top 20

TABLE 4.4

Top Overall Journals According to Google Scholar as of April 2020

Ranking	Journal
1	Nature
2	The New England Journal of Medicine
3	Science
4	The Lancet
5	Chemical Reviews
6	Nature Communications
7	Advanced Materials
8	Chemical Society Reviews
9	Cell
10	IEEE/CVF Conference on Computer Vision and Pattern Recognition
11	Journal of the American Chemical Society
12	Proceedings of the National Academy of Sciences
13	Angewandte Chemie International Edition
14	Nucleic Acids Research
15	JAMA
16	Physical Review Letters
17	Energy & Environmental Science
18	ACS Nano
19	Journal of Clinical Oncology
20	Nano Letters
21	The Lancet Oncology
22	Nature Materials
23	PLoS ONE
24	Nature Genetics
25	Renewable and Sustainable Energy Reviews

soil science and health and medical sciences journals (Table 4.3) did not show up in these eight sample searches. The exceptions were the *Journal of Environmental Quality* and *PLoS One*, which both showed up in three of the JANE searches. This leads to the conclusion that *PLoS One* shows particular promise to bridge the communications gap between SSPs and HCPs, as it targets both audiences and has published a number of papers on soils and human health topics. The *Journal of Environmental Quality* is also a good journal to go to if looking to publish on or read about soil and human health issues, and it is widely read by SSPs, but it is probably not on the reading list of the typical HCP.

Journals from the top 20 environmental science list showed up quite frequently in the JANE searches, with nine of the top 20 appearing in at least one search. More precisely, *Environmental Science & Technology* showed up in one of the eight keyword searches; *Environment International*, *Chemosphere*, and *Journal of Environmental Management* each showed up twice; *Environmental Pollution and Ecotoxicology* and *Environmental Safety* showed up three times each; *Journal of Hazardous Materials* appeared in four of the eight searches; *Environmental Science and Pollution Research* showed up five times; and *Science of the Total Environment* was in seven of the eight searches. This verified the earlier thought that environmental science might help bridge the communications gap between SSPs and HCPs because relevant papers are obviously being published in these journals. However, while SSPs are frequent readers of and contributors to these journals, HCPs are less likely to be engaged with them. There are also a number of public health journals and toxicology journals that publish papers relevant to the soils and human health issue. For example, *Environmental Health Perspectives*, a leading toxicology journal, has published a large number of

TABLE 4.5

Journals That Are Listed in the JANE Data Base When Conducting Certain Keyword Searches Involving Soils and Human Health

Keywords Searched	Ten Top Journals[a]
Soil health human health	1. Science of the Total Environment
	2. Environmental Science and Pollution Research
	3. Journal of Hazardous Materials
	4. Environment International
	5. International Journal of Environmental Research and Public Health
	6. Journal of Environmental Management
	7. Chemosphere
	8. Environmental Pollution
	9. Nature
	10. Journal of Environmental Quality
Soil human health	1. Science of the Total Environment
	2. Journal of Hazardous Materials
	3. Environmental Science and Pollution Research
	4. International Journal of Environmental Research and Public Health
	5. Journal of Environmental Management
	6. Environment International
	7. Environmental Pollution
	8. Ecotoxicology and Environmental Safety
	9. Nature
	10. Journal of Environmental Quality
Soil human nutrition	1. Science of the Total Environment
	2. Ecotoxicology and Environmental Safety
	3. Scientific Reports
	4. Plant Science
	5. Applied and Environmental Microbiology
	6. Frontiers in Plant Science
	7. Environmental Science and Pollution Research
	8. Pakistan Journal of Pharmaceutical Sciences
	9. Annals of Botany
	10. Frontiers in Microbiology
Soil human disease	1. Science of the Total Environment
	2. PLoS One
	3. Applied and Environmental Microbiology
	4. International Journal of Environmental Research and Public Health
	5. Scientific Reports
	6. PLoS Neglected Tropical Diseases
	7. Environmental Monitoring and Assessment
	8. Nature
	9. International Journal of Infectious Diseases
	10. Annual Review of Phytopathology
Soil infectious disease	1. PLoS One

(Continued)

TABLE 4.5 (*Continued*)
Journals That Are Listed in the JANE Data Base When Conducting Certain Keyword Searches Involving Soils and Human Health

Keywords Searched	Ten Top Journals[a]
	2. Prion
	3. Frontiers in Microbiology
	4. PLoS Neglected Tropical Diseases
	5. Applied and Environmental Microbiology
	6. Scientific Reports
	7. Chemosphere
	8. Science of the Total Environment
	9. Journal of Environmental Quality
	10. Vector Borne and Zoonotic Diseases
Soil allergen	1. European Annals of Allergy and Clinical Immunology
	2. Immunology and Allergy Clinics of North America
	3. Current Opinion in Allergy and Clinical Immunology
	4. Frontiers in Plant Science
	5. The Journal of Allergy and Clinical Immunology
	6. Allergy, Asthma & Immunology Research
	7. Current Allergy and Asthma Reports
	8. Allergy
	9. Journal of Environmental Health
	10. Journal of Agricultural and Food Chemistry
Soil chemical exposure	1. Environmental Pollution
	2. Science of the Total Environment
	3. Environmental Science and Pollution Research
	4. Journal of Hazardous Materials
	5. Environmental Science & Technology
	6. Pest Management Science
	7. Ecotoxicology and Environmental Safety
	8. Environmental Research
	9. Integrated Environmental Assessment and Management
	10. SAR and QSAR in Environmental Research
Soil medicine	1. Journal of Hazardous Materials
	2. Environmental Science and Pollution Research
	3. Journal of Agricultural and Food Chemistry
	4. PLoS One
	5. Journal of Asian Natural Products Research
	6. Canadian Journal of Microbiology
	7. FEMS Microbiology Ecology
	8. Bulletin of Environmental Contamination and Toxicology
	9. Science of the Total Environment
	10. Journal of Biotechnology

[a] Journals were excluded from these lists if JANE indicated they did not have an Article Influence factor.

papers that investigate various soil and human health links over a wide time span (e.g., Sorenson et al., 1974; Mielke and Reagan, 1998; Kessler, 2013; Lin et al., 2017). However, these journals are likely read more by HCPs than by SSPs.

Another possible communications outlet involves books that deal with the merger of soil and medical sciences to address human health issues. Recent books specifically on this topic include Brevik and Burgess (2013b), Singh et al. (2017), and the book that this chapter is in. In addition to these books that focus exclusively on soil–human health connections, there are broader book projects, for example, those on medical geology (Selinus at al., 2013; Centeno et al., 2016; Siegel et al., 2021) that include soils information within their pages. An advantage to using books to communicate between academic fields is the ability to have an entire collection of relevant information in a single compact package as opposed to diffuse papers spread among several journals or multiple volumes of a given journal, which makes finding the information easier. Another advantage is that publishers can market a given book to the desired audience(s), whereas most of our journals already have an established audience. One disadvantage to books is that they do not receive an impact factor, so faculty evaluation procedures at some universities, either real or perceived, may discourage some faculty, especially pretenure faculty or those hoping to find a tenure-track position, from publishing in books (McKiernan et al., 2019).

4.3.2 FACILITATING JOINT CONFERENCES

In addition to publication of manuscripts that cross disciplinary lines, conferences that deliberately engage both SSPs and HCPs in joint sessions offer another means for enhancing interprofessional communication and productively integrating soil health and human health disciplines with the aim of driving research that will improve the human condition.

In 2018, The Soil Health Institute sponsored the "Conference on Connections between Soil Health and Human Health" with the aims of (i) making relevant connections between soil and human health sciences; (ii) identifying promising research opportunities; (iii) building interdisciplinary teams to address those research needs; and (iv) proposing funding mechanisms to meet these goals (SHI, 2018). Forty speakers/panelists were invited (including four physicians) in order to provide a balanced soil health/human health perspective. More than 180 attendees from 120 different organizations (including a variety of funding agencies) were in attendance. Participants were assigned to ten separate working groups charged with developing research priorities in the technical areas of human nutrition, food safety, fate and transport, and microbiome and community health (recommendations available online; SHI, 2018). Six recommended "next steps" were identified:

1. leverage conference participants to form a working group that can promote/facilitate **creation of transdisciplinary soil health/human health research teams**;
2. promote **cross-attendance/participation** at soil science/agricultural/human health meetings;
3. convene periodic **theme-specific conferences** focused on different attributes of the soil health/human health continuum;
4. develop a web-based **information repository**;
5. advocate for **funding sources** that facilitate transdisciplinary research teams involving both soil health and human health scientists;
6. develop **critical review of literature** for publication in suitable transdisciplinary journals.

Of note, as a direct result of this conference, one of the organizer's SSPs (Wayne Honeycutt, PhD) and one of the participating HCPs (David Collier, M.D., Ph.D) were invited to copresent on the connections between soil health and human health at the 2019 Rural Health Symposium. This symposium, jointly sponsored by East Carolina University's Brody School of Medicine, the University of North Carolina's Eschelman School of Pharmacy, and Vidant Health (regional hospital system) in association with the Eastern Area Health Education Center, attracted several hundred HCPs and serves as an example of successful cross-attendance/participation. Other direct results of this conference that include collaborations between SSPs and HCPs are Brevik et al. (2020) and this chapter.

Conference planners wishing to engage HCPs (specifically physicians) should work with an organization that can certify their program and allow them to offer appropriate AMA PRA Category 1 Credits™, as this legitimizes the program in the eyes of HCPs as well as attracts HCPs who need to maintain their continuing medical education. Funding to support travel is also an important, and potentially limiting, factor for facilitating cross-attendance/participation. Finally, conferences of the International Medical Geology Association provide additional examples of joint/cross-discipline conferences. Since their founding in 2004, they have organized 8 international conferences that have brought together geoscientists and medical professionals, including 187 participants from the geological and medical communities from 30 countries at their 2017 meeting in Moscow (Farkhutdinov et al., 2020). These conferences may serve as a template for additional conferences focused on soil and human health connections.

4.3.3 University Education That Crosses Disciplines

One of the challenges in getting SSPs and HCPs to engage with one another is the lack of a commonly shared language. Every field has its own professional lingo, and it can be difficult for those who are not fluent in that lingo to professionally collaborate with someone who is, even when there is willingness by both parties to engage (Kuznetsov and Kuznetsova, 2014). In addition, even when the words used are the same, their exact meanings may vary from one discipline to another (Brevik, 2013; Rothschild, 2015). Therefore, some form of university education that provides a bridge across the disciplines may help close the communication gap. This is not advocating for soil scientists who can take out an appendix, or MDs who also make a habit of digging soil pits, but for training for members of both fields that allows them to effectively communicate with each other at a professional level so that better integration between them can occur. Keith et al. (2016) saw a large degree of overlap between the ecosystem services ideas being utilized by many natural scientists and the One Health initiative coming from the health community, and proposed soil stewardship as a way to bridge the communications gap between these groups. Soil ecosystem services, in turn, are linked to human health (Brevik et al., 2018, 2019).

There are many examples of cross-disciplinary training that has led to quite successful results. For example, the University of Saskatchewan has taught a graduate level field class that enrolled both soil science and masters of fine arts students. The assignments included detailed descriptions and paintings of soils and landscapes, as well as making the pigments used to create these paintings from natural materials gathered by the students. The soil science students benefited through the development of increased attention to detail, and the fine arts students learned some of the science behind their pigments (Van Rees, 2017). Benefits have been shown when incorporating training from both biology and chemistry faculty in graduate-level chemical biology programs as opposed to basing the training in one department or the other (Silvius, 2007). Cross-training of medical researchers and regulators has shown benefits (Eitzen et al., 2019), and Williams (1998) reported on a physician who also had training as an engineer and was able to turn his dual passion into a career designing medical devices for use during space missions. Olsen et al. (2015) reported on the benefits of multidisciplinary health field courses based on their experiences running such a course for over a decade. Therefore, numerous examples have shown the advantages of cross-disciplinary training.

The call being made in this chapter for SSPs and HCPs to engage in nontraditional cross-disciplinary training is not an isolated suggestion. Other disciplines that have connections to human health but do not typically work with health professionals are also proposing closer ties. For example, Sanchez and Khreis (2020) have called for cross-disciplinary training between transportation and human health professionals, because our transportation choices have implications for our health ranging from chronic diseases to climate change concerns, a very similar situation to the links between soil and human health. Even within disciplines that might, on first glance, appear to be those that would traditionally work together, the call has been made for more interdisciplinary cooperation. For example, the need to engage in more cross-disciplinary work between scientists and HCPs has been noted regarding infectious diseases (Borer et al., 2011) and veterinarians and HCPs regarding zoonotic diseases (Ellis, 2008). In many ways, this is the point to the One Health initiative, breaking traditional disciplinary boundaries and facilitating cooperation and communication across fields (Lebov et al., 2017), and discussions regarding the structure of medical-related training have taken place under the One Health umbrella (Mor et al., 2013; Rabinowitz et al., 2017).

Davies (2019) proposed a curriculum for a master's degree focused on medical geology. Davies' proposal included contributions from multiple academic departments, with specific coursework including mineralogy and petrology; geochemistry; human, plant, and animal physiology; fundamentals of geological fieldwork; epidemiology; sampling, analytical, and statistical techniques; and research methods. Interdisciplinary courses that explore the interactions between the geologic and biologic environments were also included. There are a number of links, both historical and modern, between soil science and geology (Landa and Brevik, 2015). Therefore, this medical geology curriculum provides a template that could be used as a starting point upon which interdisciplinary medical soil programs could be built. Both medicine and soil science are broad fields; therefore, any such curriculum should also include flexibility that would allow students to pursue particular areas of personal interest within the field.

4.4 CONCLUDING STATEMENTS

SSPs and HCPs are working on many of the same issues (climate change, pathogens, nutrient supply, exposure to toxic elements/chemicals, etc.) that affect human health, just from different perspectives. This fact provides a lot of room to benefit from cross-disciplinary cooperation between the disciplines. However, to get such cooperation, it is imperative that SSPs and HCPs engage in frequent and meaningful cross-disciplinary communication. Doing so means putting information into publication outlets that both disciplines will read, or at least periodically check for relevant information. Communication would also be facilitated by regularly scheduled conferences that include both SSPs and HCPs. This would not have to be annual conferences. There are many successful conferences that run on less than an annual schedule, but having a schedule that allows interested SSPs and HCPs to plan to participate in would be beneficial. Finally, there is a need for enough cross-education such that SSPs and HCPs can speak the same professional language in their communication efforts.

ACKNOWLEDGEMENTS

E.C. Brevik was supported by the National Science Foundation EPSCoR program under Grant Number OIA-1355466 during this project.

REFERENCES

Abbas, F., Hammad, H.M., Ishaq, W., Farooque, A.A., Bakhat, H.F., Zia, A., Fahad, S., Farhad, W., and Cerdà, A. 2020. A review of soil carbon dynamics resulting from agricultural practices. *Journal of Environmental Management* 268, 110319.

Aitta, A., El-Ramady, H., Alshaal, T., El-Henawy, A., Shams, M, Talha, N., Elbehiry, F., and Brevik, E. 2019. Ecological risk assessment and spatial distribution of soil trace elements around Kitchener drain in the Northern Nile Delta, Egypt. *Agriculture* 9, 152.

Alley, D.E., Asomugha, C.N., Conway, P.H., and Sanghavi, D.M. 2016. Accountable Health Communities – Addressing social needs through Medicare and Medicaid. *New England Journal of Medicine*. doi:10.1056/NEJMp1512532.

Barrett, B., Charles, J.W., and Temte, J.L. 2015. Climate change, human health, and epidemiological transition. *Preventive Medicine* 70, 69–75.

Baumgardner, D.J. 2012. Soil-related bacterial and fungal infections. *Journal of the American Board of Family Medicine* 25, 734–744.

Bender, S.F., and van der Heijden, M.G.A. 2015. Soil biota enhance agricultural sustainability by improving crop yield, nutrient uptake and reducing nitrogen leaching losses. *Journal of Applied Ecology* 52, 2. doi:10.1111/1365-2664.1235128-239.

Blum, W.E.H., Zechmeister-Boltenstern, S., and Keiblinger, K.M. 2019. Does soil contribute to the human gut microbiome? *Microorganisms* 7, 287.

Borer, E.T., Antonovics, J., Kinkel, L.L., Hudson, P.J., Daszak, P., Ferrari, M.J., Garrett, K.A., Parrish, C.R., Read, A.F., and Rizzo, D.M. 2011. Bridging taxonomic and disciplinary divides in infectious disease. *EcoHealth* 8, 261–267.

Boulos, M.N.K., and Le Blond, J. 2016. On the road to personalised and precision geomedicine: medical geology and a renewed call for interdisciplinarity. *International Journal of Health Geographics* 15, 5. doi:10.1186/s12942-016-0033-0.

Brady, N.C., and Weil, R.R. 2016. *Nature and Properties of Soils*, 15th Ed. Pearson, Upper Saddle River, NJ.

Brevik, E.C. 2009a. Soil health and productivity. In Verheye, W. (Ed.). *Soils, Plant Growth and Crop Production*. Encyclopedia of Life Support Systems (EOLSS), Developed under the Auspices of the UNESCO, EOLSS Publishers, Oxford, UK. http://www.eolss.net.

Brevik, E.C. 2009b. Soil, food security, and human health. In Verheye, W. (Ed.). *Soils, Plant Growth and Crop Production*. Encyclopedia of Life Support Systems (EOLSS), Developed under the Auspices of the UNESCO, EOLSS Publishers, Oxford, UK. http://www.colss.net.

Brevik, E.C. 2012. Soils and climate change: gas fluxes and soil processes. *Soil Horizons* 53(4), 12–23.

Brevik, E.C. 2013. An introduction to soil science basics. In Brevik, E.C., and Burgess, L.C. (Eds.), *Soils and Human Health*. CRC Press, Boca Raton, FL, pp. 3–28.

Brevik, E.C., and Burgess, L.C. 2013a. The 2012 fungal meningitis outbreak in the United States: connections between soils and human health. *Soil Horizons* 54. doi:10.2136/sh12-11-0030.

Brevik, E.C., and Burgess, L.C. (Eds). 2013b. *Soils and Human Health*. CRC Press, Boca Raton, FL.

Brevik, E.C., Pereg, L., Pereira, P., Steffan, J.J., Burgess, L.C., and Gedeon, C.I. 2019. Shelter, clothing, and fuel: often overlooked links between soils, ecosystem services, and human health. *Science of the Total Environment* 651, 134–142. doi:10.1016/j.scitotenv.2018.09.158.

Brevik, E.C., Pereg, L., Steffan, J.J., and Burgess, L.C. 2018. Soil ecosystem services and human health. *Current Opinion in Environmental Science & Health* 5, 87–92. doi:10.1016/j.coesh.2018.07.003.

Brevik, E.C., and Sauer, T.J. 2015. The past, present, and future of soils and human health studies. *Soil* 1, 35–46. doi:10.5194/soil-1-35-2015.2018.

Brevik, E.C., Slaughter, L., Singh, B.R., Steffan, J.J., Collier, D., Barnhart, P., and Pereira, P. 2020. Soil and human health: current status and future needs. *Air, Soil, and Water Research* 13, 1–23. doi:10.1177/1178622120934441.

Buck, B.J., Londono, S.C., McLaurin, B.T., Metcalf, R., Mouri, H., Selinus, O., and Shelembe, R. 2016. The emerging field of medical geology in brief: some examples. *Environmental Earth Sciences* 75, 449. doi:10.1007/s12665-016-5362-6.

Bultman, M.W., Fisher, F.S., and Pappagianis, D. 2013. The ecology of soil-borne human pathogens. In Selinus, O., Alloway, B., Centeno, J., Finkelman, R., Fuge, R., Lindh, U., and Smedley, P. (Eds.), *Essentials of Medical Geology*. Springer, Cham, Switzerland, pp. 477–504.

Butcher, K., Wick, A.F., DeSutter, T., Chatterjee, A., and Harmon, J. 2018. Corn and soybean yield response to salinity influenced by soil texture. *Agronomy Journal* 110, 1243–1253. doi:10.2134/agronj2017.10.0619.

Carr, P.M., Brevik, E.C., Horsley, R.D., and Martin, G.B. 2015. Long-term no-tillage sequesters soil organic carbon in cool semi-arid regions. *Soil Horizons* 56. doi:10.2136/sh15-07-0016.

Centeno, J.A., Finkelman, R.B., and Selinus, O. 2016. *Medical Geology: Impacts of the Natural Environment on Public Health*. MDPI AG, Basel, Switzerland.

Cockx, E.M., and Simonne, E.H. 2003. Reduction of the impact of fertilization and irrigation on processes in the nitrogen cycle in vegetable fields with BMPs. In *Nutrient Management of Vegetable and Row Crops Handbook*. University of Florida Agricultural Extension Service, Gainesville, FL, pp. 8–24.

Davies, T.C. 2019. A medical geology curriculum for African geoscience institutions. *Scientific African* 6, e00131.

De Pascale, S., and Barbieri, G. 1995. Effects of soil salinity from long-term irrigation with saline-sodic water on yield and quality of winter vegetable crops. *Scientia Horticulturae* 64, 145–157.

Eitzen, M.M., Jones, E.Z., McCowan, J., and Brasel, T. 2019. A cross-disciplinary training program for the advancement of medical countermeasures. *Health Security* 17, 344–351.

Ellis, K.H. 2008. One health initiative will unite veterinary, human medicine. *Thorofare* 21, 11.

Ernakovich, J.G., Evans, T.A., Macdonald, B., and Farrell, M. 2016. The effect of ecosystem engineers on N cycling in an arid agroecosystem. *Proceedings of the 2016 International Nitrogen Initiative Conference, "Solutions to Improve Nitrogen Use Efficiency for the World"*, 4–8 December 2016, Melbourne, Australia. 4 p.

Farkhutdinov, I., Farkhutdinova, L., Zlobina, A., Farkhutdinov, A., Volfson, I., and Matveenko, I. 2020. Historical aspects of medical geology. *Earth Sciences History* 39, 172–183.

Ferris, H., and Tuomisto, H. 2015. Unearthing the role of biological diversity in soil health. *Soil Biology & Biochemistry* 85, 101–109.

Filippelli, G.M., Adamic, J., Nichols, D., Shukle, J., and Frix, E. 2018. Mapping the urban lead exposome: a detailed analysis of soil metal concentrations at the household scale using citizen science. *International Journal of Environmental Research and Public Health* 15, 1531. doi:10.3390/ijerph15071531.

Finkelman, R.B. 2019. The influence of clays on human health: a medical geology perspective. *Clays and Clay Minerals* 67, 1–6.

Haahtela, T. 2019. A biodiversity hypothesis. *Allergy* 74, 1445–1456.

Hanyu, K., Tamura, K., and Mori, H., 2014. Changes in heart rate variability and effects on POMS by whether or not soil observation was performed. *Open Journal of Soil Science* 4, 36–41.

Havlin, J.L., Tisdale, S.L., Nelson, W.L., and Beaton, J.D. 2013. *Soil Fertility and Fertilizers*, 8th Ed. Pearson-Prentice Hall, Upper Saddle River, NJ.

Helmke, M.F., and Losco, R.L. 2013. Soil's influence on water quality and human health. In Brevik, E.C., and Burgess, L.C. (Eds.), *Soils and Human Health*. CRC Press, Boca Raton, FL, pp. 155–176.

Henry, J.M., and Cring, F.D. 2013. Geophagy: an anthropological perspective. In Brevik, E.C., and Burgess, L.C. (Eds.), *Soils and Human Health*. CRC Press, Boca Raton, FL, pp. 179–198.

Keith, A.M., Schmidt, O., and McMahon, B.J. 2016. Soil stewardship as a nexus between ecosystem services and one health. *Ecosystem Services* 17, 40–42.

Kemper, K.J., and Lal, R. 2017. Pay dirt! Human health depends on soil health. *Complementary Therapies in Medicine*. doi:10.1016/j.ctim.2017.04.005.

Kessler, R. 2013. Urban gardening: managing the risks of contaminated soil. *Environmental Health Perspectives* 121, A327–A333.

Khaledian, Y., Brevik, E.C., Pereira, P., Cerdà, A., Fattah, M.A., and Tazikeh, H. 2017. Modeling soil cation exchange capacity in multiple countries. *Catena* 158, 194–200. doi:10.1016/j.catena.2017.07.002.

Kuznetsov, A., and Kuznetsova, O. 2014. Building professional discourse in emerging markets: language, context and the challenge of sensemaking. *Journal of International Business Studies* 45, 583–599.

Lal, R. 2007. Soil science and the carbon civilization. *Soil Science Society of America Journal* 71, 1425–1437.

Lal, R. 2013. Enhancing ecosystem services with no-till. *Renewable Agriculture and Food Systems* 28, 102–114.

Landa, E.R., and Brevik, E.C. 2015. Soil science and its interface with the history of geology community. *Earth Sciences History* 34, 296–309.

Lazarus, J.H. 2015. The importance of iodine in public health. *Environ Geochem Health* 37, 605–618. doi:10.1007/s10653-015-9681-4.

Lebov, J., Grieger, K., Womack, D., Zaccaro, D., Whitehead, N., Kowalcyk, B., and MacDonald, P.D.M. 2017. A framework for one health research. *One Health* 3, 44–50.

Lin, C., Wang, B., Cui, X., Xu, D., Cheng, H., Wang, Q., Ma, J., Chai, T., Duan, X., Liu, X., Ma, J., Zhang, X., and Liu, Y. 2017. Estimates of soil ingestion in a population of Chinese children. *Environmental Health Perspectives*, 077002.

Lowry, C.A., Hollis, J.H., de Vries, A., Pan, B., Brunet, L.R., Hunt, J.R.F., Paton, J.F.R., van Kampen, E., Knight, D.M., Evans, A.K., Rook, G.A.W., and Lightman, S.L. 2007. Identification of an immune-responsive mesolimbocortical serotonergic system: potential role in regulation of emotional behavior. *Neuroscience* 146, 756–772.

Mbila, M. 2013. Soil minerals, organisms, and human health: medicinal uses of soils and soil materials. In Brevik, E.C., Burgess, L.C. (Eds). *Soils and Human Health*. CRC Press, Boca Raton, FL, pp. 199–214.

McGinnis, J.M., Williams-Russo, P., Knickman, J.R. 2002. The case for more active policy attention to health promotion. *Health Aff (Millwood)* 21, 78–93.

McKiernan, E.C., Schimanski, L.A., Nieves, C.M., Matthias, L., Niles, M.T., and Alperin, J.P. 2019. Use of the Journal Impact Factor in academic review, promotion, and tenure evaluations. *eLife* 8, e47338.

Mielke, H.W., and Reagan, P.L. 1998. Soil is an important pathway of human lead exposure. *Environmental Health Perspectives* 106, 217–229.

Mor, S.M., Robbins, A.H., Jarvin, L., Kaufman, G.E., and Lindenmayer, J.M. 2013. Curriculum asset mapping for One Health education. *Journal of Veterinary Medical Education* 40, 363–369.

Morgan, R. 2013. Soil, heavy metals, and human health. In Brevik, E.C., and Burgess, L.C. (Eds.), *Soils and Human Health*. CRC Press, Boca Raton, FL, pp. 59–82.

NASEM. 2017. *Soils: The Foundation of Life: Proceedings of a Workshop – In Brief*. The National Academies Press, Washington, DC. doi:10.17226/24866.

Njira, K.O.W., and Nabwami, J. 2013. Soil management practices that improve soil health: elucidating their implications on biological indicators. *Journal of Animal & Plant Sciences* 18, 2750–2760.

O'Brien, M.E.R., Saini, A., Smith, I.E., Webb, A., Gregory, K., Mendes, R., Ryan, C., Priest, K., Bromelow, K.V., Palmer, R.D., Tuckwell, N., Kennard, D.A., and Souberbielle, B.E. 2000. A randomized phase II study of SRL172 (Mycobacterium vaccae) combined with chemotherapy in patients with advanced inoperable non-small-cell lung cancer and mesothelioma. *British Journal of Cancer* 83, 853–857.

Oliver, M.A. 1997. Soil and human health: a review. *European Journal of Soil Science* 48, 473–592.

Oliver, M.A. and Gregory, P.J. 2015. Soil, food security and human health: a review. *European Journal of Soil Science* 66, 257–276.

Olsen, C., Conway, J., DiPrete-Brown, L. Hutchins, F., Poulsen, K., Solheim, K., Kraus, C., Gaus, D., and Silawan, T. 2015. Advancing integrative "One-Health" approaches to global health through multidisciplinary, faculty-led global health field courses. *The Lancet Global Health* 3, S6.

Pepper, I.L. 2013. The soil health: human health nexus. *Critical Reviews in Environmental Science and Technology* 43, 2617–2652.

Pereg, L., Steffan, J.J., Gedeon, C., Thomas, P., and Brevik, E.C. Medical geology of soil ecology. 2021. In Siegel, M., Selinus, O., and Finkelman, R. (Eds.), *Practical Applications of Medical Geology*.

Rabinowitz, P.M., Natterson-Horowitz, B.J., Kahn, L.H., Kock, R., and Pappaioanou, M. 2017. Incorporating one health into medical education. *BMC Medical Education* 17, 45.

Rayman, M.P. 2012. Selenium and human health. *The Lancet* 379, 1256–1268.

Rothschild, B. 2015. A convoluted road from biochemistry to clinical medicine to primatology. *Primatology* 4, 2.

Sanchez, K.A., and Khreis, H. 2020. The role of cross-disciplinary education, training, and workforce development at the intersection of transportation and health. In Nieuwenhuijsen, M.J., and Khreis, H. (Eds.), *Advances in Transportation and Health: Tools, Technologies, Policies, and Developments*, Elsevier, Amsterdam, pp. 423–449.

Sanchez, P.A., and Swaminathan, M.S. 2005. Hunger in Africa: the link between unhealthy people and unhealthy soils. *The Lancet* 365, 442–444.

Selinus, O., Alloway, B., Centeno, J., Finkelman, R., Fuge, R., Lindh, U., and Smedley, P. (Eds.) 2013. *Essentials of Medical Geology*, Revised Edition. Springer, Dordrecht, Netherlands.

SHI. 2018. Conference on the connections between soil health and human health. October 16–17, 2018. Silver Springs, Maryland USA. Accessed October 29, 2019. Available from: https://soilhealthinstitute.org/humanhealthconference/

Siegel, M., Selinus, O., and Finkelman, R. 2021. Practical applications of medical geology. Springer, Dordrecht.

Silvius, J. 2007. Strength in diversity: a cross-disciplinary approach to graduate training in chemical biology. *Nature Biotechnology* 25, 255–258.

Singh, B.R., McLaughlin, M.J. and Brevik, E.C. (Eds.). 2017. *The nexus of soils, plants, animals and human health*. Catena-Schweizerbart, Stuttgart, Germany.

Smith, D.G., Martinelli, R., Besra, G.S., Illarionov, P.A., Szatmari, I., Brazda, P., Allen, M.A., Xu, W., Wang, X., Nagy, L., Dowell, R.D., Rook, G.A.W., Brunet, L.R., and Lowry, C.A. 2019. Identification and characterization of a novel anti-inflammatory lipid isolated from Mycobacterium vaccae, a soil-derived bacterium with immunoregulatory and stress resilience properties. *Psychopharmacology* 236, 1653–1670.

Smith, T.E., Grattan, S.R., Grieve, C.M., Poss, J.A., Läuchli, A.E., and Suarez, D.L. 2013. pH dependent salinity-boron interactions impact yield, biomass, evapotranspiration and boron uptake in broccoli (*Brassica oleracea* L.). *Plant and Soil* 370, 541–554. doi:10.1007/sl1104-013-1653-9.

Sorenson, J.R.J., Campbell, I.R., Tepper, L.B., and Lingg, R.D. 1974. Aluminum in the environment and human health. *Environmental Health Perspectives* 8, 3–95.

Steffan, J.J., Brevik, E.C., Burgess, L.C., and A. Cerdà. 2018. The effect of soil on human health: an overview. *European Journal of Soil Science* 69, 159–171. doi:10.1111/ejss.12451.

Turmel, M.S., Speratti, A., Baudron, F., Verhulst, N., and Govaerts, B. 2015. Crop residue management and soil health: a systems analysis. *Agricultural Systems* 134, 6–16.

USDA-NRCS. 2019. Soil health. Accessed October 29, 2019. Available from: https://www.nrcs.usda.gov/wps/portal/nrcs/main/soils/health/

USDHHS. 2019. US Department of Health and Human Services, Office of Disease Prevention and Health Promotion. Social determinants of health/healthy people 2020. Accessed October 29, 2019. Available from: https://www.healthypeople.gov/2020/topics-objectives/topic/social-determinants-of-health.

USGCRP. 2016. *The Impacts of Climate Change on Human Health in the United States: A Scientific Assessment.* Crimmins, A., J. Balbus, J.L. Gamble, C.B. Beard, J.E. Bell, D. Dodgen, R.J. Eisen, N. Fann, M.D. Hawkins, S.C. Herring, L. Jantarasami, D.M. Mills, S. Saha, M.C. Sarofim, J. Trtanj, and L. Ziska, (Eds.). U.S. Global Change Research Program, Washington, DC, 312 pp. doi:10.7930/J0R49NQX.

Van Rees, K. 2017. Evolution of a soil scientist into an artist: impacts on my teaching and life. *Geophysical Research Abstracts* 19, EGU2017-3717.

Wall, D.H., Uffe, N.N., and Six, J. 2015. Soil biodiversity and human health. *Nature* 528, 69–76.

WHO. 1946. *Preamble to the Constitution of WHO as adopted by the International Health Conference*, New York, 19 June–22 July 1946; signed on 22 July 1946 by the representatives of 61 States (Official Records of WHO, no. 2, p. 100) and entered into force on 7 April 1948.

WHO. 2008. World health Organization. Social determinants of health. Accessed October 29, 2019. Available from: https://www.who.int/social_determinants/sdh_definition/en/

Williams, L.B. 2017. Geomimicry: harnessing the antibacterial action of clays. *Clay Minerals* 52, 1–24.

Williams, L.B., and Hillier, S. 2014. Kaolins and health: from first grade to first aid. *Elements* 10, 207–211.

Williams, L.S. 1998. Frustrated by lack of job opportunities in Canada, MD moves to US space-medicine program. *Canadian Medical Association Journal* 158, 384–385.

Wolf, B., and Snyder, G.H. 2003. *Sustainable Soils: The Place of Organic Matter in Sustaining Soils and their Productivity.* Haworth Press, New York.

Wrench, G.T. 1939. Health and the soil. *The British Medical Journal* 1, 276–277.

5 Soil and Cancer

Marium Husain
The Ohio State University

CONTENTS

5.1 INTRODUCTION

The idea of cancer cells spreading to other organs (metastases) was termed the "seed and soil" theory, coined by Dr. Stephen Paget in 1889. A surgeon from England hypothesized that the metastatic cancer cells ("seed") and the microenvironment of the particular organ ("soil") have an active relationship that determines metastatic spread, and not a random act (Paget 1889). The theory has evolved since then after significant research and inquiry into how cancer cells spread/metastasize. Contemporarily, the network around the cancer cells, called "reactive stroma," consists of the extracellular matrix (ECM), blood vessel development (angiogenesis), immune system modulation, signaling chemicals (cytokines), including normal tissue cells (Tsai and Yang 2013, Motz and Coukos 2011).

The analogy to the soil is interesting and tells of the importance soil plays in our culture and understanding of the world. The previous chapters have outlined the relationship between human health and soil and the various health effects from poor soil conditions. In this chapter, we introduce how soil health impacts cancer development, risk, and treatment/care. We also discuss some cancer prevention/treatment strategies from understanding the soil.

5.2 CARCINOGENESIS

Cancer is still a prevalent issue in society (Figure 5.1). Cancer development is a complex process that involves many players: cytokines (chemicals involved in cellular communication), the immune system, genetic mutations, epigenetic changes, risk factors, virus/bacterial infections, etc. (Colditz). Some cancers develop *de novo*, or without any underlying genetic abnormality. Others can develop in the context of familial genetic syndromes: Lynch, Familial Adenomatous Polyposis, Li Fraumeni, BRCA 1/2, Cowden, and Hereditary Papillary Renal Carcinoma. Cancers can start in the bone marrow and lymph nodes (leukemia and lymphoma, respectively) or a solid organ in the body (breast, pancreas, skin). There are multiple risk factors for cancer and, in particular, tobacco use, excess weight, poor diet, and inactivity account for two-thirds of all cancers in the United States (Harvard

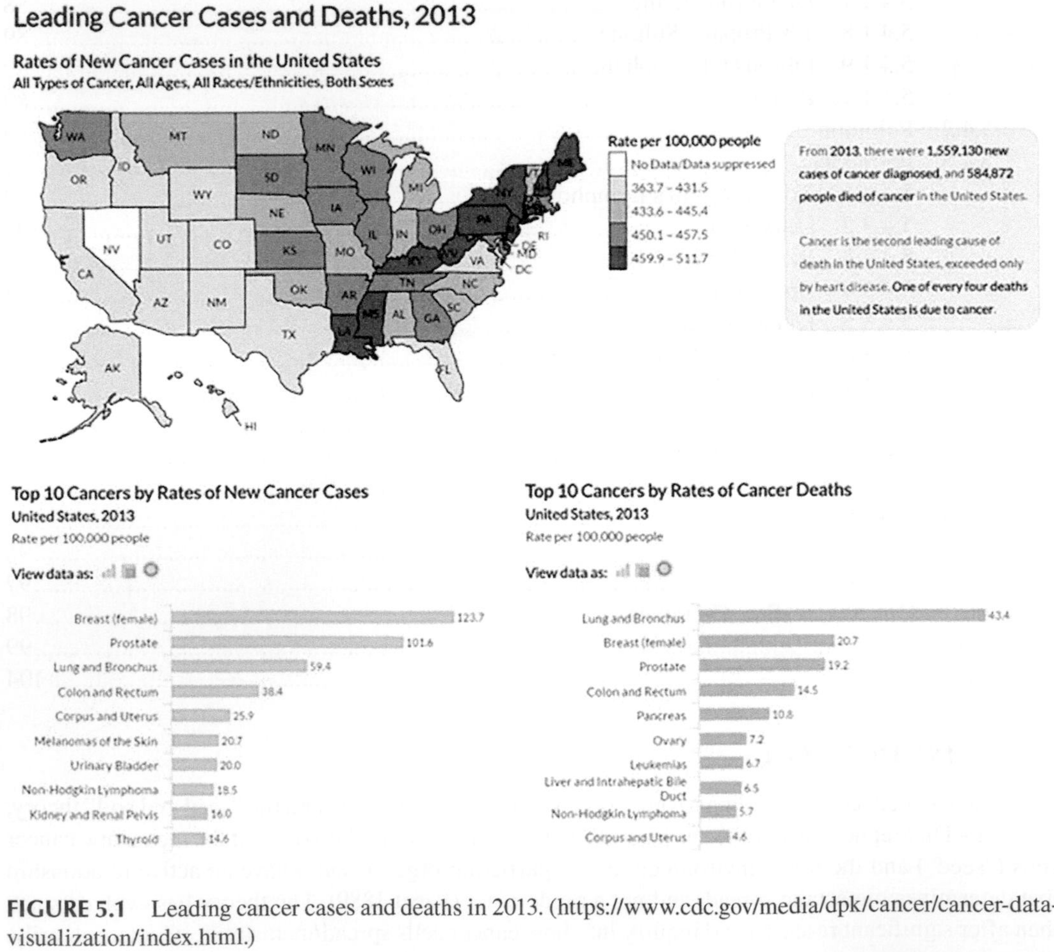

Leading Cancer Cases and Deaths, 2013
Rates of New Cancer Cases in the United States
All Types of Cancer, All Ages, All Races/Ethnicities, Both Sexes
WA
MT
ND
MN
OR
ID
SD
WI
WY
MI
NY
ME
NE
IA
PA
NV
UT
CO
KS
MO
IL
IN
OH
WV
VA
KY
DE
MD
DC
RI
CA
TN
NC
AZ
NM
OK
AR
SC
MS
AL
GA
TX
LA
FL
AK
HI
Rate per 100,000 people
No Data/Data suppressed
363.7 - 431.5
433.6 - 445.4
450.1 - 457.5
459.9 - 511.7
From 2013, there were 1,559,130 new cases of cancer diagnosed, and 584,872 people died of cancer in the United States.
Cancer is the second leading cause of death in the United States, exceeded only by heart disease. One of every four deaths in the United States is due to cancer.
Top 10 Cancers by Rates of New Cancer Cases
United States, 2013
Rate per 100,000 people
View data as:
Breast (female) 123.7
Prostate 101.6
Lung and Bronchus 59.4
Colon and Rectum 38.4
Corpus and Uterus 25.9
Melanomas of the Skin 20.7
Urinary Bladder 20.0
Non-Hodgkin Lymphoma 18.5
Kidney and Renal Pelvis 16.0
Thyroid 14.6
Top 10 Cancers by Rates of Cancer Deaths
United States, 2013
Rate per 100,000 people
View data as:
Lung and Bronchus 43.4
Breast (female) 20.7
Prostate 19.2
Colon and Rectum 14.5
Pancreas 10.8
Ovary 7.2
Leukemias 6.7
Liver and Intrahepatic Bile Duct 6.5
Non-Hodgkin Lymphoma 5.7
Corpus and Uterus 4.6

FIGURE 5.1 Leading cancer cases and deaths in 2013. (https://www.cdc.gov/media/dpk/cancer/cancer-data-visualization/index.html.)

Report 1997). Worldwide, there are 9 modifiable risk factors that account for 35% of cancer cases: smoking, alcohol use, diet low in fruit and vegetables, excess weight, inactivity, unsafe sex, urban air pollution, use of solid fuels, and contaminated injections in healthcare settings (Danaei et al. 2005).

5.3 CARCINOGENS

The International Agency for Research on Cancer (IARC) Monographs Program identifies and evaluates environmental causes of cancer in humans (IARC 2019). A working group of international experts meets to discuss the current scientific evidence on a suspected agent and using specific criteria, deliberate on evaluations of carcinogenicity to humans. The four data points evaluated are as follows:

1. The situations in which people are exposed to the agent.
2. Epidemiological studies on cancer in humans exposed to the agent (scientific evidence of carcinogenicity in humans).
3. Experimental studies on cancer in laboratory animals treated with the agent (scientific evidence of carcinogenicity in animals).
4. Studies of how cancer develops in response to the agent (scientific evidence on cancer mechanisms).

The IARC classifies agents based on the evidence as to whether an agent is capable of causing cancer, not how much risk is associated with exposure. There are four groups of classifications (IARC 2019):

1. Group 1: The agent is carcinogenic to humans
2. Group 2A: The agent is probably carcinogenic to humans
3. Group 2B: The agent is possibly carcinogenic to humans
4. Group 3: The agent is not classifiable as to its carcinogenicity to humans
5. Group 4: The agent is probably not carcinogenic to humans

To further explain these classifications, Group 2A is used when "there is limited evidence of carcinogenicity in humans and sufficient evidence of carcinogenicity in experimental animals. Limited evidence means that a positive association has been observed between exposure to the agent and cancer but that other explanations for the observations (technically termed chance, bias, or confounding) could not be ruled out." Group 2B is used when "there is limited evidence of carcinogenicity in humans and less than sufficient evidence of carcinogenicity in experimental animals." This essentially means that there is inadequate evidence. This classification can also be used when "there is some evidence that the agent could cause cancer in humans and in experimental animals but neither the evidence in humans nor the evidence in animals is convincing enough to permit a definite conclusion to be drawn." Group 3 is used when the "evidence of carcinogenicity is inadequate in humans and inadequate or limited in experimental animals." (For more information on the list of agents and the Monographs Programme: http://monographs.iarc.fr/ENG/Classification/index.php and http://monographs.iarc.fr/index.php).

The National Toxicology Program (NTP) is federally mandated by Congress to prepare the Report on Carcinogens and then present it to the Secretary of the U.S. Department of Health and Human Services. This report is cumulative and to date has listed 248 agents, substances, mixtures, and exposure circumstances that are known or reasonably anticipated to cause cancer in humans (https://ntp.niehs.nih.gov/whatwestudy/assessments/cancer/roc/index.html). The most recent report is the 14th version, released in 2016. For a listing of all carcinogens, please refer to the Appendix.

5.4 SOIL HEALTH

5.4.1 SOIL COMPOSITION

Humans can be exposed to environmental pollutants through three main mechanisms: ingestion, inhalation, and dermal absorption from soil and dust (Gabarron et al. 2017). The soil can be exposed to increased heavy metals from atmospheric particles that end up precipitating down to surface soil and accumulating over time (Acosta et al. 2015) or from industrial pollutants (Rashidi and Alavaipanah 2016). Heavy metals of particular concern that are carcinogens include cadmium, chromium, nickel, lead, and cobalt. Many studies have been done to assess the cancer risk (CR) related to exposure from these metals. One measure is lifetime average daily dose (LADD) (Li et al. 2013, Lu et al. 2014, Dehghani et al. 2017)

$$\text{LADD} = \frac{C \times EF}{AT \times PEF} \times \left(\frac{\text{InhR}_{child} \times ED_{child}}{BW_{child}} + \frac{\text{InhR}_{adult} \times ED_{adult}}{BW_{adult}} \right) \qquad (5.1)$$

where C is the concentration of metal, EF is the exposure frequency, AT is average time, PEF is the particulate emission factor, InhR is the inhalation rate, ED is the exposure duration, and BW is standard body weight.

Another risk assessment is CR, measured as the dose of dermal, inhalation, or ingestion of a compound multiplied by a slope factor (SF). There are corresponding SFs for each metal, including carcinogens, that are accessed from the Risk Assessment Information System (Gabarron et al. 2017).

$$CR = D \times SF \qquad (5.2)$$

where D is the ingestion-, dermal-, or inhalation-calculated dose and SF is the corresponding slope factor.

Gabarron displayed how these human health risk assessments can be useful to determine what areas (e.g., road dust and surface soil) expose humans to a higher risk from environmental pollutants, like heavy metals. Their study found that adults were exposed to heavy metals via inhalation routes more frequently than children, who were exposed to dermal absorption or ingestion (Gabarron et al. 2017).

Because the soil is so heterogeneous in composition, it can be statistically difficult to correlate specific components of the soil with health conditions. For example, a childhood cancer cluster was studied in Sandusky, Ohio, and the location of where these children lived overlapped with statistically significant hotspots for cadmium, chromium, and nickel (Figure 5.2). Hotspots were determined after evaluating cores from cottonwood trees. The data reflected a spatial pattern and not so much actual concentrations of these metals. However, it is difficult to completely correlate the two because of this and how further research is indicated (Garvin et al. 2019).

In addition, health-related risks of certain compounds in soil, like heavy metals, need to be assessed based on their bioavailability and not necessarily their total concentration. Datta and Sarkar found that the bioavailability of heavy metals is dependent on soil chemistry: clay content, organic carbon content, pH, cation exchange capacity, Fe, Aluminum (Al), Ca, Mg, and Phosphorus (P) concentrations (2005). The bioavailability and soil chemistry are different based on geography and soil type, in addition to external climate factors.

5.4.1.1 Selenium

Selenium is a trace mineral that is naturally occurring in soil, exists in amino acids in humans, and is present in some plants. There are different trials that were done with selenium to evaluate

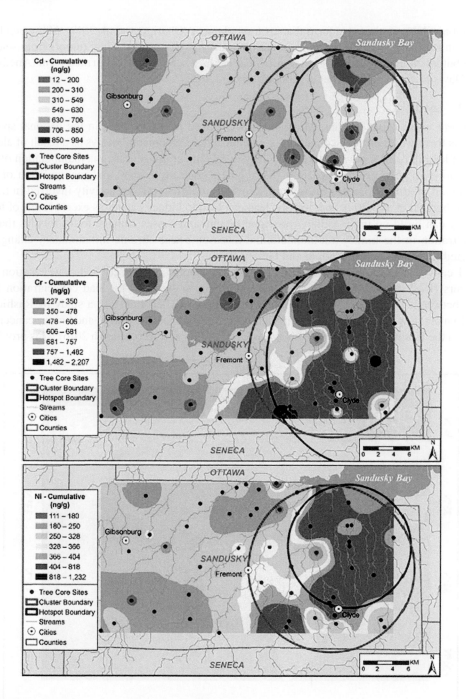

FIGURE 5.2 Childhood cancer cluster in Sandusky, Ohio.

its role in the nutritional prevention of cancers. Some trials showed no benefit in the selenium intake: Cochrane review and the SELECT trial (Vinceti 2014, Lippman et al. 2009). However, the Nutritional Prevention of Cancer Trial evaluated selenium as Se-enriched yeast, compared to the other trials that had evaluated selenium as L-selenomethionine supplement. In the yeast form, selenium was shown to significantly reduce the incidence of lung, colorectal and prostate cancer, total cancer incidence, and total cancer mortality (Clark 1996). It has been reported that selenium may be more bioavailable in the yeast form than as a supplement (Gharipour et al. 2017). Many preclinical studies in rats have shown a decrease in the mammary gland and intestinal tumors from selenium intake but when it was administered in food and not as a supplement in purified form (Abedi et al. 2018, Davis et al. 2002, Finley et al. 2000, Ip et al. 2000).

5.4.1.2 Lead

The recent water crisis in Flint, Michigan, USA, placed lead poisoning in the national spotlight again, as there was a profound concern for anemia and developmental delays as a result of elevated levels of lead in drinking water. Lead also plays a role in carcinogenesis, via manipulation of DNA (methylation in gene promoter regions that regulate gene expression) and manipulation of RNA (histone tail modifications that impact transcription and affect mRNA activity involved in translation) (Silbergeld et al. 2000). Specifically, lead can lead to kidney cancer as excess levels of lead in the body can build up in the kidney (proximal tubule) and causes damage to the tubules that help with filtration in the kidney (tubulointerstitial disease). These chronic inflammatory changes are suspected to lead to cancer (Cooper and Gaffey 1975).

Soil can be contaminated with lead via industrial pollutants, fossil fuel power stations, and fertilizers (Rashidi and Alavaipanah 2016). International standard for lead concentration in the soil is between 50 parts per million (ppm) and 150 ppm (Allen 1993). In a study by Rashidi and Alavaipanah, they showed a positive correlation between soil concentration and kidney cancer prevalence in the Isfahan province in Iran (Figure 5.3) (Rashidi and Alavaipanah 2016). This province is

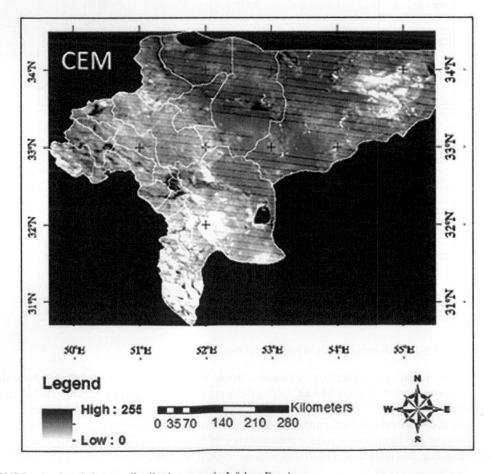

FIGURE 5.3 Lead element distribution maps in Isfahan Province.

a major agricultural area, and there is frequent use of chemical fertilizers and industrial pollutants (Rashidi and Alavaipanah 2016).

5.4.1.3   Arsenic

Arsenic contamination in soil is a particular concern around coal mining sites (Pal 2015). Pyrite in the bedrock can get oxidized to iron oxide and release arsenic and sulfates that can leak into the groundwater (Pal 2015, Carraro et al. 2015, Wang et al. 2010). Groundwater is relevant to this discussion as during dry seasons, groundwater is used for irrigating crops. Contamination of groundwater can lead to excess arsenic intake. At levels greater than 170 mg L^{-1}, there is an increased risk of skin cancer from arsenic exposure (Karagas et al. 2001); acceptable ranges are up to 100 parts per billion (WHO 1993). Long-term arsenic exposure can lead to lung, skin, and bladder cancers (Mazumder 2008, IARC 2004, Baris et al. 2016). Chattopadhyay et al. evaluated the Ganga river region in India and found unsafe levels of arsenic in the meandering position of the river and in three surrounding villages (Figure 5.4). This area is also known for intensive agricultural practices with "indiscriminate use of agrochemicals" (Chattopadhyay et al. 2019).

5.4.1.4   Trichloroethylene

Trichloroethylene (TCE) was used as a solvent, metal degreaser, and dry cleaning agent. As a solvent, it was used in different products, like electrical transformers and semiconductors (Stoiber and Naidenko 2018). TCE can be released into the groundwater from more than 400 Superfund sites. One Superfund site is near Camp LeJeune in North Carolina, and some veterans and their

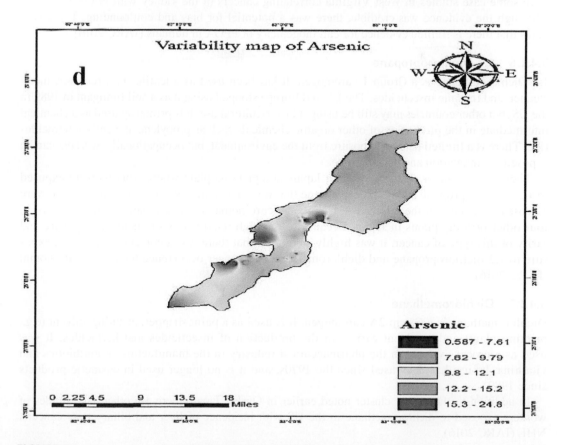

FIGURE 5.4   Ganga river region in India and the associated arsenic concentration.

families developed various forms of cancer: leukemia, bladder, kidney, liver cancer, multiple myeloma, non-Hodgkin's lymphoma (NHL). However, there is also a risk that is TCE exposure into groundwater from products made with TCE that are thrown away in landfills and other undesignated disposal sites.

In 2016, per the NTP Report on Carcinogens, TCE was added to the list of known carcinogens. Since 2000, TCE was listed as a "reasonably anticipated" human carcinogen. Multiple human studies demonstrating a causal relationship between TCE exposure and kidney cancer upgraded the status of TCE to "known carcinogen" (NTP 2016).

5.4.1.5   Perfluorooctanoic Acid

Perfluorooctanoic acid (PFOA) is a Group 2B carcinogen. It is used in the production of fluoropolymers, materials used in nonstick coatings on cookware, waterproof clothing, fire- and chemical-resistant tubing, and plumber's thread-seal tape. Some common names of these fluoropolymers are polytetrafluoroethylene and fluoroelastomers. There was a widespread use of PFOA in the 1960s–1970s: fire-fighting foam, metal cleaners, cement, leather, textile treatments, paper, and water-repellent coating in food packaging. It was also used in cosmetics, paints, and adhesives. Environmental exposure to PFOA occurs during manufacture and disposal, and as impurities or precursors from substances it was used to produce. PFOA presence in soil has occurred from surface water contamination from industrial waste mixing with soil conditioner (from agricultural use).

There are a few case studies with PFOA exposure and cancer of the testis documented, but the evidence is considered "credible and unlikely to be explained by bias or confounding." There were some case studies in West Virginia correlating cancers of the kidney with PFOA exposure. Although the evidence was credible, there was a potential for bias and confounding. The IARC notes that there is limited evidence for carcinogenicity of PFOA in humans (IARC 2016).

5.4.1.6   1,2-Dichloropropane

1,2-Dichloropropane is a Group 1 carcinogen. It has been used as a textile stain remover, metal cleaner, and in some insecticides. The US and Europe stopped using it as a soil fumigant in 1989 in the US, but other countries may still be using it in agricultural use. It is primarily used as a chemical intermediate in the production of other organic chemicals such as propylene and carbon tetrachloride. There is a limited data on exposure from the environment, but occupationally, workers can be exposed via inhalation and dermal route.

There was a cancer cluster in Osaka, Japan, at a printing plant where workers were exposed to 1,2-dichloropropane and dichloromethane that were above the international limit. There were 17 cases of cancers of the biliary tract (cholangiocarcinoma) along with seven other cases from four other printing plants in Japan. Because of the high relative risk of chemical exposure and rarity of this type of cancer, it was highly suggested that there was a correlation between exposure to 1,2-dichloropropane and dichloromethane and cancer occurrence (cholangiocarcinoma) (IARC 2016).

5.4.1.7   Dichloromethane

Dichloromethane is a Group 2A carcinogen. It is used as a paint stripper, cleaning solvent (e.g., printing plant) and a solvent carrier in the production of insecticides and herbicides. It was used as a process solvent in the pharmaceutical industry in the manufacture of antibiotics and vitamins. Its use has decreased since the 1970s, and it is no longer used in cosmetic products since 1989.

In addition to the cancer cluster noted earlier in Osaka, Japan, there are other case studies of workers exposed to dichloromethane in the US with incidence of liver and biliary tract cancers, NHL (IARC 2016).

5.4.1.8   1,3-Propane Sultone

1,3-Propane sultone is a Group 2A carcinogen. It is not widely used since the 1960s but is still used as an intermediate in the production of other chemicals: hydroxyl sulfonate surfactants for use in oil recovery, lithium ion batteries, and fungicides and insecticides.

There is one case study from the 1950s–1960s in Germany where workers have developed a range of different cancers in a plant that manufactured 1,3-propane sultone. However, it is a small sample size and there is inadequate evidence of carcinogenicity in humans. There is sufficient evidence of carcinogenicity in experimental animals (IARC 2016).

5.4.1.9   Fibrous Glaucophane

Glaucophane is an alkaline amphibole compound, essentially a mineral fiber that is naturally found in soil, particularly in blueschist metamorphic rock in the Franciscan Complex in California, USA, and along coastal California (Wakabayashi 2015). This Franciscan Complex is the most studied blueschist terrain in the world (Wakabayashi 2015). It has also been found in Brittany, France, and Piedmont, Italy. By crystal structure and chemical properties, Glaucophane is similar to asbestos and asbestos-like fibers. However, glaucophane is not classified as a carcinogen by the IARC. Di Giuseppe et al. (2019) further evaluated the chemical properties of glaucophane and its potential health hazards in the San Anselmo, Marin County, which is along the coastal region of California, as these areas undergo excavation for building/construction purposes for the commercial and residential use.

Existing at micrometer and submicrometer fibers in these areas (Erskine and Bailey 2018), the concern with glaucophane, similar to other asbestos-like fibers, is inhalation causing respiratory issues. Di Giuseppe's study found that the toxicity profile for glaucophane was similar to asbestos fibers (Di Giuseppe et al. 2019), although it is not recognized as asbestos-like mineral. Fibers from blueschist are predominantly smaller fibers, which have less toxicity/pathogenic potential (more efficient clearance by macrophages in the alveoli) (Gualtieri et al. 2017). However, glaucophane short fibers can aggregate to form long fibrous bundles. As a result, Gtuseppe suggests that the risks of glaucophane exposure are significant (Di Giuseppe et al. 2019), and it is unclear if it can be assumed that the excavation in these blueschist areas would not require occupational safety standards.

5.4.1.10   Radon

Radon is a naturally occurring gas within bedrock and soil. It is produced by the natural decay of radium, which is a byproduct when uranium and thorium decay into lead. Radon concentrations depend on the geological location. Radon can impact human health, such as if houses are built on soil containing large concentrations of radon. Radon is associated with lung cancer incidence, around 20,000 deaths a year in the United States alone. Radon detection kits can be purchased to check the level of radon accumulation. Levels higher than 4 pCi L^{-1} (picoCuries per liter) are associated with increased risk.

5.4.2   Pollution

As discussed, earlier soil composition is partially influenced by atmospheric particle concentration that is precipitated down to the surface soil. As air pollution increases, the soil is exposed to more particles, including heavy metals, like lead. Asian sand dust-particulate matter (ASD-PM) is another consequence of air pollution that leads to negative health impacts, like respiratory illnesses, particularly in several East Asian Countries (Chen et al. 2004). ASD-PM consists of soil material, but when aerosolized by wind erosion, it consists of sulfur dioxide and nitrogen dioxide, which causes health-related effects (Wong et al. 2002) but also causes oxidative damage to DNA (Kim et al. 2003, Hwang et al. 2010, Yeo et al. 2010), increases IL-6 and IL-8 levels, along with activating eosinophils because of the cytokine production (Shin et al. 2013). This suggests an elevated inflammatory response to ASD-PM. Lee et al. specifically focused on liver fibrosis and studied the relationship between ASD-PM and liver fibrosis markers, Tenascin-C (Tn-C). Tn-C is a protein that is part of the ECM and is expressed under abnormal conditions, such as chronic inflammation and tumor formation (Sato et al. 2012, Kalembeyi et al. 2003). In the study, Lee used human cancer cell lines and mouse liver tissue cells and exposed them to ASD-PM. Those cells that were exposed to ASD-PM had statistically significant increase in the levels of Tn-C and fibronectin (another component of the ECM). Tn-C has been shown to lead to liver fibrosis and liver cancer (El-Karef et al. 2007). Tn-C activates MMP-9 expression through TLR-4 (toll-like receptor 4) pathway and increases the production of ECM. This increased ECM proliferation leads to liver damage (Kuriyama et al. 2011). Lee also found that long-term exposure to ASD-PM leads to more liver damage (Lee et al. 2019).

Soil is also affected by water contamination. Researchers in Taiwan compared different regions of the country with serpentine and nonserpentine soil (Wang et al. 2020). The serpentine soil naturally has high levels of chromium and nickel. However, nonserpentine soil, soil usually used for farmland, has become contaminated with chromium and nickel as a result of industrialization (anthropogenic): electroplating industry and agricultural irrigation systems (Taiwan EPA, 1985). Another important consideration is the bioaccessibility of heavy metals in the soil. Bioaccessibility is affected by soil properties and the composition of minerals in the soil. In soil that is anthropogenically contaminated, there is an increased bioaccessibility of heavy metals compared to geologic terrains (Walraven et al. 2015). For example, lead bullets lead to more bioaccessibility of lead in the soil than urban waste containing lead (Walraven et al. 2015). In the study in Taiwan, they discovered that nickel had an increased bioaccessibility of the nonserpentine soils evaluated. Nickel is of relevance as studies have shown increased mortality from lung cancer and cancer of the nasal cavities in the nickel refinery workers. This is caused by chronic exposure to nickel-containing dust and fumes (Seilkop and Oller 2003). Many studies have been done on the health effects of illegal toxic waste dumping, which ultimately impacts water and soil contamination. As it pertains to cancer, in areas where there are more illegal practices of dumping toxic and urban waste, there have been an increased mortality rate from lung, liver, kidney, stomach (gastric), and bladder cancers (Fazzo et al. 2008). Outside of illegal dumping of waste, authorizing urban waste dumping has been associated with an increased risk of stomach cancer (Di Ciaula 2016).

The Lake Erie region in Ohio became well-known toxic algae blooms, threatening water supplies in 2014. The cyanobacterial blooms not only affect the freshwater sources but also the soil and groundwater, also affecting crops and the food supply (Lee et al. 2017). These blooms are increasing in the United States (Zhang et al. 2015) and around the world because of increasing average temperature and human activity causing runoff into water sources (Lee et al. 2017). There are toxic and nontoxic cyanobacteria that coexist (Davis et al. 2009, Hu et al. 2016). Similar to how the body's immune surveillance fails to recognize a mutation or cancer cell presence and the malignancy grows, scientists believe that climate change/global warming and eutrophication led to the growth of toxic cyanobacteria strains (Davis et al. 2009). One type of cyanobacteria, microcystins, has been associated with liver cancer and colon cancer in China (Yu et al. 2001). Cadmium is naturally found in the soil but becomes mineralized and enriched when associated with zinc or lead (Kabata-Pendias 2011, Garrett 2000). This can happen at Zn/Pb ores that are a result of human sources. Barium ores are another site where cadmium is enriched (Trefry et al. 1986, Crecelius 2007, Shahab et al. 2016). China has abundant barium ores and represents 50% of the world's production of barite (Johnson et al. 2017). Cadmium has been associated with lung, breast, and bladder cancer and is classified as a Group 1 carcinogen (Nordberg 2009). Cadmium is absorbed by plants, and rice is a significant source of cadmium intake in humans (Chen et al. 2018). Lu et al. (2019) studied the region near the Tianzhu Ba mine in the Guizhou Province of

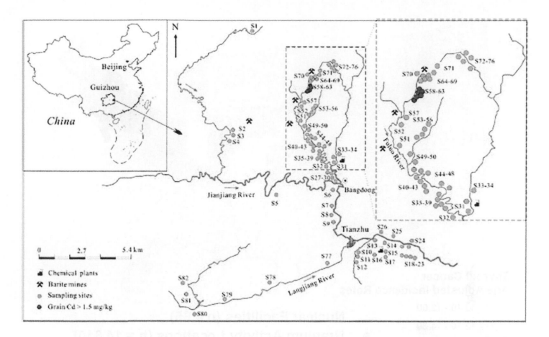

FIGURE 5.5   Tianzhu Ba mine in the Guizhou Province of southwest China.

southwest China, the world's largest barium ore mine (Figure 5.5). The local population was exposed to high cadmium levels through rice intake. Rice is the main cereal crop in that region. Soil concentrations of cadmium were as high as 91 mg/kg (baseline soil concentration levels: 0.27 mg kg^{-1}) (ATSDR 2012).

Biochars are derived from biomass (organic waste) and are carbon compounds. In recent years, biochars have shown potential in decreasing heavy metal concentrations in soil, increasing carbon retention in soil, soil water retention, and cation exchange (to decrease oxidation) (Khan et al. 2014, Atkinson et al. 2010). It is being viewed as a hopeful solution to climate change and increased CO_2 emissions. There is a concern for negative health effects from biochars, depending on the production methods and the type of biomass used. In the process of biochar production, polycyclic aromatic hydrocarbons (PAHs) can be produced which are pollutants and known to have carcinogenic, teratogenic, and mutagenic toxicities (Grimmer 1998, Perera 1997, Wang et al. 2017). Biochars can release PAHs into farmland soils in which it is being utilized, increasing the level of PAHs in the soil as well as in vegetable roots and leaves (Wang et al. 2018, Quilliam et al. 2013). There are no current standards for tolerable limits of PAHs concentrations produced in biochar manufacturing. Current practices can range from 4 to 300 mg kg^{-1} (IBI 2015, EBC 2012). Wang et al. (2019) further explored PAHs health effects and concentrations in soil. They found that there was a low risk of cancer development as PAHs' concentrations were low in their samples. However, it was noted that there are large variations in the manufacturing of biochars, and therefore, health risks can develop with higher concentrations of PAHs. Wang et al. (2019) encouraged optimizing agronomic practices, production conditions, and postmanufacturing treatment methods.

Other environmental exposures leading to soil contamination have led to the potential for adverse health effects. One study found an increase in uranium concentration in the soil and water near the location of nuclear facilities. Those same areas also had an increase in thyroid cancer incidence (Figure 5.6). Although there was no statistically significant correlation, it does suggest an increased risk of environmental uranium exposure in those areas, and a close monitoring of residents should be performed (Gerwen et al. 2020).

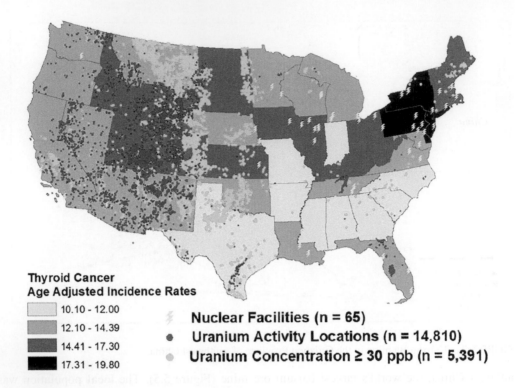

Thyroid Cancer
Age Adjusted Incidence Rates

10.10 - 12.00	
12.10 - 14.39	
14.41 - 17.30	
17.31 - 19.80	

⚡ **Nuclear Facilities (n = 65)**
● **Uranium Activity Locations (n = 14,810)**
● **Uranium Concentration ≥ 30 ppb (n = 5,391)**

FIGURE 5.6 Distributions of age-adjusted thyroid cancer incidence rates per state and sources of uranium exposure, including uranium concentrations in water.

5.4.3 PESTICIDES

There have been some studies and more case reports/case studies evaluating the connection between cancer and pesticides since 1975. It has been a controversial topic as there are case reports and case series showing associations between cancer and pesticide exposure. However, these were case studies and some studies had small sample sizes. Glyphosate is the main ingredient in the brand name pesticide, RoundUp. The International Agency for Research on Cancer (IARC) classified glyphosate as a Category 2A herbicide, a possible carcinogen in 2015 (Davoren and Schiestl 2018). The Agricultural Health Study first studied glyphosate in 2005 and concluded that there were no statistically significant associations with any cancer site, including NHL. An updated study was done in 2018 as a prospective cohort study and also found no statistically significant associations between glyphosate and NHL and solid tumors. However, this study did find an increased risk of acute myeloid leukemia (Andreotti et al. 2018). Other pesticides were evaluated in various case studies and cohort studies, like the Agriculture Health Study: malathion and diazinon.

5.4.3.1 Non-Hodgkin's Lymphoma

There has been an increased attention to pesticides leading to lymphoma with the case of RoundUp, whose main ingredient is glyphosate (Figure 5.7). Studies have been testing this relationship, and a few were found to have statistical significance. One study evaluated 155,000 farmers and found an increased risk of NHL that correlated with acres of field sprayed with pesticides (Morrison et al. 1994). Other studies have found that it is not just one class of pesticides that has been associated with NHL incidence; for example, dacamba, mecoprop, and carbamate (McDuffie et al. 2001).

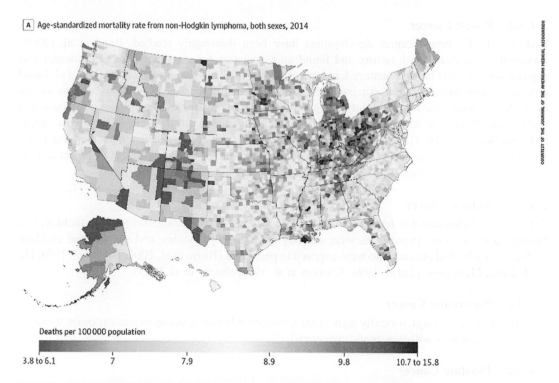

A | Age-standardized mortality rate from non-Hodgkin lymphoma, both sexes, 2014

COURTESY OF THE JOURNAL OF THE AMERICAN MEDICAL ASSOCIATION

Deaths per 100000 population

3.8 to 6.1   7   7.9   8.9   9.8   10.7 to 15.8

FIGURE 5.7 Age-standardized mortality rate from non-Hodgkin lymphoma, both sexes, 2014. (http://i2.cdn. cnn.com/cnnnext/dam/assets/170124124000-11-cancer-cluster-maps--super-169.jpg.)

A study evaluating golf course superintendents found an increased rate of NHL who were exposed to pesticides and other chemicals: fertilizer and diesel fumes (Kross et al. 1992).

The IARC reviewed four case studies from the US, Sweden, and Canada and found an increase in NHL incidence with exposure to the insecticide, malathion. Malathion is also used in the mosquito-control methods to help control the spread of malaria in less-industrialized countries. A positive exposure–response relationship was also found between diazinon and NHL (small lymphocytic leukemia, follicular lymphoma, diffuse large B-cell lymphoma) in the case control and in a large cohort study (the Agriculture Health Study) (IARC 2017).

5.4.3.2 Leukemia

When evaluating patients involved with livestock farming, there were increased rates of leukemia (Kristensen et al. 1996a, 1996b). Young children can be exposed to pesticides at different times of their development, and there is an impact on leukemia incidence. One particular study showed the critical exposure period during pregnancy that had the most impact on developing leukemia later (Ma et al. 2002). Another study found that patients who had been exposed to pesticides had similar chromosome abnormalities, bone marrow characteristics, and resistance to treatment as patients who had prior exposure to radiation and chemotherapy that also alters chemical and chromosome profiles (Cuneo 1992).

The Agriculture Health Study found a positive association between diazinon and leukemia incidence, and risk appeared to be associated with cumulative exposure to the pesticide (IARC 2017).

5.4.3.3 Brain Cancer

Bassil et al. (2007) reviewed 13 articles related to brain cancer incidence (i.e., gliomas) and found an association between pesticide exposure and cancer incidence. Of concern, many cases were found in children of parents who were exposed to pesticides.

5.4.3.4 Breast Cancer

Risk factors for breast cancer development have been thoroughly studied. Band et al. (2000) reviewed occupational risk factors and found an association between exposure to pesticides and organic solvents and breast cancer risk in both pre- and postmenopausal women. Kettles et al. found a positive association between triazine herbicides and breast cancer, although causality was unable to be determined based on the study design. What they found was that exposure to medium and high levels of triazine was associated with a statistically significant increase in the breast cancer risk (Kettles et al. 1997). Environmental exposures, like soil contamination from pesticides and contaminated water, are being further studied as risk factors for the development of breast cancer, particularly in low-income, disadvantaged communities (Natarajan et al. 2019).

5.4.3.5 Kidney Cancer

A positive relationship was found between prolonged exposure to pesticides and the incidence of kidney cancers. Some populations were directly exposed to pesticides, and some found children with cancer who had parents who were exposed to pesticides (Buzio et al. 2002, Fear et al. 1998, Hu et al. 2002, Mellemgaard et al. 1994, Ramlow et al. 1996, Sharpe et al. 1995).

5.4.3.6 Pancreatic Cancer

Ji et al. (2001) found a statistically significant association between occupational exposure to herbicides and fungicides with pancreatic cancer risk.

5.4.3.7 Prostate Cancer

Alavanja et al. (2003) studied 55,000 men who had applied pesticides and found an increased risk of prostate cancer, particularly in those men who were exposed to methyl bromide (a fumigant) and who had a family history of prostate cancer. Seven other studies were reviewed by Bassil et al. and found a positive association between pesticide exposure and prostate cancer incidence (Alavanja et al. 2003, Dich et al. 1998, Fleming et al. 1999, MacLennan et al. 2002, Mills and Yang 2003, Settimi et al. 2003, Sharma-Wagner et al. 2000).

A case–control study in Canada found an exposure–response relationship between malathion exposure and prostate cancer (IARC 2017).

Hu et al. studied the effects of the pesticide, atrazine, on prostate cancer cells. They found that atrazine may activate a signaling pathway in the body that decreases the production of reactive oxygen species (involved in oxidation), which can promote the progression of tumor cells (Hu et al. 2016). The authors suggested that prostate cancer patients should avoid exposure to atrazine and that farmers should be screened for prostate cancer (Hu et al. 2016).

5.4.3.8 Lung Cancer

The Agriculture Health Study found "a consistent and robust" finding of increased risk of lung cancer and exposure to diazinon, on a cumulative basis. Confounders, like smoking, were adjusted for, and findings were consistent based on three updates to the study in 2004, 2005, and 2015 (IARC 2017).

5.4.3.9 Stomach Cancer

The use of the pesticide, atrazine, was associated with an increased incidence of gastric cancer. The study noted atrazine contamination in the water (Van Leeuwen et al. 1999).

The Agriculture Health Study highlighted a case–control study in California that found an increased risk of stomach cancer associated with the pesticide, methyl bromide (Barry et al. 2012). Further evaluation was indicated, because although methyl bromide is decreasing in use in developed countries, it is still in use in developing countries (Barry et al. 2012).

There is a train that carries patients with cancer from the Punjab region of India to different treatment centers, called "The Cancer Train." (Pandey 2015). It has seen an increase in the number of passengers with cancer who are traveling to the Bikaner hospital for treatment since the 2010s. The increased incidence of cancer has been linked with pesticide use and growing pollution. Nitrate levels in well water were found to be significantly higher than World Health Organization (WHO) standards. One study found "disturbingly high" levels of heavy metals, such as barium, cadmium, and lead (Blaurock-Busch et al. 2010). Critics point to a lack of infrastructure in terms of water purification. The use of pesticides in this area can be traced back to the "Green Revolution," a campaign in the 1960s to utilize pesticides to help crop yields for a starving Indian population at the time of more than 1 billion people. Although the Green Revolution helped prevent mass starvation and improved economic conditions in the area, there were unintended consequences of rampant pesticide use and less-than-optimal public health management, such as cancer (Blaurock-Busch et al. 2014).

In a study of women from Mayan communities in Mexico, contamination of breastmilk with organochlorine pesticides was found in those areas with improper waste disposal. About 30% of Mayan communities consume water from water wells and sinkholes, where pesticide contamination was also found (Polanco Rodriguez et al. 2017).

Dibromochloropropane is a pesticide that has been banned by the U.S. Environmental Protection Agency (EPA) since 1979; however, it is still one of the most frequently detected pesticides in groundwater samples (Barbash and Resek 1997). It has also not shown a gradual decrease in concentration since being banned, unlike other pesticides. This means that the application rate was not an accurate estimate of groundwater contamination (Barbash and Resek 1997). As a result, there is still a risk for exposure via drinking water as well as food exposure. Its concentration in the food residue was cited to be a "major concern" and was one of the many reasons for its banning (Costle and Harwood 1979). Although human case and ecological studies had their limitations, animal studies in rats and mice did show tumors of the nasal cavity and mammary glands (NTP 1982).

Not only are study and case study results conflicting, but different organizations and decision-making bodies have conflicting statements on the carcinogenicity of glyphosate (e.g., EPA, International Life Sciences Institute, IARC). The economics of glyphosate is undoubtedly playing a role in this conflict. Pesticide use has led to a 22% increase in crop yield and a 68% increase in profits (Klümper and Qaim 2014). As glyphosate is a Category 2A herbicide, there have been many legal battles for Monsanto, the producer of glyphosate. Monsanto has lost three court battles since August 2018, and more than 13,000 claims have been lodged (Braceras 2019, Stockstad 2019).

One potential conclusion from all these studies is that there needs to be an increased education of pesticide industry workers and farmers about the risks of exposure, as some workers in developing countries can have little to no knowledge of or access to health and safety measures that need to be taken when handling pesticides (Arshad et al. 2016).

5.5 NUTRITION

There's an increasing body of research redemonstrating the link between diet and increased risk of cancer (WCRF 2015, 2018). The World Cancer Research Foundation (WCRF) highlights that the consumption of diets low in fiber, high in processed red meats, and high-nitrite diet have been associated with cancers of the colon, breast, and pancreas (WCRF 2015). Nitrates are converted to nitrites, and nitrates exist in different forms in food and in the soil. Excess use of nitrogenous fertilizers has led to increased nitrate concentration in the soil and subsequently food, like vegetables. The nitrate concentration in vegetables differs depending on various factors, such as "the amount and number of applications of nitrogen-containing fertilizers for soil fertility, growth conditions, weather conditions, season, temperature, light intensity, cultivation type (traditional versus greenhouse), harvesting time, moisture stress, plant species, plant age, soil pH, storage conditions, and postharvest storage" (Salehzadeh et al. 2020, Boroujerdnia et al. 2007, Pavlou et al. 2007).

When considering micronutrients, research recognizes the importance of diversity in food crop species in order to improve nutrition with key micronutrients (Ruel 2003). Small farms throughout the world provide a more diverse set of food crops to the people of low- and middle-income countries (Herrero et al. 2017). This is of concern for regions like the Carribean Small Island Developing States, where agriculture is negatively affected by hurricanes and natural disasters as a result of climate change. Some research of crop yields is revealing that increased temperatures will lead to a decrease in iron and zinc in C3 crops (Myers et al. 2014) and decreased protein in wheat and non-leguminous grain crops (Hatfield et al. 2011, Ainsworth and McGrath 2010). There is not enough existing research on the nutritional impacts of crop yields affected by climate change, and the concern is that more vulnerable populations will be most affected (Wheeler and von Braun 2013, Hawkes and Ruel 2006).

Another significant research area in nutrition is the microbiome of the intestines and how it interacts with and influences the immune system to modulate many disease processes, including cancer, depression, and gastrointestinal diseases (Sheehan et al. 2015, Maes et al. 2008, Balkwill et al. 2005). More research is highlighting the importance of diversity in the microbiome and how that also impacts health outcomes and the body's ability to fight off infection and cancer. The soil has its own microbiome that interplays between the soil, plants, and microorganisms. Could an integrative approach to understanding both microbiomes help researchers understand key elements in how the human body maintains balance in order to better understand and treat disease?

5.6 POSSIBLE INTERVENTIONS

Cancer treatment has evolved from high doses of toxic chemotherapies to dose adjustments in chemotherapies to targeted molecular therapies to immunotherapy. How does this relate to the soil? Some chemotherapy medications are derived from plants. Chemicals from the Pacific Yew tree led to paclitaxel and docetaxel. Trastuzumab emtasine (Kadcyla) (HER2 receptor antagonist used in breast and gastric/esophageal cancers) is derived from the Ethiopian plant, Maytenus ovatus. Etoposide (used in lung cancer, lymphoma, and neuroendocrine cancer) is chemically similar to the toxin found in Podophyllum peltatum (May Apple). Vincristine (used in leukemia and lymphoma) is derived from the Madagascar periwinkel. Eribulin (used in breast cancers and sarcoma) is an analog of halicondrin B, found in the sea sponge, Halichondria. The Cynomorium plant species are showing some potential anticancer properties, such as through downregulation of the c-myc gene and inducing autophagy (body's ability to kill cancer cells). They are found along the coasts of the Mediterranean Sea (Sdiri et al. 2018).

Cancer therapies are becoming more targeted and utilizing the body's own immune system to help kill cancer cells, which presents an exciting time in medicine for discovery and the path toward the possible cures of cancer. Here we discuss cancer-related studies on chemicals already present in the soil and novel methods for the supportive care during cancer treatment.

5.6.1 SELENIUM

As discussed earlier, selenium is a trace mineral that is naturally occurring in soil (Figure 5.8). The recommended daily dose of selenium is 55 mg day^{-1}, and the tolerable upper intake level is 400 mg day^{-1} (Navarro-Alcaron and Cabrera-Vique 2008). As selenium is already present in some plants, there has been a lot of research on biofortification of selenium, or increasing the amount of selenium that plants uptake, in order to optimize selenium levels in diets. Some studies show that selenium biofortification can increase the amount of selenoproteins, many of which function as

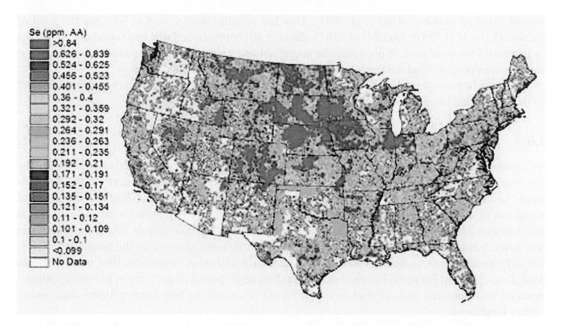

Se (ppm, AA)
- >0.84
- 0.626 - 0.839
- 0.524 - 0.625
- 0.456 - 0.523
- 0.401 - 0.455
- 0.36 - 0.4
- 0.321 - 0.359
- 0.292 - 0.32
- 0.264 - 0.291
- 0.236 - 0.263
- 0.211 - 0.235
- 0.192 - 0.21
- 0.171 - 0.191
- 0.152 - 0.17
- 0.135 - 0.151
- 0.121 - 0.134
- 0.11 - 0.12
- 0.101 - 0.109
- 0.1 - 0.1
- <0.099
- No Data

FIGURE 5.8 Map of soil selenium in United States. Red=high-selenium areas. (U.S. Department of the Interior, U.S. Geological Survey, Mineral Resources On-Line Spatial Data. https://mrdata.usgs.gov/.)

antioxidant enzymes (Puccinelli et al. 2017) which have anticancer and anti-inflammatory properties (Pandey and Rizvi 2009). It can also increase the levels of anthocyanins, a class of flavonoids that also have anti-inflammatory and potent antioxidant properties (Pandey and Rizvi 2009, Ríos et al. 2008). There is a certain synergy when biofortifying plants with selenium. Selenium can increase the level of vitamin C (Ríos et al. 2008) in certain plants. In turn, vitamin C can increase the bioavailability of iron and zinc that are already present in certain plants, both of which are essential minerals (White and Broadley 2009). Of note, there were concentrations of selenium added that did not increase anthocyanin or antioxidant levels, meaning that there is a threshold for the synergistic effects of adding selenium (Liu et al. 2017, Hawrylak 2008). Importantly, studies did show that selenium biofortification actually decreased the levels of glucosinolates and isothiocyanates, which work by activating the transcription of antioxidant enzymes (Sanchez-Pujante 2017, Ávila et al. 2013). In addition, depending on a plant species maturity, there were differences in the selenium biofortification impact (Ávila et al. 2014, 2013). For example, mature broccoli florets had a 6-fold decrease in glucosinolate content compared to broccoli sprouts after the selenium biofortification (Ávila et al. 2013).

It is important to note that depending on the plant species, selenium biofortification can have positive and negative impacts on chemopreventive compounds that impact human health. It is vital to understand which plant species and at which level of maturity they are at prior to planning the selenium biofortification.

5.6.2 Phenols

Phenols are naturally occurring compounds in vegetables and have been associated with decreased CR (Boivin et al. 2009, Mazewski et al. 2018), specifically by decreasing metastases (spread of cancer), causing apoptosis (cell death) and stopping cell proliferation (Sadeghi Ekbatan et al. 2018, Vidya Priyadarsini et al. 2010, Cilla 2009). A proposed method to increase the concentration of phenols in vegetables is by decreasing exposure to nitrogen, as the levels are usually small under current farming methods (Guo et al. 2013). This has actually been shown to increase the level of phenols (Lillo et al. 2008, Qadir et al. 2017). Zhou et al. (2019) studied this and found that there was an increased accumulation of phenolic compounds in lettuce that was exposed to less nitrogen and that this was proportional among all phenolic compounds, not just one or two.

These findings suggest that excess fertilizer use on crops could potentially have a negative effect on the nutritional composition of foods.

5.6.3 OUTSIDE-THE-BOX THERAPIES

As chemicals to prevent and treat cancer are linked to the soil, so are microorganisms in the soil that lead to antibiotics. Infectious complications are a very common part of cancer treatment. Sometimes patients may present with infections as a result of their white blood cell count being low from cancer (such as a patient with acute leukemia or a patient with any type of cancer that has extensive bone marrow involvement) or as a result of their cancer treatment which has decreased white blood cell and platelet production in the bone marrow as the treatment tries to kill off cancer cells. As humans evolve so do bacteria and other organisms that many patients have drug-resistant bacteria. Researchers are finding new species of bacteria in previously unrecognized and untested soils to find newer antibiotic therapies to help treat patients undergoing cancer treatment.

Cancer treatment is sometimes referred to as a marathon and not a race. It consists of multiple office visits, frequent blood work, and imaging, waiting for scans to see if the cancer has progressed, etc. This can take an emotional toll on patients and caregivers. Many mind-body activities are recommended to patients to help maintain a positive attitude and decrease stress which can suppress the immune system and its ability to fight off cancer and other complications. One such method is forest bathing. This is a concept of connecting back with nature by taking time to be in and walk through a forest. It can consist of feeling the stream or the bark of a tree, even lying on the ground. This idea of connecting back with the soil on an emotional and psychological level to help reduce stress is intriguing as more people are living in cities and spend time indoors (Li 2018).

Although many advancements have been made in cancer therapies, still much has to be done. As treatment becomes more personalized and targeted, revisiting humanity's connection with the surrounding environment can shine new light on the prevention and treatment methods. Perhaps, we can learn more about our ECM by learning how the soil's matrix interacts with its surroundings. By learning more about soil health and quality, we can learn about what exposures humans can face and how to prevent adverse health conditions. Turning a negative situation into a positive, Vittecoq et al. suggests collaborations between oncology and ecology to better research and understand already polluted environments and how wildlife adapts to these conditions. The idea is hormesis, where it could be possible to identify interactions between pollutants/chemicals that actually promote positive health impacts at lower doses (Vittecoq et al. 2018) and could be used as new research avenues for treatment.

For more information:

IARC (International Agency for Research on Cancer) https://www.iarc.fr/
NIEHS (National Institute of Environmental Health Sciences) https://www.niehs.nih.gov/

APPENDIX

Listing of all known carcinogens (Figure 5.9)

Substances Listed in the Fourteenth Report on Carcinogens

Bold entries indicate new or changed listings in the Fourteenth Report on Carcinogens.

Known To Be Human Carcinogens

Aflatoxins

Alcoholic Beverage Consumption

4-Aminobiphenyl

Analgesic Mixtures Containing Phenacetin (see Phenacetin and Analgesic Mixtures Containing Phenacetin)

Aristolochic Acids

Arsenic and Inorganic Arsenic Compounds

Asbestos

Azathioprine

Benzene

Benzidine (see Benzidine and Dyes Metabolized to Benzidine)

Beryllium and Beryllium Compounds

Bis(chloromethyl) Ether and Technical -Grade Chloromethyl Methyl Ether

1,3-Butadiene

1,4-Butanediol Dimethanesulfonate

Cadmium and Cadmium Compounds

Chlorambucil

1-(2-Chloroethyl)-3-(4-methylcyclohexyl)-1-nitrosourea (see Nitrosourea Chemotherapeutic Agents)

Chromium Hexavalent Compounds

Coal Tars and Coal -Tar Pitches

Coke-Oven Emissions

Cyclophosphamide

Cyclosporin A

Diethylstilbestrol

Dyes Metabolized to Benzidine (Benzidine Dye Class) (see Benzidine and Dyes Metabolized to Benzidine)

Epstein-Barr Virus (see Viruses: Eight Listings)

Erionite

Estrogens, Steroidal

Ethylene Oxide

Formaldehyde

Hepatitis B Virus (see Viruses: Eight Listings)

Hepatitis C Virus (see Viruses: Eight Listings)

Human Immunodeficiency Virus Type 1 (see Viruses: Eight Listings)

Human Papillomaviruses: Some Genital -Mucosal Types (see Viruses: Eight Listings)

Human T-Cell Lymphotrophic Virus Type 1 (see Viruses: Eight Listings)

Kaposi Sarcoma–Associated Herpesvirus (see Viruses: Eight Listings)

Melphalan

Merkel Cell Polyomavirus (see Viruses: Eight Listings)

Methoxsalen with Ultraviolet A Therapy

Mineral Oils: Untreated and Mildly Treated

Mustard Gas

2-Naphthylamine

Neutrons (see Ionizing Radiation)

Nickel Compounds (see Nickel Compounds and Metallic Nickel)

Radon (see Ionizing Radiation)

Silica, Crystalline (Respirable Size)

Solar Radiation (see Ultraviolet Radiation Related Exposures)

Soots

Strong Inorganic Acid Mists Containing Sulfuric Acid

FIGURE 5.9 Substances listed in the fourteenth report on carcinogens (https://ntp.niehs.nih.gov/ntp/roc/content/listed_substances_508.pdf)

(Continued)

Sunlamps or Sunbeds, Exposure to (see Ultraviolet Radiation Related Exposures)

Tamoxifen

2,3,7,8-Tetrachlorodibenzo-*p*-dioxin

Thiotepa

Thorium Dioxide (see Ionizing Radiation)

Tobacco Smoke, Environmental (see Tobacco -Related Exposures)

Tobacco Smoking (see Tobacco -Related Exposures)

Tobacco, Smokeless (see Tobacco -Related Exposures)

o-Toluidine

Trichloroethylene

Ultraviolet Radiation, Broad -Spectrum (see Ultraviolet Radiation Related Exposures)

Vinyl Chloride (see Vinyl Halides [selected])

Wood Dust

X-Radiation and Gamma Radiation (see Ionizing Radiation)

Reasonably Anticipated To Be Human Carcinogens

Acetaldehyde

2-Acetylaminofluorene

Acrylamide

Acrylonitrile

Adriamycin

2-Aminoanthraquinone

o-Aminoazotoluene

1-Amino-2,4-dibromoanthraquinone

2-Amino-3,4-dimethylimidazo[4,5-*f*]quinoline (see Heterocyclic Amines [Selected])

2-Amino-3,8-dimethylimidazo[4,5-*f*]quinoxaline (see Heterocyclic Amines [Selected])

1-Amino-2-methylanthraquinone

2-Amino-3-methylimidazo[4,5-*f*]quinoline (see Heterocyclic Amines [Selected])

2-Amino-1-methyl-6-phenylimidazo[4,5-*b*]pyridine (see Heterocyclic Amines [Selected])

Amitrole

o-Anisidine and Its Hydrochloride

Azacitidine

Basic Red 9 Monohydrochloride

Benz[*a*]anthracene (see Polycyclic Aromatic Hydrocarbons: 15 Listings)

Benzo[*b*]fluoranthene (see Polycyclic Aromatic Hydrocarbons: 15 Listings)

Benzo[*j*]fluoranthene (see Polycyclic Aromatic Hydrocarbons: 15 Listings)

Benzo[*k*]fluoranthene (see Polycyclic Aromatic Hydrocarbons: 15 Listings)

Benzo[*a*]pyrene (see Polycyclic Aromatic Hydrocarbons: 15 Listings)

Benzotrichloride

2,2-Bis(bromomethyl)-1,3-propanediol (Technical Grade)

Bis(chloroethyl) Nitrosourea (see Nitrosourea Chemotherapeutic Agents)

Bromodichloromethane

1-Bromopropane

Butylated Hydroxyanisole

Captafol

Carbon Tetrachloride

Ceramic Fibers (Respirable Size)

Chloramphenicol

Chlorendic Acid

Chlorinated Paraffins (C$_{12}$60% Chlorine)

Chloroform

1-(2-Chloroethyl)-3-cyclohexyl-1-nitrosourea (see Nitrosourea Chemotherapeutic Agents)

3-Chloro-2-methylpropene

FIGURE 5.9 (CONTINUED) Substances listed in the fourteenth report on carcinogens (https://ntp. niehs.nih.gov/ntp/roc/content/listed_substances_508.pdf)

(Continued)

4-Chloro-*o*-phenylenediamine

Chloroprene

p-Chloro-*o*-toluidine and Its Hydrochloride

Chlorozotocin (see Nitrosourea Chemotherapeutic Agents)

Cisplatin

Cobalt and Cobalt Compounds That Release Cobalt Ions *In Vivo* (see Cobalt -Related Exposures)

Cobalt–Tungsten Carbide: Powders and Hard Metals (see Cobalt -Related Exposures)

p-Cresidine

Cumene

Cupferron

Dacarbazine

Danthron

2,4-Diaminoanisole Sulfate

2,4-Diaminotoluene

Diazoaminobenzene

Dibenz[*a,h*]acridine (see Polycyclic Aromatic Hydrocarbons: 15 Listings)

Dibenz[*a,j*]acridine (see Polycyclic Aromatic Hydrocarbons: 15 Listings)

Dibenz[*a,h*]anthracene (see Polycyclic Aromatic Hydrocarbons: 15 Listings)

7H-Dibenzo[*c,g*]carbazole (see Polycyclic Aromatic Hydrocarbons: 15 Listings)

Dibenzo[*a,e*]pyrene (see Polycyclic Aromatic Hydrocarbons: 15 Listings)

Dibenzo[*a,h*]pyrene (see Polycyclic Aromatic Hydrocarbons: 15 Listings)

Dibenzo[*a,i*]pyrene (see Polycyclic Aromatic Hydrocarbons: 15 Listings)

Dibenzo[*a,l*]pyrene (see Polycyclic Aromatic Hydrocarbons: 15 Listings)

1,2-Dibromo-3-chloropropane

1,2-Dibromoethane

2,3-Dibromo-1-propanol

1,4-Dichlorobenzene

3,3'-Dichlorobenzidine and Its Dihydrochloride

Dichlorodiphenyltrichloroethane

1,2-Dichloroethane

Dichloromethane

1,3-Dichloropropene (Technical Grade)

Diepoxybutane

Diesel Exhaust Particulates

Di(2-ethylhexyl) Phthalate

Diethyl Sulfate

Diglycidyl Resorcinol Ether

3,3'-Dimethoxybenzidine (see 3,3' -Dimethoxybenzidine and Dyes Metabolized to 3,3' -Dimethoxybenzidine)

4-Dimethylaminoazobenzene

3,3'-Dimethylbenzidine (see 3,3' -Dimethylbenzidine and Dyes Metabolized to 3,3' -Dimethylbenzidine)

Dimethylcarbamoyl Chloride

1,1-Dimethylhydrazine

Dimethyl Sulfate

Dimethylvinyl Chloride

1,6-Dinitropyrene (see Nitroarenes [Selected])

1,8-Dinitropyrene (see Nitroarenes [Selected])

1,4-Dioxane

Disperse Blue 1

Dyes Metabolized to 3,3' -Dimethoxybenzidine (3,3' -Dimethoxybenzidine Dye Class)
(see 3,3'-Dimethoxybenzidine and Dyes Metabolized to 3,3' -Dimethoxybenzidine)

Dyes Metabolized to 3,3' -Dimethylbenzidine (3,3' -Dimethylbenzidine Dye Class)
(see 3,3'-Dimethylbenzidine and Dyes Metabolized to 3,3' -Dimethylbenzidine)

Epichlorohydrin

FIGURE 5.9 (CONTINUED) Substances listed in the fourteenth report on carcinogens (https://ntp. niehs.nih.gov/ntp/roc/content/listed_substances_508.pdf)

(*Continued*)

Ethylene Thiourea
Ethyl Methanesulfonate
Furan
Glass Wool Fibers (Inhalable), Certain
Glycidol
Hexachlorobenzene
Hexachloroethane
Hexamethylphosphoramide
Hydrazine and Hydrazine Sulfate
Hydrazobenzene
Indeno[1,2,3-*cd*]pyrene (see Polycyclic Aromatic Hydrocarbons: 15 Listings)
Iron Dextran Complex
Isoprene
Kepone
Lead and Lead Compounds
Lindane, Hexachlorocyclohexane (Technical Grade), and Other Hexachlorocyclohexane Isomers
2-Methylaziridine
5-Methylchrysene (see Polycyclic Aromatic Hydrocarbons: 15 Listings)
4,4′-Methylenebis(2-chloroaniline)
4,4′-Methylenebis(*N,N*-dimethyl)benzenamine
4,4′-Methylenedianiline and Its Dihydrochloride
Methyleugenol
Methyl Methanesulfonate
N-Methyl-*N*′-Nitro-*N*-Nitrosoguanidine (see *N*-Nitrosamines: 15 Listings)
Metronidazole
Michler's Ketone
Mirex
Naphthalene
Nickel, Metallic (see Nickel Compounds and Metallic Nickel)
Nitrilotriacetic Acid
o-Nitroanisole
Nitrobenzene
6-Nitrochrysene (see Nitroarenes [Selected])
Nitrofen
Nitrogen Mustard Hydrochloride
Nitromethane
2-Nitropropane
1-Nitropyrene (see Nitroarenes [Selected])
4-Nitropyrene (see Nitroarenes [Selected])
N-Nitrosodi-*n*-butylamine (see *N*-Nitrosamines: 15 Listings)
N-Nitrosodiethanolamine (see *N*-Nitrosamines: 15 Listings)
N-Nitrosodiethylamine (see *N*-Nitrosamines: 15 Listings)
N-Nitrosodimethylamine (see *N*-Nitrosamines: 15 Listings)
N-Nitrosodi-*n*-propylamine (see *N*-Nitrosamines: 15 Listings)
N-Nitroso-*N*-ethylurea (see *N*-Nitrosamines: 15 Listings)
4-(*N*-Nitrosomethylamino)-1-(3-pyridyl)-1-butanone (see *N*-Nitrosamines: 15 Listings)
N-Nitroso-*N*-methylurea (see *N*-Nitrosamines: 15 Listings)
N-Nitrosomethylvinylamine (see *N*-Nitrosamines: 15 Listings)
N-Nitrosomorpholine (see *N*-Nitrosamines: 15 Listings)
N-Nitrosonornicotine (see *N*-Nitrosamines: 15 Listings)
N-Nitrosopiperidine (see *N*-Nitrosamines: 15 Listings)
N-Nitrosopyrrolidine (see *N*-Nitrosamines: 15 Listings)
N-Nitrososarcosine (see *N*-Nitrosamines: 15 Listings)

FIGURE 5.9 (CONTINUED) Substances listed in the fourteenth report on carcinogens (https://ntp. niehs.nih.gov/ntp/roc/content/listed_substances_508.pdf)

(Continued)

o-Nitrotoluene

Norethisterone

Ochratoxin A

4,4'-Oxydianiline

Oxymetholone

Pentachlorophenol and By-products of Its Synthesis

Phenacetin (see Phenacetin and Analgesic Mixtures Containing Phenacetin)

Phenazopyridine Hydrochloride

Phenolphthalein

Phenoxybenzamine Hydrochloride

Phenytoin and Phenytoin Sodium

Polybrominated Biphenyls

Polychlorinated Biphenyls

Procarbazine and Its Hydrochloride

Progesterone

1,3-Propane Sultone

β-Propiolactone

Propylene Oxide

Propylthiouracil

Reserpine

Riddelliine

Safrole

Selenium Sulfide

Streptozotocin (see Nitrosourea Chemotherapeutic Agents)

Styrene

Styrene-7,8-oxide

Sulfallate

Tetrachloroethylene

Tetrafluoroethylene

Tetranitromethane

Thioacetamide

4,4'-Thiodianiline

Thiourea

Toluene Diisocyanates

Toxaphene

2,4,6-Trichlorophenol

1,2,3-Trichloropropane

Tris(2,3-dibromopropyl) Phosphate

Ultraviolet Radiation A (see Ultraviolet Radiation Related Exposures)

Ultraviolet Radiation B (see Ultraviolet Radiation Related Exposures)

Ultraviolet Radiation C (see Ultraviolet Radiation Related Exposures)

Urethane

Vinyl Bromide (see Vinyl Halides [Selected])

4-Vinyl-1-cyclohexene Diepoxide

Vinyl Fluoride (see Vinyl Halides [Selected])

FIGURE 5.9 (CONTINUED) Substances listed in the fourteenth report on carcinogens (https://ntp. niehs.nih.gov/ntp/roc/content/listed_substances_508.pdf)

REFERENCES

Abedi J, Saatloo MV, Nejati V, et al. Selenium-Enriched Saccharomyces Cerevisiae Reduces the Progression of Colorectal Cancer. *Biol Trace Elem Res*. 2018;185(2):424–432.

Acosta JA, Gabarrón M, Faz A, Martínez-Martínez S, Zornoza R, Arocena JM. Influence of population density on the concentration and speciation of metals in the soil and street dust from urban areas. *Chemosphere*. 2015;134:328–337.

Agency for Toxic Substances and Disease Registry. ToxGuide for Cadmium. 2012. Retrieved from https://www.atsdr.cdc.gov/toxguides/toxguide-5.pdf.

Ainsworth E, McGrath JM. Direct effects of rising atmospheric carbon dioxide and ozone of crop yields. *Adv Global Change Res Climate Change Food Security*. 2010;37:109–130.

Alavanja MC, Samanic C, Dosemeci M, et al. Use of agricultural pesticides and prostate cancer risk in the Agricultural Health Study cohort. *Am J Epidemiol*. 2003;157(9):800–814.

Allen HE. The significance of trace metal speciation for water, sediment and soil quality criteria and standards. *Sci Total Environ*. 1993;134:23–45.

Andreotti G, Koutros S, Hofmann JN, et al. Glyphosate use and cancer incidence in the agricultural health study. *J Natl Cancer Inst*. 2018;110(5):509–516.

Arshad M, Siddiqa M, Rashid S, et al. Biomonitoring of toxic effects of pesticides in occupationally exposed individuals. *Safety Health Work*. 2016;7:156–160.

Atkinson CJ, Fitzgerald JD, Hipps NA. Potential mechanisms for achieving agricultural benefits from biochar application to temperate soils: a review. *Plant Soil*. 2010;337:1–18.

Ávila FW, Faquin V, Yang Y, et al. Assessment of the anticancer compounds se-methylselenocysteine and glucosinolates in se-biofortified broccoli (Brassica oleracea L. var. Italica) sprouts and florets. *J Agric Food Chem*. 2013;61:6216–6223.

Ávila FW, Yang Y, Faquin V, et al. Impact of selenium supply on se-methylselenocysteine and glucosinolate accumulation in selenium-biofortified Brassica sprouts. *Food Chem*. 2014;165:578–586.

Balkwill F, Charles KA, Mantovani A. Smoldering and polarized inflammation in the initiation and promotion of malignant disease. *Cancer Cell*. 2005;7:211–217.

Band PR, Le ND, Fang R, Deschamps M, Gallagher RP, Yang P. Identification of occupational cancer risks in British Columbia. A population-based case-control study of 995 incident breast cancer cases by menopausal status, controlling for confounding factors. *J Occup Environ Med*. 2000;42(3):284–310.

Barbash JE, Resek EA. 1997. Chapter 3, overview of pesticide occurrence and distribution in relation to use. In: Gilliom RJ, ed. *Pesticides in Ground Water; Distribution, Trends, and Governing Factors*. Boca Raton, FL: Lewis Publishers, pp. 91–178.

Baris D, Waddell R, Beane Freeman LE, et al. Elevated bladder cancer in northern New England: the role of drinking water and arsenic. *J Natl Cancer Inst*. 2016;108. doi:10.1093/jnci/djw099.

Barry KH, Koutros S, Lubin JH, et al. Methyl bromide exposure and cancer risk in the Agricultural Health Study. *Cancer Causes Control*. 2012 June;23(6):807–818.

Bassil KL, Vakil C, Sanborn M, Cole DC, Kaur JS, Kerr KJ. Cancer health effects of pesticides: systematic review. *Can Fam Physician*. 2007;53(10):1704–1711.

Blaurock-Busch E, Busch YM, Friedle A, Buerner H, Parkash C, Kaur A. Comparing the metal concentration in the hair of cancer patients and healthy people living in the Malwa region of Punjab, India. *Clin Med Insights Oncol*. 2014;8:1–13.

Blaurock-Busch E, Friedle A, Godfrey M, Schulte-Uebbing, CEE. Metal exposure in the physically and mentally challenged children of Punjab, India. *Maedica*. 2010;5(2):102–110.

Boivin D, Lamy S, Lord-Dufour S, Jackson J, Beaulieu E, Côté M. Antiproliferative and antioxidant activities of common vegetables: a comparative study. *Food Chem*. 2009;112:374–380.

Boroujerdnia M, Ansari NA, Dehcordie FS. Effect of cultivars, harvesting time and level of nitrogen fertilizer on nitrate and nitrite content, yield in Romaine lettuce. *Asian J Plant Sci*. 2007;6(3):550–553.

Braceras JC. Rounding up the science behind the roundup nuisance litigation. *Independent Women's Forum*, May 28, 2019.

Buzio L, Tondel M, De Palma G, et al. Occupational risk factors for renal cell cancer. An Italian case-control study. *Medicina del Lavoro*. 2002;93:303–309.

Carraro A, Fabbri P, Giaretta A, Peruzzo L, Tateo, F, Tellini F. Effects of redox conditions on the control of arsenic mobility in shallow alluvial aquifers on the Venetian Plain (Italy). *Sci Total Environ*. 2015;532:581e594. doi:10.1016/j.scitoten.2015.06.003.

Chattopadhyay A, Singh AP, Singh SK, et al. Spatial variability of arsenic in Indo-Gangetic basin of Varanasi and its cancer risk assessment. *Chemosphere*. 2019;238:124623.

Chen H, Yang X, Wang P, Wang Z, Li M, Zhao F-J. Dietary cadmium intake from rice and vegetables and potential health risk: a case study in Xiangtan, southern China. *Sci Total Environ*. 2018;639:271–277.

Chen YS, Sheen PC, Chen ER, Liu YK, Wu TN, Yang CY. Effects of Asian dust storm events on daily mortality in Taipei, Taiwan. *Environ Res*. 2004;95, 151–155.

Cilla A, González-Sarrías A, Tomás-Barberán FA, Espín JC, Barberá, R. Availability of polyphenols in fruit beverages subjected to in vitro gastrointestinal digestion and their e_ects on proliferation, cell-cycle and apoptosis in human colon cancer Caco-2 cells. *Food Chem*. 2009;114, 813–820.

Clark LC, Combs GF, Turnbull BW, et al. Effects of selenium supplementation for cancer prevention in patients with carcinoma of the skin. A randomized controlled trial. Nutritional Prevention of Cancer Study Group. *JAMA*. 1996;276(24):1957–1963.

Colditz GA. Cancer Prevention. In: Post TW, ed. *UpToDate*. Waltham, MA: UpToDate Inc. https://www.uptodate.com (Accessed on October 10, 2019).

Cooper WC, Gaffey WR. Mortality of lead workers. *J Occup Environ Med*. 1975;17:100–107.

Costle DM, Harwood, G. Dibromochloropropane (DBCP)- suspension order and notice of intent to cancel. *Fed Reg*. 1979;44(219);65135–65179.

Crecelius E, Trefry J, McKinley J, Lasorsa B, Trocine R. 2007. Study of barite solubility and the release of trace components to the marine environment. *U.S. Department of the Interior, Minerals Management Service*, Gulf of Mexico OCS Region, New Orleans, LA, pp. 176 OC5 Study MMS 2007-061.

Cuneo A, Fagioli F, Pazzi I, et al. Morphologic, immunologic and cytogenetic studies in acute myeloid leukemia following occupational exposure to pesticides and organic solvents. *Leuk Res*. 1992;16:789–796.

Danaei G, Vander Hoorn S, Lopez AD, et al. Causes of cancer in the world: comparative risk assessment of nine behavioural and environmental risk factors. *Lancet*. 2005;366:1784.

Datta R, Sarkar D. Consideration of soil properties in assessment of human health risk from exposure to arsenic-enriched soils. *Int Environ Asses*. 2005;1(1):55–59.

Davis CD, Zeng H, Finley JW. Selenium-enriched broccoli decreases intestinal tumorigenesis in multiple intestinal neoplasia mice. *J Nutr*. 2002;132(2):307–309.

Davis TW, Berry DL, Boyer GL, Gobler CJ. The effects of temperature and nutrients on the growth and dynamics of toxic and non-toxic strains of Micro-cystis during cyanobacteriablooms. *Harmful Algae*. 2009;8:715–725.

Davoren MJ, Schiestl RH. Glyphosate-based herbicides and cancer risk: a post-IARC decision review of potential mechanisms, policy and avenues of research. *Carcinogenesis*. 2018;39(10):1207–1215.

Dehghani S, Moore F, Keshavarzi B, Hale BA. Health risk implications of potentially toxic metals in street dust and surface soil of Tehran, Iran. *Ecotoxicol Environ Saf*. 2017;136:92–103.

Di Ciaula, A. Increased deaths from gastric cancer in communities living close to waste landfills. *Int J Environ Health Res*. 2016;26:281–290.

Di Giuseppe D, Harpcr M, Bailey M, et al. Characterization and assessment of the potential toxicity/pathogenicity of fibrous glaucophane. *Environ Res*. 2019;178:108723.

Dich J, Wiklund K. Prostate cancer in pesticide applicators in Swedish agriculture. *Prostate*. 1998;34:100–112.

EBC. 2012. European biochar certificate - guidelines for a sustainable production of biochar. European Biochar Foundation (EBC), Arbaz, Switzerland http://www.europeanbiochar.org/en/download Version 8E of 1 January 2019. 10.13140/RG.2. 1.4658.7043.

El-Karef A, Yoshida T, Gabazza EC, et al. Deficiency of tenascin-C attenuates liver fibrosis in immune-mediated chronic hepatitis in mice. *J Pathol*. 2007;211(1):86–94.

Erskine BG, Bailey M. Characterization of asbestiform glaucophane-winchite in the franciscan complex blueschist, northern diablo range, California. *Toxicol Appl Pharmacol*. 2018;361:3–13.

Fazzo L, Belli S, Minichilli F, et al. Cluster analysis of mortality and malformations in the provinces of Naples and Caserta (Campania Region). *Ann 1st Super Sanità*. 2008;44:99–111.

Fear NT, Roman E, Reeves G, Pannett B. Childhood cancer and paternal employment in agriculture: the role of pesticides. *Br J Cancer*. 1998;77:825–829.

Finley JW, Davis CD, Feng Y. Selenium from high selenium broccoli protects rats from colon cancer. *J Nutr*. 2000;130(9):2384–2389.

Fleming LE, Bean JA, Rudolph M, Hamilton K. Cancer incidence in a cohort of licensed pesticide applicators in Florida. *J Occup Environ Med*. 1999;41:279–288.

Gabarrón M, Faz A, Acosta JA. Soil or dust for health risk assessment studies in urban environment. *Arch Environ Contam Toxicol*. 2017;73(3):442–455.

Garrett RG. Natural sources of metals to the environment. *Hum Ecol Risk Assess*. 2000;6(6):945–963.

Garvin MC, Schijf J, Kaufman SR, et al. A survey of trace metal burdens in increment cores from eastern cottonwood (Populus deltoides) across a childhood cancer cluster, Sandusky County, OH, USA. *Chemosphere*. 2019;238:124528.

Gerwen M, et al. Association between uranium exposure and thyroid health: a national health and nutrition examination survey analysis and ecological study. *Int J Environ Res Public Health*. 2020;17:712.

Gharipour M, Sadeghi M, Behmanesh M, Salehi M, Nezafati P, Gharpour A. Selenium homeostasis and clustering of cardiovascular risk factors: a systematic review. *Acta Biomed*. 2017;88(3):263–270.

Grimmer G. 2018. *Environmental Carcinogens: Polycyclic Aromatic Hydrocarbons*. CRC Press, Boca Raton, FL. .

Gualtieri AF, Mossman BT, Roggli VL. 2017. Towards a general model for predicting the toxicity and pathogenicity of minerals fibres. In: Gualtieri, AF. (Ed.), *Mineral Fibres: Crystal Chemistry, Chemical-Physical Properties, Biological Interaction and Toxicity*. London: European Mineralogical Union-EMU Notes in Mineralogy, pp. 501–526.

Guo X, Liu RH, Fu X, Sun X, Tang K. Over-expression of l-galactono--lactone dehydrogenase increases vitamin C, total phenolics and antioxidant activity in lettuce through bio-fortification. *Plant Cell Tissue Organ Cult*. 2013;114:225–236.

Colditz GA, Samplin-Salgado M, Ryan CT, Dart H, Fisher L, Tokuda A, Rockhill B; Harvard Center for Cancer Prevention. Harvard report on cancer prevention, volume 5: fulfilling the potential for cancer prevention: policy approaches. *Cancer Causes Control*. 2002 Apr;13(3):199–212.

Hatfield JL, Boote KJ, Kimball BA, Ziska LH, Izaurralde RC, Ort DR, Thomson AM, Wolfe, D. Climate impacts on agriculture: implications for crop production. Publications from USDA-ARS / UNL Faculty. 1350. 2011. https://digitalcommons.unl.edu/usdaarsfacpub/1350.

Hawkes C, Ruel M (editors). 2006. *Understanding the Links between Agriculture and Health*. Washington, DC: IFPRI.

Hawrylak-Nowak B. Enhanced selenium content in sweet basil (Ocimum basilicum) by foliar fertilization. *Veg Crop Res Bul*. 2008;69:63–72.

Herrero M, Thornton PK, Power B, et al. Farming and the geography of nutrient production for human use: a transdisciplinary analysis. *Lancet Planet Health*. 2017;1(1):e33–e42.

Hu C, Rea C, Yu Z, Lee J. Relative importance of Microcystis abundance and diversity in determining microcystin dynamics in Lake Erie coastal wetland and downstream beach water. *J Appl Microbiol*. 2016;120(1):138–151.

Hu J, Mao Y, White K. Renal cell carcinoma and occupational exposure to chemicals in Canada. *Occup Med (Oxf)*. 2002;52:157–64.

Hu K, et al. Atrazine promotes RM1 prostate cancer cell proliferation by activating STAT3 signaling. *Int J Oncol*. 2016;48:2166–2174.

Hwang YJ, Jeung YS, Seo MH, et al. Asian dust and titanium dioxide particles-induced inflammation and oxidative DNA damage in C57BL/6 mice. *Inhal Toxicol*. 2010;22(13):1127–1133.

IARC Monographs on the Evaluation of Carcinogenic Risks to Humans. Some chemicals used as solvents and in polymer manufacture. *IARC Monogr Eval Carcinog Risks Hum*. 2016;110:1–273.

IARC Monographs on the Evaluation of Carcinogenic Risks to Humans. Some organophosphate insecticides and herbicides. *IARC Monogr Eval Carcinog Risks Hum*. 2017;112:1–452.

IARC Working Group on the Evaluation of Carcinogenic Risks to Humans. Some drinking-water disinfectants and contaminants, including arsenic. *IARC Monogr Eval Carcinog Risks Hum*. 2004;84:1.

International Agency for Research on Cancer. 2019. IARC Monographs on the Identification of Carcinogenic Hazards to Humans. Retrieved from https://monographs.iarc.fr/wp-content/uploads/2018/07/QA_ENG.pdf.

International Biochar Initiative (IBI). 2015, November 23. Standardized product definition and product testing guidelines for biochar that is used in soil. Version 2.1. (Document Reference Code: IBI-STD-2.1).

Ip C, Birringer M, Block E, et al. Chemical speciation influences comparative activity of selenium-enriched garlic and yeast in mammary cancer prevention. *J Agric Food Chem*. 2000;48(9):4452.

Ji BT, Silverman DT, Stewart PA, et al. Occupational exposure to pesticides and pancreatic cancer. *Am J Ind Med*. 2001;39(1):92–99.

Johnson CA, Piatak NM, Miller MM. 2017. Barite (barium), chap. In: D of Schulz KJ, DeYoung Jr. JH, Seal IIR. R, Bradley DC. (Eds.), *Critical Mineral Resources of the United States-Economic and Environment Geology and Prospects for Future Supply*. U.S. Geological Survey Professional Paper 1802, pp. D1–D18. https://doi.org/10.3133/pp1802D.

Kabata-Pendias A. 2011. *Trace Elements in Soils and Plants*, 4th ed. Boca Raton, FL: CRC Press, pp. 287–301.

Kalembeyi I, Inada H, Nishiura R, Imanaka-Yoshida K, Sakakura T, Yoshida T. Tenascin-C upregulates matrix metalloproteinase-9 in breast cancer cells: direct and synergistic effects with transforming growth factor beta1. *Int J Cancer*. 2003;105(1):53–60.

Karagas MR, Stukel TA, Morris JS, Tosteson TD, Weiss JE, Spencer SK, Greenberg ER. Skin cancer risk in relation to toenail arsenic concentrations in a US population-based case-control study. *Am J Epidemiol.* 2001;153:559e565.

Kettles MK, Browning SR, Prince TS, Horstman SW. Triazine herbicide exposure and breast cancer incidence: an ecologic study of Kentucky counties. *Environ Health Perspect.* 1997;105(11):1222–1227.

Khan, S., Reid, B.J., Li, G., Zhu, Y.G., 2014. Application of biochar to soil reduces cancer risk via rice consumption: a case study in Miaoqian village, Longyan, China. *Environ Int.* 68:154–161.

Kim YH, Kim KS, Kwak NJ, Lee KH, Kweon SA, Lim Y. Cytotoxicity of yellow sand in lung epithelial cells. *J Biosci.* 2003;28(1):77–81.

Klümper W, Qaim MA. meta-analysis of the impacts of genetically modified crops. *PLoS One.* 2014;9(11):e111629.

Kristensen P, Andersen A, Irgens LM, Bye AS, Sundheim L. Cancer in offspring of parents engaged in agricultural activities in Norway: incidence and risk factors in the farm environment. *Int J Cancer.* 1996a;65(1):39–50. Kristensen P, Andersen A, Irgens LM, Laake P, Bye AS. Incidence and risk factors of cancer among men and women in Norwegian agriculture. *Scand J Work Environ Health.* 1996b;22:14–26.

Kross B, Burmeister L, Ogilvie L, Fuortes L, Fu C. Proportionate mortality study of golf course superintendants. *Am J Ind Med.* 1992;21:501–506.

Kuriyama N, Duarte S, Hamada T, Busuttil RW, Coito AJ. Tenascin-C: a novel mediator of hepatic ischemia and reperfusion injury. *Hepatology.* 2011;54(6):2125–2136.

Lee J, Lee S, Jiang X. Cyanobacterial toxins in freshwater and food: important sources of exposure to humans. *Annu Rev Food Sci Technol.* 2017;8:281–304.

Lee YH, Kim DY, Jeong SH, Hwang YJ. Effect of exposure to Asian sand dust-Particulate matter on liver Tenascin-C expression in human cancer cell and mouse hepatic tissue. *J Toxicol Sci.* 2019;44(9):633–641.

Li H, Qian X, Hu W, Wang Y, Gao H. Chemical speciation and human health risk of trace metals in urban street dusts from a metropolitan city, Nanjing, SE China. *Sci Total Environ.* 2013;456–457:212–221.

Li Q. 2018, May 1. 'Forest Bathing' is great for your health. Here's how to do it. *Time.* https://time.com/5259602/japanese-forest-bathing/. Accessed October 12, 2019.

Lillo C, Lea US, Ruo, P. Nutrient depletion as a key factor for manipulating gene expression and product formation in dierent branches of the flavonoid pathway. *Plant Cell Environ.* 2008;31:587–601.

Lippman SM, Klein EA, Goodman PJ, et al. Effect of selenium and vitamin E on risk of prostate cancer and other cancers: the Selenium and Vitamin E Cancer Prevention Trial (SELECT). *JAMA.* 2009;301(1):39–51.

Liu D, Li H, Wang Y et al. How exogenous selenium affects anthocyanin accumulation and biosynthesis-related gene expression in purple lettuce. *Pol J Environ Stud.* 2017;26(2):717–722.

Lu Q, Xu Z, Xu X, et al. Cadmium contamination in a soil-rice system and the associated health risk: an addressing concern caused by barium mining. *Ecotoxicol Environ Saf.* 2019;183:109590.

Lu X, Wu X, Wang Y, Chen H, Gao P, Fu Y. Risk assessment of toxic metals in street dust from a medium-sized industrial city of China. *Ecotoxicol Environ Saf.* 2014;106:154–163.

Ma X, Buffler PA, Gunier RB, et al. Critical windows of exposure to household pesticides and risk of childhood leukemia. *Environ Health Perspect.* 2002;110:955–960.

MacLennan PA, Delzell E, Sathiakumar N, et al. Cancer incidence among triazine herbicide manufacturing workers. *J Occup Environ Med.* 2002;44:1048–1058.

Maes, M., et al. The gut-brain barrier in major depression: intestinal mucosal dysfunction with an increased translocation of LPS from gram negative enterobacteria (leaky gut) plays a role in the inflammatory pathophysiology of depression. *Neuro Endocrinol Lett.* 2008;29:117–124.

Mazewski C, Liang K, Gonzalez de Mejia E. Comparison of the effect of chemical composition of anthocyanin-rich plant extracts on colon cancer cell proliferation and their potential mechanism of action using in vitro, in silico, and biochemical assays. *Food Chem.* 2018;242:378–388.

Mazumder DNG. Chronic arsenic toxicity and human health. *Indian J Med Res.* 2008;128:436e447.

McDuffie HH, Pahwa P, McLaughlin JR, et al. Non-Hodgkin's lymphoma and specific pesticide exposures in men: cross-Canada study of pesticides and health. *Cancer Epidemiol Biomarkers Prev.* 2001;10:1155–1163.

Mellemgaard A, Engholm G, McLaughlin JK, Olsen JH. Occupational risk factors for renal-cell carcinoma in Denmark. *Scand J Work Environ Health.* 1994;20:160–165.

Mills PK, Yang R. Prostate cancer risk in California farm workers. *J Occup Environ.* 2003;45(3):249–258.

Morrison HI, Semenciw RM, Wilkins K, Mao Y, Wigle DT. Non-Hodgkin's lymphoma and agricultural practices in the prairie provinces of Canada. *Scand J Work Environ Health.* 1994;20:42–47.

Motz GT, Coukos G . The parallel lives of angiogenesis and immunosuppression: cancer and other tales. *Nat Rev Immunol.* 2011;11:702–711.

Myers SS, Zanobetti A, Kloog I, et al. Increasing CO_2 threatens human nutrition. *Nature.* 2014;510(7503): 139–142.

Natarajan R, Aljaber D, Au D, et al. Environmental Exposures during Puberty: window of Breast Cancer Risk and Epigenetic Damage. *Int J Environ Res Public Health.* 2020;17:493.

Navarro-Alarcon M, Cabrera-Vique C. Selenium in food and the human body: a review. *Sci Total Environ.* 2008;400:115–141.

Nordberg GF. Historical perspectives on cadmium toxicology. *Toxicol Appl Pharmacol.* 2009;238 (3):192–200.

NTP. 1982. Carcinogenesis bioassay of 1,2-dibromo-3-chloropropane (CAS no. 96-12-8) in F344 rats and B6C3F1 mice (inhalation study). National Toxicology Program, Research Triangle Park, pp. 1–174.

NTP (National Toxicology Program). 2016. *Report on Carcinogens, Fourteenth Edition.* Research Triangle Park, NC: U.S. Department of Health and Human Services, Public Health Service. https://ntp.niehs.nih. gov/go/roc14.

Paget S. The distribution of secondary growths in cancer of the breast. *Lancet.* 1889;1:571–573.

Pal P. 2015. *Groundwater Arsenic Remediation: Treatment Technology and Scale UP.* Waltham, MA: Butterworth-Heinemann. https://www.sciencedirect.com/science/article/pii/B9780128012819099926

Pandey KB, Rizvi SI. Plant polyphenols as dietary antioxidants in human health and disease. *Oxidative Med Cell Longev.* 2009;2:270–278.

Pandey S. 2015, January 9. On the 'Cancer Train' of India's pesticides. Al-Jazeera. https://www.aljazeera.com/ indepth/features/2015/01/cancer-train-india-pesticides-20151411811508148.html. Accessed October 12, 2019.

Pavlou GC, Ehaliotis CD, Kavvadias VA. Effect of organic and inorganic fertilizers applied during successive crop seasons on growth and nitrate accumulation in lettuce. *Scientia Horticulturae.* 2007;111 (4):319–25.

Perera FP. Environment and cancer: who are susceptible? *Science.* 1997;278:1068–1073.

Polanco Rodriguez AG, López MI, Casillas TA, León JA, Prusty BA, Cervera FJ. Levels of Persistent Organic pollutants in breast milk of Maya women in Yucatan, Mexico. *Environ Monit Assess.* 2017;189:59.

Puccinelli M, Malorgio F, Pezzarossa B. Selenium enrichment of horticultural crops. *Molecules.* 2017;22(6):E933.

Qadir O, Siervo M, Seal CJ, Brandt, K. Manipulation of contents of nitrate, phenolic acids, chlorophylls, and carotenoids in lettuce (Lactuca sativa L.) via contrasting responses to nitrogen fertilizer when grown in a controlled environment. *J Agric Food Chem.* 2017;65:10003–10010.

Quilliam RS, Rangecroft S, Emmett BA, Deluca TH, Jones DL. Is biochar a source or sink for polycyclic aromatic hydrocarbon (PAH) compounds in agricultural soils? *GCB Bioenergy.* 2013;52:96–103.

Ramlow JM, Spadacene NW, Hoag SR, Stafford BA, Cartmill JB, Lerner PJ. Mortality in a cohort of penta-chlorophenol manufacturing workers, 1940–1989. *Am J Ind Med.* 1996;30:180–94.

Rashidi M, Alavipanah SK. Relation between kidney cancer and Soil leads in Isfahan Province, Iran between 2007 and 2009. *J Cancer Res Ther.* 2016;12(2):716–720.

Ríos JJ, Rosales MA, Blasco B, Cervilla LM, Romero L, Ruiz JM. Biofortification of se and induction of the antioxidant capacity in lettuce plants. *Sci Hortic.* 2008;116:248–255.

Ruel MT. Operationalizing dietary diversity: a review of measurement issues and research priorities. *J Nutr.* 2003;133(11 Suppl 2):3911S–3926S.

Sadeghi Ekbatan S, Li XQ, Ghorbani M, Azadi B, Kubow S. Chlorogenic acid and its microbial metabolites exert anti-proliferative e_ects, S-phase cell-cycle arrest and apoptosis in human colon cancer Caco-2 cells. *Int J Mol Sci.* 2018;19:723.

Salehzadeh H, Maleki A, Rezaee R, Shahmoradi B, Ponnet K. The nitrate content of fresh and cooked veg-etables and their health-related risks. *PLoS One.* 2020;15(1):e0227551. https:// doi.org/10.1371/journal. pone.0227551.

Sánchez-Pujante PJ, Borja-Martínez M, Pedreño MÁ, Almagro L. Biosynthesis and bioactivity of glucosino-lates and their production in plant in vitro cultures. *Planta.* 2017;246:19–32.

Sato A, Hiroe M, Akiyama D, et al. Prognostic value of serum tenascin-C levels on long-term outcome after acute myocardial infarction. *J Card Fail.* 2012;18(6):480–486.

Sdiri M, Li X, Du WW, et al. Anticancer Activity of. *Cancers (Basel).* 2018;10(10):354.

Seilkop SK, Oller AR. Respiratory cancer risks associated with low-level nickel exposure: an inte-grated assessment based on animal, epidemiological, and mechanistic data. *Regul Toxicol Pharm.* 2003;37:173–190.

Settimi L, Masina A, Andrion A, Axelson O. Prostate cancer and exposure to pesticides in agricultural set-tings. *Int J Cancer.* 2003;104:458–461.

Shahab B, Bashir E, Kaleem M, Naseem S, Rafique T. Assessment of barite of Lasbela, Balochistan, Pakistan, as drilling mud and environmental impact of associated Pb, As, Hg, Cd and Sr. *Environ Earth Sci*. 2016; 75(1115):1–10. doi: 10.1007/s12665-016-5916-7

Sharma Wagner S, Chokkalingam AP, Malker HS, Stone BJ, McLaughlin JK, Hsing AW. Occupation and prostate cancer risk in Sweden. *J Occup Environ Med*. 2000;42:517–525.

Sharpe CR, Franco EL, de Camargo B, et al. Parental exposures to pesticides and risk of Wilms' tumor in Brazil. *Am J Epidemiol*. 1995;141:210–217.

Sheehan, D., et al. The microbiota in inflammatory bowel disease. *J Gastroenterol*. 2015;50:495–507.

Shin SH, Ye MK, Hwang YJ, Kim ST. The effect of Asian sand dust-activated respiratory epithelial cells on activation and migration of eosinophils. *Inhal Toxicol*. 2013;25(11):633–639.

Silbergeld EK, Waalkes M, Rice JM. Lead as a carcinogen: experimental evidence and mechanisms of action. *American J Ind Med*. 2000;38:316–323.

Stockstad E. Costly cancer lawsuits may spur search to replace world's most common weed killer. *Science*. 2019 May:22. Available from: https://www-sciencemag-org.proxy.lib.ohio-state.edu/news/2019/05/costly-cancer-lawsuits-may-spur-search-replace-worlds-most-common-weed-killer.

Stoiber T and Naidenko O. 2018 July, 24. 'A Civil Action' carcinogen pollutes tap water supplies for 14 million Americans. EWG Children's Health Initiative. https://www.ewg.org/childrenshealth/carcinogen-pollutes-tap-water-supplies-14-million-americans/. Accessed October 11, 2019.

Taiwan Environmental Protection Administration. 1985. Survey of Heavy Metals in the Soil Samples.

Trefry JH, Trocine RP, Metz S, Sisler MA. 1986. *Forms, Reactivity and Availability of Trace Metals in Barite*. Report to the Offshore Operators Committee, Task Force on Environment Science. New Orleans, LA, pp. 1–50.

Tsai JH, Yang J. Epithelial-mesenchymal plasticity in carcinoma metastasis. *Genes Dev*. 2013;27:2192–2206.

Van Leeuwen JA, Waltner-Toews D, Abernathy T, Smit B, Shoukri M. Associations between stomach cancer incidence and drinking water contamination with atrazine and nitrate in Ontario (Canada) agroecosystems, 1987–1991. *Int J Epidemiol*. 1999;28:836–840.

Vidya Priyadarsini, R.; Senthil Murugan, R.; Maitreyi, S.; Ramalingam, K.; Karunagaran, D.; Nagini, S. The flavonoid quercetin induces cell cycle arrest and mitochondria-mediated apoptosis in human cervical cancer (HeLa) cells through p53 induction and NF-_B inhibition. *Eur J Pharmacol*. 2010;649:84–91.

Vinceti M, Dennert G, Crespi CM, Zwahlen M, Brinkman M, Zeegers MPA, Horneber M, D'Amico R, Del Giovane C. Selenium for preventing cancer. *Cochrane Database Syst Rev*. 2014;3:CD005195. doi: 10.1002/14651858.CD005195.pub3.

Vittecoq M, Giraudeau M, Sepp T, et al. Turning natural adaptations to oncogenic factors into an ally in the war against cancer. *Evol Appl*. 2018;11(6):836–844.

Wakabayashi J. Anatomy of a subduction complex: architecture of the Franciscan Complex, California, at multiple length and time scales. *Int Geol Rev*. 2015;57:669–746.

Walraven N, Bakker M, Van Os B, Klaver GT, Middelburg J, Davies G. Factors controlling the oral bioaccessibility of anthropogenic Pb in polluted soils. *Sci Total Environ*. 2015;506:149–163.

Wang J, Odinga ES, Zhang W, et al. Polyaromatic hydrocarbons in biochars and human health risks of food crops grown in biochar-amended soils: a synthesis study. *Environ Int*. 2019;130:104899.

Wang J, Xia K, Waigi MG, Gao Y, Odinga ES, Ling W, Liu J. Application of biochar to soils may result in plant contamination and human cancer risk due to exposure of polycyclic aromatic hydrocarbons. *Environ Int*. 2018;121:169–177.

Wang J, Zhang X, Ling W, Liu R, Liu J, Kang F, Gao Y. Contamination and health risk assessment of PAHs in soils and crops in industrial areas of the Yangtze River Delta region, China. *Chemosphere*. 2017;168:976–987.

Wang Y, Chun-Li SU, Xie X, Xie Z. The genesis of high arsenic groundwater: a case study in Datong basin. *Chin Geol*. 2010;37:771e780.

Wang YL, Tsou MC, Liao HT, et al. Influence of the soil properties on the bioaccessiblity of Cr and Ni in geologic serpentine and anthropogenically contaminated non-serpentine soils in Taiwan. *SSci Total Environ*. 2020:714. https://doi.org/10.1016/j.scitotenv.2020.136761.

Wheeler T, von Braun J. Climate change impacts on global food security. *Science*. 2013;341(6145):508–513.

White PJ, Broadley MR. Biofortification of crops with seven mineral elements often lacking in human diets-iron, zinc, copper, calcium, magnesium, selenium, and iodine. *New Phytol*. 2009; 182:49–84.

WHO. 1993. *Recommendation. Guideline for Drinking Water Quality*, vol. I. Geneva, Switzerland: WHO.

Wong TW, Tam WS, Yu TS, Wong AH. Associations between daily mortalities from respiratory and cardio-vascular diseases and air pollution in Hong Kong, China. *Occup Environ Med*. 2002;59(1):30–35.

World Cancer Research Fund International/American Institute for Cancer Research. Continuous Update Project Report: Diet, Nutrition, Physical Activity and Kidney Cancer. 2015. https://www.wcrf.org/sites/default/files/Kidney-Cancer-2015-Report.pdf

World Cancer Research Fund/American Institute for Cancer Research. Diet, Nutrition, Physical Activity and Cancer: A global Perspective. Continuous Update Project Expert Report 2018. https://www.wcrf.org/dietandcancer

Yeo NK, Hwang YJ, Kim ST, Kwon HJ, Jang YJ. Asian sand dust enhances rhinovirus-induced cytokine secretion and viral replication in human nasal epithelial cells. *Inhal Toxicol*. 2010;22(12):1038–1045.

Yu S, Zhao N, Zi X. The relationship between cyanotoxin (microcystin, MC) in pond-ditch water and primary liver cancer in China. *Zhonghua Zhong Liu Za Zhi*. 2001;23(2):96–99.

Zhang F, Lee J, Liang S, Shum CK. Cyanobacteria blooms and non-alcoholic liver disease: evidence from a county level ecological study in the United States. *Environ Health*. 2015;14:41.

Zhou W, Liang X, Dai P, et al. Alteration of phenolic composition in lettuce. *Int J Mol Sci*. 2019;20(17):1–15.

6 Addressing Urban Mal- and Undernourishment through Sustainable Home Gardens

Neeraj H. Tayal
The Ohio State University Medical Center

CONTENTS

6.1 INTRODUCTION

Meeting the nutritional needs of an ever-expanding human population is of monumental importance. The challenge is formidable when considering the predicted growth rate of the human population in the coming decades. Most of this population expansion will be concentrated in cities and surrounding lands, making urban agriculture of particular importance to community leaders, policymakers, educators, researchers, farmers, and, yes, home gardeners. This chapter explores the topic of home gardening as one component of urban agriculture with a focus on addressing human health through improved nutrition (Satterthwaite 2010).

Providing children and adults with nutrient-dense foods having appropriate calories, proteins, essential fats, vitamins, and mineral salts is required for good health. Providing food that is free of toxins is equally important. When people suffer from chronic malnutrition or exposure to pollutants, they can experience significant health problems, resulting in disabilities that can often last a lifetime. For children who experience poor nutrition in utero and in the first years of life, these consequences are most significant; physical and cognitive function is impaired, and there is an increased risk of death from infection (Pelletier 1995).

The aim of this article is to challenge the commonly held notion of home gardening as a hobby and pastime, which contributes negligibly to human health. We will make the case that through the production of fruits and vegetables, home gardens are uniquely equipped to provide nutrient-dense foods critical to human health. We will start with a discussion of the health impacts of under- and overnutrition experienced in both developed and developing nations. In the following, we will detail the nutrient-dense properties of fruits and vegetables and the evidence to support the health benefits of a diet high in fruits and vegetables. We will address the potential health impact of soil contamination, given that urbanized soils are more likely to have elevated levels of heavy metals and other toxins. We will finish by detailing the fruit and vegetable yields that can be expected from a home garden when managed with best practices, borrowed from the small farm and urban farm sectors of agriculture.

6.2 MALNUTRITION

Malnutrition encompasses conditions of both under- and overnutrition. Undernutrition is experienced by over 800 million people worldwide according to the United Nations. Undernutrition describes insufficient intake of calories and nutrients to maintain good health. Individuals who are undernourished are at increased risk of ailments such as stunting, infection, weak bones, cognitive dysfunction, anemia, and visual impairment. At its worst, undernutrition can result in increased child and maternal mortality (Pelletier 1995).

Overnutrition is a common experience of developed nations and is characterized by the high consumption of ultraprocessed foods. These are calorie-dense and often high in sodium, making them hyperpalatable. The high consumption of ultraprocessed foods, combined with an ever-increasing sedentary lifestyle, has resulted in a marked increase in obesity-related diseases such as diabetes, cardiovascular disease, heart failure, cancer, osteoarthritis, fatty liver, and kidney failure.

Unfortunately, the qualities making ultraprocessed foods popular in developed nations make them desirable in developing nations too, often more so. They are prepackaged with a long shelf life, generally not requiring refrigeration. This is of particular importance in areas where electricity is not consistently available. Industrial manufacturing processes such as mass production, automation, and raw material subsidies (corn, rice, soybean, wheat) result in a low cost per unit. In combination, the characteristics of being hyperpalatable, prepackaged, easily distributed, and low cost have made ultraprocessed foods easily marketable in developing nations. As a result, they are rapidly replacing traditional diets. Consequently, many medium and low-income nations are experiencing substantial increases in the incidence of obesity and its related diseases. Most importantly, just as with undernutrition, children are uniquely susceptible to the long-term consequences of overnutrition (Ezzati 2017).

The common features of undernutrition, micronutrient deficiency, and overnutrition are disability, lost productivity, increased mortality, and human suffering. Many low- and middle-income nations are experiencing what is described as the "double burden" of both under- and overnutrition in their population.

6.3 MICRONUTRIENT DEFICIENCY IN DIET AND HUMAN HEALTH

The term "malnutrition" evokes an image of a person who is thin, with little muscle mass and perhaps a bloated belly. Unfortunately, this image belies an important form of malnutrition referred to as micronutrient deficiency. Human health depends on finely tuned and complex chemical pathways. The building blocks of these pathways are the macronutrients: carbohydrates, proteins, and fats. The chemical reactions, however, are often dependent on relatively minuscule amounts of minerals and vitamins, referred to as micronutrients. Examples include but are not limited to: potassium, iron, calcium, zinc, magnesium, iodine, vitamin A (retinol), vitamin B12 (cyanocobalamin), vitamin C (ascorbic acid), vitamin D (cholecalciferol), and vitamin K (phylloquinone).

Here, we review the more consequential micronutrient deficiencies and the capacity for mitigation through diet. Deficiencies of vitamin A, iron, zinc, and iodine are common across the globe and cause substantial morbidity and mortality. More importantly, replacement of these particular micronutrients, through both dietary and artificial means, has demonstrated dramatic improvements in function and survival (Black 2003).

Vitamin A is a fat-soluble vitamin found in both animal and plant tissues. Animal products such as eggs, milk, cheese, liver, and fish oils are high in retinol, the activated form of vitamin A. The bioavailability of animal-derived retinol is high, at 70%–90%. Plants, on the other hand, are a source of the provitamins called carotenoids and have a lower bioavailability of 8%–16%. Carotenoids are found in deeply colored vegetables like dark green leaves and carrots, tubers such as yellow sweet potato, and fruits such as mango, papaya, and red palm. Chopping, cooking, and incorporating fats into the cooking process are all strategies that maximize the bioavailability of carotenoids from these fruits and vegetables.

Vitamin A deficiency has been well described and impacts the health of millions, mostly in developing nations, and mostly children and women of child-bearing age. An estimated 250 million preschool children are deficient in vitamin A, and 250–500,000 children go blind each year from this deficiency. Almost half of those die within a year of losing their vision. Vitamin A causes a spectrum of ocular diseases; night blindness and corneal xerosis are reversible with supplementation, while keratomalacia results in full thickness necrosis of the cornea and permanent vision loss. Anemia from vitamin A deficiency is caused by decreased mobilization of iron stores and suppressed bone marrow activity. People deficient in vitamin A are more susceptible to infection, specifically measles and diarrheal diseases. Growth retardation is another observed consequence. Given the impact of vitamin A deficiency on vulnerable populations, interventions are targeted to children, women of child-bearing age, and breastfeeding mothers in developing nations (Sommer and West 1996).

Iron deficiency is most common in childhood, adolescence, and during pregnancy. Demands are particularly high in adolescent boys, when the puberty-related growth spurt results in substantial muscle development. Adolescent girls experience a high iron demand related to menstrual loss. Iron requirements increase dramatically during pregnancy as the placenta and baby draw from the mother's stores. Unfortunately, the bioavailability of iron is quite low, so any limitation in access or absorption can result in decreased iron stores. The incidence is highest in developing nations such as South Asia where up to 88% of pregnant and 40% of nonpregnant women are iron-deficient. In developed nations, this rate is lower—around 20%—but still consequential.

Reduced red cell mass, anemia, is the most commonly associated disease of iron deficiency, but a number of other important illnesses result from low iron stores. These include impaired cognitive function, increased risk of infection, growth retardation, thyroid dysfunction, and an increased capacity to absorb heavy metal toxins such as cadmium and lead from contaminated foods. Iron plays an important role in the brain; deficiencies result in a host of cognitive impairments, which are most consequential in infants and children who are experiencing rapid neurologic expansion. Impacts include delayed psychomotor development and impaired cognitive performance. Infants with moderate anemia score lower on IQ tests when they reach school age. Children and adults with iron deficiency feel more fatigued and report reduced ability to concentrate, independent of blood hemoglobin levels. Iron deficiency results in higher rates of infection. Women with iron deficiency are more likely to have premature babies, and the mortality rate of their infants is higher. The mortality rate for iron-deficient mothers is also increased. Babies of iron-deficient mothers are iron-deficient at birth and are less capable of replenishing iron stores from the breast milk of mothers, who themselves remain iron-depleted (WHO 2001).

Foods highest in iron are meats and seafood. Iron in fruits and vegetables is far less concentrated and, depending on the dietary mix, may have variable absorption. Fruits and vegetables that are highest in iron include spinach, Swiss chard, green peas, cabbage, Brussel sprouts, leeks, asparagus, thyme, parsley, strawberries, raspberries, kiwis, figs, and rhubarb (InformedHealth.org 2018). Interestingly, iron absorption is significantly increased when consumed with vitamin C (ascorbic acid) which is common in fruits and vegetables such as sweet peppers, citrus, kiwi, broccoli, strawberry, Brussel sprouts, tomato, and cantaloupe (https://ods.od.nih.gov/factsheets/VitaminC-HealthProfessional/).

Zinc is a trace element in the body and functions as a cofactor to over 300 enzymes. It also stabilizes proteins involved in signal transduction and transcription. Thus, zinc plays an important role in many bodily tasks including cell growth and differentiation, connective tissue growth and maintenance, DNA synthesis, and cell division. It is vital for the proper operation of the immune system, thyroid gland, cognitive system, and crucial for wound healing, tasting, bone mineralization, blood clotting, and fetal growth. Because there is no storage mechanism for zinc, a steady consumption is required (Chasapis 2012). Deficiency of zinc in developing nations leads to increased incidence of and mortality from diarrheal disease and pneumonia. A pooled analysis of seven trials demonstrated that zinc supplementation in children significantly decreased both digestive and respiratory infections (Bhutta 1999). In a study conducted in New Delhi, zinc supplementation for low birth weight infants was associated with significantly lower mortality (Sazawal 2001). Zinc is found in high concentrations in animal meats and seafood. The content in fruits and vegetables is limited, but zinc is present in appreciable levels in chickpeas, dry beans, pumpkin seeds, and peas.

As food systems evolve in the coming decades, meeting both the macro- and micronutrient needs of the population will be critical. Micronutrient deficiencies will be managed through greater access to nutrient-dense foods including animal products, grains, fruits, and vegetables, along with food fortification and supplementation.

6.4   HEALTH BENEFITS OF FRUITS AND VEGETABLES

Fully addressing undernutrition, overnutrition, and micronutrient deficiencies is a complex undertaking; however, all three can be ameliorated with an increase in fruit and vegetable intake. Both the quantity and quality of food need to increase rapidly in poverty-stricken communities struggling with double burden. At the same time, battling overnutrition requires foods that are nutritionally dense and low in calories. The calorie density of fruits and vegetables is substantially lower than that of both carbohydrate-rich grains and meat and dairy products. Table 6.1 shows the nutrient density scores of vegetables commonly grown in the garden.

At the same time, fruits and vegetables are replete with important minerals such as potassium, iron, calcium, and zinc and are low in sodium. They are an excellent source of vitamins A and C, and of both soluble and insoluble fiber (FDA 2017). Table 6.2 displays nutritional information for common garden vegetables. Fruits and vegetables also contain a host of beneficial phytochemicals, specifically the class of compounds called polyphenols, which include flavonoids, phenolic acids, ligans, and stilbens. A growing body of evidence demonstrates the health benefits of fruits and vegetables.

Dietary guidelines from the World Health Organization (WHO) recommend fruit and vegetable intake as their topline recommendation, followed by recommendations to consume whole grains in place of refined grains and to consume lower amounts of saturated fat, refined sugars, and sodium. A study of over 90 nutritional guidelines from across the globe demonstrated that over 90% of existing guidelines align with the WHO in promoting a diet high in fruits and vegetables (Herforth 2019).

The benefits of diets rich in fruits and vegetables are most pronounced in the category of cardiovascular disease. In the International Journal of Epidemiology, investigators concluded that each additional 200-g consumption of fruits and vegetables (3 cups) resulted in a lower risk of cardiovascular disease, and a decrease in all-cause mortality. The effect was dose-dependent and sizable. At 800 g, the relative risk reduction of coronary heart disease was 24%, stroke 33%, cardiovascular disease 28%, and all-cause mortality 31% (Aune 2017). A similar analysis from the Prospective Urban Epidemiology (PURE) study evaluated over 130,000 subjects across 18 low-income, middle-income, and high-income countries (North America, Europe, South America, Middle East, South Asia, China, Southeast Asia, Africa). Investigators demonstrated a similar pattern of reduced risk for cardiovascular and all-cause mortality across both high and low-to-middle income nations (Miller 2017).

A diet rich in fruits and vegetables decreases the risk of diabetes. A meta-analysis in the British Medical Journal showed a relative risk reduction of 0.93 for each serving of fruit consumed per day. For vegetable consumption, the relative risk was 0.9 for each serving consumed per day. For green leafy vegetables, the risk reduction was 0.87 for each 0.2 serving consumed per day (Li 2014).

TABLE 6.1

Powerhouse Fruits and Vegetables ($N=41$), by Ranking of Nutrient Density Scores[a], 2014

Item	Nutrient Density Score
Watercress	100.00
Chinese cabbage	91.99
Chard	89.27
Beet green	87.08
Spinach	86.43
Chicory	73.36
Leaf lettuce	70.73
Parsley	65.59
Romaine lettuce	63.48
Collard green	62.49
Turnip green	62.12
Mustard green	61.39
Endive	60.44
Chive	54.80
Kale	49.07
Dandelion green	46.34
Red pepper	41.26
Arugula	37.65
Broccoli	34.89
Pumpkin	33.82
Brussels sprout	32.23
Scallion	27.35
Kohlrabi	25.92
Cauliflower	25.13
Cabbage	24.51
Carrot	22.60
Tomato	20.37
Lemon	18.72
Iceberg lettuce	18.28
Strawberry	17.59
Radish	16.91
Winter squash (all varieties)	13.89
Orange	12.91
Lime	12.23
Grapefruit (pink and red)	11.64
Rutabaga	11.58
Turnip	11.43
Blackberry	11.39
Leek	10.69
Sweet potato	10.51
Grapefruit (white)	10.47

Source: Data from Noia (2014).

[a] Calculated as the mean of percent daily values (DVs) (based on a 2,000 kcal day^{-1} diet) for 17 nutrients (potassium, fiber, protein, calcium, iron, thiamin, riboflavin, niacin, folate, zinc, and vitamins A, B_6, B_{12}, C, D, E, and K) as provided by 100 g of food, expressed per 100 kcal of food. Scores above 100 were capped at 100 (indicating that the food provides, on average, 100% DV of the qualifying nutrients per 100 kcal).

TABLE 6.2

Nutritional Information for Raw Vegetables

Vegetables Serving Size (gram weight/ounce weight)	Calories	Calories from Fat	Total Fat (g)	(%DV)	Sodium (mg)	(%DV)	Potassium (mg)	(%DV)	Total Carbo-Hydrate (g)	(%DV)	Dietary Fiber (g)	(%DV)	Sugars (g)	Protein (g)	Vitamin A (%DV)	Vitamin C (%DV)	Calcium (%DV)	Iron (%DV)
Asparagus 5 spears (93 g/3.3 oz)	20	0	0	0	0	0	230	7	4	1	2	8	2	2	10	15	2	2
Bell Pepper 1 medium (148 g/5.3 oz)	25	0	0	0	40	2	220	6	6	2	2	8	4	1	4	190	2	4
Broccoli 1 medium stalk (148 g/5.3 oz)	45	0	1	1	80	3	460	13	8	3	3	12	2	4	6	220	6	6
Carrot 1 carrot, 7″ long, 1 1/4″ diameter (78 g/2.8 oz)	30	0	0	0	60	3	250	7	7	2	2	8	5	1	110	10	2	2
Cauliflower 1/6 medium head (99 g/3.5 oz)	25	0	0	0	30	1	270	8	5	2	2	8	2	2	0	100	2	2
Celery 2 medium stalks (110 g/3.9 oz)	15	0	0	0	115	5	260	7	4	1	2	8	2	0	10	15	4	2

(Continued)

TABLE 6.2 (Continued)
Nutritional Information for Raw Vegetables

Vegetables Serving Size (gram weight/ounce weight)	Calories	Calories from Fat	Total Fat (g)	Total Fat (%DV)	Sodium (mg)	Sodium (%DV)	Potassium (mg)	Potassium (%DV)	Total Carbo-Hydrate (g)	Total Carbo-Hydrate (%DV)	Dietary Fiber (g)	Dietary Fiber (%DV)	Sugars (g)	Protein (g)	Vitamin A (%DV)	Vitamin C (%DV)	Calcium (%DV)	Iron (%DV)
Cucumber 1/3 medium (99 g/3.5 oz)	10	0	0	0	0	0	140	4	2	1	1	4	1	1	4	10	2	2
Green (Snap) Beans 3/4 cup cut (83 g/3.0 oz)	20	0	0	0	0	0	200	6	5	2	3	12	2	1	4	10	4	2
Green cabbage 1/12 medium head (84 g/3.0 oz)	25	0	0	0	20	1	190	5	5	2	2	8	3	1	0	70	4	2
Green onion 1/4 cup chopped (25 g/0.9 oz)	10	0	0	0	10	0	70	2	2	1	1	4	1	0	2	8	2	2
Iceberg lettuce 1/6 medium head (89 g/3.2 oz)	10	0	0	0	10	0	125	4	2	1	1	4	2	1	6	6	2	2
Leaf lettuce 1 1/2 cups shredded (85 g/3.0 oz)	15	0	0	0	35	1	170	5	2	1	1	4	1	1	130	6	2	4

(Continued)

TABLE 6.2 (*Continued*)
Nutritional Information for Raw Vegetables

Vegetables Serving Size (gram weight/ounce weight)	Calories	Calories from Fat	Total Fat (g)	(%DV)	Sodium (mg)	(%DV)	Potassium (mg)	(%DV)	Total Carbo-Hydrate (g)	(%DV)	Dietary Fiber (g)	(%DV)	Sugars (g)	Protein (g)	Vitamin A (%DV)	Vitamin C (%DV)	Calcium (%DV)	Iron (%DV)
Onion 1 medium (148 g/5.3 oz)	45	0	0	0	5	0	190	5	11	4	3	12	9	1	0	20	4	4
Potato 1 medium (148 g/5.3 oz)	110	0	0	0	0	0	620	18	26	9	2	8	1	3	0	45	2	6
Radishes 7 radishes (85 g/3.0 oz)	10	0	0	0	55	2	190	5	3	1	1	4	2	0	0	30	2	2
Summer squash 1/2 medium (98 g/3.5 oz)	20	0	0	0	0	0	260	7	4	1	2	8	2	1	6	30	2	2
Sweet corn kernels from 1 medium ear (90 g/3.2 oz)	90	20	3	4	0	0	250	7	18	6	2	8	5	4	2	10	0	2
Sweet potato 1 medium, 5″ long, 2″ diameter (130 g/4.6 oz)	100	0	0	0	70	3	440	13	23	8	4	16	7	2	120	30	4	4
Tomato 1 medium (148 g/5.3 oz)	25	0	0	0	20	1	340	10	5	2	1	4	3	1	20	40	2	4

Source: Data from FDA (2017).

Hypertension is a leading cause of cardiovascular disease. The DASH diet (Dietary Approaches to Stop Hypertension), developed to lower blood pressure through dietary means, stresses the importance of increased fruit and vegetable intake and low-fat dairy consumption. The mechanism for blood pressure improvement is primarily related to the high potassium content of fruits, vegetables, and low-fat dairy in the DASH diet. There are other components thought to play a contributing role, such as magnesium, calcium, fiber, and other bioactive constituents—again, commonly found in fruits and vegetables. Administration of the DASH diet has demonstrated an 11-point drop in systolic blood pressure in patients with hypertension and a 3.5-point drop in those without hypertension (Weaver 2013). Fruits and vegetables—especially beet greens, lima beans, Swiss chard, potato, acorn squash, spinach, pak choi, sweet potato, tomato, and zucchini—are high in potassium.

Plainly, there is significant evidence supporting the health benefits of fruits and vegetables, although one must take care not to overextend the health-benefit claims. Investigators have extensively studied whether fruits and vegetables decrease the incidence of cancer. Unfortunately, there is scant and often conflicting evidence for this assertion. One of the largest studies examining this possible association was the European Prospective Investigation into Cancer and Nutrition (EPIC), which enrolled over 500,000 participants from across ten nations. Information was collected with the intent of identifying a relationship between cancer risk and fruit, vegetable, and fiber intake. There was no observed association for cancers of the stomach, biliary tract, pancreas, cervix, endometrium, prostate, kidney, bladder, or lymphatic system. Fruit intake was associated with decreased cancers of the upper gastrointestinal tract, but this association was not observed for vegetable intake. Fruits, vegetables, and fiber were similarly associated with a reduced risk of colon cancer (Bradbury 2014). Separately, a pooled analysis of 13 studies investigating the relationship of fiber intake and colorectal cancer demonstrated a significant inverse relationship. However, once adjustments for red meat, total milk, and alcohol intake were applied, no beneficial impact remained (Park 2005).

Finally, we address the issue of how growing methods impacts the nutritional quality of fruits and vegetables. Many home gardeners deploy organic growing methods with the understanding that produce will be more nutrient dense and free of toxins. The literature is quite variable on this topic with some studies reporting nutrient density equivalent with organic and conventional methods, while others demonstrate clear superiority of organic production methods. In a large meta-analysis of 343 peer-reviewed publications, the following conclusions were drawn. Organically grown produce was substantially higher in antioxidant compounds such as polyphenolics including phenolic acids, flavinones, stilbens, flavones, and anthocyanins. These compounds are linked to lower rates of cardiovascular disease, neurodegenerative diseases, and some cancers. Pesticide residues were 75% lower in organically grown produce. The heavy metal cadmium was substantially lower in organically produced foods (Barański 2014). In another analysis, investigators determined that there were no differences in the mineral and trace element levels between organic and conventional production. No conclusions could be drawn regarding iron levels. The same was true for vitamins A, B1, B2, C, and protein content (Woese 1997).

6.5 HEAVY METAL AND CONTAMINANTS IN URBAN SOIL AND HUMAN HEALTH

When discussing home gardening, it is important to bear in mind that the soils of many urban environments are contaminated with potential toxins such as heavy metals (arsenic, lead, cadmium, chromium, mercury) and polycyclic aromatic hydrocarbons. Polycyclic aromatic hydrocarbons are created when materials such as coal, gasoline, and wood are combusted. Given the prolonged settlement of many urban environments and the resulting accumulation of these toxins, home gardeners and their family members are at an increased risk of exposure to toxins through produce consumption and direct exposure to soils. By knowing the historic use of the land and considering the distance to point source contamination—such as factories, roads, and waste dumps—gardeners can make assumptions about the contamination risk of their soil. Soil testing for these contaminants is also an option, although the cost can be $25–$75 per element (Antisari 2015).

Lead is naturally occurring in the soil, but excess levels in urban environments are common. Gasoline for automobiles was intentionally leaded from the 1920s until the early 1980s in order to improve engine performance. A transition to unleaded gas was mandated due to environmental pollution rules in the Clean Air Act Amendment of 1970. Growing evidence about the negative health impact of lead accelerated this transition. Additionally, there was a phase-out of lead-based paint, which had been commonly used in residential housing. Unfortunately, the legacy of leaded gasoline and lead-based paint persists, and primarily impacts urban soils, including where home gardens are frequently planted. Lead particulates from the combustion of leaded gasoline have settled on soils adjacent to areas of high auto-traffic. Over time, the exteriors of older homes painted with lead-based paint have been subjected to leaching, flaking, and scraping—thus creating a halo of soil contamination around residential properties. Exposure to lead in soil occurs from the consumption of vegetables grown in contaminated soil, inhalation of soil dust, and, to a lesser extent, from direct contact with soil. Lead is absorbed through the lung and intestines, along with smaller amounts through the skin. Once ingested, the internal absorption is approximately 8%–10%. Factors which increase absorption include diets low in zinc, iron, phosphorous, and calcium. Chronic exposure to lead results in increased mortality, related to cardiovascular disease (relative risk of 1.5) and cancer (relative risk of 1.69). Children can experience measureable decreases in neurocognitive function, hearing, and brain volume. Psychiatric problems such as phobic behavior, anxiety, depression, and increased hostility have been demonstrated. Lead toxicity can result in anemia. Nephrotoxicity is possible, as is male infertility through effects on sperm function (Needleman 1988).

Cadmium is an important soil contaminant in the urban setting. Causes include previous mining, pollution from smelters, sewage waste, and atmospheric emissions. Synthetic phosphate fertilizers are another source of possible cadmium contamination. Cadmium is readily absorbed by plants and then leads to important health effects in humans. As with lead absorption, specific mineral deficiencies, such as low iron stores, increase the absorption of cadmium. Excess cadmium levels can result in kidney damage. Although this damage may be subclinical and not progressive, affected individuals could be more susceptible to secondary renal insults such as hypertension and diabetes. Cadmium exposure can also lead to bone diseases like osteoporosis and associated bone fractures (Järup 1998).

Mercury is a toxin that can be found in excess levels in urban soils. However, the main source of human toxicity is the consumption of meats, specifically fish. Arsenic can be found in soils, but the main concern is acute poisoning; fruit and vegetable intake from urban environments is not considered a risk.

Soil contamination is an important consideration at any time land is used for food production. Home gardening is no exception. While contamination risk can be assessed by considering the location and historic use of land, gardeners would be best served to consider soil testing as a starting point, if cost is not prohibitive. Testing for lead and cadmium is the highest priority. Gauging the degree of contamination and the thresholds at which human health is impacted will help determine the degree of mitigation needed. Further research and guidance for home gardeners on this subject is needed (Leake 2009). Current options for mitigation include removing contaminated soil—although this would be impractical for most. Building raised beds is a commonly deployed method in urban settings, given the simplicity and relatively low cost. Other simple methods to decrease the uptake of contaminants include altering soil pH and adding organic matter. Soil can be covered with natural or synthetic products, preventing soil dust from contaminating edible plants parts like leaves and fruit. Finally, thoroughly washing fruits and vegetables before consumption will also decrease exposure to harmful soil contaminants.

6.6   HOME GARDENS FOR FRUITS AND VEGETABLES

Where does the humble home garden fit in the monumental struggle to feed humanity? Certainly, it cannot be a significant contributor to the caloric needs of a family, for this would require growing calorie-rich starchy grains such as wheat, corn, and rice. These are not the purview of the home gardener. Similarly, home gardening cannot satisfy the protein needs of a family, as this would require either the large-scale production of pulses such as legumes, beans and peas, or protein-rich grains such as quinoa. Although some gardeners do raise animals such as poultry, significant production of animal proteins would be out of reach for most. In contrast, the area home gardeners excel in is the production of fruits and vegetables and, in some communities, tree nuts. It is here that home gardening arrives as a meaningful contributor to human nutritional needs and, thus, human health.

Home gardens have been an established form of cultivation for millennia and are common among civilizations across the globe. The fruits and vegetables grown in home gardens are as varied as the climates and cultures they grow in. This variation is determined by physical factors such as temperature, seasonality, precipitation, soil makeup, pest pressures, and sun exposure. Additionally, socioeconomic factors such as tradition, income, family structure, and access to arable land are all important determinants of the garden makeup. Despite this rich variation, there are many common features to home gardens.

In the United States we say, "if you've seen one, you've seen them all." When speaking of home gardens, it would be appropriate to say, "if you've seen one home garden, you've seen one home garden." The variety of plants and garden designs is staggering. As an example, gardens in tropical areas are often multistoried with large trees, creating a canopy to protect low-growing plants from the harsh equatorial sun and the torrential rains. Plantings are dispersed, mixed, and less commonly in rows, when compared with temperate climates. Vegeculture or asexual reproduction from vegetative structures is the more common form of propagation, given the year-round growing conditions. What results is an environment imitating the tropical forest structure. Fruit production is predominant with plants such as banana, mango, and papaya. In the lower canopy and ground cover, maize, peppers, tomato, melons, and beans are common. Tubers and roots such as cassava are also grown.

In contrast to the tropics, temperate climates are relatively sun poor. Trees and bushes are spaced out with ground cover species unshaded. Instead of vegetative propagation predominating, seed propagation is more the norm—although not exclusively. Spring, summer, and fall growing seasons give way to winters with killing frosts and short daylight hours. Seasonality is the primary determinant for what grows and can be grown. Since crops must be replanted every year from seed, rotation becomes feasible and naturally leads to a more defined cropping pattern, in contrast to the mixed pattern described in tropical gardens. As sunlight is at a premium in temperate climates, plants are often spaced to maintain a more open canopy, though this does at times come at the expense of excessive weed competition.

Regardless of this vast range, there are common features for home gardens that distinguish them from market gardens and field agriculture. Table 6.3 shows the characteristics of these three basic food production systems. First and foremost, home gardens are located adjacent to the living space where planting, management, and harvest are most convenient (Figure 6.1). Home gardens have a large variety of plant species, comprised almost completely of fruits and vegetables. The food production is supplementary to the family's dietary needs. Consumption is within the home but often shared with neighbors and, sometimes, bartered for other products (Figure 6.2). Women, children, and the elderly commonly comprise the labor force, although this is by no means exclusive. Home gardeners often have a separate vocation and work in the garden part-time during the growing season. Harvest from the garden is daily and seasonal. Plantings occur at high density, with both horizontal and vertical spacing used to maximize production (Figure 6.3). Cost and inputs are minimal, as hand tools and family labor are used. Family gardens are found in both urban, periurban, and rural communities. Lastly, institutional support provided to family gardeners is minimal to none, resulting in a highly variable skill level (Niñez 1987).

TABLE 6.3

Tendency Characteristics of Three Basic Food Production Systems

Concept	Household Garden	Market Garden	Field Agriculture
Species density	High	Medium to low	Low
Species type	Staple, vegetable fruit (cultural)	Vegetable, fruit (market oriented)	Staple (subsistence agro-industrial)
Production objective	Home consumption	Market sale	Subsistence, market sale
Labor source	Family (female, elderly, children)	Family or hired (male, female)	Family, hired (male, female)
Labor requirements	Part-time	Full-time	Full-time
Harvest frequency	Daily, seasonal	(Short) seasonal	(Long) seasonal
Space utilization	Horizontal, vertical	Horizontal, vertical	Horizontal
Location	Close to dwelling	Close to urban market	Rural setting, close or distant from homestead
Cropping patterns	Irregular, row	Row	Row
Economic role	Supplementary	Major economic activity	Major economic activity
Technology	Simple hand tool	Hand tool or mechanized	Mechanized if possible, hand tool
Inputs—cost	Low	Medium to high	Medium to high
Distribution	Rural and urban	Sub-urban	Rural
Skills	Garden-horticultural	Market-horticultural	Agricultural, commercial
Assistance	None or minor	Credit	Credit, extension

Source: Data from Niñez (1987).

FIGURE 6.1 With the garden only feet away from home, harvest commonly occurs on a daily and weekly basis. Fruits and vegetables are picked when fully ripened; minimum handling prevents bruising and limits food waste.

FIGURE 6.2 The day's harvest is commonly shared with friends and family in a communal exercise.

FIGURE 6.3 Vertical supports are common in the home garden, given the premium on space. In the foreground, peppers are supported with bamboo. In the second row, pole beans are supported on trellis netting. In the background, tomatoes are supported on adjustable twine that allows vines to grow up to 20 ft in length throughout the growing season.

6.7   YIELDS OF FRUITS AND VEGETABLES IN THE HOME GARDENS

Home gardening has traditionally been a means for producing fruit and vegetables—and to a limited degree, animal protein, mostly poultry and eggs. With average skills, a gardener can expect to grow 0.5 lb per square foot or up to 500 lb of food in a 30-by-30-foot garden. Table 6.4 displays the average yields expected for a variety of vegetables. With a proper selection of fruits and vegetables and advanced gardening skills, these yields can be increased to 1.5 lb per square foot and a remarkable 1,500-lb yield in a season (Rabin 2012).

Fresh fruits and vegetables are for the most part—with the exception of root vegetables and some squashes—highly perishable products. They are more likely to be damaged with rough handling and often require specific temperatures and humidity to extend their already limited shelf life. When grown commercially, they require faster transportation methods, more expensive packaging, and more complicated distribution systems, as compared to grain crops. Even under careful management, spoilage occurs in transport, at the point of distribution or shortly after purchase by consumers. All of these factors contribute to the high relative cost of fruits and vegetables. Furthermore, growing and distributing fruits and vegetables exact a heavy toll on the environment. These circumstances position home gardening as an alternative to obtaining fresh fruits and vegetables from commercial sources.

As mentioned above, a relatively large quantity of produce can be grown on the limited land adjacent to living spaces. The start-up costs for a home gardener can be quite modest. The tools and equipment can be fairly rudimentary and low cost, with much of the work being manual. Access to water can be convenient. Home gardeners have the ability to scout their plants on a frequent basis, identifying barriers to optimal growth such as insect damage, disease pressure, weed pressure, water needs, physical stresses, and optimal harvest time. With the garden only feet away from home, harvest occurs on a daily and weekly basis. Fruits and vegetables are picked when fully ripened; minimum handling prevents bruising and limits food waste (Figure 6.1).

6.8   CONCLUDING REMARKS

The conclusion of this chapter is being written in the Spring of 2020 just as a devastating viral pandemic sweeps across the world. In this early stage, there is much talk about reinforcing healthcare systems and rescuing the economy. There is also talk about quarantines and the physical distancing required to "flatten the curve" of infection. Soon though, there will be talk of food insecurity. Even in a nation as wealthy as the United States, people are scared and worried about access to food. The experience in developing nations is likely to be far more devastating. The supply chains for food production and distribution are complex and interconnected, stretching across nations and oceans. As COVID-19 spreads, the links in these supply chains may break.

It is in this time of crisis, as it has been in times past, people search for means of self-reliance. Producing one's own food in a home garden is perhaps one of the oldest and readily available forms of self-sufficiency. Home gardeners can produce relatively large quantities of fruits and vegetables in a modestly sized yard. There are well-established methods to optimize the success of the home gardens. There are also new materials and technologies that can further increase the efficiency of home gardens (Figure 6.4). There is a large body of evidence demonstrating the health benefits of eating fruits and vegetables, particularly in reducing cardiovascular disease. In addition, fruits and vegetables contain vital micronutrients required for optimal health.

In order for home gardening to be fully leveraged, local communities should consider the following interventions. Offering educational programs will ensure that gardeners understand the best practices to optimize food production. Topics should include methods for improving soil health, efficient use of water resources, and organic methods that limit the use of pesticides and herbicides. Gardeners will need recommendations and access to quality seed, appropriate for the growing conditions. Programs should also provide guidance on safe food handling and storage. Municipal

TABLE 6.4
Yield Expectations for Mixed Stand, Small-Scale Agriculture

Vegetable	Spacing (inches)		Plants or Seed per 100 ft	Average Yield Expected 100 ft	Average Yield Expected ft^{-2}
	Rows	Plants			
Asparagus	36–48	18	66 plants or 1 oz	30 lb	0.08–0.1 lb
Beans, snap bush	24–36	3–4	½ lb	120 lb	0.4–0.6 lb
Beans, snap pole	36–48	4–6	½ lb	150 lb	0.38–0.5 lb
Beans, lima bush	30–36	3–4	½ lb	25 lb shelled	0.08–0.1 lb
Beans, lima pole	36–48	12–18	¼ lb	50 lb shelled	0.13–0.17 lb
Beets	15–24	2	1 oz	150 lb	0.75–1.2 lb
Broccoli	24–36	14–24	50–60 plants or ¼ oz	100 lb	0.33–0.5 lb
Brussels sprouts	24–36	14–24	50–60 plants or ¼ oz	75 lb	0.25–0.38 lb
Cabbage	24–36	14–24	50–60 plants or ¼ oz	150 lb	0.5–0.75 lb
Cabbage, Chinese	18–30	8–12	60–70 plants or ¼ oz	80 heads	n/a
Carrots	15–24	2	½ oz	100 lb	0.5–0.8 lb
Cauliflower	24–36	14–24	50–60 plants or ¼ oz	100 lb	0.33–0.5 lb
Celery	30–36	6	200 plants	180 stalks	n/a
Collards and kale	18–36	8–16	¼ oz	100 lb	0.33–0.67 lb
Corn, sweet	24–36	12–18	3–4 oz	10 doz	n/a
Cucumbers	48–72	24–48	½ oz	120 lb	0.2–0.3 lb
Eggplant	24–36	18–24	50 plants or 1/8 oz	100 lb	0.33–0.5 lb
Kohlrabi	15–24	4–6	½ oz	75 lb	0.38–0.6 lb
Lettuce, head	18–24	6–10	¼ oz	100 heads	n/a
Lettuce, leaf	15–18	2–3	¼ oz	50 lb	0.33–0.4 lb
Muskmelon, cantaloupe	60–96	24–36	50 plants or ½ oz	100 fruit	n/a
Okra	36–42	12–24	2 oz	100 lb	0.29–0.33 lb
Onions	15–24	3–4	400–600 sets or 1 oz	100 lb	0.5–0.8 lb
Parsley	15–24	6–8	¼ oz	30 lb	0.15–0.24 lb
Parsnips	18–30	3–4	½ oz	100 lb	0.4–0.67 lb
Peas, English	18–36	1	1 lb	20 lb	0.07–0.13 lb
Peas, Southern	24–36	4–6	½ lb	40 lb	0.13–0.2 lb
Peppers	24–36	18–24	50 plants or 1/8 oz	60 lb	0.2–0.3 lb
Potatoes, Irish	30–36	10–15	6–10 lb seed tubers	100 lb	0.33–0.4 lb
Potatoes, sweet	36–48	12–16	75–100 plants	100 lb	0.25–0.33 lb
Pumpkins	60–96	36–48	½ oz	100 lb	n/a
Radishes	14–24	1	1 oz	100 bunches	n/a
Spinach	14–24	3–4	1 oz	150 lb	0.3–0.5 lb
Squash, summer	36–60	18–36	1 oz	150 lb	0.3–0.5 lb
Squash, winter	60–96	24–48	½ oz	100 lb	0.13–0.2 lb
Tomatoes	24–48	18–36	50 plants or 1/8 oz	100 lb	0.25–0.5 lb
Turnip, greens	14–24	2–3	½ oz	50–100 lb	0.38–0.6 lb
Turnip, roots	14–24	2–3	½ oz	50–100 lb	0.38–0.6 lb
Watermelon	72–96	36–72	1 oz	40 fruit	n/a

Source: Data from Rabin et al. (2012).

FIGURE 6.4 Geotextiles are used to optimize the microclimate. During colder periods, heavier fabrics are used to increase environmental temperatures and maintain humidity. In the warmer months, shade cloths combined with water misters are used to cool sensitive crops. Fabrics are also used to protect crops from insects, mammals, and birds that might forage on crops.

resources should be made available to help identify areas of contamination and assistance with mitigation. Land and water use policies should be developed to facilitate home gardening further.

This COVID-19 global pandemic is laying bare the vulnerabilities of our modern, interconnected, and interdependent society. As the world eventually emerges from this crisis, the structures and systems of society will be under intense scrutiny. Redesigning food systems will be critical, given the sobering realization that repeated global disruptions are likely, whether they be from pandemic or from climate change. With regard to building a more resilient food production system, a movement toward growing locally will be key. In that future state, home gardening may no longer be viewed as a hobby. It may be viewed as a critical infrastructure and a national priority.

REFERENCES

Antisari, Livia Vittori, Francesco Orsini, Livia Marchetti, Gilmo Vianello, and Giorgio Gianquinto. 2015. "Heavy Metal Accumulation in Vegetables Grown in Urban Gardens." *Agronomy for Sustainable Development* 35 (3). Springer-Verlag France: 1139–47. doi:10.1007/s13593-015-0308-z.

Aune, Dagfinn, Edward Giovannucci, Paolo Boffetta, Lars T. Fadnes, Na Na Keum, Teresa Norat, Darren C. Greenwood, Elio Riboli, Lars J. Vatten, and Serena Tonstad. 2017. "Fruit and Vegetable Intake and the Risk of Cardiovascular Disease, Total Cancer and All-Cause Mortality-A Systematic Review and Dose-Response Meta-Analysis of Prospective Studies." *International Journal of Epidemiology* 46 (3): 1029–56. doi:10.1093/ije/dyw319.

Barański, Marcin, Dominika Średnicka-Tober, Nikolaos Volakakis, Chris Seal, Roy Sanderson, Gavin B. Stewart, Charles Benbrook, et al. 2014. "Higher Antioxidant and Lower Cadmium Concentrations and Lower Incidence of Pesticide Residues in Organically Grown Crops: A Systematic Literature Review and Meta-Analyses." *British Journal of Nutrition* 112 (5): 794–811. doi:10.1017/S0007114514001366.

Bhutta, Z. A., Robert E. Black, K. H. Brown, J. Meeks Gardner, S. Gore, A. Hidayat, F. Khatun, et al. 1999. "Prevention of Diarrhea and Pneumonia by Zinc Supplementation in Children in Developing Countries:

Pooled Analysis of Randomized Controlled Trials." *Journal of Pediatrics* 135 (6): 689–97. doi:10.1016/S0022-3476(99)70086-7.

Black, Robert. 2003. "Micronutrient Deficiency – An Underlying Cause of Morbidity and Mortality." *Bulletin of the World Health Organization.* World Health Organization. doi:10.1590/S0042–96862003000200002.

Bradbury, Kathryn E., Paul N. Appleby, and Timothy J. Key. 2014. "Fruit, Vegetable, and Fiber Intake in Relation to Cancer Risk: Findings from the European Prospective Investigation into Cancer and Nutrition (EPIC)." *The American Journal of Clinical Nutrition* 100 (suppl_1). American Society for Nutrition: 394S–398S. doi:10.3945/ajcn.113.071357.

Brantsæter, Anne Lise, Trond A. Ydersbond, Jane A. Hoppin, Margaretha Haugen, and Helle Margrete Meltzer. 2017. "Organic Food in the Diet: Exposure and Health Implications." *Annual Review of Public Health* 38 (1). Annual Reviews: 295–313. doi:10.1146/annurev-publhealth-031816-044437.

Chasapis, Christos T., Chara A. Spiliopoulou, Ariadni C. Loutsidou, and Maria E. Stefanidou. 2012. "Zinc and Human Health: An Update." *Archives of Toxicology* 86 (4): 521–34. doi:10.1007/s00204-011-0775-1.

Ezzati, M. 2017. "Worldwide trends in body-mass index, underweight, overweight, and obesity from 1975 to 2016: a pooled analysis of 2416 population-based measurement studies in 128.9 million children, adolescents, and adults." *Lancet* 390:2627–2642.

FDA. 2017. "Nutrition Information for Raw Vegetables." https://www.fda.gov/food/food-labeling-nutrition/nutrition-information-raw-vegetables.

Herforth, Anna, Mary Arimond, Cristina Álvarez-Sánchez, Jennifer Coates, Karin Christianson, and Ellen Muehlhoff. 2019. "A Global Review of Food-Based Dietary Guidelines." *Advances in Nutrition* 10 (4): 590–605. doi:10.1093/advances/nmy130.

InformedHealth.org [Internet]. Cologne, Germany: Institute for Quality and Efficiency in Health Care (IQWiG); 2006-. How can I get enough iron? 2014 Mar 20 [Updated 2018 Mar 22]. https://www.ncbi.nlm.nih.gov/books/NBK279618/

IQWiG (Institute for Quality and Efficiency in Health Care). 2018. "How Can I Get Enough Calcium?" Institute for Quality and Efficiency in Health Care (IQWiG). https://www.ncbi.nlm.nih.gov/books/NBK279330/.

Järup, Lars, Marika Berglund, Carl Gustaf Elinder, Gunnar Nordberg, and Marie Vahter. 1998. "Health Effects of Cadmium Exposure – A Review of the Literature and a Risk Estimate." *Scandinavian Journal of Work, Environment and Health* 24 (SUPPL. 1): 1–51. https://www.jstor.org/stable/40967243?casa_token=3M86MLw_VrAAAAAA:sWEIllIiT2HRjzvt0SNL-4qH1HJQsx1OZQ011uyJs6aD8Xj9sHZNzROMaA5XV4Zg3_SdVHELJ-GYzVaTs3TmDfVolMl1LTzdcCfuQsWZ6z0-IsEvBcU.

Leake, Jonathan R., Andrew Adam-Bradford, and Janette E. Rigby. 2009. "Health Benefits of 'grow Your Own' Food in Urban Areas: Implications for Contaminated Land Risk Assessment and Risk Management?" *Environmental Health: A Global Access Science Source* 8 (SUPPL. 1): S6. doi:10.1186/1476-069X-8-S1-S6.

Li, Min, Yingli Fan, Xiaowei Zhang, Wenshang Hou, and Zhenyu Tang. 2014. "Fruit and Vegetable Intake and Risk of Type 2 Diabetes Mellitus: Meta-Analysis of Prospective Cohort Studies." *BMJ Open* 4 (11): 5497. doi:10.1136/bmjopen-2014.

Miller, Victoria, Andrew Mente, Mahshid Dehghan, Sumathy Rangarajan, Xiaohe Zhang, Sumathi Swaminathan, Gilles Dagenais, et al. 2017. "Fruit, Vegetable, and Legume Intake, and Cardiovascular Disease and Deaths in 18 Countries (PURE): A Prospective Cohort Study." *The Lancet* 390 (10107): 2037–49. doi:10.1016/S0140-6736(17)32253-5.

Needleman, Herbert L. 1988. "The Persistent Threat of Lead: Medical and Sociological Issues." *Current Problems in Pediatrics* 18 (12): 703–44. doi:10.1016/0045-9380(88)90004-7.

Niñez, V. 1987. "Household Gardens: Theoretical and Policy Considerations." *Agricultural Systems* 3 (23): 167–86. https://www.sciencedirect.com/science/article/pii/0308521X87900643.

Noia, Jennifer Di. 2014. "Defining Powerhouse Fruits and Vegetables: A Nutrient Density Approach." *Preventing Chronic Disease* 11 (6). Centers for Disease Control and Prevention (CDC). doi:10.5888/pcd11.130390.

Park, Yikyung, David J. Hunter, Donna Spiegelman, Leif Bergkvist, Franco Berrino, Piet A. Van Den Brandt, Julie E. Buring, et al. 2005. "Dietary Fiber Intake and Risk of Colorectal Cancer: A Pooled Analysis of Prospective Cohort Studies." *Journal of the American Medical Association* 294 (22): 2849–57. doi:10.1001/jama.294.22.2849.

Pelletier, David L., Edward A. Frongillo, Dirk G. Schroeder, and Jean-Pierre Habicht. 1995. "The Effects of Malnutrition on Child Mortality in Developing Countries." *Bulletin of the World Health Organization* 73 (4). World Health Organization: 443–48.

Rabin, Jack, Gladis Zinati, and Peter Nitzsche. 2012. "Yield Expectations for Mixed Stand, Small-Scale Agriculture." *Sustainable Farming on Urban Fringe* 7 (1): 1–4. https://njaes.rutgers.edu/pubs/urbanfringe/pdfs/urbanfringe-v07n01.pdf.

Satterthwaite, David, Gordon McGranahan, and Cecilia Tacoli. 2010. "Urbanization and Its Implications for Food and Farming." *Philosophical Transactions of the Royal Society B: Biological Sciences*. Royal Society. doi:10.1098/rstb.2010.0136.

Sazawal, Sunil, Robert E. Black, Venugopal P. Menon, Pratibha Dinghra, Laura E. Caulfield, Usha Dhingra, and Adeep Bagati. 2001. "Zinc Supplementation in Infants Born Small for Gestational Age Reduces Mortality: A Prospective, Randomized, Controlled Trial." *Pediatrics* 108 (6): 1280–86. doi:10.1542/peds.108.6.1280.

Sommer, Alfred, and Keith P. West. 1996. *Vitamin A Deficiency Health, Survival, and Vision*. New York: Oxford University Press.

Weaver, Connie M. 2013. "Potassium and Health." *Advances in Nutrition* 4 (3). Oxford University Press (OUP): 368S–77S. doi:10.3945/an.112.003533.

WHO. 2001. *Iron Deficiency, Anaemia Assessment, Prevention, and Control. A Guide for Programme Managers*. Geneva, Switzerland. https://www.scirp.org/(S(351jmbntvnsjt1aadkposzje))/reference/ReferencesPapers. aspx?ReferenceID=1279392.

Woese, Katrin, Dirk Lange, Christian Boess, and Klaus Werner Bögl. 1997. "A Comparison of Organically and Conventionally Grown Foods-Results of a Review of the Relevant Literature." *Journal of the Science of Food and Agriculture* 74: 281–93.

7 Improving Human Health by Remediating Polluted Soils

Darryl D. Siemer
Retired Idaho National Laboratory

CONTENTS

7.1 INTRODUCTION

Soil contamination (or pollution) is usually considered to be caused by the presence of xenobiotics (human-made) chemicals or other alteration in the natural soil environment. It is typically due to industrial activity, agricultural chemicals, or improper disposal of waste. The concern stems primarily from health risks due to direct contact with contaminated soils, contaminants in food produced by them, and from secondary contamination of water supplies both within and underlying them. However, there are other instances in which the culprit is Mother Nature herself—some land is polluted by natural causes.

Humanity's contribution to soil contamination is both extensive and extremely expensive. For example, roughly 100,000 km² of China's cultivated land have been polluted, with contaminated water being used to irrigate a further 21,670 km² and another 1,300 km² either totally covered or destroyed by solid wastes (Qi 2007). In total, it accounts for one-tenth of China's cultivatable land and is situated mostly in economically developed areas. An estimated 12 million tonnes of grain are contaminated by heavy metals every year, causing direct losses of 20 billion yuan ($2.57 billion USD).

The number of estimated "potentially contaminated" sites in the European Union exceeds 2.5 million (Panagos et al. 2013), and the positively identified sites number around 340,000. Municipal and industrial wastes contribute the most (38%), followed by the industrial/commercial sector (34%). Mineral oil and heavy metals are the main contaminants contributing around 60% to soil contamination. In budgetary terms, the management of contaminated sites currently costs around 6 billion Euros (€) annually.

A recent review of the USA's soil/groundwater contamination issues (NRC 2013) concluded that …"Recent estimates by the U.S. Environmental Protection Agency (EPA) indicate that expenditures for soil and groundwater cleanup at over 300,000 sites through 2033 may exceed $200 billion (not adjusted for inflation), and many of these sites have experienced groundwater impacts."

While there is a great deal of both technical and popular discussion of soil pollution examples, damage mechanisms, and consequences, fewer words have been devoted to describing practical solutions to such problems.

There are three basic approaches to dealing with polluted soils, namely...

- isolation/landfill transportation which cannot be categorized as remediation, as it does not remove the pollutants from the soil, change its physicochemical characteristics, or return it to a productive state.
- solidification/stabilization: Solidification encapsulates the waste to form a solid material—again such work usually renders the soil useless. Stabilization converts contaminants to less soluble, mobile, or toxic forms.
- pollutant removal: This group of technologies can be subdivided as follows:
  - physicochemical methods which remove, extract, or transform the pollutant via physicochemical procedures.
  - thermal methods which heat contaminant up to temperatures high enough to destroy or immobilize it.
  - biological methods which utilize the living organisms (plants, fungus, and bacteria) to degrade or accumulate the contaminants.

Regarding the physicochemical methods, they can be divided into (i) physical soil washing, which exploits the differential physical properties (size, density and magnetism) of the soil itself and between the soil and the pollutants, and (ii) chemical soil washing, which employs chemical agents such as acids, bases, solvents, or surfactants. In the 1st case, the aim of the process is the reduction of the total volume of polluted soil, in the later the complete removal of the pollutants (Boente et al. 2018).

This chapter will briefly describe several "novel" ways to either prevent pollution or return soil to a sustainable and productive state.

The main thing I learned while researching this project is that like most of the mankind's other "technical" problems, bureaucratic inflexibility and other ignoble manifestations of "human nature" constitute both the primary cause of soil pollution and the most significant "barriers to science" (NAP 1996) facing anyone attempting to mitigate them. Basically, what needs to happen is that societal rules, laws, and customs must change so that "doing the right thing" becomes much easier than it is now. In this arena, that means...

1. enabling the people/industry involved in the manufacture of cement to utilize virtually all "organic" wastes (worn-out tires, plastics, solvents, hospital wastes, etc.) as fuel,
2. rendering it easy/cheap for farmers to amend their already-polluted soils with powdered natural minerals, primarily basalt,
3. incentivizing the repair/recovery/recycle of materials/things whenever doing so actually makes good sense,
4. looking for ways that two or more of the things that we must do could complement each other (e.g., a phosphate fertilizer plant could also be a cement plant).

7.2   SOIL POLLUTANT PROBLEMS

There are basically just two kinds of pollutants that might end up in soils, organic and inorganic. The former comprise "soft" inherently complex, covalently bonded materials (most antibiotics including, pesticides/herbicides, oils, solvents, plastics, etc.) consisting primarily of carbon, hydrogen, and oxygen often with smaller amounts of phosphorous, nitrogen, sodium, potassium, iron, calcium along with even lesser amounts of various trace minerals. Inorganic wastes comprise everything else: metals, aqueous solutions of salts, acids, bases, etc., and solid industrial slags and dusts. Some wastes, for example, worn-out rubber tires, computers, and paint chips, contain both organic and inorganic materials.

Organic wastes are intrinsically combustible and, if not overdiluted with too much water, will burn completely in air/oxygen with net energy release to form CO_2, N_2, and water vapor plus relatively small amounts of sterile inorganic ash; i.e., serve as a fuel. While inorganic wastes are generally incombustible, many of them are rendered far less troublesome if subjected to the same conditions that serve to burn organics. For example, toxic and/or radioactive element-containing nitrate/nitrite salt solutions and chromium-containing tanning wastes are converted to far smaller volumes of a relatively innocuous and easily sequestered or otherwise treatable mineral ash.

7.2.1 PREVENTION

7.2.1.1 Combustibles (Organics)

A couple of weeks ago, I answered a QUORA question that pretty much sums up my opinions about today's "plastic pollution crisis" (below).

There's currently a tremendous amount of handwringing going on about how plastic pollution is destroying our environment (especially the oceans) and we're not really doing anything about it.

Let's look at this "terribly difficult" problem the way that a chemist, engineer, or properly informed governmental decision maker should.

Total world plastic production is currently about 380 million tons per year, approximately one tenth of which (~40 million tons) is the relatively tough-to-burn chlorinated plastics - mostly polyvinyl chloride (PVC) [https://en.wikipedia.org/wiki/Polyvinyl_chloride]/

In 1975/1976 the Canadian government implemented a large scale chlorinated hydrocarbon (including PVC plastic) incineration demonstration at the St Lawrence Cement Co in Missaugua, Ontario

(Burning Waste Chlorinated Hydrocarbons in a Cement Kiln, a GOOGLABLE 1978 EPA report (EPA 1978)). Those materials contained up to 46 wt% chlorine. They were destroyed with >99.98% efficiency and no high molecular wt. chlorocarbons (e.g. dioxins) were detected in the off gas. As expected, the amount of kiln dust produced mostly a mix of sodium and potassium chloride salts often used as fertilizer increased in stoichiometric proportion to the amount of chlorine fed to the kiln. Chlorine fed was up 0.8 wt% of the clinker produced. Burning that waste reduced the amount of fossil fuels required to make the cement clinker and improved its quality (less alkali).

Global cement production is now about 4 billion tons [https://en.wikipedia.org/wiki/Cement]. A typical Portland cement is about 62 wt% CaO (molecular wt (MW)=56 g/mole) meaning that the limestone calcined to produce it contained about 1.95 billion tons [4*1E+9*0.62*44/56] of carbon dioxide (MW=44 g/mole) all of which was dumped into the atmosphere. A typical large, wet process rotary cement kiln, fitted with drying-zone heat exchangers, produces about 680 tons of clinker per day? and burns 0.25-0.30 tons of coal fuel per ton clinker to do so.

Assuming the highest carbon and highest heating value coal (anthracite essentially 100% C, - MJ/kg heating value), the amount of coal burned to satisfy the world's cement demand would be ~1.0 billion tons [0.25*4E+9] the combustion of which would generate 3.67 (1.0*44/12) billion tons of CO_2 which is also dumped into the atmosphere.

Most plastics exhibit about the same heating value as crude oil (~ 42 MJ/kg) meaning that if 100% of the plastics made/consumed per year were to be burned in the world's cement kilns rather than dumped into the oceans or landfills, the amount of coal burned to make cement would be reduced by nearly 50% (0.38*42/33 = 0.485 billion tons).

Because plastic has a somewhat higher heating value per carbon atom as coal, that substitution would also reduce the total amount of CO_2 generated/dumped per ton of cement.

What' so tough about deciding how we should address plastic pollution other than our all-too "human nature"?

A point not made in my "answer" is that the chief issue having to do with plastics recycling is the cost of sorting and cleaning up that sort of garbage—a cement kiln does not care about "clean" or exactly what sort of combustible stuff is being burned to generate its process heat.

An European report (EPA 1978) written 46 years ago indicated that the methods then employed for plastic disposal included…

1. illegal dumping of barrels or other containers on uncontrolled refuse dumps;
2. deposition of larger quantities in barrels on refuse dumps which are supposedly sanctioned for purpose;
3. combustion in simple facilities without hydrogen chloride scrubbing;
4. combustion of piles of barrels on remote beaches with an offshore wind;
5. dumping of barrels on the open sea;
6. dumping liquids into the sea from moving vessels;
7. separation of waste materials and recovery of useful components;
8. combustion with the recovery of hydrochloric acid; and
9. combustion on the open sea at temperatures guaranteeing almost complete pyrolysis.

That report concluded that "only the last three approaches can be considered to be not harmful for the environment." Unfortunately, the "best" ones, numbers 7 and 8, were (and still are) quite expensive and, in some cases, "unsustainable for the producer." Number 9's weaknesses include its reliance upon the "solution to pollution is dilution to a commons" principle as far as its combustion products are concerned and that combustion at sea requires "extensive observation of a variety of safety procedures."

Another study (French 1974) indicated that a variety of means both legal and illegal were being used to dispose of the especially tough-to-burn chlorinated hydrocarbon wastes. Of the illegal means, "discharge of what in general are insignificant quantities is disposed of in drums, or by tankers, into waterways, former quarries now used for other purposes, or discharged with unsupervised wastes that reason suggests should be retained."

Unfortunately, some of those means are still being employed for all sorts of wastes, especially ocean dumping.

Compared with the other disposal method that is usually considered environmentally correct for combustible wastes, cement kiln combustion differs in that…

- dedicated waste incinerators usually operate with flame temperatures of 1,200°C–1,560°C (EPA 1974) while cement kiln flame temperatures are on the order of 2,100°C–2,300°C (Weber 1964);
- both gasses and solids are retained for longer times within a cement kiln's hot zones than is normally the case with incinerators;
- in order to produce "good" cement clinker, extremely high temperatures are absolutely necessary, thereby eliminating the necessity for constant temperature monitoring as is the case with conventional waste incinerators;
- there is always more than sufficient lime (CaO) within a cement kiln to react with acidic gasses such as hydrogen chloride and thus prevent their emission to the atmosphere;
- burning wastes in a cement kiln saves fossil fuels, thereby incinerating within them rather than a typical dedicated incinerator that reduces both the total amount of fuel consumed and the amount of CO_2 emitted to the atmosphere.

A more up-to-date European report (Belgium 2019) concluded that "For the vast majority of environmental impacts, it is concluded that using industrial wastes as alternative fuel in cement production is better for the environment than treating them in waste incinerators." Its authors also observed that "Overall, more waste than fossil fuel was used in the six cement kilns considered in 2006 (in tons)."

A typical rotary kiln (Figure 7.1) consists of a roughly 100-m long, 3-m diameter tube made from thick steel plate and lined with firebrick. The tube slopes slightly (1°–4°) and slowly rotates

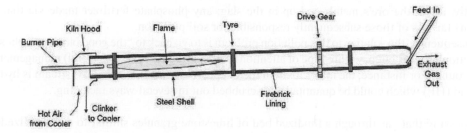

FIGURE 7.1 Cement kiln schematic (Wikipedia commons).

on its axis at between 30 and 250 revolutions per hour. Raw mix (mostly limestone ground up with a high silica clay and an iron source—typically iron ore) is fed in at the upper end, and the kiln's rotation causes it gradually to move downhill to the other end of the kiln. At that end fuel, in the form of gas, oil, or a pulverized solid fuel (most often coal), is blown in through the "burner pipe," producing a large concentric flame in the lower part of the kiln tube. As material moves under the flame, it reaches its peak temperature, before dropping out of the kiln tube into the product "clinker" cooler. Air is drawn first through the cooler and then through the kiln for combustion of the fuel. In the cooler, the air is heated by the cooling clinker, so that it may be 400°C–800°C before it enters the kiln, thus causing intense and rapid combustion of the fuel. Many such kilns also have "tire feeders" situated about halfway along the tube into which waste materials can be dropped upon each revolution.

The inorganic ash resulting from the combustion of waste within cement kilns becomes cement, which in turn ends up in concretes. Concretes are themselves often used for immobilizing toxic and radioactive waste streams because doing so represents the most affordable way to treat them in a way that protects both people and the environment (IAEA 2013). Cement's properties, both chemical and physical, make "solidification/stabilization" with it a good way to achieve the encapsulation of such wastes. Chemically, it has a high pH (e.g., its pore water) and forms hydration products which favor ionic sorption and substitution. Physically, concretes are durable solids with low permeability which facilitates safe transportation and storage. Cement is also an inexpensive and readily available material, fluid when initially cast, durable in its hardened state, which immobilizes virtually anything within almost any sort of ash. Cements have also been proved to be stable when irradiated and to act as radiation shielding which explains why cementitious materials have been employed in radioactive waste management for many decades.

7.2.1.2 Inorganic Toxins (Metals)

Phosphorus fertilizers have received a good deal of study/attention, because in addition to their nutrient elements (P with or without fixed N and K), they often contain substantial amounts of trace-level impurities including uranium, arsenic, copper, chromium, vanadium, nickel, cadmium, lead, and mercury (Oyedele et al. 2006; Alloway 1990; Singh 1994; Mortvedt et al. 1981). According to Thomas et al. (2012), "average" raw phosphate rock ore contains 11, 25, 1888, 32, 10, and 239 mg kg^{-1} of As, Cd, Cr, Cu, Pb, and Zn, respectively.

The reason for this is that most of the world's fertilizer-type phosphate is currently produced by the acid (usually sulfuric) leaching of phosphate rock, not by the old "dry" process (Swann 1922). The former dissolves up the phosphate rock's metals too which then, with the exception of calcium because it forms insoluble gypsum, accompany the phosphoric acid product throughout the rest of the process. On the other hand, the old dry process utilized an electrically heated "blast furnace" which converted a mix of powdered phosphate rock ore, iron ore, and coke into gaseous elemental phosphorous, liquid ferrophosphorus (a valuable by product), and a molten glass-like calcium silicate slag. The gaseous elemental P along with the CO, CO_2, HF, etc. was then generally mixed with air which oxidized it to solid P_2O_5 that is simple to separate from the other gasses. Because, virtually, all of the ore's metals end up in the slag; any phosphate fertilizer made via that route contains far less of those subsequently responsible for soil pollution.

Consequently, the solution to this pollution problem is to go back to "the good old days'" phosphate production process but pay more type of attention to waste (both gaseous and solid) management this time around. For instance, the major toxin in the old process' post P-removal gas stream is hydrofluoric acid (HF) which could be quantitatively scrubbed out in several ways including…

- passing that gas through a fluidized bed of limestone granules similar to those utilized to scrub SOx from coal-/oil-fired power plant off gasses (produces fluorite);
- wet scrubbing by passing it up through a sieve plate or packed absorption column countercurrent to an aqueous solution of sodium hydroxide, sodium carbonate, or sodium bicarbonate (produces sodium fluoride).

That process' solid waste, slag, should not be considered a waste because it would be easy, cheap, and sensible to consider it as "cement clinker" instead. Phosphate blast furnace slag is essentially the same thing as iron blast furnace slag which means that if it were to be treated in the properly—rapidly cooled, powdered, and mixed with an activator (e.g., lime or sodium silicate)—it could serve the same purposes as does ordinary Portland cement (Criado et al. 2017).

Of course, one of the biggest drawbacks of the "dry" process is that it requires a lot more electricity to implement than does today's sulfuric acid-based process. That is one of the reasons why it is no longer used in areas like Idaho whose demand for electricity now exceeds that which can be provided by its now-old hydroelectric dams.

7.2.1.3 Biotoxins

One of the most genuinely serious soil pollution issues is the fact that domestic animals are routinely dosed with much cheaper lots of the same new "cutting edge" antibiotics developed to treat humans. Those animals serve as test beds for bacterial evolution which results in their (the bacteria) inevitably discovering/developing/passing on biological pathways that render those antibiotics useless (Tasho and Cho 2016). Because manures and urine are often used as fertilizer, some of this natural microorganism "research and development" also occurs in soils receiving those amendments. One way to mitigate the soil-related part of that human health threat would be to convert all such animal wastes to "biochar" and "bio-oil" rather than use it as a fertilizer. Another way would be to heavily ozonate such material (pass ozone through it) to destroy the bacteria and (hopefully) the antibiotics accompanying it before applying it to the soil (EPA 1999). The former would likely destroy that wastes' value as a nitrogen fertilizer but might generate enough "biofuel" to be at least be energy-neutral. The second suggestion simply represents blue sky theorizing but might prove useful if electricity was to become cheap enough to generate the necessary ozone.

7.3 ALREADY POLLUTED SOIL REMEDIATION

7.3.1 Phytoremediation

Soils that have already been polluted are more difficult to deal with due to the fact that large-scale entropy reversal (separation or "unmixing") is intrinsically costly. Fortunately, there are two more-or-less practical approaches that can turn the trick.

The first of these is "phytoremediation"—raising some sort of crop that pulls pollutants up and out of the soil and then removing it (Violante et al. 2010). Pentavalent arsenic is a phosphate analog (AsO_4^{-3}) and is therefore readily transported in plant tissues and cells. It is bioaccumulative in many organisms, marine species in particular. In polluted areas, plant growth may be affected by the root uptake of arsenate, where in such regions, uptake of the more toxic arsenite ion found more particularly in reducing conditions like those in flooded rice paddies or any other poorly drained soil.

Rice (*Oryza sativa* L.) is the staple food for the people of arsenic endemic South (S) and South-East (SE) Asian countries. In that region, arsenic-contaminated groundwater has been used not only for drinking and cooking purposes but also for rice cultivation. Irrigation with such water results in high concentrations in the rice eaten by humans and the straw and husk consumed by their domesticated animals. Arsenic concentrations in that grain are typically an order of magnitude higher than in such water as an average ground water in those areas contains 50 ppb arsenic and the mean rice consumption by adult humans in those areas is between 400 and 650 g day^{-1}; their intake of arsenic via that route averages about at 0.20–0.35 mg day^{-1} (Azizur et al. 2008). Assuming a daily consumption of 4 L, their intake via drinking water is about 0.2 mg day^{-1}. Furthermore, cooking rice with arsenic-contaminated water also increases its arsenic burden.

Similarly, a recent Japanese study (Takahashi et al. 2016) points out that the accumulation of Cd in rice grains is a widespread problem. Phytoremediation is one of the most effective methods for reducing soil Cd levels in paddy fields, and rice is a promising candidate for doing it. In a large-scale field trial, the average Cd extraction using two varieties of rice were about 22 mg m^{-2}, and one year's worth of growth/harvesting reduced soil Cd concentrations by about 15%.

To investigate whether phytoremediation could reduce As uptake by Asia's staple food crop (rice), several studies utilizing the As "hyperaccumulator" Pteris vittata have been undertaken (Ling et al. 2011) (P. vittata, commonly known as the Chinese brake, Chinese ladder brake, or simply ladder brake, is a fern in the Pteridoideae subfamily of the Pteridaceae). It is indigenous to Asia, southern Europe, tropical Africa, and Australia and is sometimes cultivated in gardens for its attractive appearance or used in pollution control schemes. Over a 9-month period, it removed 3.5%–11.4% of the total soil As and decreased phosphate-extractable As and soil pore water As by 11%–38% and 18%–77%, respectively. Rice grown following P. vittata had significantly lower As concentrations in both straw and grain, being 17%–82% and 22%–58% of those in the control, respectively (Tu and Ma 2002). Phytoremediation also results in significant changes in As speciation in rice grain by greatly decreasing the concentration of dimethylarsinic acid. In two soils, the concentration of inorganic As in rice grain was decreased by 50%–58%. Those results demonstrate effective stripping of bioavailable As from contaminated paddy soils, thus reducing rice As uptake. Pteris vittata removed 3.5%–11.4% of the total As from five contaminated paddy soils.

Of course, as any such crops are therefore rendered "poisonous," the safest thing to do with them is to utilize them as a cement or electrical power plant's "biofuel" (if the plant ash is genuinely toxic, a cement plant would be preferable). Unfortunately, as many of the people who rely almost solely upon rice for their food are poor, they cannot afford to either stop raising it or using it as a biofuel.

A recent student-study by Pennsylvania State University (Bowe et al.?) was conducted with the intent to evaluate the remediation of abandoned mine lands within that state via utilization of several of the crops often touted for use as biofuels. It encompassed a green house and three different plant species: *Camelina sativa*, *Miscanthus sinensis*, and Switchgrass (*Panicum virgatum*). Camelina (Brassicaceae), commonly known as wild flax, is an important biofuel crop due to its 30%–40% oil by seed weight and a tolerance to cold and drought allowing it to grow on marginal agricultural lands, while increasing soil health (Janick 2007). Miscanthus is a rhizomatous plant that is another important biofuel crop due to its rapid growth, low mineral content, and its "massive" ethanol production. It also purportedly "yields more than five hundred percent of the energy" (?) which renders it able to outperform corn (*Zea mays*) in terms of biomass production (Lewandowski et al. 2000). To many biofuel enthusiasts, switchgrass is especially magical—it can apparently grow anywhere and does not need any of the things or care normally associated with crop production (maybe we could grow it on the moon where it would not compete so much with food and fiber production).

Anyway, that study concluded that "not all of the crops were suitable for metal uptake" and that…

- Miscanthus was the most suitable for phytoremediation
- Switchgrass exhibited "potential" for phytoremediation
- Camelina was the least suitable for phytoremediation

However, another conclusion that I would certainly draw from its data as published is that none of those plants removed a *significant* fraction of the metals in question (arsenic, barium, cadmium, chromium, lead, mercury, selenium, and silver) from the soil per season (growth cycle) with the possible exception of barium.

7.3.2 "TREATMENT" WITH POWDERED BASALT

The other approach is to add something to the soil that is not only apt to be beneficial in other ways but also renders toxic pollutants/ions less harmful. To me, anyway, the most suitable such amendment is simply powdered basalt.

The following statement has been cut and pasted from one of the best-written soil science papers I have seen so far (Hassan and El Toney 2014):

> *With unique 92+ essential elements, consisting of a broad range of trace minerals along with small amounts of macro elements, the soil nutrients which have been slowly lost through the ages by erosion, leaching and farming are gently regenerated. Rock dust not only improves soil vitality, but also increases plants' overall health while strengthening immunity and pest resistance. Resulting in a completely natural and disease free final product, unique in flavor and mineral composition.*

Their paper goes on to show that powdered basalt also mitigates another issue that is becoming more prevalent throughout much of the world, the oversalination of underirrigated (usually) soils. The reason for this is that freshly powdered basalt exhibits a huge ion exchange capacity that serves to "pull" excess ions (e.g., Na^+) out of soil solutions. Because most toxic metals are multivalent, their affinity to a soil's cation exchange sites tends to be much greater than that of the alkalis (most sodium) responsible for oversalination's deleterious effects upon plant growth.

Another property imparted to so-amended soils is that basalt weathering tends to raise the pH of soil solutions (Silva et al. 2005), which, in turn, tends to lower the solubility and therefore bioavailability of many toxic and radioactive elements.

Another plus for utilizing such a "natural" soil amendment (powdered basalt is much like volcanic ash) is that its weathering simultaneously retains (sequesters) plant/soil-generated CO_2 that would otherwise "respire" back to the atmosphere. The weathering of enough "Snake River Plain Basalt" [10.06 wt% CaO and 7.65 wt% MgO='s 7.35 milliequivalents $[0.1006 \times 2/(40+16)+0.0765 \times 2/(24.32+16)]$ worth of base per gram] to meet the future's agricultural phosphorous (and potassium) demand that would remove about 4.1 GT of carbon dioxide from the atmosphere per year—about 11% of today's anthropogenic CO_2 emission rate (Siemer 2020). A good deal of theorizing has been done about how the "enhanced weathering" of mafic rocks could address global warming (see Schuiling and Krijgsman 2006 and Hartmann et al. 2013), but very little realistic experimental work has been done to date. Thankfully, that subject is beginning to receive more attention (Taylor et al. 2017) and some hopefully realistic experimental studies have apparently begun (Beerling et al. 2018).

Assuming that "clean" greenhouse gas-free electricity eventually becomes as cheap as it should ne (Siemer 2019), another advantage of this scheme is that the basalt powder required to supply the crops' phosphorous, potassium, etc. (typically about 10 Mg (tons) ha^{-1} year^{-1}) will become "dirt cheap"; i.e., considerably less expensive to make/distribute than is the fixed nitrogen required to fertilize the same land (Siemer 2020).

A final advantage is that the addition of sufficient basalt powder to feed the crops would eventually replace/restore huge amounts of topsoil that is already lost to erosion due to mankind's often destructive farming practices (Pimentel and Burgess 2013).

Unlike most of the metals that are considered to be toxic, arsenic tends to be anionic in soil solutions, i.e., present as either arsenate or arsenite anions of which the latter, more reduced trivalent form, is more harmful. Both of those anions are ready absorbed by one of the most common forms of iron in nature, ferric hydroxide (rust), which suggests that another useful soil amendment in thusly affected regions would be powdered hematite (Fe_2O_3) or goethite (FeO(OH)). Drinking water passed through a bucket of sand mixed with the same minerals would be rendered "safer" to drink.

7.3.3 Desalinated Water Soil Washing

In regions where the soil is already oversalinized or contains too much of a readily leached inorganic toxin such as arsenic, pumping ground water through a desalination system and then back into the ground either directly or indirectly by irrigating fields with it would remediate it. The downside of this approach to in situ "soil washing" is that it would require a great deal of electrical power—certainly more than do most of the other schemes that have been investigated (Dermont et al. 2008). State-of-the-art desalination systems are primarily based upon reverse osmosis which typically requires about 3 kWh worth of electrical power to produce one cubic meter of deionized water. Assuming that so-washing land would require one-meter's worth of water to be spread over it, the amount of such water required per hectare would be 10,000 m³, which, in turn, would require 30,000 kWh of electrical energy to produce. If that electricity was to cost 3US cents per kWh (the approximate current wholesale cost of nuclear power), this scheme's energy cost would be $900/ha/cycle—fairly high but certainly worth it if it would render such soil capable of producing human food in a sustainable fashion.

7.4 DISCUSSION

The first thing that anyone should do to address a "technical" issue is to quit exacerbating it. For instance, since…

- fossil fuel CO_2 is responsible for global warming (Hansen 2008),
- energy-poor young people (especially males) tend to be both unhappy and war-like,
- the fossil fuels – especially petroleum – powering today's civilization (>80%) will become too rare/expensive to serve that purpose within another century,
- wind and solar power are and will remain ineluctably unreliable,
- enough batteries to render such sources sufficiently reliable to power a civilization like today's would be impossibly expensive, and…
- breeder reactors are "renewable" (Cohen 1983)…

…we should do whatever we can to quickly implement a 20–30 TWe "nuclear renaissance" utilizing breeder reactors so that we can kick our addiction to fossil fuels (Hubbert 1956). With regard to this book's topics, that means adopting policies that encourage potential polluters to see to it that their waste ends up being properly dealt with—either recycled to useful purposes or rendered innocuous—not just dumped into a river, ocean, or onto the ground. For example, individuals and businesses should be able to put anything that is burnable into a single waste container whose contents will be collected, baled up, and shipped off to the nearest cement factory. Big "contributors" to such fuel streams should be rewarded, not subjected to the sorts of paperwork nightmares and excessive costs usually associated with such hazardous (in places like the state of California, anyway) waste disposal.

This brings us to the second thing that anyone should do when faced with a "new" problem; i.e., familiarize yourself with what has already been done to deal with similar problems and is therefore already available. Most of the world's technical issues already have perfectly reasonable solutions—what has been lacking is the will to recognize, identify, and then implement them.

A television interviewer once asked me, "But what about nuclear waste? Will it not poison the whole biosphere and persist for millions of years?" I knew this to be a nightmare fantasy wholly without substance in the real world... One of the striking things about places heavily contaminated by radioactive nuclides is the richness of their wildlife. This is true of the land around Chernobyl, the bomb test sites of the Pacific, and areas near the United States' Savannah River nuclear weapons plant of the Second World War. Wild plants and animals do not perceive radiation as dangerous, and any slight reduction it may cause in their lifespans is far less a hazard than is the presence of people and their pets... I find it sad, but all too human, that there are vast bureaucracies concerned about nuclear waste, huge organizations devoted to decommissioning power stations, but nothing comparable to deal with that truly malign waste, carbon dioxide.

(Lovelock 2006)

As did I, James Lovelock started out as an analytical researcher endeavoring to push the detection limits for toxic substances ever lower. His signature invention, the electron-capture detector, permitted the detection of refrigerant-type chlorocarbons in the stratosphere—a feat that resulted in the policy changes constituting the single greatest success of the modern environmental movement—"closing the ozone hole." However, it also permitted the detection of chlorinated pesticides in foods, water, etc., at levels far lower than could possibly hurt anyone or anything (*"All things are poison, and nothing is without poison; the dosage alone makes it so a thing is not a poison.* Paracelsus 1538 AD"). The consequences of those policy changes plus his conclusions about nuclear power eventually caused him to break ranks with most of the people spearheading the world's "environmental" movements.

During the past 150 years, the Earth itself has become a "commons" undergoing a series of man-made tragedies (Hardin 1994). This chapter began with what I feel to be the proper way to go about dealing with the world's pollution problems. Like the root causes of those problems, they mostly have to do with human behavior—our culture must be changed so that people are incentivized to do things that are good for everyone both now and in the future—not just whatever immediately most-benefits each individual. (Ayn Rand was wrong – there's nothing "right" about such selfishness).

The necessary changes include the following:

1. Enable the people/industries involved in the manufacture of cement or excessively eroded or trace mineral depleted (or lime) to utilize virtually all "organic" wastes (worn out tires, plastics, solvents, hospital wastes, etc.) as kiln fuel.
2. Render it easy/cheap for farmers to amend their already-polluted soils with powdered natural minerals primarily basalt. This can only happen if reliable (not "intermittent") "clean" energy becomes much cheaper than it is now.
3. Incentivize the repair/recovery/recycle of materials/things whenever doing so actually makes good sense—not just because it it's politically correct.
4. Look for ways that two or more of the things that we do could complement each other and thereby reduce waste (e.g., a phosphate fertilizer plant could also become a blast furnace slag cement plant).
5. As the root cause of much of the world's pollution is humanity's collective failure to render fossil fuel-based energy more expensive than clean energy, we should adopt James Hansen's suggestion that carbon emissions should be heavily taxed and those funds should be rebated back to the public on a per capita basis (Hansen 2009).

A final thing that anyone facing such issues should do is to familiarize themselves with what has already been done to deal with similar problems. Perfectly reasonable solutions to most of the world's technical problems have already been identified—what has been lacking is the collective will to actually deal with them. One of the reasons for this is that most of the people employed to deal with such problems are either in the "research" (study) business (most of the scientists and engineers) or do not want to "rock the boat" (upset people) any more than they absolutely have to (most politicians and governmental program managers). Both of those human characteristics incentivize "experts" to ignore what should be obvious solutions and, instead, turn every new "technical" mole hill into a mountain requiring the performance of a great deal of cutting-edge research before any serious changes happen. This is the reason that the US Department of Energy's "legacy" nuclear fuel reprocessing waste treatment efforts at both its Idaho and Hanford sites' morphed into interminable multibillion dollar boondoggles (Siemer 2019).

Another characteristic of human nature is that especially "important" people desire to remain important. This means that when agencies established to address certain obvious safety issues have succeeded in that mission, their leadership tends to keep ratcheting up criteria in order to keep their jobs/agencies relevant. For instance, when I and my fellow analytical research chemists during the 1970s and 1980s made it possible to detect hazardous substances (in my case, toxic metals via graphite furnace atomic absorption spectrometry) at progressively lower concentrations in foods, water, or the "environment," the statutory limits for such pollutants tended to drift in the same direction, "to provide an extra margin of safety" concentration (and, of course, to keep the regulators, inspectors, researchers, and analysts busy). This tendency often goes well beyond reasonable levels and thereby generates a great deal of anxiety and, in the case of radionuclides, waste, fraud, and taxpayer-abuse.

For instance, an abandoned elemental phosphorus plant situated ten miles west of Pocatello ID is shadowed by a roughly 30 million Mg mini-mountain of dry-process phosphate rock slag that the USA's Environmental Agency (EPA) decided to deem too radioactive to utilize for any constructive purpose (Gesell 2013). Prior to that decision that slag (which is very much like slowly cooled cement clinker) had been widely used for road and parking lot construction, construction fill, railroad ballast, home foundations, driveways, and even built up-type roofing materials. The "alarm level" imposed upon the people owning, living in, or trying to sell such property (20 mrem h^{-1} or 0.00175 Sievert/a) represents ~25% of the radiation dose that an average American citizen gets from everything if he/she were to live in else he/she does and far less than he/she would experience from natural background radiation levels in cities like Ramsar (Iran) or the southwest Indian state of Kerala (Russell 2015). Sections of Kerala experience an average of 70 mSv year^{-1}, with some areas as high as 500 mSv year^{-1} and locally grown food averages five times more radioactive than that consumed in the United States. Despite its excessive background radiation levels, Kerala's cancer incidence is the same as that of greater India, about one-half that of Japan's and under a third that of Australia's. The actuaries who studied that issue stated that "Cancer experts know a great deal about the drivers of these huge differences, and radiation is not on the list."

There are a number of well-written papers and books freely available to anyone wishing to learn more about how unreasonable conservative assumptions lead to hugely destructive consequences (Henriksen 2015; Calabrese 2013; Cuttler 2014; Scott and Tharmalingam 2019; Russell 2014). It seems that everyone has read them except the officials responsible for establishing regulatory limits.

The elements responsible for most of the background radiation (uranium, thorium, and potassium) render "advanced" life on earth is possible because they generate the heat necessary for two of its unique features: plate tectonics and a still molten and therefore magnetic iron/nickel core (Ward and Brownlee 2000).

The same sort of "precautionary principle"-driven bureaucratic overreach that has succeeded in crippling the USA's nuclear power "option" has also officially rendered a good deal of its drinking water "unsafe" because a substantial fraction of its groundwater contains enough arsenic to exceed the EPA's "suggested" 10 ppb (part per billion) limit—prior to 2001 that limit had been 50 ppb. Affected regions include Southwestern states like Nevada, to the upper Midwest and New England, where a belt of arsenic-infused bedrock taints aquifers in stretches from the coast of Maine to a point midway through Massachusetts. The studies serving to rationalize that hugely impacting (too expensive) suggestion were performed in Bangladesh where well-water arsenic levels are typically higher than 50 ppb, and more importantly, such water is used to raise a single crop (rice) comprising its citizen's and their domesticated animals' staple food under conditions uniquely well suited for its uptake (Rahman and Hasegawa 2011). The people genuinely affected by that problem are too poor to quit raising rice—the USA's citizens are not so poor and have far more varied diets.

7.5 CONCLUSION

All of the problems I have discussed would be easier to prevent than remediate after they have occurred.

There is a reasonable chance that that our descendants circa 2100 AD will live in a world that has effectively addressed soil pollution as well as the other technical issues that currently seem so threatening. However, any such success will require "radical" changes that many of the world's most influential citizens, businessmen, politicians, and even some of its scientists will resist. It is up to the people who read books like this to do whatever we can to help make those changes happen, not just raise more issues and insist upon performing more studies.

REFERENCES

Alloway, B.J. 1990. *Toxic Metals in Soil-Plant Systems*. John Wiley & Sons, Chichester, UK.

Azizur, R.M, Hasegawa H, Mahfuzur Rahman M, Mazid Miah MA, Tasmin A. 2008 February. Arsenic accumulation in rice (Oryza sativa L.): human exposure through food chain. *Ecotoxicol Environ Safety*, 69(2): 317–24. Epub 2007 March 7.

Beerling, D.J., Leake, J.R., Long, S.P., et al. 2018. Farming with crops and rocks to address global climate, food and soil security. *Nature Plants*, 4: 138–147 (also see http://lc3m.org/news/).

Belgium. 2019. Waste Treatment: Tackling industrial waste Cement kilns versus Incinerators. http://www.coprocessing.info/www.coprocessing.info/en/waste-treatment/index.html.

Boente, C., Sierra, C., Rodríguez-Valdés, E., Menéndez-Aguado, J.M., Gallego, J.R. 2018. Mineral processing technologies for the, remediation of soils polluted by trace elements. *2nd International Research Conference on Sustainable Energy, Engineering, Materials and Environment (IRCSEEME)*, Mieres, Spain, 25–27 July 2018.

Bowe, T., Matthew, R., Shuler, A., Gerst, E. Phytoremediation of Camelina, Switchgrass, and Miscanthus in mine land soils of PA. (Conference presentation) no date https://harrisburg.psu.edu/content/tylerreuminelandpdf.

Calabrese, E.J. 2013. How the US National Academy of Sciences misled the world community on cancer risk assessment: new findings challenge historical foundations of the linear dose response. *Arch Toxicol*. doi:10.1007/s00204-013-1105-6. Available at: http://link.springer.com/article/10.1007/s00204-013-1105-6.

Cohen, B.L. 1983. Breeder reactors: a renewable energy source. *Am J Phys*, 51: 75.

Criado, M.L., Ke, X., Provis, J.L., Bernal, S.A. 2017. Alternative inorganic binders based on alkali-activated metallurgical slags. In *Sustainable and Nonconventional Construction Materials using Inorganic Bonded Fiber Composites*, pp. 185–220. Elsevier Academic Press Inc. https://doi.org/10.1016/B978-0-08-102001-2.00008-5

Cuttler, J.M. 2014. Remedy for radiation fear – discard the politicized science. *CNS Bull*, 34(4). https://www.ncbi.nlm.nih.gov/pmc/articles/PMC4036393/.

Dermont, G, Bergeron, M, Mercier, G, Richer-Laflèche M., 2008 March 21. Soil washing for metal removal: a review of physical/chemical technologies and field applications. *J Hazard Mater*, 152(1): 1–31. Epub 2007 October 22.

EPA. 1974. United States Environmental Protection Agency, Permit No. 730 D008C (3) to Shell Chemical Company, Inc. and Ocean Combustion Services, B.V. December 12, 1974.

EPA. 1978. Burning waste chlorinated hydrocarbons in a cement kiln. https://nepis.epa.gov/Exe/ZyPURL.cgi?Dockey=9101XE14.TXT.

EPA. 1999. Wastewater technology fact sheet ozone disinfection, PA 832-F-99-063. https://www3.epa.gov/npdes/pubs/ozon.pdf.

French. 1974. Incineration of Industrial Chlorine Wastes on the High Seas, Report from the Environmental Agency (Ministere charg de 1 'Environment) of the Pollution and Nuisance Prevention Administration (Direction de la Prevention des Pollutions et Nuisances), France. 1974.

Gesell, T. 2013 (Professor of Health Physics). *Current Implementation of the Graded Decision Guidelines (Phase 2)* March 21, 2013. http://fmcidaho.com/wp-content/uploads/Slag-Presentation-by-Dr-Tom-Gesell.pdf.

Hansen, J. 2008. Tipping point: Perspective of a climatologist. In *State of the Wild 2008–2009: A Global Portrait of Wildlife, Wildlands, and Oceans*. E. Fearn, Ed. Wildlife Conservation Society/Island Press, NewYork, NY, pp. 6–15.

Hansen, J. 2009. "Carbon Tax & 100% Dividend vs. Tax & Trade", testimony to the Committee on Ways and Means, United States House of Representatives, 25 February 2009, http://www.columbia.edu/~jeh1/2009/WaysAndMeans_20090225.pdf.

Hardin, G. 1994. The tragedy of the unmanaged commons. Trends Ecol Evol, 9(5): 199. doi:10.1016/0169-5347(94)90097-3. PMID 21236819.

Hartmann, J., West, A.J., Renforth, P., et al. 2013. Enhanced chemical weathering as a geoengineering strategy to reduce atmospheric carbon, supply nutrients, and mitigate ocean acidification. Rev Geophys, 51. https://www.researchgate.net/pub.

Hassan, M.S., El Toney, M. 2014. Increasing productivity of bean by healing salinity of soils. Asian J Agricul Food Sci, 2(1). ISSN: 2321–1571 https://pdfs.semanticscholar.org/7935/7264bbb78efa10a2b7fef59d165e8f7a6df4.pdf.

Henriksen, T. 2015. Radiation and health. The University of Oslo medical and biophysics group. http://www.efn-usa.org/2013-11-06-07-30-02/item/777-radiation-and-health-2015-thormod-henriksen-and-biophysics-group-at-uio-norway.

Hubbert, M K, Nuclear Energy and the Fossil Fuels, Shell Publication no. 95, 1956 https://web.archive.org/web/20190610010030/http://www.hubbertpeak.com/hubbert/1956/1956.pdf

IAEA. 2013. The behaviours of cementitious materials in long term storage and disposal of radioactive waste. https://www-pub.iaea.org/MTCD/Publications/PDF/TE-1701_web.pdf. Lewandowski, C.-B., Scurlock, J.M.O., Huismanst, W. 2000. Miscanthus: European experience with a novel energy crop. Biomass Bioenerg, 9: 209–227. ISSN 0961–9534. http://www.sciencedirect.com/science/article/pii/S0961953400000325.

Ling, Y., Asaduzzaman Khan, M., McGrath, S.P., Zhao, F.-J. 2011 December. Phytoremediation of arsenic contaminated paddy soils with Pteris vittata markedly reduces arsenic uptake by rice. Environ Pollut, 159(12): 3739–3743. doi:10.1016/j.envpol.2011.07.024.

Lovelock, J. 2006. The Revenge of Gaia: Why the Earth Is Fighting Back – and How We Can Still Save Humanity, Penguin Books (Some editions of this book have a different, less optimistic subtitle: "Earth's Climate Crisis and the Fate of Humanity".).

Mortvedt, J., Mays, D.A, Osborn, G. 1981. Uptake by wheat of cadmium and other heavy metal contaminants in phosphate fertilizers. J Environ Quality, 10: 193–197.

NAP. 1996. Barriers to Science: Technical Management of the Department of Energy Environmental Remediation Program. The National Academies Press. https://www.nap.edu/catalog/10229/barriers-to-science-technical-management-of-the-department-of-energy.

NRC. 2013. Alternatives for Managing the Nation's Complex Contaminated Groundwater Sites. National Research Council, National Academies Press, Washington DC 423 pages.

Oycdele, D.J., Asonugho, C., Awotoye, O. 2006. Heavy metals in soil and accumulation by edible vegetables after phosphate fertilizer application. Electron J Environ Agricult Food Chem, 5(4): 1446–1453.

Panagos, P., Liedekerke, M.V., Yigini, Y., Montanarella, L. 2013. Contaminated sites in Europe: review of the current situation based on data collected through a European network. J Environ Public Health, 2013: 1–11. doi:10.1155/2013/158764. ISSN 1687–9805. PMC 369739714.

Pimentel, D., Burgess, M. 2013. Soil erosion threatens food production. Agriculture, 3(3): 443–463. doi:10.3390/agriculture3030443. http://www.mdpi.com/2077-0472/3/3/443.

Qi, X. 2007. Facing up to "invisible pollution", Soil contamination has grown unnoticed across China's landscape. https://www.chinadialogue.net/article/show/single/en/724-Facing-up-to-invisible-pollution.

Rahman, M.A., Hasegawa, H. 2011 October 15. High levels of inorganic arsenic in rice in areas where arsenic-contaminated water is used for irrigation and cooking. Sci Total Environ, 409(22): 4645–4655. doi:10.1016/j.scitotenv.2011.07.068. Epub 2011 September 6.

Russell, G. 2015a. What can we learn from Kerala? (Guest essay posted on Professor Barry Brook's BRAVE NEW CLIMATE environmental blogsite), http://bravenewclimate.com/2015/01/24/what-can-we-learn-from-kerala/.

Russell, G. 2014. Greenjacked! The Derailing of Environmental Action on Climate Change, ISBN 9-780980-656114 (kindle book). Amazon.com.

Schuiling, R.D., Krijgsman, P. 2006. Enhanced weathering: an effective and cheap tool to sequester CO2. Climate Change, 74(1–3): 349–354. http://www.innovationconcepts.eu/res/literatuurSchuiling/enhanced.pdf.

Scott, B.R., Tharmalingam, S. 2019 March 1. The LNT model for cancer induction is not supported by radiobiological data. Chemico-Biological Interact, 301: 34–53. doi:10.1016/j.cbi.2019.01.013 (one of several papers published in this "special issue" devoted to reexamining the LNT hypothesis).

Siemer, D.D. 2019. *Nuclear Power: Policies, Practices, and the Future.* John Wiley & Sons and Scrivener Publishing LLC. (a book describing the whys and hows of a "nuclear renaissance" implemented with breeder-type reactors.).

Siemer, D.D. 2020. *Nuclear Powered Agriculture: Fertilizers and Amendments*, chapter 12 of an upcoming volume entitled, "Soil and fertilizer" in Taylor and Francis's "Soil Science and Plant Nutrition" book series, Rattan Lal, Editor.

Singh, B.T. 1994. Trace element availability to plants in agricultural soils with special emphasis on fertilizer inputs. *Environ Rev*, 2: 133–146.

Silva, J.A., Hamasaki, R., Paull, R., Ogoshi, R., Bartholomew, D.P., Fukuda, S., Hue, N.V., Uehara, G., Tsuji, G.Y. 2005. Lime, gypsum, and basaltic dust effects on the calcium nutrition and fruit quality of pineapple. *Acta Horticulturae*, 702(702). https://pdfs.semanticscholar.org/8fb5/2093d48370c8969cd41483f d74cd2ef8b323.pdf.

Swann, T. 1922 July 1. Manufacture of phosphoric acid in the electric furnace by the condensation and electric precipitation method. *J Ind Eng Chem*, 1922147630. doi:10.1021/ie50151a018.

Takahashi R., Ito, M., Katou, K., Sato, K., Nakagawa, S., Tezuka, K., Akagi, H., Kawamoto, T. 2016. Breeding and characterization of the rice (Oryza sativa L.) line "Akita 110" for cadmium phytoremediation, *Soil Sci Plant Nutr*, 62(4): ICOBTE 2015.

Tasho, R.P., Cho, J.Y. 2016 September 1. Veterinary antibiotics in animal waste, its distribution in soil and uptake by plants: a review. *Sci Total Environ*, 563–564: 366–376. doi:10.1016/j.scitotenv.2016.04.140. Epub 2016 April 30.

Thomas, E.Y., Omueti, J.A.I., Ogundayomi, O. 2012. The effect of phosphate fertilizer on heavy metal in soils and Amaranthus caudatus. *Agricul Biol J North Am*. ISSN Print: 2151–7517, ISSN Online: 2151–7525, doi:10.5251/abjna.2012.3.4.145.149© 2012, ScienceHuβ, http://www.scihub.org/ABJNA.

Tu, C., Ma, L.Q. 2002. Effects of arsenic concentrations and forms on arsenic uptake by the hyperaccumulator ladder brake. *J Environ Qual*, 31(2): 641–647.doi:10.2134/jeq2002.6410.

Violante, A., Cozzolino, V., Perelomov, L., Caporale, A.G., Pigna, M. 2010. Mobility and bioavailability of heavy metals and metalloids in soil environments. *J Soil Sci Plant Nutr*, 10(3): 268–292. doi:10.4067/ S0718-95162010000100005.

Taylor, L.L., Beerling, D.J., Quegan, S., Banwart, S.A. 2017 April. Simulating carbon capture by enhanced weathering with croplands: an overview of key processes highlighting areas of future model development. *Biol Lett*, 13(4). PMC5414688. https://www.ncbi.nlm.nih.gov/pmc.

Ward, P., Brownlee, D.E. Rare Earth: Why Complex Life Is Uncommon in the Universe is a Copernicus Publications, 2000.

Weber, P. 1964. Alkali problems and Alkali elimination in heat-economising dry-process rotary kilns (Translation). *Zement-Kalk- Gips*, 8: 335.

8 Managing Soil for Global Peace by Eliminating Famines and Pandemics

Rattan Lal
The Ohio State University

CONTENTS

8.1 INTRODUCTION

8.1.1 STATE-OF-THE-WORLD SOILS

Hunger and malnutrition affect human health, and they, along with desperateness caused by soil degradation, are serious threats to global peace and stability (Lal 2015). As much as one-third of the agricultural land area of the world is degraded (FAO and ITPS 2015), and this is aggravating the threat to world peace. The most dominant processes of soil degradation are erosion (by water and wind), compaction, depletion of soil organic carbon, nutrient/elemental imbalance, and salinization (Oldeman 1994). Proxy methods also indicate a severe problem of soil degradation (Bai et al. 2008) with strong ecological, economic, and social consequences (Nkonya, Mirzabaev, and von Braun 2016). Soil health has also strong impacts on human health (Kemper and Lal 2017). Sustainable soil management is essential to provisioning of basic ecosystem services, or ESs (i.e., supporting, provisioning, regulating, and cultural services) (MEA 2004). However, essential ESs are jeopardized by ten threats to soil functions as outlined by FAO and ITPS (2015). These ten threats (i.e., physical, chemical, biological, and ecological; Figure 8.1) are exacerbated by the interaction between land misuse/soil mismanagement and anthropogenic climate change. The primary driver of these

143

threats to soil functions is the growing human population along with its increasing affluence, the insatiable demands on natural resources, and weak governance (Figure 8.1). Indiscriminate use of agricultural inputs (i.e., excessive tillage, flood irrigation, fertilizers, pesticides) along with the use of inappropriate practices (i.e., removal or in-field burning of residues) aggravates rate and impact of these ten threats to soil functions.The permanent solution to the dilemma is returning some managed land back to nature.

8.1.2 SOIL AND HUMAN HEALTH

Prominent among the threats of soil degradation is food and nutritional insecurity, with adverse impacts on the environment (i.e., quality of soil, water, and air) and the health of livestock and human beings (Lal 2020a). Several mechanisms through which soil degradation can directly impact livestock and human health include the following (FAO and ITPS 2015): (i) entrance of toxic elements and contaminants from soils into the food chain, (ii) direct exposure to pathogens and pests, (iii) nutrient-deficient food produced from crops and livestock, and (iv) dust inhalation. Indirectly, soil degradation affects human health through hunger and hidden hunger (Figure 8.2). The severity of degradation of soil biological health may have increased with increase in the dependence on chemical fertilizers and a regressive decline in the use of organic amendments as source of plant nutrients (Figure 8.3). Degradation in soil biological health affects human health through the quantity and nutritional quality of the food produced. Degradation-induced decline in soil quality and functionality has severe consequences to human health and well-being. Crop failures, caused by extreme climate events along with other biotic and abiotic stresses, have affected human well-being by causing famines and the attendant consequences. Food based on crops and animals raised on degraded soils are often deficient in protein, micronutrients, and vitamins (Lal 2009).

The objective of this chapter is to discuss the importance of soil and its sustainable management to advance global peace and human well-being through the elimination of famines and pandemics. This viewpoint is based on the collation of literature and synthesis of the available information in a simple language that the general public, farmers, land managers, and policymakers can understand and relate to.

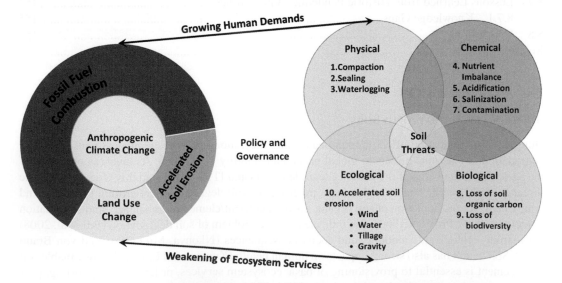

FIGURE 8.1 Ten threats to soil functions are aggravated by anthropogenic climate change which in turn is driven by fossil fuel combustion, land use conversion (deforestation), and erosion-induced emission of multiple gases (CO_2, CH_4, and N_2O).

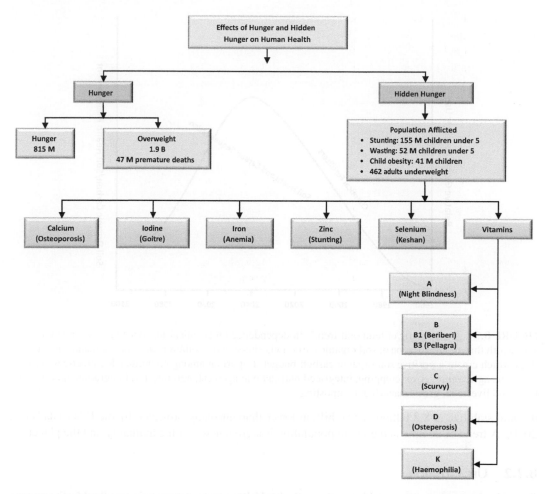

FIGURE 8.2 Some common human diseases associated with deficiencies of minerals and vitamins. (Stats and related information from FAO et al. 2019; WHO 2020; Roser and Ritchie 2020; Griffiths and Philippot 2013; Awuchi, Ikechukwu, and Echeta 2020; John Hopkins Medicine 2020.)

8.2 POPULATION

8.2.1 GLOBAL TRENDS

The world population of 7.7 B in 2019 will increase to 8.5 B by 2030 (+10%), 9.7 B by 2050 (+26%), and to 10.9 B by 2100 (+42%) (UN 2019). Almost all of the future growth in world population will occur in developing countries. The highest increase in population will occur in Sub-Saharan Africa (SSA), where the population is expected to double by 2050, and it is presently growing at the rate of 2.7% year⁻¹ compared with the growth rate of 1.2% year⁻¹ for South Asia and 0.9% year⁻¹ for Latin America. Developing countries of Asia, Africa, and Latin America account for 80% of the present world population in 2020, and they represent 99% of the global growth (PRB 2000). Furthermore, the highest growth rates of the population are observed in the poorest countries, and especially those in SSA (PRB 2000). Almost all of this annual population growth of 70 million is happening in the less developed countries of Asia, SSA, and Latin America, where the soil resources are already degraded, water resources are finite, and there is a large prevalence of undernourished and malnourished people (FAO et al. 2019). However, a recent report has indicated that the expected growth in population at the end of the 21st century may be lower than projected (Cilluffo and Ruiz 2019), and

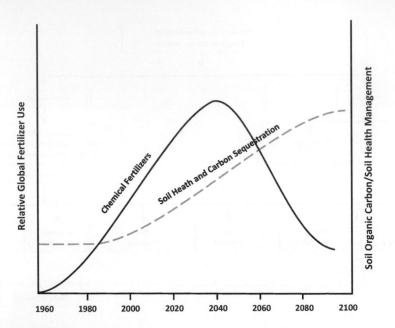

FIGURE 8.3 A schematic of temporal trends in dependence on chemical fertilizer vis-à-vis restoration of soil health through restoration of soil organic carbon via site-specific soil/crop/animal/tree management practices which create a positive soil organic carbon budget. Important among recommended practices are conservation agriculture, cover cropping, integrated nutrient management, agroforestry, integration of crops with trees and livestock, green manuring, composting, etc.

it could stabilize at 8.8 billion or two billion lower than currently projected by the U.N. (Gladstone 2020). A trend of decline in the world population is a great news for the humanity and the planet.

8.2.2 URBANIZATION

Several urban centers appeared in southern (Indus Valley) and western Asia and the Mediterranean regions prior to the Christian era. However, organized urbanized societies developed during the 9th and 12th centuries (Davis 1955). The rate of urbanization has been rapid globally since the 1800s and has not yet peaked (Table 8.1). Nonetheless, both the 20th and 21st centuries are the eras of urbanization, which is the driving force of the Anthropocene and its ramifications. Urbanization has strong implications to food and nutritional security, as well as on future demands for energy, water,

TABLE 8.1
World Total, Urban, and Rural Population

	Population (B)			Population (%)	
Year	Total	Urban	Rural	Urban	Rural
1950	2.54	0.75	1.79	29.6	70.4
1970	3.70	1.35	2.35	36.6	63.4
1990	5.33	2.29	3.04	43.0	57.0
2018	7.63	4.22	3.41	55.3	44.7
2030	8.55	5.17	3.38	60.4	39.6
2050	9.77	6.68	3.09	68/4	31.6

Source: Adapted from UN (2019).

and land area needed for habitat and infrastructure. Urban encroachment will strongly compete with agriculture for land and water resources.

In comparison with 30% of the world population living in urban centers in 1930, 55% is urbanized in 2018, and this is projected to be 68% by 2050 (Table 8.2, UN 2019). The most urbanized regions in 2018 include 82% for North America, 81% for Latin America and the Caribbean, 74% for Europe, and 68% for Oceania (UN 2019). The degree of urbanization in 2018 is 50% for Asia and 43% for Africa. However, Asia and Africa have the highest rate of urbanization. Almost 90% of the future growth in urbanization, 2.5 B people by 2050, will occur in Asia and Africa. Projected increases in the urban population between 2018 and 2050 will be 416 M in India, 255 M in China, and 189 M in Nigeria (UN 2019).

Megacities, cities with a population of ≥ 10 million people, have been phenomena since the middle of the 20th century. In 2014, there were 30 megacities in the world, housing 558 M people (8.2% of the population) compared with 261 M (6.4% of the global population) in 1975. By 2010, 460 million people lived in megacities (6.7% of the world population lived in 27 megacities) (Kennedy et al. 2015) and 757 M people (11% of the world population) resided in 101 largest cities of the world (Hoornweg and Pope 2016). The number of megacities was 2 in 1950 (London and New York), 4 in 1960, 8 in 1970, 15 in 1980, 20 in 1990, 28 in 2000 (Guest 1994), and they are projected to be 42 in 2030 (U.N. 2014). Tremendous amount of resource use in megacities, comprising 9% of global electricity, 10% of gasoline, and 13% of solid waste, pose a major environmental challenge (Kennedy et al. 2015). A megacity of 10 million requires 6,000 tons of food per day. The list of the 20 largest cities by 2100 (Table 8.3) shows that all but two (Manila and Cairo) are situated in South Asia and SSA, where natural resources (i.e., soil, water, air, vegetation, and biodiversity) are degraded, the environment polluted, and the incidence of hunger and malnutrition is the highest.

8.3 FAMINES

8.3.1 GLOBAL HISTORY

Famines, lack of access to an adequate amount of safe and nutritious food resulting in adverse effects on human health and often leading to premature and unnatural death, have plagued humanity throughout the recorded history. Simply put, famine refers to a widespread scarcity of nutritional food (Kelly 1992). A famine is defined as a food crisis that causes an elevated mortality over a specific period of time. It is called a "great famine" when it causes 100,000 or more deaths and a "catastrophic famine" when the death toll is 1 M or more (Devereux 2009). It entails a widespread

TABLE 8.2

Number of Years It Took to Reach 1 Billion for World Total and Urban Population

World Population (B)	Year When It Reached	Years to Increase by 1 B	Urban Population	Year When It Reached	Years to Increase by 1 B
1	1804	-	1	1959	-
2	1927	123	2	1985	26
3	1959	32	3	2002	17
4	1974	15	4	2015	13
5	1986	12	5	2025	13
6	1998	12	6	2041	13
7	2010	12			
8	2023	13			
9	2037	14			

Source: Adapted from UN (2019).

TABLE 8.3
Population of 20 Largest Cities by 2100

Location	Population by 2100 (M)
Lagos, Nigeria	88.3
Kinshasa, DRC	83.5
Dar Es Salaam, Tanzania	73.7
Mumbai, India	67.2
Delhi, India	57.3
Khartoum, Sudan	56.6
Niamey, Niger	56.1
Dhaka, Bangladesh	54.3
Kolkata, India	52.4
Kabul, Afghanistan	50.3
Karachi, Pakistan	49.1
Nairobi, Kenya	46.7
Lilongwe, Malawi	41.4
Blantyre City, Malawi	40.9
Cairo, Egypt	40.5
Kampala, Uganda	40.1
Manila, Philippines	40.0
Lusaka, Zambia	37.7
Mogadishu, Somalia	36.4
Addis Ababa, Ethiopia	35.8

Source: Adapted from Desjardins (2018).

---

lack of food leading directly to excess mortality from starvation or hunger-induced malnutrition (Geber and O'Donnabhain 2020).

The first famine during the Roman empire was recorded during 441 BCE. In another era, drought killed millions of Mayan people due to famine and thirst in Mesoamerica during 800–1000 A.D. The Kanki famine from 1230 to 1231 killed 2 M in Japan. The Great Famine of 1315–1317 killed 7.5 M in Europe. The Russian famine of 1601–1603 killed 2 M (Wikipedia 2020).

There is a long history of famine in India. The famous epic of Ramayana (5th century BCE) narrates a drought in Ayodhya (Northern India) that involved plowing of a field by King Janak and leading to discovery of a baby girl in an urn. The baby, goddess Sita (also called Sia), was the heroine of the Ramayana and married Prince Rama, which was avatar of the Lord Vishnu. The history of famines in India and elsewhere in Asia includes several important events: The Deccan famine (afflicting southern India) of 1702–1704 that killed 2 M, the Great Bengal Famine of 1769–1773 that killed 10 M, the Chalisa famine (1783–1784), and the Deji Bara or Skull Famine (1789–1793) that killed 11 M each (Wikipedia 2020). There is an equally tragic history of recurring famines in China.

In Europe, the decade of 1845–1850 was termed the "Hungry Forties" (Vanhaute, Paping, and Gráda 2006). The decade also included the Great Irish Famine of 1848. Prominent famines since 1900 are shown in Table 8.4. There were man-made famines in Europe during the 1940s. Chronological records of famines since 1860 are presented by the World Peace Foundation (2020) and vividly described by Hasell and Roser (2017). Whereas famines have mostly disappeared from Europe and Asia, they have persisted in Africa, especially in Ethiopia, Malawi, and Niger during the 2000s. The annual rate of people dying from famine (Table 8.5) mostly reflects the problem of poor governance and civil strife.

TABLE 8.4

Prominent Famines since 1900 (CNN Editorial Research 2020)

Year	Country	Deaths (10^6)
1921–1922	Soviet Union	9
1927	China (Northwest)	3–6
1929	China (Hunan)	2
1932–1933	Soviet Union (Ukraine)	7–8
1943	China (Henan)	3–5
1943	India (Bengal)	2.1–3
1946–1947	Soviet Union	2
1959–1961	China	5–30
1974	Bangladesh	1.5
1975–1979	Cambodia	1.5–2
1984–1985	Ethiopia	1
1991–1993	Somalia	3 (affected)
1995–1999	North Korea	2.5
1998–2011	Sudan, Darfur, S. Sudan	>2 (affected)
2008	Somalia	3.7 (affected)
2011 (20 July)	Somalia (Southern)	
2011 (5 September)	Somalia	
2013 (2 May)	Somalia	0.26
2017 (20 February)	South Sudan	4.9 (affected)

TABLE 8.5

Annual Rate of People Dying from Famine Per Decade between 1860 and 2010

Decade	Total Famine Victims (10^6)	Annual Rate of People Dying (per 100,000)
1860s	4.1	30.0
1870s	20.4	142.2
1880s	3.0	19.5
1890s	9.9	60.4
1900s	5.3	31.2
1910s	2.9	14.9
1920s	16.0	82.1
1930s	12.2	56.5
1940s	18.6	78.6
1950s	8.9	32.2
1960s	16.8	50.0
1970s	3.4	8.4
1980s	1.3	2.8
1990s	1.5	2.6
2000s	2.8	4.3
2010–2016	0.26	0.5

Source: Adapted from Hasell and Roser 2013 to 2017.

Grada, C.O. 2007. Making Famine History. *J. Economic Literature* 45(1): 5–38.

8.3.2   Causes of Famines

Most famines are preventable and are attributed to human-related causes and poor governance. Preventing famine is also in accord with Sustainable Development Goal (SDG) #2 of the United Nations or the Agenda 2030. The World Peace Foundation (2019, 2020) outlined the following steps to avert global famines: (i) minimizing risks of wars, both between and within nations; (ii) improving and sustaining humanitarian assistance; (iii) strengthening international law to make the right-to-food inviolable; (iv) eliminating poverty as per SDG #1 of the U.N.; (v) reducing wealth inequality within and between nations; (vi) decreasing vulnerability to climate change and its impact on agronomic production; (vii) balancing the world's food systems in favor of food staples; and (viii) making famines and mass starvation politically intolerable, morally toxic, ethically unthinkable, and humanely unacceptable.

Historically, wars and totalitarianism have been responsible for the most famines (Gráda 2007). On the contrary, economic growth and advances in medical science have reduced the threat of mass starvation since the middle of the 20th century. Incidences of famine have also been reduced by food markets and the globalization of disaster relief (Gráda 2007). Above all, a dramatic improvement in agronomic production since the 1960s (the Green Revolution Era) has increased global per capita food production despite the rapid growth in population. It is this agronomic success story that is responsible for "making famine history," and it is popularly called the "Borlaug Effect."

Recent trends in hunger shown in Table 8.6 indicate that despite the global quantum jump in agricultural production, there are pockets of low agronomic yields in regions prone to drought and those where scientific knowledge has not been translated into action. Anthropogenic perturbations and changes in human values have adversely affected human health and well-being (Figure 8.4). Prevalence of undernourishment (Tables 8.7 and 8.8) indicates that regions where research has not been translated into action include Sub-Saharan Africa (SSA), South Asia (SA), Central America and the Caribbean, and parts of the Andean regions. Famines and mass starvation can be averted by improving agronomic productivity through adoption of proven scientific technologies. Political unrest, civil strife, and war are important factors that aggravate the risks of famine in Africa. Kiros and Hogan (2001) observed that child mortality is the highest among those born to illiterate parents and that parental education is critical to reducing child mortality during famine periods.

TABLE 8.6
Temporal Trends in Global Hunger between 1990 and 2020

Year	Global Hunger[a]	
	Millions	% of the Total Population
1990	1011	19.0
1995	989 (5,278.6)	
2000	900	14.5
2005	945 (6,513)	
2010	821	11.8
2015	784	10.6
2016	804 (7,426)	
2017	821	10.8
2018	822	10.8
2019	821 (7,577)	
2020	690	

Source: Adapted from FAO et al. (2019, 2020) and The World Counts (2020).
[a]   COVID-19 could push an additional 83-132 million into chronic hunger.

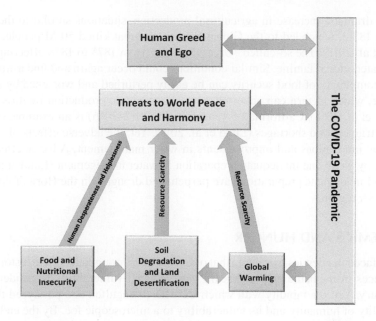

FIGURE 8.4 Human-induced challenges facing humanity.

TABLE 8.7

Prevalence of Undernourished People in the World (CNN Editorial Research 2020; FAO et al. 2019, 2020)

Year	Total (10^6)	% of the World Population
1990–1992	1,010.6	18.6
2005	947.2	14.5
2010	822.3	11.8
2015	785.4	10.6
2016	796.5	10.7
2017	811.7	10.8
2018	821.6	10.8
2020	690	8.9

TABLE 8.8

Percentage of Population Affected by Undernourishment (CNN Editorial Research 2020; FAO et al. 2019)

Year	Africa	Asia	Latin America and Caribbean	Oceania	Other
1990–1992	27.6	23.6	14.7	15.7	< 2.5
2005	21.2	17.4	9.1	5.5	< 2.5
2010	19.1	13.6	6.8	5.2	< 2.5
2015	18.3	11.7	6.2	5.9	< 2.5
2016	19.2	11.5	6.3	6.0	< 2.5
2017	19.8	11.4	6.5	6.1	< 2.5
2018	19.9	11.3	6.5	6.2	< 2.5

Despite the dramatic increase in agricultural production, situations similar to the 19th century drought (1876–1878), which led to the Global Famine and that killed 50 M people, could happen again (Singh et al. 2018). The so-called "Great Drought" from 1875 to 1878 affecting Asia, Brazil, and Africa created global famine. Similar conditions could occur again and undermine global food security. The complexity of food security can be easily perturbed and worsened by anthropogenic climate change, which in turn can adversely impact agronomic production by direct and indirect factors (Yadav et al. 2018). California's 5-year drought (2012–2016) is an example of such a risky event that can trigger food shortages (Lund et al. 2018). Yet, the adverse effects of severe drought were averted by innovations and improvements in water management. Adverse effects of drought are aggravated by weak and inadequate preparation in water management (Lund et al. 2018). Such weaknesses and inadequate preparation have perpetuated droughts in the Horn of Africa and elsewhere in the SSA region.

8.4   PANDEMICS AND HUNGER

Humanity has faced pandemics, similar to that of COVID-19, throughout human history (Figure 8.5). Modern advances in medicine and healthcare have reduced vulnerability to pandemics since the mid-20th century. Yet, the rapidity with which COVID-19 engulfed and paralyzed the world indicates the fragility of humanity and its vulnerability to a microscopic foe. By the end of September 2020, more than 30 million had been infected by COVID-19, and ~1 million had died globally. Yet, the tragedy is on the increase not only in developing countries (e.g., India and Brazil) but also in the US, and there is no end in sight. It is evident, however, that vulnerability of a community to pandemics depends on the prevalence of hunger and malnutrition, as is sadly the case in developing countries even in 2020 (FAO et al. 2019, Kretchmer 2020).

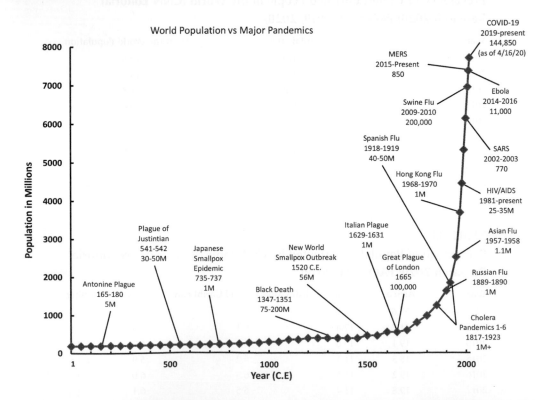

FIGURE 8.5   Relation between the frequency of world pandemics and increase in human population. (Adapted from Wikipedia (2020) and Roser, Ritchie, and Ortiz-Ospina (2020).)

Despite advances in agriculture and increases in food production, hunger and hidden hunger remain the biggest risks to human health worldwide (Table 8.5). The absolute number and the percentage of total population that is prone to hunger dropped between 1990 and 2015, but there is an increase in prevalence of hunger since 2015 (Table 8.7). Around 9 million people die every year of hunger and hunger-related diseases (The World Counts 2020). A child dies from hunger every 10 s with a total child mortality caused by hunger estimated at 3.1 million/year (The World Counts 2020). Achieving the Sustainable Development Goal (SDG) #2 (Zero Hunger) by 2030 is a daunting challenge. The adverse effects of pandemics (i.e., COVID-19) are aggravated by the poor standards of living and the lack of adequate hygienic facilities. The adverse effects of pandemics on one hand is related to lack of resources on the other, and both are mutually reinforced.

8.4.1   Global History

Global pandemics (e.g., AIDS, bubonic plague, influenza, malaria, smallpox, and the Spanish flu) have reportedly killed 300–500 million people over the last 12,000 years. The COVID-19 (coronavirus disease of 2019), started in December 2019 in Wuhan, China, infected >30 million by the end of September 2020 and killed ~1 million. The ten most serious global pandemics are outlined in Table 8.9. Whereas infectious diseases have existed ever since the evolution of humans, epidemics/pandemics were aggravated through the development of communities (e.g., urbanization) made possible by the onset of settled agriculture about ~12,000 years ago. Several pandemics have also changed the course of world history (History.com Editors 2020), because diseases struck them as these civilizations arose. In addition to those listed in Table 8.9, other pandemics (History.com Editors 2020) were a fever similar to typhoid that struck Athens, Greece 430 B.C., Cyprian Plague in 250 A.D., Leprosy in Europe in the 11th century A.D., and the 1492 Columbian diseases that were passed on by the Europeans to the native populations. Pandemics of the 19th century included the First Cholera Pandemic (1817), the Third Plague Pandemic (1855), Fiji Measles Pandemic (1875), SARS (2003), the Swine Flu or H1N1 (2009–2010), West African Ebola (2014–2016), and the Zika virus (2015) (Jarus 2020). Just as the onset of settled agriculture and the development of organized communities (i.e., urbanization) created epidemics, easy intercontinental transport has aggravated the frequency and intensity of the pandemics. Rapid spread of COVID-19 is an example of the impact of high mobility across intercontinental borders. Population mobility increases virus spread quickly. Urbanization and air travel are fueling the COVID-19 pandemic.

TABLE 8.9
Ten Most Deadly Historic Pandemics

Circa (A.D.)	Pandemic	Cause	Estimated Deaths (M)
165	Antonine Plague	Unknown	5
541–542	Plague of Justinian	Bubonic Plague	25
1346–1353	The Black Death	Bubonic Plague	200
1852–1860	Third Cholera Pandemic	Cholera	1
1889–1890	Flu Pandemic	Influenza	1
1910–1890	Sixth Cholera Pandemic	Cholera	0.8
1918	Flu Pandemic	Influenza	20-50
1956–1958	Asian Flu	Influenza	2
1968	Flu Pandemic	Influenza	1
2005–2012	HIV/AIDS	HIV/AIDS	36
2019–?	COVID-19	Coronavirus	?

Source: Adapted from MPHOnline (2020)

8.4.2 POPULATION AND PANDEMICS

The 21st century is the era of rapid urbanization (Tables 8.1–8.3). Urbanization and the attendant demographic changes can strongly increase vulnerability of human beings to pandemics such as COVID-19. Urbanization also affects the epidemiology of infectious diseases (Neiderud 2015). Urbanization, and the attendant increased demand for food and other victuals, is leading to a rapid intensification of agriculture, socio-economic alterations, and ecological fragmentation (Hassell et al. 2017). These drastic transformations in landscape and the environment impact on epidemiology of infectious diseases have disastrous consequences. Increase in human population is a major factor aggravating the intensity and rapidity of the spread of COVID-19. Urbanization, which is more rapid now in developing countries, also changes the epidemiology of emerging wildlife-borne zoonoses because of evolving wildlife–livestock–human interfaces. Hassell et al. (2017) argue that these interfaces form the critical point for cross-species transfer of pathogens into the human population. Urban planners must consider all options to minimize the human–wildlife interaction, and the wildlife habitat must never be encroached upon. Urbanization and commuter mobility are identified as key factors affecting the bimodality of influenza pandemic in Australia (Zachreson et al. 2018; Lal 2020b). Tian et al. (2018) described a complex relationship between the diffusion of zoonotic pathogens and urbanization. They observed that the number of urban immigrants in China is strongly correlated with the human incidence over time of the hantavirus epidemic. Ginter and Simko (2013) observed that rapid urbanization, aging population, obesity, and decreasing physical activity are aggravating the increase in the incidence of Type 2 diabetes mellitus (T2Dm). Dalziel et al. (2018) reported that the surges of seasonal flu in the US occurred in smaller cities with less residential densities and lower household incomes. In contrast, large, more densely populated cities had rather diffused epidemics.

8.5 ACHIEVING ZERO HUNGER AND MALNUTRITION BY 2030

The world is faced with a dilemma in the coexistence of record agricultural production and the widespread problem of hunger and hidden hunger on the one hand and the threat of COVID-19 pandemic on the other. Almost 690 M were under-nourished in 2019, and the COVID-19 pandemic may increase that number between 83 million and 132 million in 2020 (Kretchmer 2020). The problem of malnourishment is severe, both in developed and developing countries (Webb et al. 2018; Ruel, Quisumbing, and Balagamwala 2017), and the vulnerability of a community to the pandemic is also related to the prevalence of hunger and hidden hunger. Malnutrition is caused by the deficiency of protein and micronutrients, leading to overweight and obesity because of inappropriate diet. The problem goes beyond the production agronomy, which involves the knowledge of nutritious diet and quality of food. It is also affected by political unrest, civil strife, and war. Access to an adequate amount and distribution of food are important factors. Therefore, judicious government, at all levels, is critically important. The confounding factor that has emerged since December 2019 is the rapid spread of the COVID-19 pandemic which has also disrupted the global food supply chain (Lal 2020c). The complex and highly interactive issues of food/nutritional security, rapid urbanization, and high mobility of the population across continents are strongly confounded by the daunting challenges posed by the COVID-19. In the final analyses, the solution lies in how humanity will share finite resources (land, water, and biodiversity) with other coinhabitants of the planet so that zoonotic interactions are minimized while meeting the demands of the humanity for the basic victuals.

The Sustainable Development Goal #2 (Zero Hunger) by 2030 has been derailed by the COVID-19 pandemic. The challenges of meeting the food and nutritional security are more now than ever before. Food and nutritional insecurity, urbanization, rapid intercontinental mobility, the COVID-19 pandemic, and other health related issues are strongly interrelated. The solution to the humungous problem lies in the sustainable management of agriculture and natural resources. Because of the overlapping nature of several SDGs, the sustainable management of soil and agriculture will also advance other SDGs especially #1 (End Poverty), #3 (Clean Water and Sanitation), #13 (Climate Action), and #15 (Life on Land) (Lal, Horn, and Kosaki 2018).

8.6 THE SOIL–PEACE–FAMINE–PANDEMIC NEXUS

The widely recognized soil–peace nexus (Lal 2015) is threatened and confounded by the ever-growing greed fueled by insatiable demands for natural resources and the risks of pandemics (e.g., COVID-19) exacerbated by the growing interaction between human and wildlife caused by the encroachment of the habitats by urbanization (Figure 8.4).There is a strong need for education about human values and life style, healthy diet, sustainable management of soil and agricultural ecosystems, waste reduction and management, and respect for the nature. Global flashpoints, indicated by hostility among neighboring countries, soil and climate refugees, and breakdown of law and order within and across nations, are triggered by human greed and lack of tolerance. The viable solution to the paucity of natural resources lies in sustainable management, restoration of degraded ecosystems, and not taking food and natural resources for granted.

8.7 LESSONS LEARNED FROM HISTORIC PANDEMICS AND FAMINES

Humanity has faced epidemics ever since the evolution of Homo sapiens. However, the pandemics have emerged since the urbanization, and more so since the increase in mobility within and across nations circa the 1950s with increase in aviation. Rapid and ad hoc urbanization, as is the case in developing countries without adequate development of basic facilities, is an important determinant of the speed and intensity of the COVID-19 pandemic. Not only should the attempt be made to slow down the rate of population growth in Africa and elsewhere in developing countries, education of all, especially women, is critical to bringing about the much-needed paradigm shift in human values and thinking. Planned urbanization, management of green space, and development of innovative urban agriculture to strengthen and improve the resilience of local food production systems are of high priority.

Above all, the tragedy of the COVID-19 pandemic necessitates an objective and critical review of the manner in which we produce, transport, store, process, distribute, cook and consume food; and dispose of the food wastes. The entire food production and supply chain system must be critically examined.

8.7.1 KNOWLEDGE GAPS

Scientific advances since the mid-20th Century are unprecedented, and new break-throughs are happening more frequently than ever before. However, the global issues of 2020 necessitate a different kind of science—science that respects nature, restores processes which strengthen its functions and ecosystem services of critical importance to nature (not just for humans), that protects natural resources for nature, and the one that increases symbiosis and harmony between Homo sapiens and nature. The objective of science is not to conquer nature but to protect it, and enable humanity to live in harmony with it. Returning some land back to nature is the best option for humanity.

Focus on nature-centric science, which also recognizes the rights-of-nature, must also be supported by an appropriate education. There is a strong need to revise and update the curricula, with the focus on courses which teach environmental and natural resources (e.g., soil, water, climate, and biodiversity) courses at all levels from primary school to the graduate education. Respect for nature and natural resources must be integral to curricula of education at all levels.

Translating science into action is another aspect that needs to receive an increasing attention, especially in developing countries. This is also both science and education, and an important unknown which must be effectively addressed. A scientific project is completed, not with publication of the data in a peer reviewed journal but only when it is translated into action for addressing societal issues at local, regional, national, and global levels.

8.8 CONCLUSIONS

There is a strong interaction between natural resources (soil, water, climate, biota), their sustainable management, and the frequency and intensity of pandemics. Transformation of epidemics faced by humanity since the evolution of Homo sapiens to the pandemics since the 1950s is attributed to high mobility and rapid urbanization. The frequency and intensity of pandemics necessitate an objective thinking about the interaction between human and nature, and the need to return more land resources for nature conservation. Not only is the need for the restoration of degraded soils and vegetation more now than ever before but also the need for proper planning of urbanization and the management of resource fluxes through the megacities is urgently critical now than ever during the human history. There is also a strong need for the education of general public and policymakers regarding the sustainable management and stewardship of natural resources. The education curricula at all levels must focus on the "One Health" concept stating that the "health of soil, plants, animals, people, and the environment is one and indivisible." The COVID-19 pandemic also teaches us a lesson about the need to strengthening local food production systems and making them resilience against disruptions by strengthening home gardening and urban agriculture.

REFERENCES

Awuchi, C.G.I., V.S.A. Ikechukwu, and C.K. Echeta. 2020. Health Benefits of Micronutrients (Vitamins and Minerals) and Their Associated Deficiency Diseases: A Systematic Review. *International Journal of Food Sciences* 3, no. 1 (January 7): 1–32.

Bai, Z.G., D.L. Dent, L. Olsson, and M.E. Schaepman. 2008. Proxy Global Assessment of Land Degradation. *Soil Use and Management* 24, no. 3 (September 1): 223–234. doi:10.1111/j.1475-2743.2008.00169.x.

Cilluffo, A., and N.G. Ruiz. 2019. World's population is projected to nearly stop growing by the end of the century. *Pew Research Center*. https://www.pewresearch.org/fact-tank/2019/06/17/worlds-population-is-projected-to-nearly-stop-growing-by-the-end-of-the-century/

CNN Editorial Research. 2020. Famine Fast Facts. *CNN: World*. https://www.cnn.com/2013/08/20/world/famine-fast-facts/index.html.

Dalziel, B.D., S. Kissler, J.R. Gog, C. Viboud, O.N. Bjørnstad, C.J.E. Metcalf, and B.T. Grenfell. 2018. Urbanization and Humidity Shape the Intensity of Influenza Epidemics in U.S. Cities. *Science* 362, no. 6410 (October 5): 75–79. http://science.sciencemag.org/content/362/6410/75.abstract.

Davis, K. 1955. The Origin and Growth of Urbanization in the World. *American Journal of Sociology* 60, no. 5 (March 1): 429–437. doi:10.1086/221602.

Desjardins, F. 2018. By 2100 None of the World's Biggest Cities Will Be in China, the U.S. or Europe. *World Economic Forum*. https://www.weforum.org/agenda/2018/07/by-2100-none-of-the-worlds-biggest-cities-will-be-in-china-the-us-or-europe/.

Devereux, S. 2009. Why Does Famine Persist in Africa? *Food Security* 1, no. 1: 25–35. doi:10.1007/s12571-008-0005-8.

Earth Institute. 2018. Researchers Say an 1800s Global Famine Could Happen Again. Columbia University. Earth Institute. State of the Planet. https://blogs.ei.columbia.edu/2018/10/12/the-worst-ever-known-drought-could-happen-again-says-study

FAO, IFAD, UNICEF, WFP, and WHO. 2019. *The State of Food Security and Nutrition in the World 2019. Safeguarding against economic slowdowns and downturns*. Rome, Italy: FAO. http://www.fao.org/3/ca5162en/ca5162en.pdf.

FAO, and ITPS. 2015. *Status of the World's Soil Resources (SWSR) – Main Report*. Rome, Italy: Food and Agriculture Organization of the United Nations and Intergovernmental Technical Panel on Soils. http://www.fao.org/3/a-i5199e.pdf

FAO, IFAD, UNICEF, WFP and WHO. 2020. *The State of Food Security and Nutrition in the World 2020. Transforming food systems for affordable healthy diets*. Rome, FAO. https://doi.org/10.4060/ca9692en.

Geber, J., and B. O'Donnabhain. 2020. "Against Shameless and Systematic Calumny": Strategies of Domination and Resistance and Their Impact on the Bodies of the Poor in Nineteenth-Century Ireland. *Historical Archaeology* 54, no. 1 (January 14): 160–183.

Ginter, E., and V. Simko. 2013. Type 2 Diabetes Mellitus, Pandemic in 21st Century. *Advances in Experimental Medicine and Biology* 771: 42–50.

Gladstone, R. 2020. World Population Could Peak Decades Ahead of U.N. Forecast, Study Asserts. *The New York Times*. July 14, 2020. https://www.nytimes.com/2020/07/14/world/americas/global-population-trends.html

Gráda, C.Ó. 2007. Making Famine History. *Journal of Economic Literature* 45, no. 1: 5–38. http://www.jstor.org/stable/27646746.

Griffiths, B.S., and L. Philippot. 2013. Insights into the Resistance and Resilience of the Soil Microbial Community. *FEMS Microbiology Reviews* 37, no. 2 (March): 112–129.

Guest, P. 1994. The Impact of Population Change on the Growth of Mega-Cities. *Asia-Pacific Population Journal / United Nations* 9, no. 1 (March): 37–56.

Hasell, J., and M. Roser. 2017. Famines. *Our World in Data*. https://ourworldindata.org/famines

Hassell, J.M., M. Begon, M.J. Ward, and E.M. Fèvre. 2017. Urbanization and Disease Emergence: Dynamics at the Wildlife–Livestock–Human Interface. *Trends in Ecology & Evolution* 32, no. 1: 55–67. http://www.sciencedirect.com/science/article/pii/S0169534716301847.

History.com Editors. 2020. Pandemics That Changed History. *History.Com*. https://www.history.com/topics/middle-ages/pandemics-timeline.

Hoornweg, D., and K. Pope. 2016. Population Predictions for the World's Largest Cities in the 21st Century. *Environment and Urbanization* 29, no. 1 (September 24): 195–216. doi:10.1177/0956247816663557.

Jarus, O. 2020. 20 of the Worst Epidemics and Pandemics in History. *Live Science*. https://www.livescience.com/worst-epidemics-and-pandemics-in-history.html.

John Hopkins Medicine. 2020. Nutritional Deficiencies. *Digestive Weight Loss Center*. https://hopkinsmedicine.org/digestive_weight_loss-center_/conditions/post_bariatrick_surgery_/vitamins_mineral_deficiency.html.

Kelly, J. 1992. Scarcity and Poor Relief in Eighteenth-Century Ireland: The Subsistence Crisis of 1782–4. *Irish Historical Studies* 28, no. 109: 38–62. https://www.cambridge.org/core/article/scarcity-and-poor-relief-in-eighteenthcentury-ireland-the-subsistence-crisis-of-17824/080D12CD29FB99082150F27F825BDF60.

Kemper, K.J., and R. Lal. 2017. Pay Dirt! Human Health Depends on Soil Health. *Complementary Therapies in Medicine* 32 (June): A1–A2.

Kennedy, C.A., I. Stewart, A. Facchini, I. Cersosimo, R. Mele, B. Chen, M. Uda, et al. 2015. Energy and Material Flows of Megacities. *Proceedings of the National Academy of Sciences of the United States of America* 112, no. 19 (May): 5985–5990.

Kiros, G.E., and D.P. Hogan. 2001. War, Famine and Excess Child Mortality in Africa: The Role of Parental Education. *International Journal of Epidemiology* 30, no. 3 (June): 447–455.

Kretchmer, H. 2020. Global hunger fell for decades, but it's rising again. *World Economic Forum*. https://www.weforum.org/agenda/2020/07/global-hunger-rising-food-agriculture-organization-report

Lal, R. 2009. Soils and World Food Security. *Soil and Tillage Research* 102, no. 1 (January): 1–4.

Lal, R. 2015. The Soil–Peace Nexus: Our Common Future. *Soil Science and Plant Nutrition* 61, no. 4 (July): 566–578.

Lal, R. 2020a. Integrating Animal Husbandry with Crops and Trees. *Frontiers in Sustainable Food Systems* 4:113. doi: 10.3389/fsufs.2020.00113.

Lal, R. 2020b. Soil Science beyond COVID-19. *Journal of Soil and Water Conservation* 75, no. 4: 1–3.

Lal, R. 2020c. Home Gardening and Urban Agriculture for Advancing Food and Nutritional Security in Response to the COVID-19 Pandemic. *Food Security* 12: 871–876. doi:10.1007/s12571-020-01058-3

Lal, R., R. Horn, and T. Kosaki. 2018. *Soil and the Sustainable Development Goals*. Ed. R. Lal, R. Horn, and T. Kosaki. Stuttgart, Germany: Catena-Scheizerbart.

Lund, J., J. Medellin-Azuara, J. Durand, and K. Stone. 2018. Lessons from California's 2012–2016 Drought. *Journal of Water Resources Planning and Management* 144, no. 10 (October 1): 4018067. doi:10.1061/(ASCE)WR.1943-5452.0000984.

MEA. 2004. *Ecosystems and Human Well-Being: A Framework for Assessment*. Ed. Millenium Ecosystem Assessment. *Choice Reviews Online*. Washington, DC: Island Press.https://islandpress.org/books/ecosystems-and-human-well-being?prod_id=474

MPH Online Staff. 2020. Outbreak: 10 of the Worst Pandemics in History. *MPH Online*. https://www.mphonline.org/worst-pandemics-in-history/.

Neiderud, C.-J. 2015. How Urbanization Affects the Epidemiology of Emerging Infectious Diseases. *Infection Ecology & Epidemiology* 5 (June 24): 27060. https://pubmed.ncbi.nlm.nih.gov/26112265.

Nkonya, E., A. Mirzabaev, and J. von Braun. 2016. Economics of Land Degradation and Improvement: An Introduction and Overview BT - Economics of Land Degradation and Improvement – A Global Assessment for Sustainable Development. In ed. E. Nkonya, A. Mirzabaev, and J. von Braun, 1–14. Cham, Switzerland: Springer International Publishing. doi:10.1007/978-3-319-19168-3_1.

Oldeman, L. 1994. The Global Extent of Land Degradation. In *Land Resilience and Sustainable Land Use*, ed. D.J. Greenland and I. Szabolcs, 99–118. Wallingford, England: CABI.

PRB. 2000. 9 Billion World Population by 2050. *Population Reference Bureau.* https://www.prb.org/9billion worldpopulationby2050/.

Roser, M., and H. Ritchie. 2020. Hunger and Undernourishment. *Ourworldindata.Org.* https://ourworldindata. org/hunger-and-undernourishment?utm_campaign=The Preface&utm_medium=email&utm_source=Revue newsletter.

Roser, M., H. Ritchie, and E. Ortiz-Ospina. 2020. World Population Growth. *Our World in Data.* https:// ourworldindata.org/world-population-growth.

Ruel, M.T., A.R. Quisumbing, and M. Balagamwala. 2017. Nutrition-Sensitive Agriculture: What Have We Learned and Where Do We Go from Here? In *IFPRI Discussion Paper 1681*, Vol. 80. Washington, DC: International Food Policy Research Institute (IFPRI). http://ebrary.ifpri.org/cdm/ref/collection/ p15738coll2/id/131461.

Singh, D., R. Seager, B.I. Cook, M. Cane, M. Ting, E. Cook, and M. Davis. 2018. Climate and the Global Famine of 1876–78. *Journal of Climate* 31, no. 23 (October 4): 9445–9467.

The World Counts. 2020. Global Challenges: People and Poverty: Hunger and Obesity. *The World Counts.* https:// www.theworldcounts.com/challenges/people-and-poverty/hunger-and-obesity/how-many-people- die-from-hunger-each-year.

Tian, H., S. Hu, B. Cazelles, G. Chowell, L. Gao, M. Laine, Y. Li, et al. 2018. Urbanization Prolongs Hantavirus Epidemics in Cities. *Proceedings of the National Academy of Sciences of the United States of America* 115, no. 18 (May): 4707–4712.

U.N. 2014. *World Urbanization Prospects: The 2014 Revision, Highlights (ST/ESA/SER.A/352).* New York: United Nations, Department of Economic and Social Affairs, Population Division. https://population. un.org/wup/publications/files/wup2014-highlights.pdf.

UN. 2019. *World Population Prospects 2019: Highlights (ST/ESA/SER.A/423).* Rome, Italy: United Nations, Department of Economic and Social Affairs, Population Division. https://population.un.org/wpp/ Publications/Files/WPP2019_Highlights.pdf.

Vanhaute, E., R. Paping, and C. Gráda. 2006. The European Subsistence Crisis of 1845–1850. A Comparative Perspective. In *European Working Papers, IHES 2006*, Vol. 123, 32. Helsinki, Finland. http://www. helsinki.fi/iehc2006/papers3/Vanhaute.pdf

Webb, P., G.A. Stordalen, S. Singh, R. Wijesinha-Bettoni, P. Shetty, and A. Lartey. 2018. Hunger and Malnutrition in the 21st Century. *BMJ (Online)* 361 (June 13): k2238. http://www.bmj.com/content/361/ bmj.k2238.abstract.

WHO. 2020. Malnutrition. *Fact Sheets.* https://www.who.int/news-room/fact-sheets/detail/malnutrition.

Wikipedia. 2020. List of Famines. *Wikipedia.* https://en.wikipedia.org/wiki/List_of_famines.

WPF. 2019. *What Everyone Should Know about Famines. World Peace Foundation.* Somerville, MA: World Peace Foundation. https://sites.tufts.edu/wpf/files/2019/03/What-Everyone-Should-Know-about-Famine- final-20190226.pdf.

WPF. 2020. *Famine Trends Dataset, Tables and Graphs. World Peace Foundation.* Somerville, MA: World Peace Foundation. https://sites.tufts.edu/wpf/famine/.

Yadav, S., V. Hegde, A. Habibi, M. Dia, and S. Verma. 2018. Climate Change, Agriculture and Food Security. In *Food Security and Climate Change*, ed. S.S. Yadav, R.J. Redden, J.L. Hatfield, A.W. Ebert, and D. Hunter, 1–24. Hoboken, NJ: Wiley Blackwell.

Zachreson, C., K.M. Fair, O.M. Cliff, N. Harding, M. Piraveenan, and M. Prokopenko. 2018. Urbanization Affects Peak Timing, Prevalence, and Bimodality of Influenza Pandemics in Australia: Results of a Census- Calibrated Model. *Science Advances* 4, no. 12 (December 1): eaau5294. http://advances.sciencemag.org/ content/4/12/eaau5294.abstract.

9 Illustrating a Disjoint in the Soil–Plant–Human Health Nexus with Potassium

Kaushik Majumdar, T. Scott Murell,
Sudarshan Dutta, and Robert L. Mikkelsen
African Plant Nutrition Institute
Mohammed VI Polytechnic University

Ashwin Kotnis
All India Institute of Medical Sciences

Shamie Zingore, Gavin Sulewski, and Thomas Oberthür
African Plant Nutrition Institute
Mohammed VI Polytechnic University

CONTENTS

9.1   INTRODUCTION

Human nutritional security is a complex goal that connects food supply to undernourishment and prosperity. Countries with lower caloric supplies tend to experience a higher incidence of undernourishment. Food supplies increase with economic growth of countries until they eventually slow at higher income levels and then plateau with further economic prosperity (Roser and Ritchie 2020). While differences in caloric supply between countries across the world are expected, dietary inequality within countries is typically highest in Sub-Saharan Africa.

Dietary deficiencies of nutrients can be addressed through a number of approaches, including dietary diversification, food fortification, supplementation, or by genetic and agronomic biofortification (Broadley et al. 2010). Agronomic management practices, including soil management and crop nutrition, provide an opportunity to improve nutritional security. Of the approximately 29 elements considered essential for human life, 18 are also either essential or beneficial to plants, and most of the other elements can be taken up from the soil by plants. Approximately 78% of the average per capita calorie consumption worldwide comes from crops grown directly in soil, and another nearly 20% comes from terrestrial food sources that rely indirectly on soil.

Soils that provide a healthy, nutrient-rich growth medium for plants will result in plant tissues that contain most of the elements required for human life when plants are consumed (Combs 2005; Committee on Minerals and Toxic Substances in Diets and Water for Animals, National Research Council 2005). Population growth, increasing urbanization, and high requirements for food and energy are putting enormous pressure on global soil resources and exploiting soil fertility that has developed over thousands of years (Hansjürgens et al. 2018). Nutrient-deficient soils are limited in their ability to supply sufficient nutrients to crops. Jones et al. (2013) note that nitrogen (N) removed by crops from soils is often replenished but other nutrients are not, and this is particularly prevalent in Sub-Saharan Africa (Drechsel et al. 2001). In their seminal work, Stoorvogel and Smaling (1990) demonstrated such significant nutrient depletion in Africa across various scales.

To combat the prevalence of malnutrition, the emphasis on soil health and plant nutrition must be improved, especially in developing countries where a significant part of the population is dependent on its own food production. Responsible plant nutrition is therefore considered as a fundamental platform to prevent nutrient depletion and enable soils to support crop-based solutions that contribute to improved human nutritional security. Building a strong evidence-based link between soil, plant, and health and human nutrition is required to demonstrate the potential of responsible and sustainable crop nutrition as a promising pathway to improve human health.

In this review, we aim to demonstrate that agriculture, and specifically crop nutrition, is a key entry point for positive human health outcomes. We hypothesize that crop nutrition-based solutions are currently an inefficiently used pathway to address human malnutrition. We do this on only one component of soil, plant, and human health: potassium (K). We further hypothesize that K taken up by plants from soils moves up the food chain to create beneficial health outcomes in humans. Better soil K management can translate into better plant health and human health. To demonstrate this, we review key concepts for K behavior in soils and its importance to the functioning of plants and humans. We conclude by providing opportunities where these three areas of health could be better integrated when considering interventions that focus on K nutrition.

9.2   DEFINITIONS

Soil, plant, and human health each have their own definitions. The World Health Organization (2006) defines human health as "…a state of complete physical, mental and social well-being and not merely the absence of disease or infirmity." Doran and Zeiss (2000) defined soil health as "…the capacity of soil to function as a vital living system, within ecosystem and land-use boundaries, to sustain plant and animal productivity, maintain or enhance water and air quality, and promote plant and animal health." A peer-reviewed, published definition of plant health could not be located, but the United States Department of Agriculture (2020) states that "Healthy plants are able to grow and produce in the face of environmental stress, pests, and competition."

A shortcoming of all of these definitions is that they lack operationalization. Stucki et al. (2020) discussed the requirements of operationalization for human health. They proposed that an operational definition of human health needs to (i) "…make it possible to consistently and uniformly describe and measure health" and (ii) be in accord with our basic intuition about why health matters. They proposed that the operationalization of health be "…suitable for the scientific description, measurement, and explanation of the lived experience of health from the person's perspective." The WHO (2013) created "functioning" as the basis for evaluating health, which allows comparison across studies.

Environmental scientists have yet to fully operationalize definitions of soil and plant health. Until they do so, quantifying the impacts of interventions in the soil, plant, and human health continuum will be unstandardized, and subject to bias.

9.3 POTASSIUM AS AN INDICATOR FOR HEALTH OUTCOMES

Potassium is the second-most abundant element in plant tissue, and the K concentration required for optimum growth ranges from 20 to 50 g kg^{-1} dry weight (Hawkesford et al. 2012). Plants require a large quantity of K to complete their lifecycle, more than any other nutrient except N. Consequently, the K status of soils is a key soil health parameter (https://soilhealthinstitute.org/north-american-project-to-evaluate-soil-health-measurements/). In humans, K is the principal intercellular cation and is the third most abundant macromineral after calcium (Ca) and phosphorus (P) (Lindshield 2018). Potassium is physiologically important in many functions of the human body. These include the functional integrity of the cardiovascular, respiratory, digestive, endocrine, renal, and neurological systems (Kotnis et al. 2017; Weaver 2013). Potassium has been selected for this review, because it is a mineral element that originates in the soil and the K nutrition of plants and humans largely rely on the transition of K from the soil to the plants and human via the food chain. And, unlike many other nutrients, K provides a readily available knowledge base for an in-depth review across the soil–plant–human health continuum.

9.4 REVIEW

9.4.1 SEARCH PROCESS

The review was conducted using publicly available literature databases to identify articles related to the interrelations between K and soil, plant, and human health, and their outcomes. The approach was keyword-based using a step-wise process with various independent combinations of keywords. The initial search focused on individual components, K, and soil health; K and plant health; and K and human health. The next round narrowed the focus on the extent of K deficiencies in the continuum, and evidences of specific health outcomes when the K supply is reduced in plants and humans. We concluded with searches linking soil, plant health, and human health, specifically focusing on the bioavailability of K while using potato as an example.

9.4.2 SOIL HEALTH AND POTASSIUM

Several studies have analyzed the rise and fall of ancient human civilizations with the quality of soils they thrived on (Katsuyuki 2009). Evidence presented in several monographs on the impacts of human activities on soil and the environment identifies soil degradation as a major factor in the collapse of ancient civilizations. Decline in soil fertility due to erosion led to the decline of civilizations in the Middle East, Greece, Rome, and Mesoamerica (Montgomery 2007). The destruction of the Roman civilization and the decline of the Mesopotamian civilization were also accelerated by soil degradation (Carter and Dale 1974; Katsuyuki 2009). The collapse of Easter Island can be seen as an instance of deforestation and its accompanying soil erosion. Similarly, the Mayan civilization in the Copán valley suffered a decline when they cut forests covering the hills, which promoted soil erosion (Katsuyuki 2009). Soil supplies humanity with most of its food, and soil degradation leads to the collapse of civilization when it fails to supply enough nutritious food to support the population. In 1912, Alexis Carrel (1873–1944), the French Nobel Laureate in Physiology and Medicine, stated:

> Since soil is the basis for all of human life, our only hope for a healthy world rests on re-establishing the harmony in the soil we have disrupted by our modern methods of agronomy. All of life will be either healthy or unhealthy according to the fertility (potential) of the soil.

(Tompkins and Bird 1989; Katsuyuki 2009)

Kibblewhite et al. (2008) identified plant nutrients as a controlling input to the soil system that support the processes for all the major services delivered by the soil. Nutrient concentrations, solubility, and their biogeochemical transformations are critical to soil health. After carbon (C), the cycling of nutrients to, from, and within the soil system most affects its dynamics and the delivery of ecosystem services, including agricultural production. Soil health is gradually compromised when nutrients removed from the field during harvest are not replenished. The limitation imposed by inadequate nutrient status strips the soil of its "capacity to function or perform" as adequate availability of nutrients in the soil is critical for sustained soil health (Sanyal et al. 2014).

Manipulation of plant nutrient supplies to increase crop outputs has been one of the keystones of agriculture for millennia. Agricultural strategies were first based on additions of animal manures and later supplemented with mineral fertilizers to counter gradual nutrient depletion with the aim of restoring and sustaining soil health (Ciceri et al. 2015). Kihara et al. (2020) associated improved soil health with appropriate synchrony between nutrient supply and crop demand, thereby minimizing problems of leaching and emissions.

There are numerous examples that underpin the importance of integrated and balanced nutrient management for increased crop productivity and quality, as well as improved soil health parameters (Benbi and Brar 2009; Bandyopadhyay et al. 2010; Ayuke et al. 2011). Increased richness and diversity of soil microbes has been observed in maize-bean rotations in Kenya with the use of farmyard manure and its combination with mineral fertilizers (Kibunja et al. 2010). Similarly, a large body of experimental evidence is now available that exhibits underperformance of soils when soil fertility levels are downgraded due to overextraction and underapplication of nutrients (Chauhan et al. 2012; Ladha et al. 2003). The review by Geisseler and Skow (2014) showed that mineral N application increased microbial biomass and soil organic carbon compared with an unfertilized control. The use of both mineral fertilizers and the return of crop residues to the soil produced more crops and sequestered much more soil C than under subsistence agriculture in Zimbabwe (Zingore et al. 2005).

Global studies provide compelling evidence to emphasize soil fertility as a crucial component of soil health. It allows the soil to produce adequate and nutritious food, one of the most essential services provided by the soil.

Potassium is a component of many soil minerals, especially mica and feldspar. The presence of mica and feldspar in soils provides an abundant *in-situ* source of this nutrient. During soil formation, these minerals transform to secondary minerals (on weathering) and partially release some of their structural K into solution (Majumdar et al. 2017). Once released from the mineral structure, plants can take up K^+ from the soil solution. A healthy soil with adequate K-supplying capacity supports healthy crops that produce adequate and nutritious feed for animals, and food for the human family (Figure 9.1). McLean and Watson (1985) estimated that only 5% of the K requirement of a crop is present in the soil solution at any one time.

9.4.2.1 Potassium Status in Global Croplands

Although soil nutrient deficit is recognized as a potential threat to soil health and sustainability of crop production, a comprehensive estimation of K fertility status in global cropland is rare. Smil (1999) and Manning (2015) estimated that replenishment of K in agricultural soils is only between 10% and 35% of the crop K removed. Blanchet et al. (2017), while summarizing several studies, found that despite the natural abundance of K in the soils in certain regions of the world, such as Australia, China and Iran, crop K deficiencies are prevalent over large areas due to particular pedo-climatic conditions or chronic underfertilization of K (Figure 9.2). Other authors reported that large agricultural areas of the world are deficient in K, including three-fourths of the paddy soils of China and two-thirds of the wheat belt of Southern Australia (Römheld and Kirkby 2010). Sardans and Peñuelas (2015) stated that at a global level, K is as limiting as N and P for plant productivity in terrestrial ecosystems. These authors observed some degree of K limitation in up to 70% of all studied terrestrial ecosystems. Tan et al. (2005), using data for four major crops, (i.e., rice wheat, maize, and barley) estimated the global soil K deficit at 38.8 kg K ha^{-1} year^{-1}. These authors stated that 283 million hectares of crop land in developing countries and 31 million hectares in less-developed

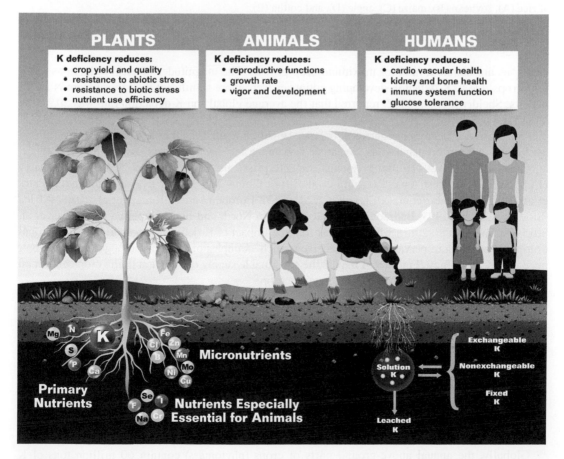

FIGURE 9.1 Inadequate supply of soil K can lead to important negative impacts on key indicators of plant, animal, and human health.

FIGURE 9.2   Visual symptoms of potassium deficiency (clockwise order from top left) observed in the fields of rice (A), soybean (B), maize (C), apple (D), and cotton (E).

countries are deficient in K for maximum crop growth. The magnitudes of yield reductions of all four crops due to K deficit in developing countries were more significant than those from N and P deficits. Sheldrick et al. (2002) reported that the average global soil K depletion rate was 20.2 kg K ha^{-1} year^{-1} in 1996.

Potassium deficiency, or negative K balance (input< output), is an indicator of increasing impoverishment of K in cropland soils. Negative K balance has been reported in different soil types, agro-ecosystems, crops/cropping systems, or enterprises in Australia (Edis et al. 2012; Christy et al. 2015); Southeast Asia (Pasuquin et al. 2014); China (Liu et al. 2017a, 2017b); India (Patra et al. 2017; Majumdar et al. 2017; Dutta et al. 2013); Europe (Andrist-Rangel et al. 2010); Africa (Stoorvogel et al. 1993; Badraoui et al. 2003); and in North and South America (Wingeyer et al. 2015; Barbazán et al. 2012), which suggest that the impact of K deficiency on soil health in global croplands is far more pervasive than perceived. Such examples of widespread negative soil K balance and K deficiency (Figure 9.3) impair the sustainable supply of nutrition to the crops grown on it, and impact the health and performance of the crop.

9.4.3   PLANT HEALTH AND POTASSIUM

Potassium is among the most abundant elements in plant tissues. It is involved in numerous biochemical and physiological processes crucial to growth, performance, quality, and stress tolerance (Oosterhuis et al. 2014). Adequate K nutrition is associated with greater crop yield and fruit size, increases in soluble solids and ascorbic acid concentrations, improved market quality of crops, and higher concentration of phytonutrients of vital importance for human health (Mikkelsen 2018; Salih et al. 2016; Lester et al. 2010). A balanced supply of K improves the plant's efficiency to use other essential nutrients, including N and P (Brar et al. 2011).

Globally, the annual above-ground parts of crops (phytomass) contain 60 million tons of K (Römheld and Kirkby 2010). The amount of K accumulated by the crop varies depending on the crop species, the environmental conditions during the growing season, and the management practices

FIGURE 9.3   Map of predicted exchangeable K for African soils. The range between light-shaded K-deficient soils to dark-shaded soils with higher K concentrations is reported as 0.01–2.4 cmol+kg^{-1}. (Modified from Hengl et al. (2015).)

(Mullins et al. 1997). Potassium is taken up by plants against its concentration gradient through K$^+$ transporters and channels located in the plasma membrane of root cells (Ashley et al. 2006). Potassium can be stored, both in the cell cytoplasm and the vacuole, and the distribution among these two locations is the major factor for determining the K function in the plant (Marschner 1995). Plant uptake of K usually precedes dry matter production (Kant et al. 2005), and about 95% of the total accumulated K has been absorbed by the time of 50% dry matter accumulation in maize (Welch and Flannery 1985). In contrast to annuals, the K uptake by longer-duration crops such as sugarcane and cassava is very slow at early stages of growth and steadily increases until harvest (Filho 1985; Howler 1985).

9.4.3.1   Potassium for Yield Building

Potassium regulates the biosynthesis, conversion, and allocation of metabolites in plants, and the integrative effects of K on different physiological processes have profound influence on crop growth and development (Hasanuzzaman et al. 2018). The role of K application toward yield improvement is well-established, especially in regions with low or depleted soil K. Countless studies have shown that appropriate K fertilizer application can significantly increase the yields of cereals (Jiang et al. 2018), cropping systems (Singh et al. 2018); oilseeds and pulses (Majumdar and Govil 2013); vegetables and fruits (Lester et al. 2009; Mitra and Dhaliwal 2009); and plantation crops

(Ruan et al. 2013). Hasanuzzaman et al. (2018) summarized several studies across regions and crops providing compelling evidence that K application is directly or indirectly responsible for higher crop yield. Besides its direct effect on crop yields, balanced application of K improves the uptake and use efficiency of other essential nutrients that support high yields.

One of the more visually obvious consequences on plant growth from insufficient K is a reduction in plant stature, with consequent reduction in leaf area, solar radiation interception, and photosynthesis per unit leaf area that lead to a decreased photosynthetic assimilate pool produced in the leaves (Pettigrew 2008). Restricted K uptake affected plant height, aerial stem number, and leaf number per plant in potato (Zelelew et al. 2016), plant dry matter production, leaf area, and internode size in cotton (Gerardeaux et al. 2010), and rhizobial nodulation in legumes (Divito and Sadras 2014).

9.4.3.2   Potassium for Quality Improvement

Along with increased crop production, improved quality of harvested products is crucial for human nutritional security (Cakmak 2010). Potassium exerts significant influence on increasing many human health-related quality compounds in grain crops, fruits, and vegetables (Asaduzzaman and Asao 2018; Cakmak 2010; Kausar and Gull 2014; Mikkelsen 2018; Ray et al. 2019).

Since quality parameters differ among crops, only general examples are given here. With citrus, benefits from adequate K are manifested in the thickness of the peel and vitamin C concentration; for apples, sugar concentrations; while for tomatoes, the development of uniformly red fruit rich in lycopene (Mikkelsen 2018). Lester et al. (2010) presented an exhaustive review of published abstracts that consistently demonstrated the positive impacts of specific K fertilizer forms, in combination with specific application regimes, on fruit quality attributes. Recently, Ray et al. (2019) reported that the omission of K fertilizer application in maize had the most inhibitory effect on maize oil quality compared with N and P. Positive correlation between K and soybean yield and quality is also well-documented in reviews (Imas and Magen 2007). Potassium fertilizer application increased the major isoflavones—genistein, daidzein, and glycitein in soybean and the effect was more in the soil with low plant-available K (Vyn et al. 2002). The quality of harvested tea shoots was improved by K fertilizer application as measured by increased concentrations of free amino acids, water-extractable dry matter, and total polyphenols (Ruan et al. 2013). Several hydroponic studies showed that adequate K in the nutrient solutions increased tomato fruit production, total soluble solid content, and lycopene concentration of tomato compared with plants growing in low K conditions. Conversely, a low K concentration in the root zone resulted in melon fruits with lower K content when supply of K was inadequate (Asaduzzaman and Asao 2018). Similarly, Khan et al. (2012) reported improved specific gravity, starch, sugar, and vitamin C contents of potato tubers at higher K application rates over the control.

Besides playing a critical role in many of the metabolic processes that enhance the quality, nutrition, flavor, and appearance, ensuring an adequate K supply during crop growth enhances longevity of fresh food crops. Many other examples of the influence of an adequate K supply on many aspects of crop quality have been documented. The enhanced quality of crops, supported by an adequate K supply from the soil, not only contributes to better human nutrition but also adds to the marketability and revenue stream of the farmers toward better livelihoods.

9.4.3.3   Potassium for Biotic and Abiotic Stress Management

Potassium is an essential nutrient that affects most of the biochemical and physiological processes that help in the survival of plants exposed to various biotic and abiotic stresses (Wang et al. 2013). The stress-signaling and stress-mitigating roles of K are important for plants, particularly under drought, salinity, and upon pathogenic infection (Cakmak 2005; Amtmann et al. 2008).

9.4.3.3.1   Potassium for Abiotic Stress Management

In the face of significant food and nutritional insecurity, agricultural production continues to be constrained by a variety of abiotic (e.g., drought, salinity, cold, acidity, and compaction) stresses that can significantly reduce the quantity and quality of crops. Yield losses from abiotic stress were estimated at 66% for maize, 82% for wheat, 69% for soybeans, and 54% for potatoes (Wang et al. 2013). Potassium plays a vital role in protecting plants against abiotic stresses through its role in maintaining ion homeostasis, cellular integrity, and enzymatic activities (Hasanuzzaman et al. 2018).

Drought is the single-most critical threat to world food security affecting vast areas of agricultural lands. Potassium helps crops use water more effectively. Adequate K availability helps plants develop deeper roots to access water at greater soil depth, creates higher osmotic gradient within cells for faster water uptake, and speeds the closure of the crop canopy that results in more water accessible to the crop. Hasanuzzaman et al. (2018) reported that an adequate K supply helped rice, wheat, maize, rapeseed, and cotton to produce more of organic osmolytes (e.g., Pro) that protect plants when exposed to drought stress. A long-term study in a maize–soybean rotation showed that yields and profits were highest in the plots receiving K application in the year with water stress (IPNI 1998). Srinivasarao et al. (2009) reiterated the special role of K in plant adaptation to water stress in rain-fed production systems. Amanulaah et al. (2016) also reported improved growth and yield of maize with K application in water-stressed semi-arid climates of Pakistan. Bhattacharyya et al. (2018), studying the role of K in water-stressed maize, reported that water-soluble and exchangeable soil K emerged as the most significant factors influencing yield. Many studies have elucidated the role of K in triggering drought-resistance mechanisms within plant species (Hosseini et al. 2016; Jákli et al. 2017). There is increasing evidence that drought-affected plants have a larger internal requirement for K (Anschütz et al. 2014).

Salinity affects over hundreds of millions of hectares of agricultural land worldwide (Rengasamy 2010). Salinity inhibits seed germination and plant growth, affects the leaf anatomy and physiology of plants, and thereby disrupts photosynthesis, water relations, protein synthesis, energy production, and lipid metabolism (Parida and Das 2005). Hasanuzzaman et al. (2018) summarized the beneficial effect of K application under salinity stress and highlighted the essential roles of K related to osmotic adjustment, maintaining turgor and regulating the membrane potential, cytoplasmic homeostasis, protein synthesis, and enzyme activation under salt stress. For example, barley growth and salt tolerance were sharply reduced when exposed to a combination of salt stress and K deficiency, and low K conditions significantly increased salt sensitivity and impaired photosynthesis (Horie et al. 2009).

Low K status might induce the formation of harmful reactive oxygen species and related cell damage under saline conditions, which was attributed to the effects of K^+ deficiency and/or Na^+ toxicity on stomatal closing and the inhibition of photosynthetic activity that ultimately inhibits plant growth and reduces crop production (Gong et al. 2011). The plant's ability to increase K^+ fluxes and reduce Na^+ fluxes in the presence of salinity to attain a higher K^+/Na^+ selectivity ratio is essential for salt tolerance. The addition of K to a saline culture solution has been found to increase K concentrations in the plant tissue with corresponding decreases in Na, resulting in increased plant growth and salt tolerance. Cytosolic K^+ retention is essential to confer salinity stress tolerance in plants (Wang et al. 2013) and a strong positive correlation between the ability of plant roots to retain K after exposure to NaCl, and a genotype's salinity tolerance was reported for a wide range of crop species (Anschütz et al. 2014).

Cold stress can inhibit plant growth and productivity. It affects plants by directly inhibiting metabolic reactions and indirectly influencing cold-induced osmotic, oxidative, and other stresses (Wang et al. 2013). Higher tissue K concentrations reduced chilling damage and increased cold resistance, ultimately increasing crop yield (Kant and Kafkafi 2002). Haque (1988) reported that spikelet sterility in rice, induced by low temperature, was reduced with increased K supply and the increased K/N ratio in the rice leaves. Oats that were well supplied with K survived late frost without obvious damage, whereas much of the crop that was grown on K-deficient soil did not survive

(Bogdevitch 2000). Frost damage is inversely related to K concentrations, and the increased K soil supply improved the plant frost resistance by the increase in phospholipids, membrane permeability, and improvement in the biophysical and biochemical properties of cells (Hakerlerler et al. 1997).

Soil compaction restricts root growth and plant uptake of nutrients. Soil compaction is one of the reasons why positive yield responses to K occur more frequently in no-till and ridge-till cropping systems compared with conventional tillage (IPNI 1998). The impact of soil compaction can be reduced in low-to-medium K fertility soils by the application of K. Anaerobic conditions resulting from soil compaction increase K fixation by some clay minerals, making K less accessible to plants and possibly requiring higher K applications to meet plant demands. Roots also depend on oxygen and are less active in taking up K when soils become anaerobic.

9.4.3.3.2 Potassium for Biotic Stress Management

Global crop production is significantly restricted by biotic stresses (Wang et al. 2013). Weeds caused the highest potential loss (32%), followed by animal pests (18%), fungi and bacteria (15%), and viruses (3%) to attainable production of wheat, rice, maize, barley, potatoes, soybeans, sugar beets, and cotton between 1996 and 1998 (Oerke and Dehne 2004).

Adequate application of K fertilizers is widely reported to decrease insect infestation and disease incidence in many host plants (Prabhu et al. 2007). Potassium deficiency manifestations, such as thin cell walls, weakened stalks and stems, smaller and shorter roots, sugar accumulation in the leaves, and accumulation of unused N, encourage disease infection. Each of these maladies reduces the ability of the plant to resist entry and infection by fungal, bacterial, and viral disease organisms.

A healthy plant, free from stress, is much more resistant to disease attack. Although the disease resistance of a plant is related to its genetic traits, the natural disease resistance mechanism can be enhanced by balanced nutrition, especially adequate K (IPNI 1998). Proper K nutrition impacts a number of physiological and biochemical processes to induce a plant's resistance against pathogens and insects (Amtmann et al. 2008; Zorb et al. 2014). The large impact of K fertilizer on the reduction of stalk, leaf, or root diseases was attributed to responses that promoted enzyme activity and induced abundant natural compounds (Gao et al. 2018). Gao et al. (2018) demonstrated that the application of K fertilizer increased the plant's inherent defense potential against soybean cyst nematode by increasing the root exudation of phenolic acids and plant pathogen-related gene expression. The release of cinnamic, ferulic, and salicylic acids was significantly enhanced by K application, and each of the three acids can dramatically constrain soybean cyst nematode (*Heterodera glycines*) in vitro. Potassium-deficient plants were more susceptible to nematode infection than those with an adequate supply of K. Similarly, rice borer infestation was greatest with a low soil K supply, but damage decreased rapidly as the K application rate increased from zero to 60 kg K_2O ha^{-1} (Sarwar 2012). Williams and Smith (2001) also reported negative correlations between K in rice leaf blades and disease severity incidence of stem rot and aggregate sheath spot. The importance of K fertilization in reducing the severity of *Alternaria* leaf spot disease was seen in cotton on a soil testing low in K. This disease organism can cause significant yield reduction with premature plant defoliation (IPNI 1998). Perrenoud (1990) reviewed 2,449 studies and found K use significantly decreased the incidence of fungal diseases by 70%, bacteria by 69%, insects and mites by 63%, viruses by 41%, and nematodes by 33%.

9.4.4 HUMAN HEALTH AND POTASSIUM

9.4.4.1 Dietary Potassium Requirement in Humans

For adults, a daily adequate intake (AI: the average daily level of intake assumed to ensure nutritional adequacy) of 3,500 mg (or 90 mmol) of K is recommended (WHO 2012a). Bellows and Moore

(2013) provided AI values for K (mg day^{-1}) as 400 for infants 0–6 months; 700 for 7–12 months; 3,000 for 1–3 years; 4,500 for 4+ years; 4,700 for 19+ years; and 5,100 while breastfeeding (Prasad et al. 2016).

Weaver et al. (2018) reviewed the basis of deciding the adequate rate of K intake. The first published reports of K guidelines were produced by the Scientific Committee for Food (SCF) and a joint committee convened by the WHO and Food and Agriculture Organization. The SCF proposed a Population Reference Intake (equivalent to North American Recommended Dietary Allowance (RDA)) for K of 3,500 mg day^{-1} (90 mmol day^{-1}) to lower blood pressure and assist with Na excretion. Similarly, the WHO/Food and Agriculture Organization Joint Commission recommended K intakes of 70–80 mmol day^{-1} (2,750–3,100 mg day^{-1}) to lower blood pressure and protect against cardiovascular disease (CVD).

Later, at the request of member states, the WHO examined the evidence linking K intake to noncommunicable diseases. The key aims were (i) to identify health outcomes correlated with increased levels of K intake compared with lower levels of intake and (ii) to compare K intakes of less than 90–120, 120–155, and greater than 155 mmol day^{-1} (<3,500, 3,500–4,700, 4,700–6,000, and >6,000 mg day^{-1}, respectively) in relation to health outcomes. The discussions led to two documents: the first contained the official recommendations of the WHO (2012b), and the second one detailed the systematic review and meta-analysis upon which the recommendations were based (Aburto et al. 2013). After evaluating the evidence, WHO made a strong recommendation for increasing K intake to reduce blood pressure, CVD, stroke, and coronary heart disease. In addition, it made a conditional recommendation to consume at least 90 mmol day^{-1} (3,500 mg day^{-1}) of K to achieve these benefits. At the request of the European Commission, European Food Safety Administration expanded upon the meta-analysis conducted by the WHO and considered evidence for potassium's effect on additional health outcomes. Subgroup analyses revealed that the effect was greater for hypertensive individuals not on any other antihypertensive treatments and that the effect was greatest for those consuming 3,500–4,700 mg K day^{-1} (90–120 mmol day^{-1}). Although current data precluded the setting of an average requirement (equivalent to the United States and Canadian EAR), they determined that there were sufficient data to set the AI at 3,500 mg day^{-1} (90 mmol day^{-1}). A chronology of the development of the global K recommendations is given in Table 9.1.

TABLE 9.1
Summary of Global Potassium Recommendation

Organization	Year Established	Recommended Daily K Intake[a]	Strength of Evidence	Reference
SCF	1993	3,100 mg day^{-1} (80 mmol day^{-1})	Population reference intake	Scientific Committee on Food
WHO/FAO	2002	70–80 mmol day^{-1} (2,750–3,100 mg day^{-1})	Convincing	World Health Organization
IOM	2005	4,700 mg day^{-1} (120 mmol day^{-1})	Adequate intake	Food and Nutrition Board, Institute of Medicine
WHO	2012	90 mmol day^{-1} (3,500 mg day^{-1})	Conditional recommendation	World Health Organization
EFSA	2016	3,500 mg day^{-1} (90 mmol day^{-1})	Adequate intake	European Food Safety Administration

Source: Weaver et al. (2018).

Abbreviations: EFSA, European Food Safety Administration; IOM, Institute of Medicine (now National Academies of Medicine); SCF, Scientific Committee for Food; WHO, World Health Organization; WHO/FAO, Joint World Health Organization/Food and Agriculture Organization Expert Consultation.

[a]Target intake for healthy adults aged 18–65 years.

9.4.4.2 Food Sources for Dietary Potassium

Potassium is found in a wide variety of plant and animal foods and in beverages. Many fruits and vegetables are excellent sources, as are some legumes (e.g., soybeans) and potatoes. Meats, poultry, fish, milk, yogurt, and nuts also contain abundant K. Among starchy grains, whole wheat flour and brown rice are much higher in K than their refined counterparts, white wheat flour and white rice. Milk, coffee, tea, other nonalcoholic beverages, and potatoes are the top sources of K in the diets of U.S. adults. Among children in the United States, milk, fruit juice, potatoes, and fruit are the top K sources. It is estimated that the body absorbs about 85%–90% of dietary K. The forms of K in fruits and vegetables include K phosphate, sulfate, citrate, and others, but not K chloride. The following table provides a list of food sources of K (https://ods.od.nih.gov/factsheets/Potassium-HealthProfessional/). An extended list is available at https://www.nal.usda.gov/sites/www.nal.usda.gov/files/potassium.pdf (Table 9.2).

TABLE 9.2
Selected Food Sources of Potassium

Food	Milligrams(mg) per Serving	PercentDV[a]
Apricots, dried, ½ cup	1,101	23
Lentils, cooked, 1 cup	731	16
Prunes, dried, ½ cup	699	15
Squash, acorn, mashed, 1 cup	644	14
Raisins, ½ cup	618	13
Potato, baked, flesh only, 1 medium	610	13
Kidney beans, canned, 1 cup	607	13
Orange juice, 1 cup	496	11
Soybeans, mature seeds, boiled, ½ cup	443	9
Banana, 1 medium	422	9
Milk, 1%, 1 cup	366	8
Spinach, raw, 2 cups	334	7
Chicken breast, boneless, grilled, 3 ounces	332	7
Yogurt, fruit variety, nonfat, 6 ounces	330	7
Salmon, Atlantic, farmed, cooked, 3 ounces	326	7
Beef, top sirloin, grilled, 3 ounces	315	7
Molasses, 1 tablespoon	308	7
Tomato, raw, 1 medium	292	6
Soymilk, 1 cup	287	6
Yogurt, Greek, plain, nonfat, 6 ounces	240	5
Broccoli, cooked, chopped, ½ cup	229	5
Cantaloupe, cubed, ½ cup	214	5
Turkey breast, roasted, 3 ounces	212	5
Asparagus, cooked, ½ cup	202	4
Apple, with skin, 1 medium	195	4
Cashew nuts, 1 ounce	187	4

(Continued)

TABLE 9.2 (*Continued*)

Selected Food Sources of Potassium

Food	Milligrams(mg) per Serving	PercentDV[a]
Rice, brown, medium-grain, cooked, 1 cup	154	3
Tuna, light, canned in water, drained, 3 ounces	153	3
Coffee, brewed, 1 cup	116	2
Lettuce, iceberg, shredded, 1 cup	102	2
Peanut butter, 1 tablespoon	90	2
Tea, black, brewed, 1 cup	88	2
Flaxseed, whole, 1 tablespoon	84	2
Bread, whole-wheat, 1 slice	81	2
Egg, 1 large	69	1
Rice, white, medium-grain, cooked, 1 cup	54	1
Bread, white, 1 slice	37	1
Cheese, mozzarella, part skim, 1½ ounces	36	1
Oil (olive, corn, canola, or soybean), 1 tablespoon	0	0

[a] DV, Daily value.

9.4.4.3 Dietary Potassium Intake in Different Countries/Regions

Globally, the adverse health outcomes of inadequate intake of K have received attention because of a clear role of K in addressing the global burden of diseases. This has led to significant efforts at country- or global scales to assess the dietary K intake in active populations.

In 2010, a study involving 21 countries shockingly revealed that none of the studied countries met the adequate intake for K (van Mierlo et al. 2010). The authors used comprehensive evidence from national population-based dietary surveys from PubMed, conducted from 1990 to 2009, that reported data on K intake in more than 1,000 adults. They also contacted health authorities worldwide. Using the recommended level of K intake at 4.7 g day[-1], based on the Dietary Reference Intakes from the Institute of Medicine (2004), the authors concluded that in 21 countries spread across North America, Europe, Asia, and Oceania, the mean K intake ranged from 1.7 g day[-1] (China) to 3.7 g day[-1] (Finland, the Netherlands, and Poland) (Figure 9.4).

Other large-scale multicountry studies (O'Donnell et al. 2014; Mente et al. 2014) also pointed toward insufficient K intake, with evidence of mean K intakes as low as 2,100 mg day[-1]—well

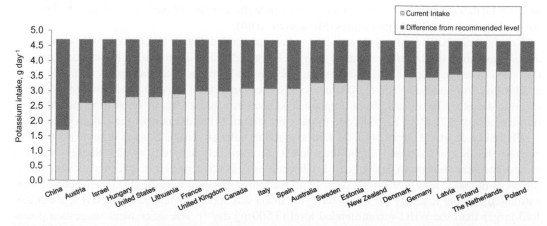

FIGURE 9.4 Potassium intakes and differences from the recommended level for 21 countries. (Modified from van Mierlo et al. (2010).)

below the WHO recommendation of at least 3,500 mg day^{-1}. Cogswell et al. (2012) estimated the distributions of usual daily K intakes by sociodemographic and health characteristics and concluded that overall <2% of US adults and ~5% of US men consumed ≥4,700 mg K day^{-1}, meeting the recommend intake for K. In comparison with their counterparts without these conditions, usual K intake was statistically significantly lower among US adults with diabetes or chronic kidney disease but not obesity or hypertension. In 2009–2010, the average dietary K intake of the U.S. population aged 2 years and older was 2,640 mg day^{-1} (Hoy and Goldman 2012).

Studies conducted in Chile (Cornejo et al. 2014) and Brazil (Pereira et al. 2015) also showed K consumption below the recommendation. Carrillo-Larco et al. (2018), using 24 h urine samples of a population-based study in a semi-urban area in Peru, reported a low K (2,000 mg day^{-1}; SD: 1,200 mg day^{-1}) consumption, which matched earlier studies. Less than one in ten met the K intake recommendation.

Welch et al. (2009) examined K intakes from 10 European countries and found that for both men and women, Greece had the lowest average intakes at 3,536 and 2,730 mg day^{-1}, respectively. The highest K intakes were found in Spain where the average intake for men was 4,870 mg day^{-1} and for women was 3,723 mg day^{-1}. Campanozzi et al. (2015) assessed the dietary K intake of 1,625 subjects aged 6–18 years from 10 Italian regions and reported much lower intake than the age-specific adequate intakes (2,800 mg day^{-1} between age 7 and 10 and 3,900 mg day^{-1} above age 10 years) in both boys and girls, in all the age categories and in all the regions surveyed. Athanasatou et al. (2018) stated that average K intake was 2,264 mg day^{-1} in 163 adults surveyed in Greece. The K intake remained lower than recommendations and only 7.4% of subjects met WHO recommendations for K intake.

Bolton et al. (2019) used 24-h urine and dietary recall data to assess K intake in a cross-section of Australian adults and concluded that the participants in the study were not consuming enough K to reduce the risk of chronic diseases. Another study (Grimes et al. 2017) that examined Na and K urinary excretion in a sample of Australian school children revealed that children aged 4–12 years were consuming too much Na and not enough K, which has important implications for cardiovascular health. In New Zealand, McLean et al. (2015) reported inadequate K intakes, consistent with K intake estimates from the 2008/2009 New Zealand Adult Nutrition Survey. The survey using 24-h diet recall revealed inadequate intakes of K (median usual daily intake of K of 3,449 mg for males and 2,757 mg for females, below the AI of 3,800 mg day^{-1} for men and 2,800 mg day^{-1} for women).

Du et al. (2020) used data on 29,926 adults aged ≥20 years between 1991 and 2015 from an ongoing cohort in China (the China Health and Nutrition Survey) and collected detailed diet data with the use of weighing methods with 3 consecutive 24-h dietary recalls. Intake was 1,700 mg K day^{-1} in 1991 and 1,500 mg K day^{-1} in 2015, below half of that recommended by the WHO. The most at-risk populations lived in China's central region and rural areas were middle-aged, had lower educations, or were farmers. Intakes were higher in Korea where the average intake was 2,900 mg K day^{-1} yet still did not meet the WHO guidelines (Stone et al. 2016).

Kotnis et al. (2017) used population data and recommended dietary K requirement to estimate an annual human K intake requirement of 1.96 Mt K for India. The authors used agricultural production data of major crops in India and average K content in marketable plant parts to assess the K availability from consumption of different food sources in India. The results indicated that the total estimated dietary K available from all plant food sources was 1.24 Mt K, suggesting an additional requirement of 0.72 Mt of K from food to meet the overall K requirement of the Indian population.

There is a dearth of information on K intake in Sub-Saharan Africa. Studies in some populations suggest that K consumption may be insufficient. A cross-sectional study of 325 black, white, and mixed-ancestry subjects in South Africa (Charlton et al. 2005) reported that K intakes were below the recommended level of 90 mmol day^{-1} in all the ethnic groups. Interestingly, Yawson et al. (2018) estimated that the K supply per capita day^{-1} in Ghana was approximately 9,086 mg day^{-1}, about 2.6-fold larger than the WHO-recommended level (3,500 mg day^{-1}). The assessment suggests a potentially large risk of excess dietary K supply at both individual and population levels and the need for assessing options for managing K excess as part of food security and public health strategies.

9.4.4.4 Potassium and Disease Incidence

The above section revealed lower than required dietary intake of K across regions and countries. Until recently, humans consumed a diet rich in K. However, with the increasing consumption of processed food, combined with a reduction in the consumption of fruits and vegetables, K intake has decreased in most developed countries. Dietary intake of Americans now averages around 2,290–3,026 mg K day^{-1}, i.e., less than one-third of our evolutionary intake (Palmer and Clegg 2016). White and Broadley (2005) reviewed the composition of vegetables, fruits, and nuts from the UK (1930s vs.1980s) and the US (1930s vs. 2004) and reported a 10% and 13% decrease in the average K concentration, respectively. Ekholm et al. (2007) reported a significant decrease in the K concentration of cereals and a smaller decrease in other food crops in Finland. Some of this decrease is due to a dilution effect where grain yields are increased but the mineral content remains static (McGrath 1985). Similarly, Balboa et al. (2018) documented that between 1930 and 2010, soybean yields increased almost 250% (from 1.3 to 3.2 Mg ha^{-1}), while the seed K concentration decreased by 13%. Marles (2017) reviewed the changes in the mineral composition of food and concluded that a diet rich in the recommended vegetables, fruits, and whole grains still provides the nutrient-dense foods recommended for good dietary health.

Generally, low K intake combined with a high salt intake causes a rise in blood pressure (He and MacGregor 1999) and increases the risk of CVD (Khaw and Barrett-Connor 1987), renal disease (Tobian et al. 1984), and bone demineralization (New et al. 1997). The health benefits of adequate dietary intake of K have been reported in many studies and have been shown to be instrumental in combating noncommunicable as well as infectious diseases, which are major contributors to global mortality and morbidity (WHO 2003, 2012a). The following sections review the existing evidences connecting K to human health outcomes.

9.4.4.4.1 Hypertension

An estimated 26% of the world's population (972 million people) has hypertension, and the prevalence is expected to increase to 29% by 2025, driven largely by increases in economically developing nations (Kearney et al. 2005). The high prevalence of hypertension exacts a tremendous public health burden and is preventable. As a primary contributor to heart disease and stroke, the first and third leading causes of death worldwide, respectively, high blood pressure was the top modifiable risk factor for disability-adjusted life-years lost worldwide in 2013 (Forouzanfar et al. 2015). Approximately, 32% of all the deaths in the world (17.8 million) occurred due to CVDs in 2017 (https://www.who.int/elena/titles/potassium_cvd_adults/en/, accessed on 8th May, 2020).

There are several mechanisms by which K reduces blood pressure. In acute conditions, enhanced plasma K brings about endothelium-dependent vasodilation by the stimulation of Na$^+$-K$^+$ ATPase pumps and the opening of K channels in vascular smooth muscle cells and adrenergic nerve receptors. In the long term, K induces increases in the number of Na$^+$-K$^+$ ATPase pumps in basolateral cell membranes and also increases the trans-epithelial voltage (Rodan 2017). Many epidemiological studies have shown an inverse association between K intake and blood pressure (INTERSALT 1986; Khaw and Rose 1982; Dyer et al. 1994). The large international study of electrolytes and blood pressure (INTERSALT) confirmed that K intake, as measured by 24-h urinary K excretion, was an important independent determinant of population blood pressure. Similarly, a meta-analysis of 32 randomized K supplementation trials showed that an increase of 53 mmol day^{-1} in K intake resulted in a fall in blood pressure of 4.4/2.5 and 1.8/1.0 mm Hg in hypertensive and normotensive individuals, respectively (Whelton et al. 1997). Increased intake of K via fruits and vegetables is also beneficial for arterial pressure by increasing Na urinary excretion (Cogswell et al. 2016). Adequate K intake was the most influential dietary component in lowering blood pressure for decreasing the incidence of hypertension in Americans by 17% and increasing the life span by 5.1% (Roger et al. 2012). In a cross-sectional study of 4,716 participants, frequency of high systolic blood pressure was reported more in adolescents with K intake <700 mg day^{-1} than those consuming >2,800 mg day^{-1}

of K (Chmielewski and Carmody 2017). The Dietary Approaches to Stop Hypertension (DASH) reports that it is certain that lowering Na in the diet lowers the blood pressure, but the effect is amplified when the diet of the same individual is enriched with K. Hence, it is important to consider the Na-to-K ratio in the discussion of K homeostasis (McDonough and Youn 2017). It is, however, important to understand the bioavailability of K and the specific influence of various compounds and processes on its movement in order to better understand its role in human health (Stone et al. 2016).

9.4.4.4.2 Cardiovascular Diseases

In a 12-year prospective study, Khaw and Barrett-Connor (1987) showed that an increase of 10 mmol K day^{-1} intake was associated with a 40% reduction in the risk of stroke-associated mortality. This association was independent of other dietary variables and also independent of other known cardiovascular risk factors including age, sex, blood pressure, blood cholesterol level, obesity, fasting blood glucose level, and cigarette smoking. A community-based prospective study concluded that community-dwelling elders with low-to-normal serum K (3.5–3.8 mmol L^{-1}) were more prone to cardiovascular mortality rates compared with those with mid-normal values (3.9–4.4 mmol K L^{-1}) (Lai et al. 2015). Randomized trials have shown that increasing fruit and vegetable consumption with a subsequent increase in 24-h urinary K excretion lowers blood pressure to a similar extent as K supplementation. As elevated blood pressure throughout the range is the major cause of CVD (Lewington et al. 2002), the blood pressure-lowering effect of K is likely to be a major mechanism that accounts for the protective effect of fruit and vegetables against CVD. Further, an experimental study found a causative link between the reduced dietary K and vascular calcification in atherosclerosis by unveiling the pathogenic mechanisms which integrate the increased intracellular Ca influx, activated CREB signaling, and elevation of autophagy resulting in initiation and progression of vascular calcification (Sun et al. 2017). Similarly, a Japanese prospective study has shown that the incidence of cardiac complications in Type II diabetes mellitus patients with normal renal function is lower in those with higher urinary K excretion than those with lower K excretion (Araki et al. 2015). Increasing evidence from epidemiological studies in humans and experimental studies in animals also suggests that K may have a direct beneficial effect on the cardiovascular system, which may be independent of, but additive to, its effect on blood pressure (He and MacGregor 2001). For instance, experimental studies in animals suggest that K may have inhibitory effects on free radical formation, vascular smooth muscle proliferation, and arterial thrombosis (Young and Ma 1999). It has also been shown that K may reduce macrophage adherence to the vascular wall and thereby contribute to the reduction of vascular lesions (Ishimitsu et al. 1995). In an outcome trial of K-enriched salt on cardiovascular mortality in 1981, elderly veterans, randomized into two groups, were given either K-enriched salt (experimental group) or regular salt (control group) for approximately 31 months (Chang et al. 2006). They showed that with the intervention, there was a 76% increase in K intake and a 17% reduction in Na intake in the experimental group, 60% reduction in CVD mortality, participants lived 0.3–0.9 years longer, and spent significantly less (approximately US$426 per year) for in-patient care for CVD after controlling for age and previous hospitalization expenditures.

9.4.4.4.3 Kidney and Bone Disorders

Many studies have shown that an increase in dietary K intake reduces urinary Ca excretion and causes a positive Ca balance (Lemann et al. 1991, 1993). Increased K intake plays an important role in the management of hypercalciuria (Osorio and Alon 1997). By reducing urinary Ca excretion, high K intake reduces the risk of kidney stone formation, as Ca is the main component of most urinary stones. A 4-year prospective study of 45,619 men aged 40–75 years showed that K intake was inversely related to the risk of kidney stones (Curhan et al. 1993). In humans, several case reports suggest that chronic hypokalemia is related to renal lesions (Riemenschneider and Bohle 1983). Experimental studies in hypertensive rats have shown that high K intake prevents the development of renal vascular, glomerular, and tubular damage, independent of its effect on blood pressure (Tobian et al. 1984).

Numerous population-based studies underline the beneficial effects of dietary K or fruit and vegetables on bone health through its effect on acid-base balance (Macdonald et al. 2005; New et al. 2000). For example, the cross-sectional study by New et al. (1997) in 994 healthy premenopausal women aged 45–49 years showed that with increasing K intake, there was a significant increase in bone mineral density in the lumbar spine and femoral neck after the adjustment of confounding factors. Treatment trials have shown that supplementation with K bicarbonate (Sebastian et al. 1994) or K citrate (Marangella et al. 2004) reduced the bone turnover in post-menopausal women. In an ancillary study to the DASH-Sodium trial, Lin et al. (2003) showed that, compared with the usual American diet, the DASH diet reduced bone turnover significantly, as indicated by decreased serum osteocalcin and C-terminal telopeptide of type 1 collagen. Importantly, the effect was significant at all three levels of salt intake. The study also suggested that the DASH diet and reduced salt intake may have complementary and beneficial effects on bone health.

9.4.4.4.4 Glucose Intolerance

Blood glucose levels are stringently regulated within the range of 70–100 mg dL^{-1}, failing which diabetes mellitus results. Potassium is crucial in the blood glucose control by regulating insulin secretion from the pancreas. Glucose intolerance often occurs in clinical conditions where there is severe hypokalemia and a deficit in K balance, such as primary or secondary aldosteronism or after prolonged treatment with diuretics. An analysis of 59 clinical trials of thiazide diuretics showed a strong relationship between hypokalemia and glucose intolerance (Zillich et al. 2006). This study, as well as others, suggests that the treatment of thiazide-induced hypokalemia with K supplementation or K-sparing diuretics could lessen the glucose intolerance and possibly prevent the development of diabetes. Further evidence in support of the role of K in glucose tolerance comes from experimental studies in healthy individuals (Rowe et al. 1980), where K depletion, e.g., induced by a low K diet, caused glucose intolerance associated with impaired insulin secretion. One prospective cohort study (84,360 US women aged 34–59 years followed up for 6 years) showed that a high K intake was associated with a lower risk of developing Type II diabetes (Colditz et al. 1992). Similarly, other such large-scale studies have shown that dietary K intake is significantly associated with diabetes risk (Chatterjee et al. 2011).

9.4.4.4.5 Cardiac Arrhythmias

The relationship between intracellular and extracellular K plays a major role in determining the electrophysiological properties of cardiac-conducting tissue. Hypokalemia can cause prolonged repolarization, the pathogenic factor in the genesis of torsade de pointes, particularly in those at risk, i.e., patients with ischemic heart disease, heart failure, and left ventricular hypertrophy. Several studies suggested that administration of K to raise serum K levels could improve repolarization in patients with inherited or acquired long QT syndrome (Compton et al. 1996; Tan et al. 1999). For instance, the Multiple Risk Factor Intervention Trial where 1,403 hypertensive men were on diuretics showed that for a decrease of 1 mM in serum K, there was a 28% increase in the number of ventricular premature complexes (Cohen et al. 1987). In patients with acute myocardial infarction, low serum K is related to a higher risk of severe ventricular arrhythmias. A recent meta-analysis of cohort studies showed that patients with acute myocardial infarction who developed primary ventricular fibrillation had an average of 0.27 mM lower serum K at admission (Gheeraert et al. 2006). Other studies have shown a concentration-dependent risk between hypokalemia and ventricular arrhythmias in patients with acute myocardial infarction, and the relationship was independent of diuretic usage (Gettes 1992). The major sympathetic response to acute myocardial infarction with an elevation of circulating catecholamines, particularly adrenaline by stimulating b2-adrenergic receptor, causes an influx of K into cells, thereby causing a reduction in serum K. Beta-blockers lessen hypokalemia in acute myocardial infraction, and this may partially explain their benefit (Nordrehaug et al. 1985a, 1985b). In patients with heart failure, hypokalemia is very common which may be, in part, because of the diuretics but may also be because of activation of the renin–angiotensin system and the sympathetic nervous system. Low serum K in chronic heart failure increases the likelihood of arrhythmias, particularly those related to digoxin therapy. Correction of the serum K reduces the frequency and complexity of ventricular arrhythmias and may possibly prevent the subsequent occurrence of sudden cardiac death (Nolan et al. 1998).

9.4.4.5 Potassium in Immune System Disorders

The human body has an intricate and efficient defense system comprising cells and biochemicals that protect from invading pathogens. Potassium plays a crucial role in the function of the human immune system by regulating ion channels of the macrophages, T-cells, and B cells involved in the immune system.

Among these channels, the voltage-gated Kv1.3 and the Ca-activated KCa3.1 channels are the two essential ones. The function of Kv1.3 is to maintain a negative membrane potential and to support Ca^{2+} influx, whereas the Ca-activated KCa3.1 channel sustains more intense Ca^{2+} signaling events in the activated human T-cells and induces the production of different cytokines, namely, IL-2, IFN-ϒ, and IL-17 and also induces memory T-cell differentiation (Feske et al. 2015; Hasso-Agopsowicz et al. 2018). Memory B cells show high expression of Kv1.3. It is shown that Kv1.3 and KCa3.1 blockers inhibit the production of the memory B cells and is observed in patients with primary biliary cirrhosis (Feske et al. 2015). The K-regulated KCa3.1 channel is the predominant K^+ channel on Th1 and Th2 cells and may be useful to treat disease mediated by these cells. Thus, they can be a major target in the treatment of disease like Crohn's disease and ulcerative colitis (Feske et al. 2015).

9.4.4.6 Potassium in Cancer and Autoimmune Diseases

It has been known that continual T-cell recognition signaling and progressive cellular differentiation occur in cancers due to persistent antigen exposure to T-cells which leads to the exhaustion of tumor-infiltrating lymphocytes. T-cells depend on the extracellular nutrients to remain robust. It was observed that elevated extracellular K reduced the uptake and consumption of local nutrients by antitumor T-cells, thus inducing a state of caloric restriction. Therefore, increased level of extracellular K restricts T-cell effector function by limiting nutrient uptake, thereby inducing autophagy and reduction of histone acetylation at effector and exhaustion loci which ultimately produces CD8 + T-cells with enhanced in vivo persistence, multipotency, and tumor clearance (Baixauli et al. 2019).

In a large-scale study among US women, it was found that dietary intake of K, but not Na, was inversely associated with Crohn's disease (Khalili et al. 2016) by modulating the IL-23/T_H17 pathway showing a genetic×environment interaction in disease pathogenesis. Potassium was reported not only to enhance the induction of Foxp3 induced by TGF-β1 but also to reinforce Foxp3 expression in TH17 cells, suggesting a possible anti-inflammatory function of K via induction of Foxp3-mediated T-cells tolerance.

9.4.4.7 Potassium and Infectious Diseases

SARS-CoV-2 has recently caused a series of COVID-19 infections globally. Hypokalemia is prevalent in patients with COVID-19. SARS-CoV-2 binds to ACE2 of the renin-angiotensin system and causes dominant hypokalemia. Of the 175 cases of COVID-19 reported in the Wenzhou hospital of China, 93% of the severely and critically ill patients had hypokalemia. The COVID-19 patients with severe hypokalemia responded favorably when administered supplements at doses of 3,000 mg K day^{-1}. The correction of hypokalemia is challenging because of the continuous renal K loss resulting from the degradation of ACE2 (Chen et al. 2020). Apart from SARS-CoV-2, *Mycobacterium tuberculosis* also shows unique transcriptional response to change in K levels (MacGilvary et al. 2019).

9.5 INTERACTIONS AMONG HEALTH COMPONENTS

The importance of maintaining an adequate supply of mineral nutrients in the human and animal food supply has been recognized for many years. The lack of sufficiently diverse diets in many low-income families is exacerbated by their reliance on staple foods (such as wheat, rice, and maize) that do not provide sufficient quantities of many nutrients. Perhaps, the attention on the mineral nutrients has mostly been focused on increasing the concentrations of selenium (Se), zinc (Zn), iron (Fe), and iodine (I) to overcome the human dietary shortages.

There are several examples of successful intervention to increase the mineral concentration of staple foods. One of the first attempts to increase the mineral concentration of crops was done in Finland. Selenium is essential for animals and humans but is not required for plant growth. In 1984, the Finnish government implemented a strategy to fortify agricultural fertilizers with selenite to increase the concentration of Se in their population. The Se-enhancement program has been very successful (Alfthan et al. 2015). Another successful example of boosting the nutritional quality of food is the introduction of I to soil, crops, or irrigation water (Qiang et al. 2008; Mao et al. 2014). A large-scale effort was launched in 2006 by HarvestPlus to increase the concentrations of mineral nutrients in staple foods, especially Fe and Zn (harvestplus.org).

Traditional soil fertilization approaches to boost the mineral content of food must overcome the challenges presented by the chemical environment in the soil and in the rhizosphere. For example, when Zn or Fe fertilizers are added to the soil, they go through a series of reactions that reduce their solubility and availability for plant uptake. The unique soil mineralogy and environment in each field will determine the extent of these reactions and the processes that regulate plant nutrient availability. Similarly, addition of selenite (SeO_3^{2-}) to soil results in far less plant Se recovery than an application of selenate (SeO_4^{2-}) in most soils (Mikkelsen et al. 1989). The method and form of fertilizer applied, the specific soil environment, and the crop rooting characteristics all need to be considered when attempting to manipulate the mineral concentrations of crops.

Applications of K fertilizer result in far fewer competing soil and microbial reactions that reduce plant availability than for many other nutrients. Following the addition of K fertilizer to soil, there are very few competing microbial processes (compared with N) that reduce the soluble K concentration, and the cation exchange reactions primarily dominate the initial chemical processes. The direct application of K fertilizer to the soil is a practical and effective nutrient delivery mechanism. The low environmental impact of K fertilization also provides some flexibility in its management, compared with N and P fertilizers.

Potassium fertilizer is almost always applied with the expectation of an economic return in increased yield, so the additional benefits are easily overlooked. The advantages associated with K fertilization and enhanced crop quality and improved resistance to biotic and abiotic stress are sometimes appreciated, but these are not primary factors in the fertilizer-decision-making process.

Even less valued and recognized are the benefits for human nutrition from the harvested crops when they are enriched in K. A number of healthful components of food are boosted by the application of nutrients, in particular K (Bruulsema 2000). A large volume of nutrition research has recently been focused on potential benefits of functional foods. This term generally refers to food that contains significant amounts of naturally occurring, biologically active molecules that have health benefits beyond the basic mineral nutrients (Jitendra and Amit 2015). Boeing et al. (2012) and Dillard and German (2000) chronicle the multiple health benefits of a diet rich in phytochemicals. There is evidence that plant nutrition will influence the accumulation of beneficial phytochemicals. Bruulsema et al. (2000) showed the relationship between K fertilization and the concentration of phytochemicals (e.g., isoflavone) in soybeans. The link between mineral nutrition and various functional foods has been demonstrated for a number of crops. For example, lycopene content increased by 67% with K fertilization (Serio et al. 2007). However, the health benefit of additional phytochemicals resulting from K fertilization remains largely unexplored. The nutritional benefits from K fertilization are not valued because consumers cannot easily measure the content in their food, and farmers rarely receive financial rewards for producing crops with above-average nutritional value.

9.5.1 Case Study

To explore opportunities and challenges with integrating soil, plant, and human health, we focus on one food and three nutrients to perform a small case study: K, Na, and vitamin C (ascorbic acid, ascorbate). Potato is a staple crop, ranking fourth in consumption worldwide after maize, wheat, and rice. Potato products delivered to consumers may include tubers, chips (crisps), frozen potatoes, dehydrated potatoes, and flour. Besides being a source of carbohydrates that provides satiety, potatoes are among the highest K-containing foods (Weaver 2013). Potatoes are also an important source of vitamin C (Love and Pavek 2008).

9.5.1.1 Potato as a Source of Nutrients and Energy

Potassium, Na, and vitamin C are important to human health outcomes. The relationship between higher consumption of dietary K and reduced incidence of noncommunicable diseases has been demonstrated previously; however, for benefits to be realized, higher levels of K need to be accompanied by lower intakes of Na (Weaver 2013). Ascorbate is an electron donor for eight enzymes in the human body. Inadequate intake of vitamin C leads to anemia and scurvy.

To avoid these diseases, recommended daily intake levels have been established for both K and ascorbic acid. These requirements vary by group (Table 9.3). The recommended daily intake is 3,510 mg K day^{-1} for adults, and the maximum Na intake is 2,000 mg day^{-1}, leading to a minimum K:Na ratio of 1.76:1. The WHO (2012a, 2012b) recommends that intakes of K and Na for children and adolescents should be reduced based on the lower energy requirements of those groups; however, no quantitative guidance is given. We decided to use the proportion of energy in the children's

---

TABLE 9.3
Recommended Intakes of Daily Energy, Potassium (K), Sodium (Na), and Ascorbic Acid for Various Groups

	Girls 4–5	Girls 8–9	Girls 11–12	Adolescent Women 14–15	Adolescent Women 16–17	Pregnant Women	Lactating Women	Adult Women 19–65
Energy (kcal)[a]	1,241	1,698	2,149	2,449	2,503	3,296	5,624	2,800
Potassium (mg K)[b]	1,556	2,129	2,694	3,070	3,138	3,510	3,510	3,510
Sodium (mg Na)[c]	886	1,213	1,535	1,749	1,788	2,000	2,000	2,000
Ascorbic acid (mg)[d]	30	35	35	40	40	55	70	45

[a] Data are from FAO (2004). The energy requirement of adult women (2,800 kcal day^{-1}) was that for a 65 kg woman, 18–29.9 years, and an active lifestyle (2.20×BMR). The additional energy requirements for pregnant and lactating women were added to this value. The additional energy requirement for pregnant women (496 kcal day^{-1}) was the estimated additional energy cost of pregnancy for the third trimester (the most energy costly). The additional energy requirement of 2,824 kcal day^{-1} was added for lactating women at 6 months postpartum.

[b] Data are from WHO (2012a). The required K intake was 3,510 mg day^{-1} and adjusted downward by the proportion of the energy requirement of a given group (kcal day^{-1}) to that of adult women (2,800 kcal day^{-1}). A maximum of 3,510 mg day^{-1} was used.

[c] Data are from WHO (2012b). The maximum allowable Na intake was 2,000 mg day^{-1} and adjusted downward by the proportion of the energy requirement of a given group (kcal day^{-1}) to that of adult women (2,800 kcal day^{-1}). A maximum of 20,000 mg day^{-1} was used.

[d] Data are from WHO and FAO (2004).

and adolescent's groups compared with that of the adult group as the fraction of K recommended. For each group, the ratio of K:Na equaled 1.76:1. Quantitative recommendations for ascorbic acid already existed for various age groups (WHO and FAO 2004), requiring no interpolation.

To quantify the extent to which potatoes can contribute K and vitamin C to the human diet while keeping Na intake low, we combined recommended daily intakes and potato nutrient contents. Table 9.4 presents a listing of average energy and nutritional contents in 100 g of raw, white potato with skin (USDA 2020). Potatoes are about 82% water and supply 69 kcal of energy, 407 mg K, and 9.1 mg ascorbic acid (100 g fresh weight). Additionally, fat content is low (0.1 g total lipids). Most of the carbohydrates (about 86%) are starch, while sugars represent about 7.3%. The K:Na ratio in white potatoes is 25:1, making it a high K, low Na food. Per unit of energy (1 kcal), potatoes deliver 5.9 mg K and 0.13 mg ascorbic acid. Such nutrient density makes it a good choice for obtaining K and ascorbic acid in reduced calorie, low fat, and low Na diets.

It is worth noting some key aspects of the data in Table 9.4. In the authors' experiences accessing these data from the USDA in the past, only the amounts of nutrients were available for given serving sizes. New additions have been the number of observations used to derive each amount (n) and the minimum (Min) and maximum (Max) of the range. Providing additional meta-data, such as the date the amounts were last updated and whether or not the amounts were measured (analytical) or calculated provides needed transparency to the end user. Similar meta-data and transparency do not currently exist for nutrient contents reported in agricultural information intended for farmers and their consultants.

9.5.1.2 Linking Soil and Potato Plant Health to Human Health

Managing soil and plant health can have meaningful impacts on human nutrition. We examined the following outcomes of K fertilization in potato production: (i) increasing the quantity of nutrients produced on a given area of land; (ii) changing the nutritional profile of potatoes.

9.5.1.2.1 Increasing the Quantity of Nutrients Produced

A key functional outcome of soil and plant health is producing needed quantities of energy and nutrients. Yield is generally the primary measure of performance in agricultural systems. For instance, research conducted in Rwanda (Haverkort and Rutayisire 1986) found that K fertilization

---

TABLE 9.4

Energy and Nutrients in 100 g Raw White Potato, Flesh and Skin (USDA 2020)[a]

Component	Derived by	Amount	n	Min	Max	Last Updated
Water (g)	Analytical	81.6	3	80.2	83.9	5/1/2002
Energy (kcal)	Calculated	69	-	-	-	8/1/2010
Energy (kJ)	Calculated	288	-	-	-	8/1/2010
Protein (g)	Analytical	1.68	3	1.54	1.86	5/1/2002
Total lipid (fat) (g)	Analytical	0.1	3	0.08	0.14	5/1/2002
Carbohydrate (g)	Calculated	15.7	-	-	-	8/1/2010
Fiber, total dietary (g)	Analytical	2.4	1			3/1/2006
Sugars, total including NLEA (g)	-	1.15	-	-	-	8/1/2010
Starch (g)	Calculated	13.5	-	-	-	5/1/2002
Potassium (mg K)	Analytical	407	3	379	433	5/1/2002
Sodium (mg Na)	Analytical	16	3	5	23	8/1/2010
Ascorbic acid (mg)	Analytical	9.1	3	7.6	11.7	2/1/2002
Folate, food (mcg)	Analytical	18	3	9	34	8/1/2010

[a] *Database entry:* potatoes, white, flesh and skin, raw; portion: 100 g; data type: SR Legacy; food category: vegetables and vegetable products; FDC ID: 170028; NBD Number: 11354; FDC published: 4/1/2019; accessed 5/1/2020.

up to 166 kg K ha^{-1} significantly increased average yields by 1.5–3.3 Mg ha^{-1} during the dry season and 1.1–6.6 Mg ha^{-1} during the rainy season.

To understand the importance of increasing yield for human nutrition, we calculated how much of the annual requirements of one person would be met with an additional quantity of 1 Mg ha^{-1} of potatoes. We multiplied the quantities of nutritional components in potatoes in Table 9.2 by 1 Mg ha^{-1}. The result was 6.9×10^5 kcal ha^{-1}, 4.07×10^6 mg K ha^{-1}, and 9.1×10^4 mg ascorbic acid ha^{-1}. Additionally, 1 Mg of potatoes is equivalent to about 5,000 medium-sized (200 g) potatoes. The quantities of nutrients were then put in terms of percentages of the annual requirements of one person that 1 Mg would meet, based on the daily requirements multiplied by 365 days. Percentages less than 100 met less than a year's requirement and percentages over 100 met more than a year's, with every 100% increase representing an additional year or an additional person in the same year.

Table 9.3 shows the ramifications of potato being a low-energy, high-nutrient density food. The most intensive energy requirements are for lactating women, and 1 Mg of potatoes would meet only a third of their individual annual energy requirement. The lowest energy-requiring group in Table 9.3 is children aged 4–5 years, and 1 Mg would provide over 1 year worth of energy for one child. For K, 1 Mg of potatoes provide enough K to provide annual requirements for 3 adolescents and adults, 4 girls aged 11–12, 5 girls aged 8–9, and 7 girls aged 4–5.

We calculated the fresh weight (81.6% moisture) of potatoes needed to meet the requirements in Table 9.3. The results are presented in Table 9.5. To meet the greatest K requirements, 863 g of potatoes would be needed per day (about 4.5 medium potatoes). The minimum requirements, those of girls ages 4–5, could be met with 383 g of potatoes (about 2 medium). Lower quantities of potatoes are needed to be consumed to meet ascorbic acid requirements. Tables 9.5 and 9.6 demonstrate that potatoes are important sources of K and ascorbic acid and that simply producing sufficient quantities of these nutrients per unit land area is a key function of soil and plant health.

In Africa, eight countries produce over 1 Mt of potatoes annually: Algeria, Egypt, Kenya, Malawi, Morocco, Nigeria, South Africa, and the United Republic of Tanzania (FAO 2020).

TABLE 9.5

Percentages of Per-person Annual Requirements for Energy, Potassium, or Ascorbic Acid Provided by 1 Mg ha^{-1} Potato Yield (Fresh Weight)

Requirement	Girls 4–5	Girls 8–9	Girls 11–12	Adolescent Women 14–15	Adolescent Women 16–17	Pregnant Women	Lactating Women	Adult Women 19–65
Energy (kcal)[a]	152	111	88	77	76	57	34	68
Potassium (mg K)[b]	717	524	414	363	355	318	318	318
Ascorbic acid (mg)[c]	831	712	712	623	623	453	356	554

Based on Table 9.1, 1 Mg ha^{-1} of potatoes is estimated to provide 6.9×10^5 kcal, 4.07×10^6 mg K, and 9.1×10^4 mg ascorbic acid, and the equivalent of 5,000 medium (200 g) potatoes.

[a] Data are from FAO (2004). The energy requirement of adult women (2,800 kcal day^{-1}) was that for a 65 kg woman, 18–29.9 years, and an active lifestyle ($2.20 \times$ BMR). The additional energy requirements for pregnant and lactating women were added to this value. The additional energy requirement for pregnant women (496 kcal day^{-1}) was the estimated additional energy cost of pregnancy for the third trimester (the most energy costly). The additional energy requirement of 2,824 kcal day^{-1} was added for lactating women at 6 months postpartum.

[b] Data are from WHO (2012a). The required K intake was 3,510 mg day^{-1} and adjusted downward by the proportion of the energy requirement of a given group (kcal day^{-1}) to that of adult women (2,800 kcal day^{-1}). A maximum of 3,510 mg day^{-1} was used.

[c] Data are from WHO (2012b). The maximum allowable Na intake was 2,000 mg day^{-1} and adjusted downward by the proportion of the energy requirement of a given group (kcal day^{-1}) to that of adult women (2,800 kcal day^{-1}). A maximum of 20,000 mg day^{-1} was used.

TABLE 9.6
Fresh Weight (g) of Potatoes Required to Meet Daily Intake Requirements for Potassium or Ascorbic Acid. None of the Weights Exceed Na Intake Maxima

Requirement	Girls 4–5	Girls 8–9	Girls 11–12	Adolescent Women 14–15	Adolescent Women 16–17	Pregnant Women	Lactating Women	Adult Women 19–65
Potassium (mg K)[a]	383	524	662	755	772	863	863	863
Ascorbic acid (mg)[b]	330	385	385	440	440	605	770	495

Based on value in Tables 9.1 and 9.2. A medium potato weighs approximately 200 g.

[a] Data are from WHO (2012a). The required K intake was 3,510 mg day^{-1} and adjusted downward by the proportion of the energy requirement of a given group (kcal day^{-1}) to that of adult women (2,800 kcal day^{-1}). A maximum of 3,510 mg day^{-1} was used.

[b] Data are from WHO (2012b). The maximum allowable Na intake was 2,000 mg day^{-1} and adjusted downward by the proportion of the energy requirement of a given group (kcal day^{-1}) to that of adult women (2,800 kcal day^{-1}). A maximum of 20,000 mg day^{-1} was used.

Of those countries, only four had officially reported data for production, yield, and harvested area (Table 9.7). Striking is the comparison of Kenya potato production with Morocco. Both produce about the same annual quantities of potatoes. However, Morocco's yields are 3.5 times higher and produced on less than a third of the area.

Quantitative soil health indexes have yet to be applied to potato-production areas. Such indexes integrate physical, chemical, and biological factors into an overall, quantitative score of soil health. Examples of these indexes include SMAF—the Soil Management Assessment Framework (Andrews et al. 2004) and CASH—the Comprehensive Assessment of Soil Health (Moebius-Clune et al. 2016). CASH incorporates exchangeable K into the index, using a "more is better" score. Higher K fertility, therefore, is thought to contribute to improved soil health.

As Wulff et al. (1998) point out, some soils, particularly those with poor K-holding capacity, cannot accumulate a large reserve of soil K and potatoes must be fertilized each season. Therefore, soil test K concentrations cannot be the sole indicator of the K component of soil health, and more complete indicators of the contribution of K will need to be developed.

Although potatoes are an important food crop worldwide, a few studies have pointed out possible impacts on some components of soil health. Intensive potato production can lead soils to have lower microbial biomass carbon (Nelson et al. 2009), lower earthworm abundance and mass (Nelson et al. 2009), lower residue on the soil surface (Hills et al. 2020, Zebarth et al. 2019), and a risk of soil compaction (Hills et al. 2020). For example in Egypt, one of the world's major potato-producing countries, mono-cropping potatoes is a common practice. As a result, soil health has been declining along with potato productivity (El-Azeim et al. 2020). Mono-cropping is not a recommended practice potato production for a number of reasons.

TABLE 9.7
Total Production, Yield, and Harvested Area of Potatoes for the Four Countries with Complete Sets of Unimputed Data and Producing over 1 Mt of Potatoes Annually (FAO 2020)

Country	Total Production (Mg)	Total Harvested Area (ha)	Yield (Mg ha^{-1})
Algeria	4,653,322	149,665	31.1
Kenya	1,870,375	217,315	8.6
Malawi	1,125,874	68,133	16.5
Morocco	1,869,149	62,033	30.1

Lambert et al. (2005) reviewed the impacts of crop nutrition on several potato diseases. Their study revealed that little work has been done to isolate the effects of K on disease incidence and severity specifically for potatoes.

9.5.1.2.2 Changing the Nutritional Profile of Potatoes

Potassium fertilization can create changes in the nutrient content of potatoes that improves their suitability for human consumption, especially on soils that have health problems. An example comes from Algerian research conducted on a highly saline soil with an electrical conductivity of 5.9 dS m^{-1} (Oustani et al. 2015). In Algeria, salinity is one of the main constraints to potato production. On this saline soil, poultry manure was applied. Compared with the check, the highest rate of manure application increased potato yield from 20.5 to 43.6 t ha^{-1} (Table 9.8). With no poultry manure applied, the K concentration in tubers was only 291 mg 100 g^{-1}, only 71% of the USDA reference; however, with poultry manure the K concentration increased to a little over twice the USDA reference. In addition, the Na content dropped from 589 to 412 mg Na 100 g^{-1}. Without the manure application, tuber Na concentrations are so high and K concentrations are so low that 72% of the dietary intake need of K would be left unmet with potato consumption after reaching the Na dietary intake limits. At this Na limit, potato consumption would only meet 28% of the dietary K needs. Where this very large quantity of manure was applied, all dietary K needs could be met without exceeding the Na intake limits. The higher yield, higher K content, and lower Na content resulting from the manure application yielded a total of 3.71×10^8 mg K ha^{-1}—enough to safely meet the annual dietary K intake needs of 289 adult women.

Albeit scare, but there is some evidence that K fertilization can increase ascorbic acid content of potato tubers. Mondy and Munshi (1993) found that high application rates of K (520 kg K ha^{-1}) increased tuber ascorbic acid concentration by 4–5 mg 100 g^{-1} FW (fresh weight). Hamouz et al. (2007) also observed 6% increases in ascorbic acid with increases in K fertilization rates. Similar increases were shown by Naumann et al. (2020). Although there appears to be a relationship, more work is needed to quantify these responses and their probability.

Although potatoes contain beneficial nutrients and vitamins, they also contain or can form compounds that may pose health risks. There are two types of compounds of concern in potatoes: glycoalkaloids and acrylamide.

Potassium alone or in combination with other nutrients has been shown to increase the glycoalkaloid concentration in potatoes (Wszelaki et al. 2005). Glycoalkaloids are present in all members of the Solanaceae family. Friedman (2006) provided a comprehensive review of glycoalkaloids in potatoes, and a few of his key points follow. The primary glycoalkaloids in potato tubers are α-chaconine and

---

TABLE 9.8

Statistically Significant Changes in Yield and Composition of Tubers After Applying 60 t ha^{-1} Chicken Manure (Oustani et al. 2015)

Potato Tuber Parameter	Check	60 t ha^{-1} Manure	USDA Reference[b]
Yield (t ha^{-1})	20.5	43.6	-
mg K (100 g FW[a])$^{-1}$	291	852	407
mg Na (100 g FW)$^{-1}$	589	412	16
K:Na ratio	0.49	2.07	25.4
% of recommended K intake after reaching Na limit[c]	−72	18	1,348

[a] FW = fresh weight at 81.6 % moisture.

[b] From Table 9.4.

[c] Percent of recommended K intake (Table 9.3) unmet (negative) or oversupplied (positive) after reaching the maximum Na intake limits in Table 9.3. The percent of recommended K intake presented is the same for all groups listed in Table 9.3.

α-solanine. Potatoes produce glycoalkaloids to guard against attacks by bacteria, fungi, viruses, insects, animals, and humans. These compounds are much more concentrated in the skin of the tuber than in the flesh, as the skin is the first line of defense against attack by soil pathogens and animals. While K fertilization may help the plant stave off attacks, ingestion of glycoalkaloids can cause a number of human health issues, adversely affecting the intestinal tract and nervous system and in large doses, poisoning and death (Nie et al. 2018). Of the two primary glycoalkaloids, α-chaconine is more toxic than α-solanine. The recommended glycoalkaloid threshold in potato products is 20 mg 100 g^{-1} FW.

Glycoalkaloids may also have health benefits (Friedman 2006). They may help lower cholesterol, reduce inflammation, reduce blood sugar levels, and provide antibiotic effects. There is much yet to learn about these compounds, and studies have typically relied on pure forms of these compounds, rather than on food sources.

Acrylamide is another compound of concern since it is a known carcinogen and intake should be limited. Potatoes do not contain acrylamide; however, its precursors are present in raw potatoes: reducing sugars (glucose and fructose) and free asparagine. Potassium fertilization can decrease levels of both these precursors, and bringing down sugar levels is most effective for minimizing acrylamide production (Gerendás et al. 2007; Rosen et al. 2018).

In the presence of high heat, reducing sugars react with asparagine to form acrylamide through the Maillard as well as other reactions (Camire et al. 2009). Both French fries and crisps have the highest levels of acrylamide of the processed potato products (Camire et al. 2009).

9.5.1.3 Postharvest Fate of Tuber Nutrients

The discussion has been focused on the impacts of K fertilization on the nutrient composition of raw potato tubers (Step 1, Figure 9.5); however, there are many steps that exist between harvest and human consumption. The final fate of potato K is unknown as it moves through multiple steps from the field to the consumer plate. Since we could find no studies that traced these processes for a given batch of raw potatoes, we assembled representative information on each.

9.5.1.3.1 Downstream Value Chain Processes (Step 2, Figure 9.5)

Downstream value chain processes include steps that create potato products that are delivered to the wholesaler or retailer. This process may begin during the initial potato grading, washing, and possible months-long storage.

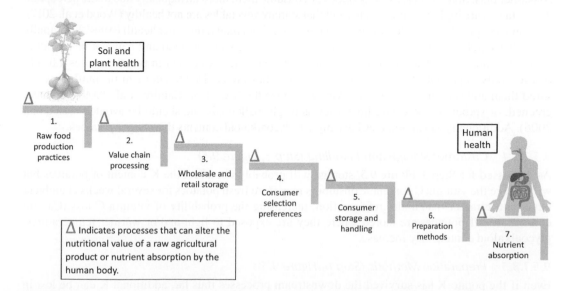

FIGURE 9.5 Diagram of processes that affect the content and postharvest fate of tuber nutrients.

It is likely that processed potatoes contain less K than their raw precursors. Fresh produce delivered to wholesalers or retailers likely has K concentrations near that present at harvest. Removing the skin, and also cutting and washing during processing, decreases the K content. The addition of potassium sorbate as a preservative may increase the K content.

Peeling the skin to a depth of 3–4 mm removes nearly all of the glycoalkaloids (Friedman 2006). When potatoes are fried in oil, glycoalkaloids diffuse from the tuber flesh into the oil. Repeated batches of potatoes in the same oil can saturate oil with glycoalkaloids. When this happens, glycoalkaloids will diffuse from the oil back into the potato flesh. The frequency with which oil is changed, the glycoalkaloid content of the potatoes being fried, whether or not the potatoes have been peeled, and the number of batches between oil changes all affect the glycoalkaloid content in fried products (Friedman 2006).

Most processes (cutting, washing, cooking) decrease the vitamin C content of the final potato products (Camire et al. 2009; Love and Pavek 2008), with dehydrated products having the lowest content (Augustin et al. 1979). Preventing enzymatic browning during processing is important in the production of potato products. Ascorbic acid may be added in washing solutions to reduce enzymatic browning, which increases the ascorbic acid content in the products (Bobo-Garcia et al. 2019).

9.5.1.3.2 Wholesaler/Retailer Storage and Handling (Step 3, Figure 9.5)

Potassium fertilization can increase the shelf life of whole potatoes (Moinuddin and Umar 2004). Storage is not expected to change the K content. However, ascorbic acid content can decrease steadily with storage time (Love and Pavek 2008). If $CaCl_2$ was added to the washing solution (for peeled potatoes), ascorbic acid levels are maintained (Hägg et al. 1998).

Storage temperature at 8°C can reduce "cold sweetening"—a process where concentration of reducing sugars increases, as observed at a lower storage temperature of 4°C (Rosen et al. 2018). After cold sweetening, more acrylamide can be formed during high-temperature cooking. Light and/or heat exposure during storage, as well as mechanical injury, stimulates glycoalkaloid synthesis in the tuber (Friedman 2006).

9.5.1.3.3 Consumer Selection Preferences (Step 4, Figure 9.5)

Regardless of how much K is present in potatoes, some consumers choose not to buy them for a variety of reasons. This breaks the chain of soil health to plant health and to human health. One reason is consumer bias. Some people avoid potatoes or consume them more infrequently due to the perception that potatoes are high in carbohydrates and that starchy vegetables are not healthy (Wood et al. 2017). For some people, the choice to not buy potatoes arises from managing other health issues. The rapidly digestible starch in potatoes quickly increases glucose levels in the blood and may not be a healthy choice for those with diabetes (Camire et al. 2009). The high K content in potatoes poses a health risk for dialysis patients. When preparing potatoes, they are advised to cut them in small pieces or shred them and soak them prior to cooking to reduce the K content (Camire et al. 2009). Blighted, greened, or sprouted tubers have higher levels of glycoalkaloids and should be avoided (Friedman 2006). Additionally, small potatoes have higher glycoalkaloid contents than larger potatoes.

9.5.1.3.4 Customer Storage and Handling (Step 5, Figure 9.5)

As discussed for Step 3, Figure 9.5, storage will probably not affect the K content of potatoes but will reduce the vitamin C content. Consumers may store fresh potatoes for several weeks at ambient temperature rather than under refrigeration, increasing the probability of vitamin C loss (Camire et al. 2009). If potatoes are stored where they are exposed to light and/or warmer temperatures, glycoalkaloid content may increase.

9.5.1.3.5 Preparation Methods (Step 6, Figure 9.5)

Even if the potato K has survived the downstream processes thus far, additional K can be lost in the preparation prior to consumption, particularly for fresh-market potatoes. Removing the skin reduces K content, as does cutting and boiling. Highest K retention occurs with roasting and frying when prepared with the skins left on (Camire et al. 2009). The K concentration also varies within the potato tuber. The K concentration at the bud end of the potato is 50% higher than at the stem end (LeRiche et al. 2009). Consequently, the quantity of K ingested may depend in part on what portion of the potato was consumed. Vitamin C content decreases with moisture, heat, and exposure to air (Love and Pavek 2008). Removing the skin, though it reduces the K content, effectively removes glycoalkaloids (Friedman 2006). Cooking under high heat until browning forms acrylamide (Rosen et al. 2018).

9.5.1.3.6   Nutrient Absorption (Step 7, Figure 9.5)

Only a portion of each nutrient in food is absorbed and utilized by the human body after consumption. The amount absorbed depends on

> (i) the state of the food (e.g., raw or processed), (ii) particle size, (iii) mode of size reduction (e.g., cell rupture or separation), (iv) digestive enzymes and products of digestion, (v) composition of the meal (i.e., fat/carbohydrate/protein content), (vi) the presence of bile acids/salts, and (vii) time.

(Stahl et al. 2002)

Thus, nutrient absorption can be equal to or much less than the quantity consumed. Potato-derived K is well-absorbed (96%) by the body. The processes discussed in Figure 9.5 demonstrate that there are many opportunities for nutrient losses to occur during the postharvest processes. Therefore, a potato grown in a way that improved its potential nutritional value may not have that potential fully realized if the downstream processes chip away its nutrient content.

9.6   CONCLUSIONS

Concepts of soil, plant, and human health need to be operationalized and done so in a collaborative manner. A common theme to all three concepts is functionality. To be considered healthy, soils must function properly to provide quality products as well as many other ecosystem services. For our bodies to function properly, we rely on food with nutritional profiles that match our needs and are safe to eat. Scientists in soil health have proposed some operational definitions, as have scientists in human health. Future efforts need to ensure that the definitions within each discipline integrate well with those of the others.

We also see opportunities to quantify the fate of nutrients, or more broadly, food quality, from harvest to consumption. This will require a multidisciplinary effort to trace food composition of single batches of raw food as it passes through each of the steps from the farm to the consumer. Quantifying component losses, gains, and transformations through downstream processes is critical if we are to understand how changes made at the field level translate to changes in food consumed by families and individuals and their eventual health. We envision disciplines coming together to understand the flow of essential minerals and trace elements from the soil to specific food-end products and value chains to create operational plans for quality testing at each stage.

Many opportunities exist for seemingly diverse disciplines to come together to create rich data sets that serve multiple purposes. For instance, agriculture typically limits its testing to mineral content and feed value. Food and nutrition science also test for many of the same nutrients but analyzes many more components. Agreeing on units and the moisture content upon which concentrations are based are clear needs. More broadly, interdisciplinary standards for chemical analyses, data reporting, terminology (controlled vocabularies), and adherence to FAIR data compliance (findable, accessible, interoperable, reusable) will need to be accomplished.

Operational health definitions, nutrient fate analyses, and interdisciplinary standards are key steps to integrate soil, plant, and human health and quantify their interactions. Associating nutrient compositions in food with soil health parameters has, to our knowledge, not been done. Pieces of those stories exist in the scientific literature, but a full integration is lacking. Comprehensive data sets capturing the fates of nutrients and other quality parameters offer promise to deepen our understanding of how soil, plant, and human health interact to meet the energy and nutritional needs of humans to create desired health outcomes in each system.

REFERENCES

Aburto, N. J., S. Hanson, H. Gutierrez, L. Hooper, P. Elliott, and F. P. Cappuccio. 2013. Effect of increased potassium intake on cardiovascular risk factors and disease: systematic review and meta-analyses. *Br Med J* 346:f1378.

Alfthan, G., M. Eurola, P. Ekholm, E. R. Venäläinen, T. Root, K. Korkalainen, H. Hartikainen, P. Salminen, V. Hietaniemi, P. Aspila, A. Aro. 2015. Effects of nationwide addition of selenium to fertilizers on foods, and animal and human health in Finland: from deficiency to optimal selenium status of the population. *J Trace Elem Med Biol* 31:142–7.

Amanulaah, A. Iqbal, Irfanullah, and Z. Hidayat. 2016. Potassium management for improving growth and grain yield of maize (Zea mays L.) under moisture stress condition. *Sci Rep* 6:34627. doi:10.1038/srep34627.

Amtmann, A., S. Troufflard, and P. Armengaud. 2008. The effect of potassium nutrition on pest and disease resistance in plants. *Physiologia Plantarum* 133:682–691.

Andrews, S. S., D. L. Karlen, and C. A. Cambardella. 2004. The soil management assessment framework: a quantitative soil quality evaluation method. *Soil Sci Soc Am J* 68:1945–1962.

Andrist-Rangel, Y., S. Hillier, I. Öborn, A. Lilly, W. Towers, A. C. Edwards, and E. Paterson. 2010. Assessing potassium reserves in northern temperate grassland soils: a perspective based on quantitative mineralogical analysis and aqua-regia extractable potassium. *Geoderma* 158 (3–4):303–314.

Anschütz, U., B. Dirk, and S. Sergey. 2014. Going beyond nutrition: regulation of potassium homoeostasis as a common denominator of plant adaptive responses to environment. *J Plant Physiol* 171:670–687.

Araki, S., M. Haneda, D. Koya, K. Kondo, S. Tanaka, H. Arima, S. Kume, J. Nakazawa, M. Chin-Kanasaki, S. Ugi, and H. Kawai. 2015. Urinary potassium excretion and renal and cardiovascular complications in patients with type 2 diabetes and normal renal function. *Clin J Am Soc Nephrol* 10(12):2152–2158. doi:10.2215/CJN.00980115.

Asaduzzaman, M. and T. Asao. 2018. Introductory chapter: potassium in quality improvement of fruits and vegetables. In *Potassium - Improvement of Quality in Fruits and Vegetables through Hydroponic Nutrient Management*. Asaduzzaman, M. and A. Toshiki eds. pp. 1–7. IntechOpen, London, UK. doi:10.5772/intechopen.75654.

Ashley, M. K., M. Grant, and A. Grabov. 2006. Plant responses to potassium deficiencies: a role for potassium transport proteins. *J Exp Bot* 57(2):425–436.

Athanasatou, A., A. Kandyliari, O. Malisova, A. Pepa, and M. Kapsokefalou. 2018. Sodium and potassium intake from food diaries and 24-h urine collections from 7 days in a sample of healthy Greek adults. *Front Nutr* 5:13. doi:10.3389/fnut.2018.00013.

Augustin, J., B. G. Swanson, S. F. Pometto, C. Teitzel, W. E. Artz, and C. P. Huang. 1979. Changes in nutrient composition on dehydrated potato products during commercial processing. *J Food Sci* 44(1):216–219.

Ayuke, F. O., L. Brussaard, B. Vanlauwe, J. Six, D. K. Lelei, C. N. Kibunja, and M. M. Pulleman. 2011. Soil fertility management: impacts on soil macrofauna, soil aggregation and soil organic matter allocation. *Appl Soil Ecol* 48(1):53–62.

Badraoui, M., M. Agbani, R. Bouabid, and A. Houssa. 2003. Potassium status in soils and crops: Fertilizer recommendations and present use in Morocco. In *Potassium and Water Management in West Asia and North Africa*. Johnston, A. E. ed. pp. 161–167. The regional workshop of the International Potash Institute, Amman, Jordan, 5–6 November 2001, Accessed on 16 May, 2020.

Baixauli, F., M. Villa, and E. L. Pearce. 2019. Potassium shapes antitumor immunity. *Science* 363(6434):1395–13966.

Balboa, G. R., V. O. Sadras, and I. A. Ciampitti. 2018. Shifts in soybean yield, nutrient uptake, and nutrient stoichiometry: a historical synthesis-analysis. *Crop Sci* 58:43–54.

Bandyopadhyay, K. K., A. K. Misra, P. K. Ghosh, and K. M. Hati. 2010. Effect of integrated use of farmyard manure and chemical fertilizers on soil physical properties and productivity of soybean. *Soil Till Res* 110(1):115–125.

Barbazán, M., C. Bautes, L. Beux, M. Bordoli, A. Califra, J. D. Cano, A. del Pino, O. Ernst, A. García, F. García, S. Mazzilli, and A. Quincke. 2012. Soil potassium in Uruguay: current situation and future prospects. *Better Crops* 96(4):21–23.

Bellows, L. and R. Moore. 2013. *Potassium and the Diet*. Colorado State University Extension, Food and Nutrition Series/Health Fact Sheet 9.355, Fort Collins, CO.

Benbi, D. K. and J. S. Brar. 2009. A 25-year record of carbon sequestration and soil properties in intensive agriculture. *Agron Sustain Deve* 29:257–265.

Bhattacharyya, K., T. Das, K. Ray, S. Dutta, K. Majumdar, A. Pari, and H. Banerjee. 2018. Yield of and nutrient-water use by maize exposed to moisture stress and K fertilizers in an inceptisol of West Bengal, India. *Agril Water Manage* 206:31–41.

Blanchet, G., Z. Libohova, S. Joost, N. Rossier, A. Schneider, B. Jeangros, and S. Sinaj. 2017. Spatial variability of potassium in agricultural soils of the canton of Fribourg, Switzerland. *Geoderma* 290:107–121.

Bobo-Garcia, G., C. Arroqui, G. Merino, and P. Virseda. 2019. Antibrowning compounds for minimally processed potatoes: a review. *Food Rev Intl*. doi:10.1080/87559129.2019.1650761.

Boeing, H., A. Bechthold, A. Bub, S. Ellinger, D. Haller, A. Kroke, E. Leschik-Bonnet, M. J. Muller, H. Oberritter, M. Schulze, P. Stehle, and B. Watzl. 2012. Critical review: vegetables and fruit in the prevention of chronic Diseases. *Eur J Nutr* 51:637–663.

Bogdevitch, I. 2000. *IPI Internal Report*. International Potash Institute: Basel, Switzerland.

Bolton, K. A., K. Trieu, M. Woodward, C. Nowson, J. Webster, E. K. Dunford, B. Bolam, and C. Grimes. 2019. Dietary intake and sources of potassium in a cross-sectional study of Australian adults. *Nutrients* 11:2996. doi:10.3390/nu11122996.

Brar, M. S., Bijay-Singh, S. K. Bansal, and C. Srinivasarao. 2011. Role of potassium nutrition in nitrogen use efficiency in cereals. *e-ifc* No. 29, December 2011. International Potash Institute, Basel, Switzerland.

Broadley, M. R., J. Alcock, J. Alford, P. Cartwright, I. Foot, S. J. Fairweather- Tait, D. J. Hart, R. Hurst, P. Knott, S. P. McGrath, and M. C. Meacham. 2010. Selenium biofortification of high-yielding winter wheat (*Triticum aestivum* L.) by liquid or granular Se fertilisation. *Plant Soil* 332:5–18.

Bruulsema, T. W. 2000. Functional food components: a role for mineral nutrients? *Better Crops* 84(2):4–5.

Bruulsema, T. W., C. J. C. Jackson, I. Rajcan, and T. J. Vyn. 2000. Functional food components: a role for potassium? *Better Crops* 84(2):6–7.

Cakmak, I. 2005. The role of potassium in alleviating detrimental effects of abiotic stresses in plants. *J Plant Nutri Soil Sci* 168:521–530.

Cakmak, I. 2010. Potassium for better crop production and quality. *Plant Soil* 335:1–2. doi:10.1007/s11104-010-0534-8.

Camire, M. E., S. Kubow, and D. Donnelly. 2009. Potatoes and human health. *Crit Rev Food Sci Nutr* 49(10):823–840.

Campanozzi, A., S. Avallone, A. Barbato, R. Iacone, O. Russo, G. De Filippo, G. D'Angelo, L. Pensabene, B. Malamisura, G. Cecere, and M. Micillo. 2015. High sodium and low potassium intake among italian children: relationship with age, body mass and blood pressure. *PLoS One* 10(4):e0121183. doi:10.1371/journal.pone.0121183.

Carrillo-Larco, R. M., L. Saavedra-Garcia, J. J. Miranda, K. A. Sacksteder, F. Diez-Canseco, R. H. Gilman, and A. Bernabe-Ortiz. 2018. Sodium and potassium consumption in a semi-urban area in peru: evaluation of a population-based 24-hour urine collection. *Nutrients* 10(2):245. doi:10.3390/nu10020245.

Carter, V. G. and T. Dale. 1974. *Topsoil and Civilization*. pp. 262. University of Oklahoma Press. ISBN 13: 9780806103327.

Chang, H. Y., Y. W. Hu, C. S. Yue, Y. W. Wen, W. T. Yeh, L. S. Hsu, S. Y. Tsai, and W. H. Pan. 2006. Effect of potassium-enriched salt on cardiovascular mortality and medical expenses of elderly men. *Am J Clin Nutr* 83:1289–1296.

Charlton, K. E., K. Steyn, N. S. Levitt, J. V. Zulu, D. Jonathan, F. J. Veldman, and J. H. Nel. 2005. Diet and blood pressure in South Africa: intake of foods containing sodium, potassium, calcium, and magnesium in three ethnic groups. *Nutrition* 21(1):39–50.

Chatterjee, R., H. C. Yeh, D. Edelman, and F. Brancati. 2011. Potassium and risk of type 2 diabetes. *Expert Rev Endocrinol Metab* 6(5):665–672. doi:10.1586/eem.11.60.

Chauhan, B. S., G. Mahajan, V. Sardana, J. Timsina, and M. L. Jat. 2012. Productivity and sustainability of the rice-wheat cropping system in the Indo-Gangetic plains of the Indian subcontinent: problems, opportunities, and strategies. *Adv Agron* 117:315–369.

Chen, D. X. Li, Q. Song, C. Hu, F. Su, J. Dai, Y. Ye, J. Huang, and X. Zhang. 2020. Hypokalemia and clinical implications in patients with coronavirus disease 2019 (COVID-19). *Medixv* doi:10.1101/2020.02.27.2 0028530.

Chmielewski, J. and J. B. Carmody. 2017. Dietary sodium, dietary potassium, and systolic blood pressure in US adolescents. *J Clin Hypertens* 19:904–909. doi:10.1111/jch.13014.

Christy, B., A. Clough, P. Riffkin, R. Norton, J. Midwood, G. O'Leary, K. Stott, A. Weeks, and T. Potter. 2015. Managing crop inputs in a high yield potential environment - HRZ of southern Australia. State of Victoria Department of Economic Development, Jobs, Transport and Resources. Melbourne. doi:10.13140/RG.2.1.3224.6884.

Ciceri, D, D.A.C. Manning, and A. Allanore. 2015. Historical and technical developments of potassium resources. *Sci Total Environ* 502:590–601. doi:10.1016/j.scitotenv.2014.09.013.

Cogswell, M. E., K. Mugavero, B. A. Bowman, and T. R. Frieden. 2016. Dietary sodium and cardiovascular disease risk – measurement matters. *N Engl J Med* 375(6):580–586.

Cogswell, M. E., Z. Zhang, A. L. Carriquiry, J. P. Gunn, E. V. Kuklina, S. H. Saydah, Q. Yang, and A. J. Moshfegh. 2012. Sodium and potassium intakes among US adults: NHANES 2003–2008. *Am J Clin Nutr* 96:647–657.

Cohen, J. D., J. D. Neaton, R. J. Prineas, and K. A. Daniels. 1987. Diuretics, serum potassium and ventricular arrhythmias in the multiple risk factor intervention trial. *Am J Cardiol* 60:548–554.

Colditz, G. A., J. E. Manson, M. J. Stampfer, B. Rosner, W.C. Willett, and F. E. Speizer. 1992. Diet and risk of clinical diabetes in women. *Am J Clin Nutr* 55:1018–1023.

Combs Jr, G. F. 2005. Geological impacts on nutrition. In *Essentials of Medical Geology*. Selinus, O. ed. pp 161–177. Elsevier, Amsterdam.

Committee on Minerals and Toxic Substances in Diets and Water for Animals, National Research Council. 2005. *Mineral Tolerance of Animals*, 2nd revised ed. The National Academies Press, Washington, DC.

Compton, S. J., R. L. Lux, M. R. Ramsey, K. R. Strelich, M. C. Sanguinetti, L. S. Green, M. T. Keating, and J. W. Mason. 1996. Genetically defined therapy of inherited long-QT syndrome. Correction of abnormal repolarization by potassium. *Circulation* 94:1018–1022.

Cornejo, K., F. Pizarro, E. Atalah, and J. E. Galgani. 2014. Evaluación de la ingesta dietética y excreción urinaria de sodio y potasio en adultos. *Rev Méd Chile* 142:687–695.

Curhan, G. C., W. C. Willett, E. B. Rimm, and M. J. Stampfer. 1993. A prospective study of dietary calcium and other nutrients and the risk of symptomatic kidney stones. *N Engl J Med* 328:833–838.

Dillard, C. J. and J. B. German. 2000. Phytochemicals: nutraceuticals and human health. *J Sci Food Agric* 80:1744–1756.

Divito, G. A. and V. O. Sadras. 2014. How do phosphorus, potassium and sulphur affect plant growth and biological nitrogen fixation in crop and pasture legumes? A meta-analysis. *Field Crops Res* 156:161–171.

Doran, J.W. and M. R. Zeiss. 2000. Soil health and sustainability: managing the biotic component of soil quality. *Appl Soil Ecol* 15:3–11.

Drechsel, P., D. Kinze, and F. P. de Vries. 2001. Soil nutrient depletion and population growth in Sub-Saharan Africa: a Malthusian nexus? *Popul Environ* 22:411–423.

Du, S., H. Wang, B. Zhang, B. M. Popkin. 2020. Dietary potassium intake remains low and sodium intake remains high, and most sodium is derived from home food preparation for Chinese adults, 1991–2015 trends. *J Nutri* 150(5):1230–1239.

Dutta, S., K. Majumdar, H. S. Khurana, G. Sulewski, V. Govil, T. Satyanarayana, and A. Johnston. 2013. Mapping potassium budgets across different states of India. *Better Crops South Asia* 7(1):28–31.

Dyer, A. R., P. Elliott, and M. Shipley. 1994. Urinary electrolyte excretion in 24 hours and blood pressure in the INTERSALT Study. II. Estimates of electrolyte-blood pressure associations corrected for regression dilution bias. The INTERSALT Cooperative Research Group. *Am J Epidemiol* 139(9):940–951.

Edis, R., R. M. Norton, and K. Dassanayake. 2012. Soil nutrient budgets of Australian natural resource management regions. In *Proceedings of the 5th Joint Australian and New Zealand soil Science Conference: Soil Solutions for Diverse Landscapes*. Burkitt, L. L. and L. A. Sparrow. eds. p. 11. Australian Society of Soil Science Inc, Hobart.

Ekholm, P., H. Reinivuo, P. Mattila, H. Pakkala, J. Koponen, A. Happonen, J. Hellström, and M. L. Ovaskainen. 2007. Changes in the mineral and trace element contents of cereals: fruits and vegetables in Finland. *J Food Compost Anal* 20:487–495.

El-Azeim, M. M. A., M. A. Sherif, M. S. Hussein, and S. A. Haddad. 2020. Temporal impacts of different fertilization systems on soil health under arid conditions of potato monocropping. *J Soil Sci Plant Nutr* 20:322–334. doi:10.1007/s42729-019-00110-2.

FAO (Food and Agriculture Organization of the United Nations). 2004. Human energy requirements: report of a joint FAO/WHO/UNU expert consultation. Available at www.fao.org/3/y5686e/y5686e00.htm. Accessed on May 16, 2020.

FAO (Food and Agriculture Organization of the United Nations). 2020. FAOSTAT. Available at http://www.fao.org/faostat/en/#home. Accessed on May 16, 2020.

Feske, S., H. Wulff, and E. Y. Skolnik. 2015. Ion channels in innate and adaptive immunity. *Annu Rev Immunol* 33:291–353.

Filho, J. O. 1985. Potassium nutrition of sugarcane. In *Potassium in Agriculture*. Munson, R. D. ed. pp. 1045–1062. American Society of Agronomy, Madison, WI.

Forouzanfar, M. H., L. Alexander, H. R. Anderson, V. F. Bachman, S. Biryukov, M. Brauer, R. Burnett, D. Casey, M. Coates, and A. Cohen. 2015. Global, regional, and national comparative risk assessment of 79 behavioural, environmental and occupational, and metabolic risks or clusters of risks in 188 countries, 1990–2013: a systematic analysis for the Global Burden of Disease Study 2013. *Lancet* 386(10010):2287–323.

Friedman, M. 2006. Potato glycoalkaloids and metabolites: roles in the plant and in the diet. *J Agric Food Chem* 54:8655–8681.

Gao, X., S. Zhang, X. Zhao, and Q. Wu. 2018. Potassium-induced plant resistance against soybean cyst nematode via root exudation of phenolic acids and plant pathogen-related genes. *PLoS One* 13(7):e0200903. doi:10.1371/journal.pone.0200903.

Geisseler, D. and K. M. Scow. 2014. Long-term effects of mineral fertilizers on soil microorganisms: a review. *Soil Biol Biochem* 75:54–63.

Gerardeaux, E., L. Jordan-Meille, J. Constantin, S. Pellerin, and M. Dingkuhn. 2010. Changes in plant morphology and dry matter partitioning caused by potassium deficiency in *Gossypium hirsutum* L. *Environ Exp Bot* 67:451–459.

Gerendás, J., F. Heuser, and B. Sattelmacher. 2007. Influence of nitrogen and potassium supply on contents of acrylamide precursors n potato tubers and on acrylamide accumulation in French fries. *J Plant Nutr* 30(9):1499–1516.

Gettes, L. S. 1992. Electrolyte abnormalities underlying lethal and ventricular arrhythmias. *Circulation* 85(1 Suppl):I-70–I-76.

Gheeraert, P. J., M. L. De Buyzere, Y. M. Taeymans, T. C. Gillebert, J. P. Henriques, G. De Backer, and D. De Bacquer. 2006. Risk factors for primary ventricular fibrillation during acute myocardial infarction: a systematic review and meta-analysis. *Eur Heart J* 27:2499–2510.

Gong, X., L. Chao, M. Zhou, M. Hong, L. Luo, L. Wang, W. Ying, C. Jingwei, G. Songjie, and H. Fashui. Oxidative damages of maize seedlings caused by exposure to a combination of potassium deficiency and salt stress. *Plant Soil* 340:443–452.

Grimes, C. A., L. J. Riddell, K. J. Campbell, K. Beckford, J. R. Baxter, F. J. He, and C. A. Nowson. 2017. Dietary intake and sources of sodium and potassium among Australian schoolchildren: results from the cross-sectional Salt and Other Nutrients in Children (SONIC) study. *BMJ Open* 7:e016639. doi:10.1136/bmjopen-2017-016639.

Hägg, M., U. Häkkinen, J. Kumpulainen, R. Ahvenainen, and E. Hurme. 1998. Effects of prepration procedures, packaging and storage on nutrient retention in peeled potatoes. *J Sci Food Agric* 77:519–526.

Hakerlerler, H., M. Oktay, N. Eryuce, B. Yagmur. 1997. Effect of potassium sources on the chilling tolerance of some vegetable seedlings grown in hotbeds. In *Food Security in the WANA Region, the Essential Need for Balanced Fertilization*. Johnston, A. E. ed. pp. 353–359. International Potash Institute, Basel, Switzerland.

Hamouz, K., J. Lachman, P. Dvorák, O. Dušková, and M. Čížek. 2007. Effect of conditions of locality, variety and fertilization on the content of ascorbic acid in potato tubers. *Plant Soil Environ* 53(6):252–257.

Hansjürgens, B., A. Lienkamp, and S. Möckel. 2018. Justifying soil protection and sustainable soil management: creation-ethical, legal and economic considerations. *Sustainability* 10:3807. doi:10.3390/su10103807.

Haque, M. Z. 1988. Effect of nitrogen, phosphorus and potassium on spikelet sterility induced by low temperature at the reproductive stage of rice. *Plant Soil* 109:31–36.

Hasanuzzaman, M., M. H. M. B. Bhuyan, K. Nahar, M. S. Hossain, J. A. Mahmud, M. S. Hossen, A. A. C. Masud, Moumita and M. Fujita. 2018. Potassium: a vital regulator of plant responses and tolerance to abiotic stresses. *Agronomy* 8(3):31. doi:10.3390/agronomy8030031.

Hasso-Agopsowicz, M., T. J. Scriba, W. A. Hanekom, H. M. Dockrell, and S. G. Smith. 2018. Differential DNA methylation of potassium channel KCa3.1 and immune signalling pathways is associated with infant immune responses following BCG vaccination. *Sci Reports* 8:13086. doi:10.1038/s41598-018-31537-9.

Haverkort, A. J. and C. Rutayisire. 1986. Utilisation des engrais chimiques sous conditions tropicales. 2. Effet de l'application d'azote, phosphore et potasse sur la relation entre la radiation interceptée et le rendement de la pomme de terre en Afrique centrale. *Potato Res* 29:357–365.

Hawkesford, M., W. Horst, H. Lambers, J. Schjoerring, I. S. Møller, and P. White. 2012. Functions of macronutrients. In *Marschner's Mineral Nutrition of Higher Plants*. Marschner, P. ed. (Kindle version). Chapter 6. ISBN 978-0-12-384905-2.

He, F. J. and G. A. MacGregor. 1999. Potassium intake and blood pressure. *Am J Hypertens* 12:849–851.

He, F. J. and G. A. MacGregor. 2001. Fortnightly review: beneficial effects of potassium. *BMJ* 323:497–501.

Hengl, T., G.B.M. Heuvelink, B. Kempen, J.G.B. Leenaars, M.G. Walsh, K.D. Shepherd, A. Sila, R.A. MacMillan, J. Mendes de Jesus, L.T. Desta, J.E. Tondoh. 2015. Mapping soil properties of africa at 250 m resolution: random forests significantly improve current predictions. *PLoS One* 10(6). doi:10.1371/journal.pone.0125814.

Hills, K., H. Collins, G. Yorgey, A. McGuire, and C. Kruger. 2020. Improving soil health in Pacific Northwest potato production: a review. *Am J Potato Res* 97:1–22.

Horie, T., F. Hauser, J. I. Schroeder. 2009. HKT transporter-mediated salinity resistance mechanisms in Arabidopsis and monocot crop plants. *Trends Plant Sci* 14:660–668.

Hosseini, S. A., M. R. Hajirezaei, C. Seiler, N. Sreenivasulu, and N. von Wirén. 2016. A potential role of flag leaf potassium in conferring tolerance to drought-induced leaf senescence in barley. *Front Plant Sci* 7:206. doi:10.3389/fpls.2016.00206.

Howler, R. H. 1985. Potassium nutrition of cassava. In *Potassium in Agriculture*. Munson, R. D. ed. pp. 891–841. American Society of Agronomy, Madison, WI.

Hoy, M. K. and J. D. Goldman. 2012. Potassium intake of the U.S. population: what we eat in America, NHANES 2009–2010. Food Surveys Research Group Dietary Data Brief No. 10. September 2012. Available at: https://www.ars.usda.gov/ARSUserFiles/80400530/pdf/DBrief/10_potassium_intake_0910.pdf. Accessed on May 16, 2020.

Imas, P. and H. Magen. 2007. Role of potassium nutrition in balanced fertilization for soybean yield and quality – global perspective. In *Proceedings of Regional Seminar on Recent Advances in Potassium Nutrition Management for Soybean Based Cropping System*. Imas, P. and A. K. Vyas eds. p. 126. National Research Centre for Soybean, Indore, India.

INTERSALT. 1986. Intersalt Cooperative Research Group. Intersalt: an international study of electrolyte excretion and blood pressure. Results for 24 hour urinary sodium and potassium excretion. *BMJ* 297:319–328. doi:10.1136/bmj.297.6644.319.

IPNI. 1998. *Better Crops with Plant Food*. Amstrong, D. L. ed. IPNI, Norcross, GA. http://www.ipni.net/publication/bettercrops.nsf/0/E90E04A957EA624285257980007CD63C/$FILE/BC-1998-3.pdf. Accessed on May 18, 2020.

Ishimitsu, T., L. Tobian, K. Sugimoto, and J. M. Lange. 1995. High potassium diets reduce macrophage adherence to the vascular wall in stroke-prone spontaneously hypertensive rats. *J Vasc Res* 32:406–412.

Jákli, B., E. Tavakol, M. Tränkner, M. Senbayram, and K. Dittert. 2017. Quantitative limitations to photosynthesis in K deficient sunflower and their implications on water-use efficiency. *J Plant Physiol* 209:20–30.

Jiang, W., X. Liu, Y. Wang, Y. Zhang, and W. Qi. 2018. Responses to potassium application and economic optimum K rate of maize under different soil indigenous K supply. *Sustainability* 10:2267. doi:10.3390/su10072267.

Jitendra, K. and P. Amit. 2015. An overview of prospective study on functional food. *Int J Recent Sci Res* 6:5497–5500.

Jones, D. L., P. Cross, P. J. A. Withers, T. H. De-Luca, D. A. Robinson, R. S. Quilliam, I. M. Harris, D. R. Chadwick, and G. Edwards-Jones. 2013. Review: nutrient stripping: the global disparity between food security and soil nutrient stocks. *J Appl Ecol* 50:851–862.

Kant, S. and U. Kafkafi. 2002. Potassium and abiotic stresses in plants. In *Potassium for Sustainable Crop Production*. Pasricha, N. S. and S. K. Bansal. eds. pp. 233–251. Potash Institute of India, Gurgaon, India.

Kant, S., P. Kant, and U. Kafkafi. 2005. Potassium uptake by higher plants: from field application to membrane transport. *Acta Agron Hungarica* 53(4):443–459.

Katsuyuki, M. 2009. Soil and humanity: culture, civilization, livelihood and health. *Soil Sci Plant Nutri* 55(5):603–615.

Kausar, A. and M. Gull. 2014. Effect of potassium sulphate on the growth and uptake of nutrients in wheat (*Triticum aestivum* L.) under salt stressed conditions. *J Agric Sci* 6(8):101–112.

Kearney, P. M., M. Whelton, K. Reynolds, P. Muntner, P. K. Whelton, and J. He. 2005. Global burden of hypertension: analysis of worldwide data. *Lancet* 365(9455):217–23.

Khalili, H., S. Malik, A. N. Ananthakrishnan, J.J. Garber, L.M. Higuchi, A. Joshi, J. Peloquin, J.M. Richter, K.O. Stewart, G.C. Curhan, and A. Awasthi. 2016. Identification and characterization of a novel association between dietary potassium and risk of crohn's disease and ulcerative colitis. *Frontiers Immunol* 7:554. doi:10.3389/fimmu.2016.00554.

Khan, M. Z., M. E. Akhtar, M. Mahmood-ul-Hassan, M. Masud Mahmood, and M. N. Safdar. 2012. Potato tuber yield and quality as affected by rates and sources of potassium fertilizer. *J Plant Nutri* 35:664–677.

Khaw, K. T. and E. Barrett-Connor. 1987. Dietary potassium and stroke-associated mortality. A 12-year prospective population study. *N Eng J Med* 316:235–240.

Khaw, K. T. and G. Rose. 1982. Population study of blood pressure and associated factors in St Lucia, West Indies. *Int J Epidemiol* 11:372–377.

Kibblewhite, M. G., K. Ritz and M. J. Swift. 2008. Soil health in agricultural systems. *Philos Trans R Soc Lond B Biol Sci* 363(1492):685–701.

Kibunja, C. N., F. B. Mwaura, and D. N. Mugendi. 2010. Long-term land management effects on soil properties and microbial populations in a maize-bean rotation at Kabete, Kenya. *Afr J Agric Res* 5(2):108–113.

Kihara, J., P. Bolo, M. Kinyua, J. Rurinda, and K. Piikki. 2020. Micronutrient deficiencies in African soils and the human nutritional nexus: opportunities with staple crops. *Environ Geochem Health*. doi:10.1007/s10653-019-00499-w.

Kotnis, A., T. Satyanarayana, and K. Majumdar. 2017. Vulnerability of human health due to inadequate dietary potassium is associated with potassium nutrition of crops. *Ind J Fert* 13(11):16–23.

Ladha, J. K., D. Dawe, H. Pathak, A. T. Padre, R. L. Yadav, B. Singh, Y. Singh, Y. Singh, P. Singh, A. L. Kundu, R. Sakal, N. Ram, A. P. Regmi, S. K. Gami, A. L. Bhandari, R. Amin, C. R. Yadav, E. M. Bhattarai, S. Das, H. P. Aggarwal, R. K. Gupta, and P. R. Hobbs. 2003. How extensive are yield declines in long-term rice-wheat experiments in Asia? *Field Crops Res* 81:159–180.

Lai, Y., H. Leu, W. Yeh, H. Chang, and W. Pan. 2015. Low-normal serum potassium is associated with an increased risk of cardiovascular and all-cause death in community-based elderly. *J Formosan Med Assoc* 114:517–525. doi:10.1016/j.jfma.2015.01.001.

Lambert, D. H., M. L. Powelson, and W. R. Stevenson. 2005. Nutritional interactions influencing diseases of potato. *Am J Potato Res* 82:309–319.

Lemann Jr, J., J. A. Pleuss, and R. W. Gray. 1993. Potassium causes calcium retention in healthy adults. *J Nutr* 123:1623–1626.

Lemann Jr, J., J. Pleuss, R.W. Gray, and R. G. Hoffmann. 1991. Potassium administration reduces and potassium deprivation increases urinary calcium excretion in healthy adults [corrected]. *Kidney Int* 39: 973–983.

LeRiche, E. L., G. Wang-Pruski, and V. D. Zheljazkov. 2009. Distribution of elements in potato (*Solanum tuberosum* L.) tubers and their relationship to after-cooking darkening. *Hort Sci* 44(7):1866–1873.

Lester, G. E., J. L Jifon, and D. J. Makus. 2010. Impact of potassium nutrition on postharvest fruit quality: melon (*Cucumis melo* L) case study. *Plant Soil* 335:117–131.

Lewington, S., R. Clarke, N. Qizilbash, R. Peto, and R. Collins. 2002. Age-specific relevance of usual blood pressure to vascular mortality: a meta-analysis of individual data for one million adults in 61 prospective studies. *Lancet* 360:1903–1913.

Lin, P. H., F. Ginty, L. J. Appel, M. Aickin, A. Bohannon, P. Garnero, D. Barclay, and L. P. Svetkey. 2003. The DASH diet and sodium reduction improve markers of bone turnover and calcium metabolism in adults. *J Nutr* 133:3130–3136.

Lindshield, B. L. 2018. Kansas State University Human Nutrition (FNDH 400) Flexbook. NPP eBooks. 19. Available at: http://newprairiepress.org/ebooks/19 (accessed on 15 October 2019).

Liu, S. X., X. G. Mo, Z. H. Lin, Y. Q. Xu, J. J. Ji, G. Wen, J. Richey, T. R. Green, Q. A. Yu, and L.W. Ma. 2010. Crop yield responses to climate change in the Huang-Huai-Hai Plain of China. *Agricult Water Manage* 97:1195–1209. doi:10.1016/S2095-3119(19)62585-2.

Liu, Y. X., J. C. Ma, W. C. Ding, W. T. He, Q. L. Lei, Q. Gao, and P. He. 2017a. Temporal and spatial variation of potassium balance in agricultural land at national and regional levels in China. *PLoS One* 12(9):e0184156. doi:10.1371/journal.pone.0184156.

Liu, Y. X., J. Y. Yang, W. T. He, J. C. Ma, Q. Gao, Q. L. Lei, P. He, H. Y. Wu, S. Ullah, F. Q. Yang. 2017b. Provincial potassium balance of farmland in China between 1980 and 2010. *Nutr Cycl Agroecosys* 107(2):247–264. doi:10.1007/s10705-017-9833-2.

Love, S. L. and J. J. Pavek. 2008. Positioning the potato as a primary food source of vitamin C. *Am J Potato Res* 85:277–285.

Macdonald, H. M., S. A. New, W. D. Fraser, M. K. Campbell, and D. M. Reid. 2005. Low dietary potassium intakes and high dietary estimates of net endogenous acid production are associated with low bone mineral density in premenopausal women and increased markers of bone resorption in postmenopausal women. *Am J Clin Nutr* 81:923–933.

MacGilvary, N. J., Y. L. Kevorkian, and S. Tan. 2019. Potassium response and homeostasis in Mycobacterium tuberculosis modulates environmental adaptation and is important for host colonization. *PLoS Pathog* 15(2):e1007591. doi:10.1371/journal.ppat.1007591.

Majumdar, K. and V. Govil. 2013. Potassium response and fertilizer application economics in oilseeds and pulses in India. *Better Crops* 7(1):23–25.

Majumdar, K., S. K. Sanyal, V. K. Singh, Sudarshan Dutta, T. Satyanarayana, and B. S. Dwivedi. 2017. Potassium fertiliser management in Indian agriculture: current trends and future needs. *Ind J Fert* 13(5):20–30.

Manning, D. A. 2015. How will minerals feed the world in 2050? *Proc Geol Assoc* 126:14–17.

Mao, H., J. Wang, Z. Wang, Y. Zan, G. Lyons, and C. Zou. 2014.Using agronomic biofortification to boost zinc, selenium, and iodine concentrations of food crops grown on the loess plateau in China. *J Soil Sci Plant Nutr* 14:459–470.

Marangella, M., M. Di Stefano, S. Casalis, S. Berutti, P. D'Amelio, and G. C. Isaia. 2004. Effects of potassium citrate supplementation on bone metabolism. *Calcif Tissue Int* 74:330–335.

Marles, R. J. 2017. Mineral nutrient composition of vegetables, fruits and grains: the context of reports of apparent historical declines. *J Food Comp Anal* 56:93–103.

Marschner, H. 1995. Functions and mineral nutrients [Chapter 8]. In *Mineral Nutrition of Higher Plants.* pp. 229–312. Academic Press, London, UK. Second edition.

McDonough, A. A. and J. H. Youn. 2017. Potassium homeostasis: the knowns, the unknowns, and the health benefits. *Physiology (Bethesda)* 32(2):100–111. doi:10.1152/physiol.00022.2016.

McGrath, S. P. 1985. The effects of increasing yields on the macro- and microelement concentrations and offtakes in the grain of winter wheat. *J Sci Food Agric* 36:1073–1083.

McLean, E. O. and M. E. Watson. 1985. Soil measurements of plant available potassium. In *Potassium in Agriculture.* Munson, R. D. ed. pp. 277–308. American Society of Agronomy, Madison, WI.

McLean, R., J. Edmonds, S. Williams, J. Mann, and S. Skeaff. 2015. Balancing sodium and potassium: estimates of intake in a New Zealand adult population sample. *Nutrients* 7:8930–8938.

Mente, A., M. J. O'Donnell, S. Rangarajan, M. J. McQueen, P. Poirier, A. Wielgosz, H. Morrison, W. Li, X. Wang, C. Di, and P. Mony. 2014. Association of urinary sodium and potassium excretion with blood pressure. *N Engl J Med* 371:601–611.

Mikkelsen, R. 2018. Quality: potassium management is critical for horticultural crops, *Better Crops* 102:24–26.

Mikkelsen, R. L., A. L. Page, and F. T. Bingham. 1989. Uptake and accumulation of selenium by agricultural crops. In *Selenium in Agriculture and the Environment.* Jacobs, L.W. ed. pp. 65–94. Soil Sci Soc Amer Special Pub. 23, Madison, WI.

Mitra, S. K. and S. S. Dhaliwal. 2009. Effect of potassium on fruit quality and their storage life. In *Potassium Role and Benefits in Improving Nutrient Management for Food Production, Quality and Reduced Environmental Damages.* Brar, M. S. and S. S. Mukhopadhaya. eds. pp. 327–342 IPI, Horgen, Switzerland and IPNI, Norcross.

Moebius-Clune, B. N., D. J. Moebius-Clune, B. K. Gugino, O. J. Idowu, R. R. Schindelbeck, A. J. Ristow, H. M. van Es, J. E. Thies, H. A. Shayler, M. B. McBride, K. S. M. Kurtz, D. W. Wolfe, and G. S. Abawi. 2016. Comprehensive assessment of soil health: the Cornell framework. https://soilhealth.cals.cornell.edu/. Accessed on May 16, 2020.

Moinuddin, and S. Umar. 2004. Influence of combined application of potassium and sulfur on yield, quality, and storage behavior of potato. *Commun Soil Sci Plant Anal* 35(7–8):1047–1060.

Mondy, N. I. and C. B. Munshi. 1993. Effect of type of potassium fertilizer on enzymatic discoloration and phenolic, ascorbic acid, and lipid contents of potatoes. *J Agric Food Chem* 41(6):849–852.

Montgomery, D. R. 2007. Soil erosion and sustainability. *PNAS* 104(33):13268–13272.

Mullins, G. L., C. H. Burmester, D. W. Reeves. 1997. Cotton response to in-row subsoiling and potassium fertilizer placement in Alabama. *Soil Till Res* 40:145–154.

Naumann, M., M. Koch, H. Thiel, A. Gransee, and E. Pawelzik. 2020. The importance of nutrient management for potato production part II: plant nutrition and tuber quality. *Potato Res* 63:121–137.

Nelson, K. L., D. H. Lynch, and G. Boiteau. 2009. Assessment of changes in soil health throughout organic potato rotation sequences. *Agr Ecosyst Environ* 131:220–228.

New, S. A., C. Bolton-Smith, D. A. Grubb, and D. M. Reid. 1997. Nutritional influences on bone mineral density: a cross-sectional study in premenopausal women. *Am J Clin Nutr* 65:1831–1839.

New, S. A., S. P. Robins, M. K. Campbell, J. C. Martin, M. J. Garton, C. Bolton-Smith, D. A. Grubb, S. J. Lee, and D. M. Reid. 2000. Dietary influences on bone mass and bone metabolism: further evidence of a positive link between fruit and vegetable consumption and bone health? *Am J Clin Nutr* 71:142–151.

Nie, X, G. Zhang, L.V. Shidong, and H. Guo. 2018. Steroidal glycoalkaloids in potato foods as affected by cooking methods. *Int. J Food Prop* 21(1):1875–1887. doi:10.1080/10942912.2018.1509346.

Nolan, J., P. D. Batin, R. Andrews, S. J. Lindsay, P. Brooksby, M. Mullen, W. Baig, A. D. Flapan, A. Cowley, R. J. Prescott, and J.M. Neilson. 1998. Prospective study of heart rate viability and mortality in chronic heart failure: results of the United Kingdom heart failure evaluation and assessment of risk trial (UK Heart). *Circulation* 98(15):1510–1516.

Nordrehaug, J. E., K. A. Johannessen, G. von der Lippe, and O. L. Myking. 1985a. Circulating catecholamine and potassium concentrations early in acute myocardial infarction: effect of intervention with timolol. *Am Heart J* 110: 944–948.

Nordrehaug, J. E., K. A. Johannessen, G. von der Lippe, M. Sederholm, P. Grottum, and J. Kjekshus. 1985b. Effect of timolol on changes in serum potassium concentration during acute myocardial infarction. *Br Heart J* 53:388–393.

O'Donnell, M., A. Mente, S. Rangarajan, M.J. McQueen, X. Wang, L. Liu, H. Yan, S.F. Lee, P. Mony, A. Devanath, and A. Rosengren. 2014. Urinary sodium and potassium excretion, mortality and cardiovascular events. *N Engl J Med* 371:612–623.

Oerke, E. C. and H. W. Dehne. 2004. Safeguarding production-losses in major crops and the role of crop protection. *Crop Prot* 23:275–285.

Oosterhuis, D. M., D. A. Loka, E. M. Kawakami, and W. T. Pettigrew. 2014. The physiology of potassium in crop production. *Adv Agron* 126:203–234.

Osorio, A. V. and U. S. Alon. 1997. The relationship between urinary calcium, sodium, and potassium excretion and the role of potassium in treating idiopathic hypercalciuria. *Pediatrics* 100:675–681.

Oustani, M., M. T. Halilat, and H. Chenchouni. 2015. Effect of poultry manure on the yield and nutriments uptake of potato under saline conditions of arid regions. *Emir J Food Agric* 27(1):106–120.

Palmer, B. F. and D. J. Clegg. 2016. Achieving the benefits of a high-potassium, paleolithic diet, without the toxicity. *Mayo Clin Proc* 91(4):496–508.

Parida, A. K. and A. B. Das. 2005. Salt tolerance and salinity effects on plants: a review. *Ecotox Environ Safe* 60:324–349.

Pasuquin, J. M., M. F. Pampolino, C. Witt, A. Dobermann, T. Oberthür, M. J. Fisher, and K. Inubushi. 2014. Closing yield gaps in maize production in Southeast Asia through site-specific nutrient management. *Field Crop Res* 156:219–230.

Patra, A. K., S. K. Dutta, P. Dey, K. Majumdar, and S. K. Sanyal. 2017. Potassium fertility status of Indian soils: National Soil Health Card database highlights the increasing potassium deficit in soils. *Ind J Fert* 13(11):28–33.

Pereira, T. S. S., I. J. M. Benseñor, J. G. V. Meléndez, C. P. D. Faria, N. V. Cade, J. G. Mill, and M. D. C. B. Molina. 2015. Sodium and potassium intake estimated using two methods in the Brazilian Longitudinal Study of Adult Health (ELISA-Brasil). *Sao Paulo Med J* 133:510–516.

Perrenoud, S. 1990. *Potassium and Plant Health.* pp. 8–10. International Potash Institute, Bern, Switzerland.

Pettigrew, W. T. 2008. Potassium influences on yield and quality production for maize, wheat, soybean and cotton. *Physiol Plant* 133:670–81.

Prabhu, A. S., N. K. Fageria, and D. M. Huber. 2007. Potassium nutrition and plant diseases. In *Mineral Nutrition and Plant Disease.* Datnoff, L. E., W. H. Elmer, and D. M. Huber. eds. pp 57–78. The American Phytopathological Society Press, Saint Paul, MN.

Prasad, R., K. Majumdar, Y. S. Shivay, and U. Kapil. 2016. *Minerals in Plant and Human Nutrition and Health.* International Plant Nutrition Institute, Peachtree Corners, GA. ISBN: 978-0-9960199-5-8.

Qiang, R., J. Fan, Z. Zhizhong, Z. G. Xiaoying, and R. DeLong. 2008. An environmental approach to correcting iodine deficiency: supplementing iodine in soil by iodination of irrigation water in remote areas. *J Trace Elem Med Biology* 22:1–8.

Ray, K., H. Banerjee, S. Dutta, A. K. Hazra, and K. Majumdar. 2019. Macronutrients influence grain yield and oil quality of hybrid maize (*Zea mays* L.). *PLoS One* 14(5):e0216939.

Rengasamy, P. 2010. Soil processes affecting crop production in salt-affected soils. *Functional Plant Biol* 37(7):613–620.

Riemenschneider, T. and A. Bohle. 1983. Morphologic aspects of low-potassium and low-sodium nephropathy. *Clin Nephrol* 19:271–279.

Rodan, A. R. 2017. Potassium: friend or foe? *Pediatr Nephrol* 32(7):1109–1121.

Roger, V. L., A. S. Go, D. M. Lloyd-Jones, E. J. Benjamin, J. D. Berry, W. B. Borden, D. M. Bravata, S. Dai, E. S. Ford, C. S. Fox, H. J. Fullerton, C, Gillespie, S. M. Hailpern, J. A. Heit, V. J. Howard, B. M. Kissela, S. J. Kittner, D. T. Lackland, J. H. Lichtman, … M. B. Turner; American Heart Association Statistics Committee and Stroke Statistics Subcommittee. Heart disease and stroke statistics – 2012 update: a report from the American Heart Association. *Circulation.* 125:e2–e220. doi:10.1161/CIR.0b013e31823ac046.

Römheld, V. and E. A. Kirkby. 2010. Research on potassium in agriculture: needs and prospects. *Plant Soil* 335:155–180.

Rosen, C., N. Sun, N. Olsen, M. Thornton, M. Pavek, L. Knowles, and N. R. Knowles. 2018. Impact of agronomic and storage practices on acrylamide in processed potatoes. *Am J Potato Res* 95:319–327.

Roser, M. and H. Ritchie. 2020. Food supply. Published online at OurWorldInData.org. Retrieved from: https://ourworldindata.org/food-supply [Online Resource]. Accessed on May 16, 2020.

Rowe, J. W., J. D. Tobin, R. M. Rosa, and R. Andres. 1980. Effect of experimental potassium deficiency on glucose and insulin metabolism. *Metabolism* 29:498–502.

Ruan, J., L. Ma, and Y. Shi. 2013. Potassium management in tea plantations: its uptake by field plants, status in soils, and efficacy on yields and quality of teas in China. *J Plant Nutr Soil Sci* 176:450–459.

Salih, R. F., K. Abdan, A. Wayayok, N. Hashim, and K. A. Rahman. 2016. Improve quality and quantity of plant products by applying potassium nutrient (A Critical Review). *J Zankoy Sulaimani* 18(2):197–208.

Sanyal, S. K., V. K. Singh, and K. Majumdar. 2014. Nutrient management in Indian agriculture with special reference to nutrient mining – a relook. *J Ind Soc Soil Sci* 62(4):307–325.

Sardans, J. and J. Peñuelas. 2015. Potassium: a neglected nutrient in global change. *Glob Ecol Biogeogr* 24:261–275.

Sarwar, M. 2012. Effects of potassium fertilization on population build up of rice stem borers (lepidopteron pests) and rice (Oryza sativa L.) yield. *J Cereals Oilseeds* 3:6–9.

Sebastian, A., S. T. Harris, J. H. Ottaway, K. M. Todd, and R. C. Morris Jr. 1994. Improved mineral balance and skeletal metabolism in postmenopausal women treated with potassium bicarbonate. *N Engl J Med* 330:1776–1781.

Serio, F., J. J. Leo, A. Parente, and P. Santamaria. 2007. Potassium nutrition increases the lycopene content of tomato fruit. *J Hort Sci Biotech* 82:941–945.

Sheldrick, W. F., J. K. Syers, and J. Lingard. 2002. A conceptual model for conducting nutrient audits at national, regional, and global scales. *Nutri Cyc Agroecosys* 62:61–72.

Singh, V. K., B. S. Dwivedi, Yadvinder-Singh, S. K. Singh, R. P. Mishra, A. K. Shukla, S. S. Rathore, K. Sekhawat, K. Majumdar, and M. L. Jat. 2018. Effect of tillage and crop establishment, residue management and K fertilization on yield, K use efficiency and apparent K balance under rice-maize system in north-western India. *Field Crops Res* 224:1–12.

Smil, V. 1999. Crop residues: agriculture's largest harvest. *Biosci* 49:299–308.

Srinivasarao, Ch., K. P. R. Vittal and B. Venkateswarlu. 2009. Role of potassium in water stress management in dry land agriculture. In *Potassium Role and Benefits in Improving Nutrient Management for Food Production, Quality and Reduced Environmental Damages*. Brar, M. S. and S. S. Mukhopadhaya. eds. IPI, Horgen, Switzerland and IPNI, Norcross.

Stahl, W., H. van den Berg, J. Arthur, A. Bast, J. Dainty, R. M. Faulks, C. Gärtner, G. Haenen, P. Hollman, B. Holst, F. J. Kelly, M. C. Polidori, C. Rice-Evans, S. Southon, T. van Vliet, J. Viña-Ribes, G. Williamson, S. B. Astley. 2002. Bioavailability and metabolism. *Mol Aspects Med* 23(1–3):39–100.

Stone, M. S., L. Martyn, and C. M. Weaver. 2016. Potassium intake, bioavailability, hypertension, and glucose control. *Nutrients* 8:444. doi:10.3390/nu8070444.

Stoorvogel, J. J. and E. M. A. Smaling. 1990. *Assessment of Soil Nutrient Depletion in Sub-Saharan Africa: 1983–2000*. Vol. II. Nutrient balances per crop and per Land Use System. Report 28. The Winand Staring Centre, Wageningen, The Netherlands.

Stoorvogel, J. J., E. M. A. Smaling, and B. H. Janssen. 1993. Calculating soil nutrient balances in Africa at different scales. *Fert Res* 35(3):227–235.

Stucki, G., S. Rubinelli, and J. Bickenbach. 2020. We need an operationalization, not a definition of health. *Disabil Rehabil* 42(3):442–444.

Sun, Y., C. H. Byon, Y. Yang, W. E. Bradley, L. J. Dell'Italia, P. W. Sanders, A. Agarwal, H. Wu, and Y. Chen. 2017. Dietary potassium regulates vascular calcification and arterial stiffness. *JCI Insight* 2(19):e94920. doi:10.1172/jci.insight.94920.

Tan, H. L., M. Alings, R. W. Van Olden, and A. A. Wilde. 1999. Long-term (subacute) potassium treatment in congenital HERG-related long QT syndrome (LQTS2). *J Cardiovasc Electrophysiol* 10:229–233.

Tan, Z. X., R. Lal, and K. D. Wiebe. 2005. Global soil nutrient depletion and yield reduction. *J Sustainable Agri* 26(1):123–146.

Tobian, L., D. MacNeill, M. A. Johnson, M. C. Ganguli, and J. Iwai. 1984. Potassium protection against lesions of the renal tubules, arteries, and glomeruli and nephron loss in salt loaded hypertensive Dahl S rats. *Hypertension* 6:I170–I176.

Tompkins, P. and C. Bird. 1989. *The Secret Life of Plants*. p. 444. Penguin, London. ISBN 0670835617.

USDA (United States Department of Agriculture). 2020. *FoodData Central*. Available at https://fdc.nal.usda. gov/index.html. Accessed on May 16, 2020.

van Mierlo, L. A. J., A. Greyling, P. L. Zock, F. J. Kok, and J. M. Geleijnse. 2010. Suboptimal Potassium intake and potential impact on population blood pressure. *Arch Intern Med* 170(16):1501–1502.

Vyn, T. J., X. Yin, T. W. Bruulsema, C. C. Jackson, I. Rajcan, and S. M. Brouder. 2002. Potassium fertilization effects on isoflavone concentrations in soybean [Glycine max (L.) Merr.]. *J Agric Food Chem* 50(12):3501–3506.

Wang, M., Q. Zheng, Q. Shen, and S. Guo. 2013. The critical role of potassium in plant stress response. *Int J Mol Sci* Apr 14(4):7370–7390.

Weaver, C. M. 2013. Potassium and health. *Adv Nutr* 4:368S–377S. doi:10.3945/an.112.003533.

Weaver, C. M., M. S. Stone, A. J. Lobene, D. P. Cladis, and J. K. Hodges. 2018. What is the evidence base for a potassium requirement? *Nutri Today* 53(5):184–195.

Welch, A. A., H. Fransen, M. Jenab, M.C. Boutron-Ruault, R. Tumino, C. Agnoli, U. Ericson, I. Johansson, P. Ferrari, D. Engeset, and E. Lund. 2009. Variation in intakes of calcium, phosphorus, magnesium, iron and potassium in 10 countries in the European prospective investigation into cancer and nutrition study. *Eur J Clin Nutr* 63:S101–S121.

Welch, L. F. and R. L. Flannery. 1985. Potassium nutrition of corn. In *Potassium in Agriculture*. Munson, R. D. ed. pp. 647–664. American Society of Agronomy, Madison, WI.

Whelton, P.K., J. He, J. A. Cutler, F. L. Brancati, L. J. Appel, D. Follmann, and M. J Klag. 1997. Effects of oral potassium on blood pressure. Meta-analysis of randomized controlled clinical trials. *J Am Med Assoc* 277:1624–1632.

White, P. J. and M. R. Broadley. 2005. Historical variation in the mineral composition of edible horticultural products. *J Hortic Sci Biotechnol* 80:660–667.

WHO. 2003. *Prevention of Recurrent Heart Attacks and Strokes in Low and Middle Income Populations: Evidence-based Recommendations for Policy Makers and Health Professionals*. World Health Organization (WHO), Geneva, Switzerland. Retrieved 18 November 2017 from: http://www.who.int/cardiovascular_diseases/resources/ pub0402/en/.

WHO (World Health Organization). 2012a. Guideline: potassium intake for adults and children. Available at http://www.fao.org/nutrition/requirements/minerals/en/ (Accessed 1 May 2020).

WHO (World Health Organization). 2012b. Guideline: sodium intake for adults and children. Available at http://www.fao.org/nutrition/requirements/minerals/en/ (Accessed 1 May 2020).

WHO (World Health Organization). 2013. How to use the ICF: a practical manual for using the International Classification of Functioning, Disability and Health (ICF). Exposure draft for comment. Geneva, WHO. Available at https://www.who.int/classifications/drafticfpracticalmanual2.pdf?ua=1 (Accessed on 1 May 2020).

WHO and FAO (World Health Organization and Food and Agriculture Organization of the United Nations). 2004. *Vitamin and mineral requirements in human nutrition*, 2nd ed. Available at http://www.fao.org/nutrition/requirements/vitamins/en/ (Accessed 1 May 2020).

Williams, J. and S. G. Smith. 2001. Correcting potassium deficiency can reduce rice stem diseases. *Better Crops* 85:7–9.

Wingeyer, A. B., T. J. C. Amado, M. Pérez-Bidegain, G. A. Studdert, C. H. P. Varela, F. O. García, and D. L. Karlen. 2015. Soil quality impacts of current South American agricultural practices. *Sustainability* 7(2):2213–2242.

Wood, K., J. Carragher, and R. Davis. 2017. Australian consumers' insights into potatoes – nutritional knowledge, perceptions, and beliefs. *Appetite* 114:169–174.

Wszelaki, A. L., J. F. Delwiche, S. D. Walker, R. E. Liggett, J. C. Scheerens, and M. D. Kleinhenz. 2005. Sensory quality and mineral and glycoalkaloid concentrations in organically and conventionally grown redskin potatoes (*Solanum tuberosum*). *J Sci Food Agric* 85:720–726.

Wulff, F., V. Schulz, A. Jugk, and N. Claasen. 1998. Potassium fertilization on sandy soils in relation to soil test, crop yield and K-leaching. *Z Pflanzenernahr Bodenk* 161:591–599.

Yawson, D. O., M. O. Adu, B. Ason, F. A. Armah, E. Boateng, and R. Quansah. 2018. Ghanaians Might be at risk of excess dietary intake of potassium based on food supply data. *J Nutri Meta* Article ID 5989307, 9 pages. doi:10.1155/2018/5989307.

Young, D. B. and G. Ma. 1999. Vascular protective effects of potassium. *Semin Nephrol* 19:477–486.

Zebarth, B. J., A. N. Cambouris, I. Perron, D. L. Burton, L. P. Comeau, and G. Moreau. 2019. Spatial variation of soil health indices in a commercial potato field in eastern Canada. *Soil Sci Soc Am J* 83:1786–1798.

Zelelew, D. Z., S. Lal, T. T. Kidane, and B. M. Ghebreslassie. 2016. Effect of potassium levels on growth and productivity of potato varieties. *Am J Plant Sci* 7:1629–1638.

Zillich, A. J., J. Garg, S. Basu, G. L. Bakris, and B. L. Carter. 2006. Thiazide diuretics, potassium, and the development of diabetes: a quantitative review. *Hypertension* 48:219–224.

Zingore, S., C. Manyame, P. Nyamugafata, and K. E. Giller. 2005. Long-term changes in organic matter of woodland soils cleared for arable cropping in Zimbabwe. *Euro J Soil Sci* 56:727–736.

Zorb, C., M. Senbayram, and E. Peiter. 2014. Potassium in agriculture–status and perspectives. *J Plant Physiol* 171:656–669.

10 Soil Aquaphotomics for Understanding Soil–Health Relation through Water–Light Interaction

Jelena Muncan
Kobe University

Balkis Aouadi, Flora Vitalis, and Zoltan Kovacs
Szent István University

Roumiana Tsenkova
Kobe University

CONTENTS

10.1 INTRODUCTION

In our everyday life, we usually view soil as something of a mere material similar to how we would see plastic, or stone or wood, and we often disregard the fact that it is a vital, living ecosystem, and the starting link in a chain that sustains human life as we know it. In a similar manner, water is often viewed as only a medium for biochemical reactions in living cells and organisms, a passive player in the milieu of life. And the opposite is true—water is one of the most intricate substances on earth, a complex molecular system, connecting the elements present in it, and influenced by the physical fields, with a very active and crucial role—a biomolecule in its own right (Ball 2008a, 2008b).

Due to our misguided beliefs about the water and soil, which stem from the scarce knowledge and limited awareness, today we are facing serious challenges to make our world sustainable for future generations. And our first task is to seek and acquire new knowledge, understand better our resources and their potential; in simple words to shift our perspective and change the basic paradigms which shape the strategies we adopt to solve the problems at hand. Instead of thinking of soil quality as a mere property defined by some set of physico-chemical-biological parameters, we might better use the term "soil health" (SH) to describe the functionality of soil to sustain plants, animals, and humans. We should also have in mind that soil not only provides habitat for living organisms but, together with other soil elements, they are all acting synergistically to provide necessary functionality.

The intention of this chapter is to introduce the novel science—aquaphotomics (Tsenkova 2009; Muncan and Tsenkova 2019; van de Kraats, Munćan, and Tsenkova 2019), which, as you will see, brings the paradigm shift of seeing water as a passive, inert molecule. Instead, we will introduce new findings about water revealed by spectroscopic methods, and how these findings served as basis to develop new strategies for dealing with the problems of food and water quality monitoring, classification, and better understanding of microorganisms, and also diagnostics of diseases in plant, animal, and human organisms. The water will be explained as a complex, interconnected molecular system composed of many molecular species with different functionalities, strongly influenced by all the components in it, and physical fields, which makes it a perfect sensor to capture and describe the state of the aqueous or biosystem. Light–water interaction then reveals what water sensor measured in a form of spectra which is the foundation stone of innovative measurement technology. Water spectral pattern (WASP) thus becomes an integrative marker or biomarker describing the state of the examined aqueous or biological system and which is directly related to its functionality.

The aquaphotomics measurement technology allowed non-destructive, real-time water and food quality control, applications in the fields of non-invasive diagnostics, and other health-related applications for different biological organisms and made important new discoveries on which many more applications could follow.

The ideas presented here will, hopefully influence the readers to also see the soil as a complex system, of which water is an intrinsic part, influencing and in turn being influenced by all the other chemical, physical, and biological components. In a similar way, aquaphotomics uses the water as a sensor for the examination of other aqueous and biological systems; in soil science, it can be used to describe in an integrative way the state of the soil and relate it directly to the state of its health/quality/functionality or in other words, the ability of soil to perform its function. Adequate and readily available technology for SH assessment would provide much needed management tools and aid for farming and agriculture and would also give the feedback needed to design measures which would enable moving toward sustainability for future generations.

10.2 FOUNDATIONS OF AQUAPHOTOMICS

In this chapter section, we will briefly define core aquaphotomics terms and its methodological approach in order to present the most basic foundations of aquaphotomics.

Aquaphotomics (aqua-photo-omics) or the study of water–light interactions has emerged as a complementary discipline to conventional "omics" disciplines (Tsenkova et al. 2018). Since its advent, efforts have been shifted from solely focusing on macro-scale water measurements (moisture content, water potential, water activity, etc.) to a more holistic, integrative approach. In the latter, the dynamic behavior of water, within aqueous and biosystems, is thoroughly examined, and valuable insights about the structure of water, its surrounding molecules, and the state of the system could be unraveled. In a sense, water serves as a reflector mimicking the energetic and molecular changes within the studied matrix. However, extracting accurate information from this constantly transient state of water calls for a meticulously designed methodology.

A first pivotal step to optimizing the outcome of aquaphotomics-based research is to elucidate its founding terminology and principles. As a spectroscopy-based technique that is majorly focused on the study of water, defining the electromagnetic range that accentuates the spectral patterns of varying water molecular conformations is of utmost importance.

In this regard, both early and recent studies have applied the near-infrared (NIR) spectral region 1,300–1,600 nm, which has been attributed to the 1st overtone of OH stretching band and contains several water absorbance bands (WABs) (Smith et al. 2005). The NIR 680–2,500 nm range has been equally reported as quite informative with respect to water studies (Büning-Pfaue 2003).

Combining rapidity, non-invasiveness, and considerable penetration depth, NIR has already been ascertained as effective in evaluating biosystems while producing distinctive spectra depending on the perturbing conditions. Still, studying aqueous samples in the NIR range has not always been effortless. In fact, for the most part, it implied measurement inaccuracies, signal interferences, and extra preparation steps (Jamrógiewicz 2012). However, what other scientific approaches considered as a constraint, aquaphotomics found as essential. These very same perturbing factors, when looked at from this novel perspective, get translated into the spectra via the action of newly constructed or broken hydrogen bonds through what is termed as WASP (Tsenkova 2009).

WASPs are thus considered as integrative markers of even the most subtle structural changes and descriptors of the system as a whole. Typically, these WASPs comprise combinations of major WABs. Interestingly, existing literature have reported an extensively rich database of essential absorbance bands exceeding 500 WABs in the 750–2,500 nm range (Tsenkova, Kovacs, and Kubota 2015). In studying numerous biological systems, in the range of the 1st overtone of water vibrational frequencies, these bands were predominantly located at 12 specific spectral regions, which were later on defined as water matrix coordinates (WAMACs) (Tsenkova 2009) (Figure 10.1).

Through empirical and theoretical studies, the correlation between these WAMACs and their corresponding water molecular species was established (Chatani et al. 2014; Headrick et al. 2005; van de Kraats, Munćan, and Tsenkova 2019; Tsenkova 2009). Thus, C1 (1,336–1,348 nm) was mainly assigned to ν_3 H_2O asymmetric stretching vibration while C2 was attributed to the water solvation shell [OH –$(H_2O)_n$] with n being equal to 1, 2, or 4 molecules in hydroxylated water cluster. C3 on the other hand was ascribed to $\nu_1+\nu_3$, combination of H_2O stretching and asymmetric stretching vibration of protonated water clusters. C4 corresponded to the water solvation shell [OH –$(H_2O)_n$; $(n=1, 4)$] and superoxide hydrates [O_2 $(H_2O)_n$ $(n=4)$] whilst trapped water and free water molecules were the major water species at C5. At the 6th WAMAC, the dominance of H-OH band and molecules involved in hydration was marked. WAMACs C7, C9, C10, and C11 represent water molecules with one, two, three, and four hydrogen bonds, respectively. The authors attributed

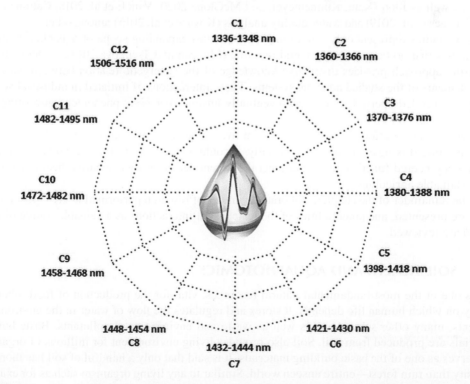

FIGURE 10.1 The 12 WAMACs in the spectral range of the 1st overtone of water (1,300–1,600 nm).

water solvation shell [OH $-(H_2O)_n$; n=4,5] to C8 and strongly bound water (ν_1, ν_2), to the C12 WAMAC. Each system under perturbation can be described with a specific spectral pattern based on WAMACs within the ranges described above.

Next, of prime importance to this research approach is finding a link between the spectral patterns and potential functionalities of the studied aqueous-biosystems. A recommended graphical representation of data, an "aquagram" (Tsenkova 2010; Tsenkova et al. 2018), provides insights into this structure–function relationship and facilitates, merely by visual inspection, the comparison not only between different samples but also different sets of the same sample under different conditions.

Aquagrams are, by definition, an illustration of normalized absorbance values (A') at relevant WABs. The below equation showcases how normalized absorbance values are computed:

$$A' = \frac{(A_\lambda - \mu_\lambda)}{\sigma_\lambda} \tag{10.1}$$

While A_λ is the absorbance obtained after scatter correction, μ_λ stands for the mean of the spectra and σ_λ is the respective absorbance's standard deviation, all of which are obtained at the wavelengths (λ) activated by the studied perturbation (Tsenkova et al. 2018).

This conception of water as a multi-element system which can be characterized by multidimensional spectra has opened new prospects in spectroscopy-based research. It contributed to overcoming common shortcomings of conventional analytical tools, namely, the need to identify and eliminate perturbing factors, which is no longer a prerequisite.

This systematic endeavor has already spurred the interest of several researchers in various fields and, and, successfully, the inspection of WASPs provided insightful findings with applications encompassing plant health (Jinendra et al. 2010; Jinendra 2011; Kuroki et al. 2019), human health (Chatani et al. 2014; Sakudo et al. 2005; Muncan et al. 2016), animal health (Tsenkova 2007b; Tsenkova et al. 2009; Meilina et al. 2009; Tsenkova and Atanassova 2002; Kinoshita et al. 2012, 2016), as well as food (Kaur, Künnemeyer, and McGlone 2020; Vanoli et al. 2018; Cattaneo et al. 2016; Kovacs et al. 2019) and water quality analysis (Kovacs et al. 2015) among others.

These studies represent only few examples of the fast-expanding scope of aquaphotomics into applications that go beyond fundamental research (Muncan and Tsenkova 2019). Undeniably, this promising approach provides distinctive knowledge of the synergetic relation between water and other elements of the studied aqueous systems. These interactions, if initiated in industrial settings through controlled "perturbations", can eventually unveil other basic phenomena governing vital functionalities.

Targeting future efforts toward the study of even more intricate matrices would substantially help enriching this field and most importantly, it could address most of this century's challenges through engineered functionalities that meet the urgent needs for food security, health assurance, and sustainable agriculture.

In the remainder of the chapter, elaborate examples of research performed in cross-disciplinary fields are presented, and potentialities of the water–light interactions as a valuable source of information are reviewed.

10.3 SOIL HEALTH AND AQUAPHOTOMICS

Soil is one of the most fundamental natural resources, vital for the production of food, fiber, and energy on which human life depends. It stores and regulates the flow of water in the environment, nutrients, many other substances, as well as waste and environmental pollutants. Basic building materials are produced from soil. Soil also provides living environment for millions of organisms and serves as one of the basic building materials. It is said that only a handful of soil has more biodiversity than rain forest—entire unseen world. Similar to any living organism such as for example, human body, the soil is a dynamic, living, natural system.

All of the diverse soil functions depend on the chemical and biological composition, its structure, and physical properties.

In general, major components of soil are around 40%–45% inorganic matter (inorganic mineral matter—sand, silt, clay particles), 25% water, 25% gases and air, and 5% organic matter (living organisms such as worms, insects, nematodes, fungi, bacteria, algae). Chemical, physical, and biological processes are happening at all times in the soil creating a dynamic soil system in which there is a continual interplay between molecules in solid, liquid, and gaseous state (Doran and Parkin 1994).

There are many definitions of soil quality. It is usually defined as an ability to perform its functions which represents a composite of physico-chemical-biological properties (Doran and Parkin 1994). The need for sustainability extended the definition of soil quality as a capacity of efficient function, not only at present but in future also. Definition of function encompasses sustaining biological productivity, maintaining quality of the environment, and promoting plant and animal health (Doran and Parkin 1994), to which we would also add human health.

From the previous definitions of soil "quality" in the terms of its functions, naturally the focus is placed on identification, measurements, and monitoring of the soil elements responsible for ensuring the performance of this desired functionality. As basic soil quality indicators, a myriad of physical, chemical, and biological factors have been identified; however, it is still unclear what exact combination constitutes a good soil. It is agreed that this perception may vary depending on what is the prioritized soil function (Doran and Parkin 1994). Since many chemical, physical, and biological indicators of quality have been over the years identified and proposed, it resulted in development, application, and adaptations of a large number of soil quality measurement techniques. Without going into the details of these, we will focus on one which started to be used around 1970, visible–NIR spectroscopy (Vis-NIR), which has a huge, but still not well explained, and enough recognized potential.

Vis-NIR spectroscopy is a non-destructive, reagent-free, environmentally friendly method which can provide rapid analysis of physical and chemical properties of the soil. The measurements are performed using light in the Vis-NIR region (400–2,500 nm). A large number of scientific papers have shown the advantages of NIR spectroscopy (NIRS) applications in soil quality assessment and soil science, which in combination with chemometric and multivariate analysis methods was mostly used for quantification of many physical and chemical, even biological parameters. Vis-NIRS in the reflectance mode has become a widely used measurement tool for rapid and accurate assessments of soil quality parameters in laboratory conditions (Dor, Ong, and Lau 2015; Stenberg et al. 2010), but with the development of portable spectral technology and hyperspectral imaging (HSI) systems also measurements have become possible to perform directly in field conditions and from air (remote sensing) (Ben-Dor et al. 2009; Cécillon et al. 2009). The two latter techniques allow proximal spectral acquisition *in situ* (for example by hand-held devices (Kooistra et al. 2001), devices mounted on agricultural machinery (Shonk et al. 1991; Sudduth and Hummel 1993; Mouazen et al. 2007, etc.)), and mapping of soil properties from the air using remote HSI sensors (Lagacherie et al. 2013; Gomez, Lagacherie, and Coulouma 2012; Chabrillat et al. 2019; Palacios-Orueta and Ustin 1998), respectively.

Since the first applications in the field of soil science, and especially over the past two decades, use of Vis-NIRS and imaging modalities have been growing rapidly both in number and complexity of measured parameters. Table 10.1 provides an overview of some of the research works dealing with such applications that illustrates this increasing trend over the years and the obvious shift from laboratory measurements to onsite measurements and soil mapping (Table 10.1).

The first works primarily dealt with measurements of soil organic matter, mainly soil organic carbon (SOC), and total nitrogen (N), followed by soil minerals, soil texture (particle size distribution), nutrients and nutrient availability, and then rapidly expanding to more complex ones, such as structure, soil type discrimination, microbial activity, and fertility. While first applications were concerned with measurements of individual attributes, it is evident that for successful utilization of

TABLE 10.1

Soil Physical, Chemical, and Biological Characteristics, Basic Indicators of Soil Quality, Using NIR Spectroscopy and/or NIR Hyperspectral Imaging

Range (nm)	Measured Properties (Physical, Chemical, Biological)	Reference	Year
400–2,600	Clay content (CC), organic matter (OM)	Al-Abbas, Swai, and Baumgardner	1972
400–2,400	OM content	Krishnan et al.	1980
400–2,400	Moisture content (MC), organic C, total N	Dalal and Henry	1986
1,500–2,400	Soil texture, OM, total N, and nitrogen mineralization potential	Meyer	1989
1,000–2,500	CC, specific surface area, cation-exchange capacity (CEC), hygroscopic moisture, carbonate and OM content	Ben-Dor and Banin	1995
400–2,500	Iron and OM content	Palacios-Orueta and Ustin	1998
1,800–2,500	Texture, soil nitrate content, total N	Ehsani et al.	1999
1,300–2,500	Sand, silt, total C, total N, moisture, (CEC), Mehlich III extractable Ca	Chang et al.	2001
1,300–2,500	Particle size distribution: clay, salt, sand; aggregation , total C, total N, moisture, pH, Mehlich extractable metals: Ca; Cu, Fe, K, Mn, Zn and P	Chang et al.	2001
400–2,500	Organic carbon, total N, and their potential mineralization	Fystro	2002
1,000–2,500	OM content	Fidêncio et al.	2002
400–2,500	CEC, exchangeable Ca and Mg, pH and Ca:Mg ratio, organic C and exchangeable sodium percentage (ESP), pH.	Dunn et al.	2002
350–2,500	Exchangeable Ca, effective cation-exchange capacity (ECEC), exchangeable Mg, organic C, CC, sand content, soil pH	Shepherd and Walsh	2002
400–2,500	Ca, Mg, Fe, Mn and K	Udelhoven, Emmerling, and Jarmer	2003
400–2,500	MC, OM, CC	Kooistra et al.	2003
400–2,500	Content of silt, sand, clay, iron (Fe), copper (Cu), manganese (Mn) and zinc (Zn)	Moron and Cozzolino	2003
400–2,500	Silt, sand, clay, calcium (Ca), potassium (K), sodium (Na), magnesium (Mg), copper (Cu) and iron (Fe)	Cozzolino and Morón	2003
350–2,500	Organic C and Inorganic C	Brown, Bricklemyer, and Miller	2005
400–700 700–2,500 2,500–2,500	pHCa, pHw, lime requirement (LR), organic C , CC, silt, sand, CEC, exchangeable calcium (Ca), exchangeable aluminium (Al), nitrate–nitrogen (NO3–N), available phosphorus (PCol), exchangeable potassium (K) and electrical conductivity (EC)	Viscarra Rossel et al.	2006
350–2,500	Organic C, total N, CEC, CC; soil fertility index (SFI)	Vågen, Shepherd, and Walsh	2006
400–2,500	Sand content, pH, clay, organic C, inorganic C, dithionate–citrate extractable Fe (FEd), CEC	Brown et al.	2006
950–1,650	Gravimetric moisture, OM, pH, Mehlich 1 extractionable K, Ca, Mg, Mn, Zn, Carbon, Lime etc	Christy	2008
300–2,500	Clay content , organic C; influence of rewetting on accuracy of measurements	Stenberg	2010
400–2,500	Contents of thermolabile organic carbon (C375 °C), the inert organic C fraction (Cinert) and the sum of both (total soil organic carbon, OCtot)	Vohland et al.	2011
400–2,450	Mapping soil surface properties at different depths (CC, sand content and CEC)	Lagacherie et al.	2013

(Continued)

TABLE 10.1 (*Continued*)

Soil Physical, Chemical, and Biological Characteristics, Basic Indicators of Soil Quality, Using NIR Spectroscopy and/or NIR Hyperspectral Imaging

Range (nm)	Measured Properties (Physical, Chemical, Biological)	Reference	Year
350–2,500	Prediction of soil organic content and influence of moisture on accuracy	Nocita et al.	2013
900–1,700	Soil discrimination on the basis of soil taxonomic classes	Todorova et al.	2015
400–1,100			
908–1,676	Total C, total N content, organic C carbon mapping, identification of soil types	Mura et al.	2019
350–2,500	Soil salinity (soil degradation)	Wang et al. a	2018
400–2,498	Mehlich-3 (M3), DTPA, and water-soluble (H2O) soil heavy metals	St. Luce et al.	2017
350–2,500	Total arsenic (As) and five different solid As phases (Mg, PO4, Ox, HCl and org pools)	Chakraborty et al.	2017
350–2,500	Total bacteria, gram positive bacteria, gram negative bacteria, OM	Weindorf	2018
400–2,500	Water holding capacity (as a difference of permanent wilting point (PWP) and volumetric water content at field capacity (FC))	Blaschek et al.	2019
325–1,075	Cadmium quantification and mapping of pollution	Chen et al.	2015
400–2,500	Classification of soil based on soil taxonomy	Vasques et al.	2014

measurement technologies it is of vital importance to define what functional soil is, i.e., to recognize which attributes need to be measured.

As we can see from Table 10.1, the functionality of the soil, or the concept of soil quality has significantly shifted from description in the terms of several, individual attributes to the integrative quality index.

For example, Vagen, Shepherd, and Walsh developed soil fertility index integrating ten well-recognized indicators of soil fertility and then developed prediction model based on the obtained Vis-NIR spectra of the soil, which they further applied for generating spatial maps of soil fertility using remote satellite imagery (Vågen, Shepherd, and Walsh 2006). Another trend is formation of soil spectral databases across the globe—taking soil samples from various locations, land uses, covering different climatic regions and soil types and conditions (Rinot et al. 2019; Dor, Ong, and Lau 2015). The samples, in addition to the acquisition of spectral information in laboratory conditions following usually standardized protocols, are also being measured using various other techniques for biological, physical, and chemical attributes. The creation of spectral libraries is an excellent step forward from the aspect of building models for prediction of soil attributes; increasing the number of samples collected from each soil location as well as the variety of soil locations, led to a large increase in the prediction accuracy (Shepherd and Walsh 2002; Christy 2008). It is expected that soil databases combined with spectral imaging can provide modeling, assessment, and mapping of key soil properties (Ben-Dor et al. 2009), which opens up new possibilities for precision agriculture, environmental, and engineering applications (Shepherd and Walsh 2002).

We can see the large body of evidence in the scientific literature testifying to the fact that Vis-NIR spectra are a rich source of information about the soil. In addition to physical information that can be obtained about soil, such as texture, particle size and other, this range is mostly used, as already mentioned for moisture content, soil organic content, and mineral constituents. The basis of obtaining the chemical information is in the fact that overtones and combinations of the fundamental absorption bands of many soil components (–OH, –SH, –CH, –NH) are occurring in the

NIR region (~780–2,500 nm), while in the visible region (400–780 nm) electronic absorptions are giving rise to distinctive colors (Schwertmann 1993). On the other hand, the perceived limitations of Vis-NIRS when focusing on single elements are weakness of the signals due to a low content of components of interest or simply that components are not Vis-NIR active (Brown et al. 2006; Viscarra Rossel et al. 2006).

Clearly, with the introduction of aquaphotomics, this limitation is not an issue any more. Water is a strong absorber in Vis-NIR region, exhibiting very strong absorbance peaks around 1,450 and 1,940 nm, attributed to the 1st overtone of the OH stretching band ($2\nu_{1,3}$) and a combination of the OH stretching and OH bending band ($2\nu_{1,3}+\nu_2$), respectively, and two smaller ones located approximately around 970 and 1,190 nm attributed to the 2nd overtone of the OH stretching band ($3\nu_{1,3}$), and a combination of the 1st overtone of the OH stretching and OH bending band ($2\nu_{1,3}+\nu_2$) (Luck 1974). Under these broad peaks, there are numerous WABs of various water species; in the 400–2,500 nm range more than 500 WABs can be found (Tsenkova, Kovacs, and Kubota 2015), which proves that water actually absorbs light in the entire Vis-NIR region. Even in the case of low-moisture systems, for example agricultural products which typically have water content up to 15%, the water despite not being dominant component, profoundly influences all the components of the system in which they are brought together, contributes to their behavior and shapes of the resultant spectra in the Vis-NIR region (Williams 2009). Similarly, the water is an intrinsic matrix of the soil, and its influence is apparent even when the soil is dried, changing its absorption pattern when physical or chemical variations occur (Stenberg 2010). Due to this property of water—that is influenced by all the components of the system and its environment, usually in the aquaphotomics works, it is referred to as "water-mirror" or water-sensor (Tsenkova 2009; Muncan and Tsenkova 2019; Tsenkova 2007a, 2008; Kovacs et al. 2015; Bázár et al. 2015). It does not matter whether a component in the system is NIR active or not, it influences the water matrix of the system, and indirectly its influence is revealed in the spectra of the system. Many pioneering works in NIR spectroscopy demonstrated these phenomena by measuring the effects of salts on the water spectra (Grant, Davies, and Bilverstone 1989; Hirschfeld 1985). Gowen et al. (2015) performed measurements of different salts in water, demonstrating clearly that substances which do not absorb NIR light can be measured using aquaphotomics' water-mirror approach, i.e., indirectly by utilizing the effects they exert on the water structure. The rearrangement of the water molecular structure, which happens under the influence of solutes, for example, also helped quantification of solutes even when they are present in very low concentrations, such as proteins or sugars (Bázár et al. 2015; Tsenkova 2008). Works in soil science also demonstrated this fact. For example, one of the works used Vis-NIRS for detection and assessment of soil salinity and several salts and they reported the following characteristic absorbance bands: for determination of NaCl: 1,930 nm, KCl: 1,430 nm, $MgSO_4$: 1,480 nm, Na_2SO_4: 1,825 nm, and $MgCl_2$: 1,925 nm (Farifteh et al. 2008).

Similarly, in another work, the absorbance bands found to show the strongest correlation with the iron content and organic matter were at 1,400 and 1,900 nm; as the authors themselves noticed, the absorption features for iron and organic matter do not occur at these bands, and they also noticed that the intensity of the bands (more specifically the depth, since they worked with the reflectance spectra) is in correlation (Palacios-Orueta and Ustin 1998). The mentioned bands are the two most strong absorbance bands of water, influenced by both organic matter and iron, and of course, in correlation because they originate from the same absorber.

One of the works also reported that rewetting of the soil prior to the NIR measurements leads to great improvements in the prediction of SOC (Stenberg 2010). Although the explanation for the mechanism behind increased accuracy of quantification was not specifically identified, as authors reported, the WABs were involved in improved calibrations, suggesting exactly the role of water as a sensor in this phenomenon.

However, the most powerful potential of utilizing the water as a sensor does not lie in the possibility of enabling measurements of individual compounds with increased accuracy. After all, many compounds and many attributes of the soil are important and describe its functionality. The true power of water in soil is in the fact that it is a matrix, bringing the soil elements together, and as such, its spectrum is an integrative property of the soil which can be related to its functionality. Stenberg et al. (2010) showed the awareness that spectra of the soil is its integrative property, and that it can be used directly in soil mapping, monitoring, relating to its function and quality, and he also envisioned the possibility of better ways to use it.

The latest publications concerned with the soil quality evaluation express the need for better understanding of soil as a system and of synergy between the soil elements, which should bring about the holistic (integrative) approach to the characterization of soil functions (Rinot et al. 2019). Cecillon et al. (2009) suggested NIRS for this purpose long time ago. The integrative approach is on the basis of aquaphotomics which views every aqueous system and/or biological system as a synergy of its elements connected by the water matrix. This is the reason why very often aquaphotomics approach is called "holistic" (Kovacs et al. 2015) or "integrative" (van de Kraats, Munćan, and Tsenkova 2019; Muncan and Tsenkova 2019)—it relates the spectral signature of the system directly to its functionality, or, in the case of biological systems, to their health (or the absence of it). Adoption of similar approach in soil science led to redefining what is traditionally viewed as "soil quality" to "soil health" showing the shift from materialistic, reductionistic approach which views soil as a mere medium to integrative, all-encompassing approach which understands soil as a living system (Rinot et al. 2019). Within this framework, SH is defined as "the capacity of soil to function as a vital living system, within ecosystem and land-use boundaries, to sustain plant and animal productivity, maintain or enhance water and air quality, and promote plant and animal health" (Rinot et al. 2019). Introduction of the novel concept of soil and SH required a "holistic index" of soil well-being related to its functionality, proposed to be derived by multivariate analysis from many soil attributes (Figure 10.2).

As Rinot et al. (2019) proposed, the SH index development should be accomplished through three steps: sampling a large variety of soil and measurement of large number of attributes (database construction), which would be subjected to the statistical methods of data analysis in order to find most relevant attributes through quantification and scoring, and finally by integration of selected

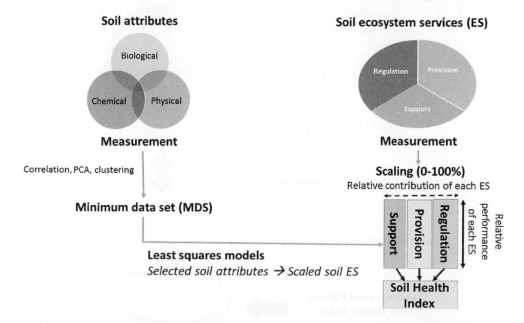

FIGURE 10.2 Novel concept of soil as a living system and redefinition of its functionality through the prism of soil health (SH), instead of outdated, materialistic approach of "soil quality." SH index as a holistic index of ecosystem services (ES) (Rinot et al. 2019).

attributes resulting in the construction of the SH index. This approach was devised based on a similar model created for assessment of the global ocean health index (Halpern et al. 2012).

In the light of what we presented about aquaphotomics, it can be deduced that aquaphotomics and Vis-NIRS fit perfectly within the envisioned concept, and what aquaphotomics already provided in other areas of life science by introducing WASP as a holistic/integrative marker either of quality (for example, in water or food quality monitoring) or as a biomarker (in biomedical fields) (Tsenkova et al. 2018; van de Kraats, Munćan, and Tsenkova 2019; Kinoshita et al. 2012; Tsenkova 2006, 2007b; Kovacs et al. 2015; Slavchev et al. 2017). WASP is derived from the spectra of the aqueous/biosystem using the aquaphotomics multivariate analysis approach (Tsenkova et al. 2018), and it is related directly to the system functionality. Soil Vis-NIR libraries already are powerful tools in soil science, allowing analysis and storage of large body of data and information (McBratney, Minasny, and Viscarra Rossel 2006). There are also pioneering aquaphotomics works in soil science, which related the WASPs as holistic descriptors of particular soil types or particular functionalities (Mura et al. 2019). With already existing databases, and an aquaphotomics multivariate analysis approach, it can be possible to develop a database of soil spectral patterns, what is called in aquaphotomics an aquaphotome (Tsenkova 2009), and identify and map soil across the globe.

The advantage of this approach is obvious—there is only one measurement technique needed—NIRS, which is rapid, non-destructive, cost-effective, and already heavily in use in soil science (Cécillon et al. 2009). Implementing aquaphotomics approach could provide a perfect diagnostic tool for SH (Figure 10.3) and continuous monitoring, just like it did in other areas of applications (which will be presented in next chapter sections).

However, since aquaphotomics is a very novel approach in soil science, there is a great need for more research, and especially in standardizing the procedure for soil spectral measurements

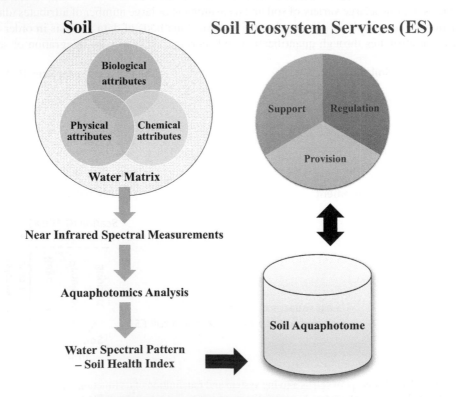

FIGURE 10.3 Aquaphotomics for SH diagnostics and monitoring: WASP of soil as an integrative marker of SH. [Developed based on Rinot et al. (2019) and Tsenkova et al. (2018).]

and smart building of soil spectral libraries. The first step toward the better management of SH assessment is in the creation of better awareness of the knowledge and potential of aquaphotomics.

10.4 AQUAPHOTOMICS IN WATER QUALITY ASSESSMENT AND MONITORING

Qualitative characterization of water, in both of its most intricate and basic forms, has long been a subject of interest to the research community, that is on the constant look up for cutting-edge techniques offering rapid, accurate, precise, and sensitive measurements. In soil science, the characterization of water by aquaphotomics can provide not only information about the water as an intrinsic part of the soil, but through utilization of water properties as a sensor, provide the information about the soil as well, in direct relation to its functions.

In water science, similar to the soil science, focus is placed on measuring individual attributes in order to provide health/quality assessment. If we take a look at a typical water quality report, we can see that everything but the water itself is being measured, as if the water is everything but H_2O. The water quality assessments have been greatly limited by the search for measurement techniques focused on measurements of individual physico-chemical-biological parameters. Thus, up to this day continuous monitoring is still a great challenge.

However, as one of the aquaphotomics studies showed, by measuring water spectra, one can use WASP as an integrative, holistic marker of quality, which will change when the content of the water is changed, signaling the need for more thorough analysis (Kovacs et al. 2015). By not limiting ourselves to measurements of individual solutes, aquaphotomics NIRS provided a real-time quality monitoring tool, tested in a real-life ground system (Kovacs et al. 2015), that can complement traditionally used techniques. In a study similar to the previously mentioned, Muncan et al. (2020) used the same approach for discerning different levels of purification of water produced by a typical purification system intended for the household use (Figure 10.4).

Measurements of individual components in water are also possible with the aquaphotomics water-mirror approach. One of the aquaphotomics studies addressed one of the most common issues nowadays—pesticides pollution (Gowen et al. 2011). Commonly used in agricultural fields, pesticides have

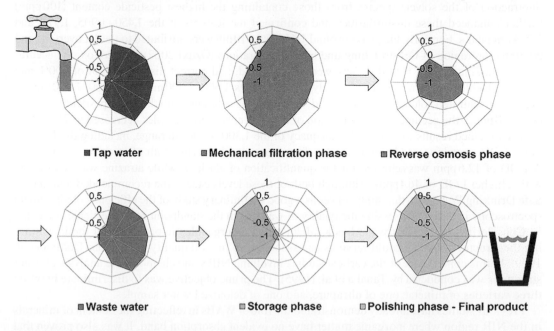

■ Tap water　　　■ Mechanical filtration phase　　　■ Reverse osmosis phase

■ Waste water　　　□ Storage phase　　　□ Polishing phase - Final product

FIGURE 10.4 WASP as an integrative marker allows tracking the changes in water composition indirectly. [Adapted based on work of Muncan et al. (2020), under the terms of CC-BY license.]

become one of the most common pollutants, particularly when drifting into water streams. Moreover, it is very well documented that they can have far-reaching adverse effects on human health.

One of the proposed alternatives to lengthy and costly conventional quantification methods is aquaphotomics NIR spectroscopy, which was the method of choice for Gowen et al. (2011) when attempting to quantify pesticides in aqueous solutions.

Research has already demonstrated that metal–OH interaction is worth investigating because despite the absence of absorbance of metals in the NIR range, the studies showed that very low concentrations of metal ions can be measured in aqueous solutions (Putra et al. 2010, 2017; Sakudo et al. 2006). Similarly, the quantification of pesticides was something of interest due to environmental pollution (Gowen et al. 2011). The particularity of this study, however, lies in the direct quantification in liquid solutions, which was not addressed by former researchers. Two of the frequently used herbicides were chosen for the purpose of the study: atrazine and alachlor.

After being dissolved in methanol/water solutions [$1CH_3OH: 1H_2O$ (v/v)], serial dilutions were prepared to obtain concentrations ranging from 1 to $50\,mg\,L^{-1}$ (ppm). Spectral acquisition was performed in the NIR range of $400–2,500\,nm$ at a temperature of $28°C \pm 1°C$ with constant monitoring of the temperature of solutions and control measurements of the solvent and air.

The pre-treated spectra were subjected to multivariate analysis by means of principal component analysis (PCA) coupled with partial least squares discriminant analysis (DA) and partial least squares regression (PLSR). As per the estimation of the limit of detection (LOD), the following equation was used:

$$LOD = mean\ blank + 1.645\ SD\ blank + 1.645\ SD\ low \tag{10.2}$$

with blank and low referring to samples without and with low pesticide content ($1.25–2.5\,ppm$), respectively (Armbruster and Pry 2008).

When comparing the mean spectra, the studied analytes were markedly overlapped with peaks majorly located at 1,450, 1,940, and 2,270 nm. The differences were highlighted when subtracting their spectra and mainly corresponded to varying spectral features near 1,420 and 1,900 nm. Subtraction of the solvent spectra from those containing the highest pesticide content (100 ppm) further enhanced these dissimilarities and confirmed the features at the 1,450, 1,905, 1,974, and 2,274 nm peaks. Formerly, the peaks around 1,455 or 1,910 nm were attributed to the 1st overtone and combination region of OH stretching and bending vibrations (Ozaki 2002). Further on, the accuracies of classification of both data sets were investigated, and correct discrimination of 100% and 85% of alachlor and atrazine, respectively, in the ranges 1,600–1,900 nm and 1,300–1,600 nm for highly concentrated solutions were reported. When testing solutions with the lower pesticides content (1.25–20 ppm), the accuracy of classification of alachlor containing samples decreased to 71%.

The obtained results confirmed the adequacy of the 1,300–1,900 nm range, including the 1st overtone of OH stretching and bending modes, for accurately identifying both of atrazine and alachlor. A LOD of 12.6 ppm was achieved for the quantification of alachlor, while atrazine was determined with a higher LOD of 46.4 ppm. Although both of these levels exceed the recommended amounts of Safe Drinking Water Act, this study serves as a good preliminary study of the efficiency of vibrational spectroscopy in terms of assessing the quality of water from the standpoint of the pesticide content.

Considering the fact that water has a salient effect on organoleptic properties of food, examining the structure of water could serve as an essential step in food quality evaluation (DeMan et al. 1999). In this context, one of the earliest studies involving NIRS and chemometric analysis of water structures was conducted by Tanaka et al. (1995). The prime objective was to discriminate between three varieties of mineral, one of ultrapure, and one of deionized water samples.

Previous studies have already demonstrated the role of WABs in reflecting the content of minerals in the NIR region where inorganic matter have no evident absorption band. It was also proven that the water structure is oriented by the existing ions (Cattaneo et al. 2011; Eisenberg and Kauzmann 1997; Tanaka et al. 1995).

In this particular study, all NIR spectra were acquired in the 1,100–1,800 nm range while keeping the samples at a temperature of 25°C. The resulting spectra along with the 2nd derivative were subjected to analysis by chemometric tools. With the absence of distinctive spectral features of the five evaluated water samples, both of PCA and DA were performed. As per the determination of the cation concentrations serving as a reference for subsequent data analysis, ion chromatography was applied. To locate the wavelengths where the effect of ion hydration is most prominent, three different mineral waters were artificially prepared so as to contain the cations' ratio of the commercially obtained mineral waters.

The results showed that, except for minor differences at the combination band ($v_1 + v_3$) of O-H stretching modes, a similarity between the 2nd derivative spectra prevented the visual discrimination of the waters. On the other hand, a good separation of all water samples was obtained by PCA. The peaks 1,378, 1,386, 1,396, and 1,404 nm of the 1st overtone of water were the ones contributing to the separation of the five waters. DA models yielded 100% and 89% accurate discrimination of the different waters (pure, commercial mineral, and deionized) based on the raw and 2nd derivative spectra, respectively.

When testing only mineral waters, the spectra (raw and 2nd derivative) enabled a 100% accuracy of the DA models. Interestingly, the very same previously stated peaks were significantly discriminatory suggesting their role as reflective of the hydration of the ions since these are the only different components of the waters.

This study highlights the suitability of NIR as a rapid characterization technique capable of pinpointing minor composition differences between marketed mineral waters namely in the region corresponding to the 1st water overtone.

10.5   AQUAPHOTOMICS IN MICROBIOLOGY

Microorganisms are integral part of the soil and they mediate many critical processes in the soil ecosystem (Carney and Matson 2005). Evaluations of microbial biomass is an essential part of the soil science and can provide the information about which microorganisms contribute to the soil functions and how (Blagodatskaya and Kuzyakov 2013). Aquaphotomics with NIRS can significantly contribute to this aspect of characterization as well, which we will show by several illustrative examples.

Remagni et al. (2013), for instance, explored the potential of aquaphotomics for the classification of bacterial strains based on their characteristics. The methodology consisted of performing a molecular profiling and spectroscopic typing of selected strains. The latter being constituted of 35 *Lactobacillus plantarum* strains originating from varying sources (dairy, olive, and grape products), six *Lactobacillus delbrueckii* subsp. *bulgaricus strains* and one strain of *Lactobacillus gasser*, both of which being isolated from probiotic yoghurt. To obtain the corresponding molecular profiles, an extraction of the DNA followed by amplification by polymerize chain reaction (PCR) were effected. Subsequently, electrophoresis and Pearson's correlation coefficient served as an assessment tool of the similarities and differences between the generated profiles. For the spectroscopic typing, centrifuged (7,000 rpm – 15 min) and water-dissolved cultures were optically analyzed in the spectral range 600–1,100 nm. To decipher the features specific to each of the studied samples, soft independent modeling of class analogy (SIMCA) was the qualitative method of choice. Initially, the data set was split into two equally sized groups: one for building the models, the other for the validation. PCA was then run to obtain the models for each group of the data set. Finally, based on the Mahalanobis distance, the data points were attributed to their corresponding classes of formerly computed PCA models.

What the results showed, is that the matrix from which the strains were isolated conferred a noticeable intraspecific heterogeneity to the strains. Additionally, all interclass distances of preliminary SIMCA models built to monitor the microbial growth were higher than 3, which indicates their robustness (Morita et al. 2014). SIMCA models developed on aqueous bacterial solutions further highlighted the differences between the strains and markedly separated those within the same biotope. The very same strains, when analyzed solely by molecular genotyping, were found to belong to the same cluster. Out of the 42 samples, 39 were correctly classified (95.1%). Together, these findings highlighted the suitability of NIRS with regard to tracking the microbial growth, distinguishing strains of same and different genotypes and more importantly, the study demonstrated the role of the aqueous matrices in reinforcing the distinctive features of the bacterial cultures. Aquaphotomics can be, therefore, an alternative to cumbersome, costly screening methods used in industrial settings.

Targeting research efforts toward determining water structures that are most reflective of cell functionality is crucial to swiftly pinpoint and select bacteria of interest. In this respect, Kovacs et al. (2019) have resorted to NIRS' contemporary approach: aquaphotomics.

In a comparative study of spectrally generated data to those obtained by reference methods, 18 bacterial strains of varying probiotic abilities were monitored (Slavchev et al. 2015). Probiotic and non-probiotic strains were either isolated from yoghurt and other marketed products or supplied by well-established provider. Calculated aquagrams (Figure 10.5) aided in the visualization of the spectral regions with prominent absorbance values of each of the probiotic, non-probiotic, and moderately probiotic strains. Thus, in the range 1,365–1,426 nm characterized by protonated water clusters, free water molecules, and weak hydrogen bonded water clusters, probiotic bacteria were the most predominant while moderate ones absorbed mostly in the 1,476–1,512 nm range, including essentially tight hydrogen bonded water. Cognizant of the correspondence between these wavelengths and the varying molecular structures of water (Segtnan, Isaksson, and Ozaki 2001;

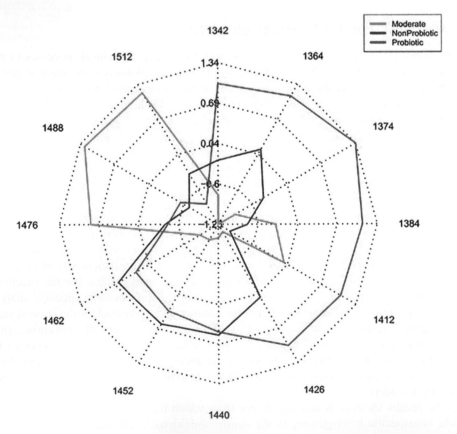

FIGURE 10.5 Aquagram, pinpointing the absorbance values of the studied bacterial groups (probiotic, moderate, and non-probiotic) at the different WAMACs (Slavchev et al. 2015).

Tsenkova et al. 2004), these findings allow a better grasp of the functionality of each of these water structural configurations.

All of these results highlight the high efficiency of aquaphotomics as a novel holistic approach with applications in the screening of microorganisms of technological interest and the determination of other potential functionalities. The screening can especially become important in food safety applications.

Food, as a water-rich system, can be amongst the most favored habitats for microorganisms, posing irrefutable risks to human health. Most of the techniques employed currently to determine microbiological quality of food are either time-consuming or cumbersome. For this purpose, Nakakimura et al. (2012) have proposed aquaphotomics NIR spectroscopy as a rather fast and non-invasive alternative. The NIRS as a technique has already been proven efficient for online monitoring of biomass during fermentation processes (Arnold et al. 2002). Nakakimura et al. (2012), however, aimed to show the feasibility of quantifying lower concentrations which are in compliance with the food safety criteria.

10.6 AQUAPHOTOMICS AND PLANT–SOIL SYSTEM HEALTH

NIRS is a fast and non-destructive technique, with diversified applications, and the aquaphotomics methodology further extends this. The water-mirror approach underlying aquaphotomics research can be effectively applied especially when examining herbaceous plants (Tsenkova 2009). The soil embraces a significant part of the abiotic (inorganic particles) and biotic (e.g., soil organisms, humus) factors essential for the evolution of plants, therefore, consideration is ineluctable (Pepper 2013). In this chapter section, we will briefly present some of the recent studies on plant health, with special emphasis on stress tolerance and quality of crops.

Stress is defined as a stressful condition that leads to divergences from normal behavior in living organisms. Stressor refers to the environmental impact. Living organisms can be affected by a variety of stressors, and these stress factors can be categorized in several ways: they can be abiotic or biotic, or anthropogenic or naturally occurring (Harrach 2009). Substantial changes in metabolism occur as a result of stress in plants. Some of these represent alterations in metabolic pathways, connections, and regulations that exist without stress, while others represent a series of alternative biochemical processes following the activation of new genes. Therefore, the stress demonstrably changes the water structure of the plant (Jinendra et al. 2010; Kuroki et al. 2019).

The plants develop in a dynamically fluctuating environment, which has varying, even harmful effects on it. The concept of abiotic stress covers all the lifeless, inorganic factors that have an adverse effect on plants. This encompasses, for example, extreme climatic conditions (cold or hot), water (drought or flood), soil nutrient deficiencies, salinity, toxic metal (arsenic, cadmium, chromium, lead, and mercury) stresses.

Water stress is one of the most common constraints on productivity in natural and agronomic crops, hence a major goal of remote sensing research is to accurately determine the symptoms of water stress in greenery by spectral reflection measurements. The aim of Peñuelas et al. (1993) research was to reveal the relationship between plant water status (e.g., leaf relative water content—RWC, potential, transpiration rates) and spectral reflectance, particularly in the 950–970 nm wavelength range (Penuelas et al. 1993). Gerbera plant desiccation and rehydration, pepper and bean cultivation under different irrigation conditions, and detached bean leaf dehydration were examined. Their results highlighted that in the 80%–90% RWC range, the reflectance changes were slight, on the other hand, significant differences occurred when the RWC was smaller than 80%–85%, as a result of the water stress (Penuelas et al. 1993). Extending their previous results—the NIR 950–970 nm region was proved to be effective in the quantification of water in plants or canopies—Peñuelas et al. (1997) used ground-based reflectance measurements and water index (R900/R970) to predict plant water content that was examined in nine Mediterranean adult plants throughout an annual cycle and in potted seedlings submitted to gradual desiccation. They observed that, due to dehydration, the wavelength analogous with the water absorption band (970–980 nm) moved to lower wavelengths (930–950 nm) (Penuelas et al. 1997).

Numerous plant physiological research studies demonstrated that water loss in the plant tissues results in significant morphological and structural realignment, caused by complex biochemical alterations at cellular and molecular levels. Depending on the different stages of water loss, different defensive mechanisms operate (Hoekstra, Golovina, and Buitink 2001). One recent study in plant biology looked for the cause of extreme desiccation tolerance of *Haberlea rhodopensis* by mapping its physiological water status during dehydration and rehydration processes by NIRS combined with the aquaphotomics methodology. The results obtained were compared with a relative but non-resurrection plant species, *Deinostigma eberhardtii* (Kuroki et al. 2019). Analyzing the entire data set of the cycle, six different water conformations were identified and their changes uncovered. There were particularly large differences in the ratio of free and bonded water molecules. While the resurrection plant drastically reduced water content, it maintained the ratios of certain water species on the same rate, the non-resurrection plant responded with only fluctuations (Muncan et al. 2019) (Figure 10.6). Another major difference happened for the plants in the dried state—the resurrection plant showed massive accumulation of dimers, increase in number of water molecules with

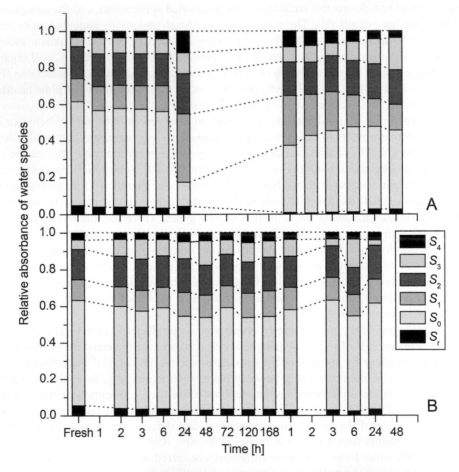

FIGURE 10.6 Dynamics of different water species during dehydration and rehydration of *Haberlea rhodopensis* and *Deinostigma eberhardtii*. Relative absorbance of water species in *Haberlea rhodopensis* (a) and *Deinostigma eberhardtii* (b) during desiccation and subsequent rehydration (Sr—protonated water clusters, S0—free water molecules, S1—water dimers, S2, S3, and S4—water molecules with 2, 3, and 4 hydrogen bonds, respectively). [Adapted from Kuroki et al. (2019), under the terms of CC-BY license.]

four hydrogen bonds, and drastically reduced free water molecules content. In the non-resurrection plant, this kind of behavior was absent.

The biotic stress factors include microbial pathogens—the prokaryotic (bacteria), eukaryotic (fungi and protozoa) organisms and viruses—vermin, insects, and herbivorous animals. The pathogens may originate and live in soils (geo-indigenous) or be introduced allogenic and become inactivated (geo-treatable) (Pepper 2013). In the last years, a trend can be observed in detection of infections in plants—instead of conventionally used destructive and expensive ELISA and PCR assays, non-invasive and real-time spectroscopy methods are becoming more widely spread. The spectral characteristics alter according to the metabolic activities and pathological conditions of plants (Mahlein 2016), prior to the visible appearance of any symptoms (Zhang et al. 2018). These reflectance changes have been used to detect, among others, the infection of maize (Ausmus and Hilty 1971) and tomato mosaic virus (Polischuk et al. 1997).

Purcell et al. (2009) applied at-field NIRS and chemometrics (PCA, PLSR) to predict sugar-cane clonal efficiency concerning Fiji leaf gall viral disease. They observed a significant decrease in the intensity between 5,205–5,393 cm^{-1} (1,854.26–1,921.23 nm) and in the OH stretching overtone (1,405.68 nm) with a shift to the higher wavelengths. The coefficients of determination and standard errors, describing the model fit, were 0.97 and 0.98 for the validation and 0.88 and 1.20 for the prediction (Purcell et al. 2009). Naidu et al. (2009) performed the spectral analysis of Grapevine leafroll-associated virus-3 infected and healthy leaves of Cabernet Sauvignon and Merlot wine grape cultivars (Naidu et al. 2009). Differences were discovered in the spectral properties of leaves near 550, 900, 1,600, and 2,200 nm. In the discrimination of sound and non-symptomatic infected leaves, the 970 nm wavelength, as variable, played a distinguished role (Naidu et al. 2009), as an indicator of plant water status (Penuelas et al. 1993).

In another prominent research, water-specific NIR spectra were used for rapid and *in vivo* detection of soybean mosaic virus even before symptom development (Jinendra et al. 2010). In data analysis, after PCA and spectral pre-treatments, SIMCA was applied to discriminate and predict different health conditions of plants. In the diagnosis, the achieved highest accuracy was 91.6% for the infected and 95.8% for the non-infected samples, respectively. Similarly to the previous observations of Peñuelas et al. (1993, 1997), the authors recognized the same peak shift in diseased leaves. Besides, multiple peaks were observed configuring the water absorbance pattern. The wavelengths 910, 936, and 970 nm proved to be the most valuable for the disease detection. In the case of infected samples, the highest absorbance was around the 2nd overtone of the OH stretching band (970 nm) (Büning-Pfaue 2003), while the healthy plants showed the smallest absorbance here (Jinendra et al. 2010).

Besides virologic assays, spectral analyzes have been successfully used to detect bacterial infections. Wang et al. (2010) developed a NIR HSI method, in the 950–1,650 nm spectral region, for early detection of onions infected by sour skin disease (*Burkholderia cepacia*). Based on the PCA (extraction of spectral signatures), 1,070 nm (center band of PC2 loading global peak) and 1,400 nm (center band of PC1 loading global peak) were selected to create band ratio images, which made it possible to distinguish more precisely the intact and the infected onions. The back-propagation neural network models allowed a 95% classification accuracy. In the aquaphotomics approach, one of the regions of interest for sour-skin detection (1,370–1,420 nm) coincides with three WAMACs (Tsenkova 2009).

Fungal diseases have been extensively studied with HSI in the Vis-NIR range. Bauriegel et al. (2011) aimed to determine the most appropriate wavelengths and grain development stage for *Fusarium* head blight disease detection in wheat. According to PCA, the wavelength ranges of pigments (Devred et al. 2013), such as carotenoids (500–533 nm) and chlorophylls (560–675 nm and 682–733 nm) could be identified for their high potential to distinguish intact and diseased wheat ears. Diminished absorption was observed in the chlorophyll region. Additionally, the spectral fluctuation in the 927–931 nm range was a result of differences in tissue water contents as a consequence of infection (Bauriegel et al. 2011). In the case of citrus, the mold rot is a frequently arising problem in the post-harvest technologies. The researchers also used HSI, in the 325–1,100 nm wavelength range, for the automatic detection of the early symptoms of decay caused by *Penicillium digitatum* fungus in navel oranges. In the examined range, 575, 698, 810, and 969 nm were selected as the characteristic wavelengths for the discrimination of healthy and damaged fruits. These results also support the above-described changes in water absorption owing to deterioration (Li et al. 2016).

10.7 IMPLICATIONS TO HUMAN HEALTH AND FUTURE PERSPECTIVES

Contemporary science has been established as a main stream goal of the "-omics" disciplines to deepen the knowledge and understanding of biomolecules which contribute about 30% to the living world. For the coming years, knowing that the rest of 70% of the body and the cell, and the surface of the earth is water which interacts very well with light, we propose to deepen the knowledge and understanding of the water molecular network as a scaffold of all these systems utilizing spectroscopy and big data mining.

This chapter discussed in parallel the current approaches in measurements and definition of SH parameters and offered a novel, complex way of defining the SH using one integrative biomarker—WASP of the soil. Water as an intrinsic matrix of the soil, is an incredible source of information about its structural, physical, chemical, and biological properties, and upon interaction with light, this information can be extracted non-destructively, rapidly, and on the spot and be related directly to its functionality. The novel approach to define and characterize the SH in this way, originated in the novel scientific field of aquaphotomics, which in the last two decades have reshaped the way we are assessing health of microorganisms, plants, animals, human organisms, and how we characterize water, food, materials, and nanomaterial functions. This chapter showed how the findings and applications of aquaphotomics can be applied in soil science as well, and SH assessed using an integrative marker—WASP.

NIRS is a fast, non-destructive measurement technology which can provide environmentally friendly and cost-effective solution for quality-related measurements and real-time monitoring. Combined with HSI, it also allows mapping of the large spatial areas. The aquaphotomics added to the NIRS by showing that utilization of properties of water as an intrinsic matrix of any aqueous or biological system allows not only simultaneous measurement of many chemical, physical, and biological properties of interest but that the WASP is an integrative marker of the system, which can be directly related to its functionality.

This chapter showed how using WASP as an integrative marker brings a rapid, non-destructive assessment of the functionality of different systems which are in interdependent relationship with soil and all of which in the end contribute to the human health as well. Through the examples of how the aquaphotomics technology works in the areas of water quality, microbiology, plant health and others, a parallel can be drawn that the same is possible for SH, and the same is already proven possible for human health monitoring (Muncan and Tsenkova 2019). Real-time monitoring of WASPs could provide early diagnosis of developmental stages and abnormalities on all levels from the water basins and soil to living systems and food and taking timely, appropriate counter measures. Further on, it has been proven that WASP of milk from dairy cows is directly related to specific feed in the diet, which illustrates the possibility of using WASP as a connective functionality element between two systems. Knowing the WASP of the soil could lead to prediction of most appropriate plants to be grown in it. Further on, the animals to be fed with these plants could be defined. Soil, plants, animals, and humans make again a system whose denominator is water, and the respective WASPs are the connections between its elements. Hence, the WASP of soil is directly related to humans and human health depends on it. As a future perspective, aquaphotomics envisage the use of water under perturbations to correct the spectral pattern of the respective system toward normal as a strategy with huge potential for sustainable development.

LIST OF ABBREVIATIONS

HCA—Hierarchical cluster analysis
HSI—Hyperspectral imaging
LDA—Linear discriminant analysis
NIR—Near infrared
NIRS—Near infrared spectroscopy
PCA—Principal component analysis
PCR—Principal component regression
PLS—DA—Partial least squares—Discriminant analysis
PLSR—Partial least squares regression
RMSE—Root mean square error
RMSECV—Root mean square error of cross validation
RMSEP—Root mean square error of prediction
SD—Standard deviation
SIMCA—Soft independent modeling of class analogies
Vis—NIR—Visible—Near infrared
WABs—Water absorbance bands
WAMACs—Water matrix coordinates
WASP—Water spectral pattern

REFERENCES

Al-Abbas, A., P.H. Swai, and M.F. Baumgardner. 1972. "Relating Organic Matter and Clay Content to the Multispectral Radiance of Soil." *Soil Science* 114 (6): 477–85.

Armbruster, D. A., and T. Pry. 2008. "Limit of Blank, Limit of Detection and Limit of Quantitation." *The Clinical Biochemist. Reviews* 29 Suppl 1 (August): S49–52.

Arnold, S. A, R. Gaensakoo, L. M. Harvey, and B. McNeil. 2002. "Use of At-Line and in-Situ near-Infrared Spectroscopy to Monitor Biomass in an Industrial Fed-Batch Escherichia Coli Process." *Biotechnology and Bioengineering* 80 (4): 405–13. doi:10.1002/bit.10383.

Ausmus, B. S, and J. W. Hilty. 1971. "Reflectance Studies of Healthy, Maize Dwarf Mosaic Virus-Infected, and Helminthosporium Maydis-Infected Corn Leaves." *Remote Sensing of Environment* 2: 77–81. Elsevier.

Ball, P. 2008a. "Water as an Active Constituent in Cell Biology." *Chemical Reviews* 108 (1): 74–108. doi:10.1021/CR068037A.

Ball, P. 2008b. "Water as a Biomolecule." *ChemPhysChem* 9 (18): 2677–85. John Wiley & Sons, Ltd. doi:10.1002/cphc.200800515.

Bauriegel, E., A. Giebel, M. Geyer, U. Schmidt, and W. B. Herppich. 2011. "Early Detection of Fusarium Infection in Wheat Using Hyper-Spectral Imaging." *Computers and Electronics in Agriculture* 75 (2): 304–12. Elsevier B.V. doi:10.1016/j.compag.2010.12.006.

Bázár, G., Z. Kovacs, M. Tanaka, A. Furukawa, A. Nagai, M. Osawa, Y. Itakura, H. Sugiyama, and R. Tsenkova. 2015. "Water Revealed as Molecular Mirror When Measuring Low Concentrations of Sugar with near Infrared Light." *Analytica Chimica Acta* 896 (October): 52–62. Elsevier. doi:10.1016/J. ACA.2015.09.014.

Ben-Dor, E., and A. Banin. 1995. "Near-Infrared Analysis as a Rapid Method to Simultaneously Evaluate Several Soil Properties." *Soil Science Society of America Journal* 59 (2): 364–72. Wiley. doi:10.2136/sss aj1995.03615995005900020014x.

Ben-Dor, E., S. Chabrillat, J.Al.M. Demattê, G.R. Taylor, J. Hill, M.L. Whiting, and S. Sommer. 2009. "Using Imaging Spectroscopy to Study Soil Properties." *Remote Sensing of Environment* 113: S38–55.

Blagodatskaya, E., and Y. Kuzyakov. 2013. "Active Microorganisms in Soil: Critical Review of Estimation Criteria and Approaches." *Soil Biology and Biochemistry*. Pergamon. doi:10.1016/j.soilbio.2013.08.024.

Blaschek, M., P. Roudier, M. Poggio, and C. B. Hedley. 2019. "Prediction of Soil Available Water-Holding Capacity from Visible near-Infrared Reflectance Spectra." *Scientific Reports* 9 (1): 1–10. Nature Publishing Group. doi:10.1038/s41598-019-49226-6.

Brown, D. J., R. S. Bricklemyer, and P. R. Miller. 2005. "Validation Requirements for Diffuse Reflectance Soil Characterization Models with a Case Study of VNIR Soil C Prediction in Montana." *Geoderma* 129 (3–4): 251–67. Elsevier. doi:10.1016/j.geoderma.2005.01.001.

Brown, D. J., K. D. Shepherd, M. G. Walsh, M. Dewayne Mays, and T. G. Reinsch. 2006. "Global Soil Characterization with VNIR Diffuse Reflectance Spectroscopy." *Geoderma* 132 (3–4): 273–90. doi:10.1016/j.geoderma.2005.04.025.

Büning-Pfaue, H. 2003. "Analysis of Water in Food by near Infrared Spectroscopy." *Food Chemistry* 82 (1): 107–15. Elsevier. doi:10.1016/S0308-8146(02)00583-6.

Carney, K. M., and P. A. Matson. 2005. "Plant Communities, Soil Microorganisms, and Soil Carbon Cycling: Does Altering the World Belowground Matter to Ecosystem Functioning?" *Ecosystems* 8 (8): 928–40. Springer. doi:10.1007/s10021-005-0047-0.

Cattaneo, T.M.P., M. Vanoli, M. Grassi, A. Rizzolo, and S. Barzaghi. 2016. "The Aquaphotomics Approach as a Tool for Studying the Influence of Food Coating Materials on Cheese and Winter Melon Samples." *Journal of Near Infrared Spectroscopy* 24 (4): 381–90. SAGE Publications Sage UK: London, England. doi:10.1255/jnirs.1238.

Cattaneo, T.M.P., S. Vero, E. Napoli, and V. Elia. 2011. "Influence of Filtration Processes on Aqueous Nanostructures by NIR Spectroscopy." *Journal of Chemistry and Chemical Engineering* 5 (11): 1046–52.

Cécillon, L., B. G. Barthès, C. Gomez, D. Ertlen, V. Genot, M. Hedde, A. Stevens, and J. J. Brun. 2009. "Assessment and Monitoring of Soil Quality Using Near-Infrared Reflectance Spectroscopy (NIRS)." *European Journal of Soil Science* 60 (5): 770–84. John Wiley & Sons, Ltd. doi:10.1111/j.1365–2389.2009.01178.x.

Chabrillat, S., E. Ben-Dor, J. Cierniewski, C. Gomez, T. Schmid, and B. van Wesemael. 2019. "Imaging Spectroscopy for Soil Mapping and Monitoring." *Surveys in Geophysics*. Springer Netherlands. doi:10.1007/s10712-019-09524-0.

Chakraborty, S., B. Li, S. Deb, S. Paul, D. C. Weindorf, and B.S. Das. 2017. "Predicting Soil Arsenic Pools by Visible near Infrared Diffuse Reflectance Spectroscopy." *Geoderma* 296 (June): 30–37. Elsevier B.V. doi:10.1016/j.geoderma.2017.02.015.

Chang, C-W., D. A. Laird, M. J. Mausbach, and C. R. Hurburgh. 2001. "Near-Infrared Reflectance Spectroscopy-Principal Components Regression Analyses of Soil Properties." *Soil Science Society of America Journal* 65 (2): 480–90. Wiley. doi:10.2136/sssaj2001.652480x.

Chatani, E., Y. Tsuchisaka, Y. Masuda, and R. Tsenkova. 2014. "Water Molecular System Dynamics Associated with Amyloidogenic Nucleation as Revealed by Real Time near Infrared Spectroscopy and Aquaphotomics." *PLoS One* 9 (7): e101997. doi:10.1371/journal.pone.0101997.

Chen, T., Q. Chang, J. G.P.W. Clevers, and L. Kooistra. 2015. "Rapid Identification of Soil Cadmium Pollution Risk at Regional Scale Based on Visible and Near-Infrared Spectroscopy." *Environmental Pollution* 206 (July): 217–26. Elsevier Ltd. doi:10.1016/j.envpol.2015.07.009.

Christy, C. D. 2008. "Real-Time Measurement of Soil Attributes Using on-the-Go near Infrared Reflectance Spectroscopy." *Computers and Electronics in Agriculture* 61 (1): 10–19. Elsevier. doi:10.1016/j.compag.2007.02.010.

Cozzolino, D., and A. Morón. 2003. "The Potential of Near-Infrared Reflectance Spectroscopy to Analyse Soil Chemical and Physical Characteristics." *Journal of Agricultural Science* 140 (1): 65–71. Cambridge University Press. doi:10.1017/S0021859602002836.

Dalal, R. C., and R. J. Henry. 1986. "Simultaneous Determination of Moisture, Organic Carbon, and Total Nitrogen by Near Infrared Reflectance Spectrophotometry." *Soil Science Society of America Journal* 50 (1): 120–23. Wiley. doi:10.2136/sssaj1986.03615995005000010023x.

DeMan, J. M., J. W. Finley, W. J. Hurst, and C. Y. Lee. 1999. *Principles of Food Chemistry*. 3rd ed. Gaithersburg, MD: Aspen Publishers.

Devred, E., K. R. Turpie, W. Moses, V. V. Klemas, T. Moisan, M. Babin, G. Toro-Farmer, M. H. Forget, and Y. H. Jo. 2013. "Future Retrievals of Water Column Bio-Optical Properties Using the Hyperspectral Infrared Imager (Hyspiri)." *Remote Sensing* 5 (12): 6812–37. doi:10.3390/rs5126812.

Dor, E. B., C. Ong, and I. C. Lau. 2015. "Reflectance Measurements of Soils in the Laboratory: Standards and Protocols." *Geoderma* 245–246: 112–24. Elsevier B.V. doi:10.1016/j.geoderma.2015.01.002.

Doran, J. W., and T. B. Parkin. 1994. "Defining and Assessing Soil Quality." *Defining Soil Quality for a Sustainable Environment* 35: 3–21. *Proceedings of Symposium, Minneapolis, MN, 1992*, October. SSSA/ASA; Special Publication. doi:10.2136/sssaspecpub35.c1.

Dunn, B. W., H. G. Beecher, G. D. Batten, and S. Ciavarella. 2002. "The Potential of Near-Infrared Reflectance Spectroscopy for Soil Analysis – A Case Study from the Riverine Plain of South-Eastern Australia." *Australian Journal of Experimental Agriculture* 42 (5): 607–14. CSIRO PUBLISHING. doi:10.1071/EA01172.

Ehsani, M. R., S. K. Upadhyaya, D. Slaughter, S. Shafii, and M. Pelletier. 1999. "A NIR Technique for Rapid Determination of Soil Mineral Nitrogen." *Precision Agriculture* 1 (2): 217–34. Springer. doi:10.1023/A:1009916108990.

Eisenberg, D., and W. Kauzmann. 1997. *The Structure and Properties of Water.* London, UK: Oxford University Press.

Farifteh, J., F. van der Meer, M. van der Meijde, and C. Atzberger. 2008. "Spectral Characteristics of Salt-Affected Soils: A Laboratory Experiment." *Geoderma* 145: 196–206.

Fidêncio, P.H., R. J. Poppi, J. C. De Andrade, and H. Cantarella. 2002. "Determination of Organic Matter in Soil Using Near-Infrared Spectroscopy and Partial Least Squares Regression." *Communications in Soil Science and Plant Analysis* 33 (9–10): 1607–15. Taylor & Francis Group. doi:10.1081/CSS-120004302.

Fystro, G. 2002. "The Prediction of C and N Content and Their Potential Mineralisation in Heterogeneous Soil Samples Using Vis-NIR Spectroscopy and Comparative Methods." *Plant and Soil* 246 (2): 139–49. Springer. doi:10.1023/A:1020612319014.

Gomez, C., P. Lagacherie, and G. Coulouma. 2012. "Regional Predictions of Eight Common Soil Properties and Their Spatial Structures from Hyperspectral Vis-NIR Data." *Geoderma* 189–190: 176–85. Elsevier B.V. doi:10.1016/j.geoderma.2012.05.023.

Gowen, A. A., F. Marini, Y. Tsuchisaka, S. De Luca, M. Bevilacqua, C. O'Donnell, G. Downey, et al. 2015. "On the Feasibility of near Infrared Spectroscopy to Detect Contaminants in Water Using Single Salt Solutions as Model Systems." *Talanta* 131 (January). Elsevier: 609–18. doi:10.1016/J.TALANTA.2014.08.049.

Gowen, A.A., Y. Tsuchisaka, C. O'Donnell, and R. Tsenkova. 2011. "Investigation of the Potential of near Infrared Spectroscopy for the Detection and Quantification of Pesticides in Aqueous Solution." *American Journal of Analytical Chemistry* 2: 53–62. doi:10.4236/ajac.2011.228124.

Grant, A., A. M. C. Davies, and T. Bilverstone. 1989. "Simultaneous Determination of Sodium Hydroxide, Sodium Carbonate and Sodium Chloride Concentrations in Aqueous Solutions by near-Infrared Spectrometry." *The Analyst* 114 (7): 819. The Royal Society of Chemistry. doi:10.1039/an9891400819.

Halpern, B.S., C. Longo, D. Hardy, K.L. McLeod, J.F. Samhouri, S.K. Katona, K. Kleisner, et al. 2012. "An Index to Assess the Health and Benefits of the Global Ocean." *Nature* 488: 615–20.

Harrach, B. D. 2009. *Abiotic and Biotic Stress Effects on Barley and Tobacco Plants (in Hungarian).* Budapest, Hungary: Corvinus University.

Headrick, J.M., E.G. Diken, R.S. Walters, N.I. Hammer, R.A. Christie, J. Cui, E.M. Myshakin, M.A. Duncan, M.A. Johnson, and K.D. Jordan. 2005. "Spectral Signatures of Hydrated Proton Vibrations in Water Clusters." *Science* 308 (5729): 1765–69. American Association for the Advancement of Science. doi:10.1126/science.1113094.

Hirschfeld, T. 1985. "Salinity Determination Using NIRA." *Applied Spectroscopy* 39 (4): 740–41. Society for Applied Spectroscopy. https://www.osapublishing.org/as/abstract.cfm?uri=as-39-4-740.

Hoekstra, F A, E A Golovina, and J Buitink. 2001. "Mechanisms of Plant Desiccation Tolerance." *Trends in Plant Science* 6 (9): 431–38. http://www.ncbi.nlm.nih.gov/pubmed/11544133.

Jamrógiewicz, M. 2012. "Journal of Pharmaceutical and Biomedical Analysis Application of the Near-Infrared Spectroscopy in the Pharmaceutical Technology." *Journal of Pharmaceutical and Biomedical Analysis* 66: 1–10. Elsevier B.V. doi:10.1016/j.jpba.2012.03.009.

Jinendra, B. 2011. "Near Infrared Spectroscopy and Aquaphotomics: Novel Tool for Biotic and Abiotic Stress Diagnosis of Soybean. PhD Thesis." Kobe University, Kobe, Japan.

Jinendra, B., K. Tamaki, S. Kuroki, M. Vassileva, S. Yoshida, and R. Tsenkova. 2010. "Near Infrared Spectroscopy and Aquaphotomics: Novel Approach for Rapid in Vivo Diagnosis of Virus Infected Soybean." *Biochemical and Biophysical Research Communications* 397 (4): 685–90. Kobe University, Kobe, Japan: Elsevier Inc. doi:10.1016/j.bbrc.2010.06.007.

Kaur, H., R. Künnemeyer, and A. McGlone. 2020. "Investigating Aquaphotomics for Temperature-Independent Prediction of Soluble Solids Content of Pure Apple Juice." *Journal of Near Infrared Spectroscopy*, January, 096703351989889. doi:10.1177/0967033519898891.

Kinoshita, K., N. Kuze, T. Kobayashi, E. Miyakawa, H. Narita, M. Inoue-Murayama, G. Idani, and R. Tsenkova. 2016. "Detection of Urinary Estrogen Conjugates and Creatinine Using near Infrared Spectroscopy in Bornean Orangutans (Pongo Pygmaeus)." *Primates* 57 (1): 51–59. doi:10.1007/s10329-015-0501-3.

Kinoshita, K., M. Miyazaki, H. Morita, M. Vassileva, C. Tang, D. Li, O. Ishikawa, H. Kusunoki, and R. Tsenkova. 2012. "Spectral Pattern of Urinary Water as a Biomarker of Estrus in the Giant Panda." *Scientific Reports* 2 (1): 856. Nature Publishing Group. doi:10.1038/srep00856.

Kooistra, L., J. Wanders, G. F. Epema, R. S.E.W. Leuven, R. Wehrens, and L. M.C. Buydens. 2003. "The Potential of Field Spectroscopy for the Assessment of Sediment Properties in River Floodplains." *Analytica Chimica Acta* 484 (2): 189–200. Elsevier. doi:10.1016/S0003–2670(03)00331-3.

Kooistra, L., R. Wehrens, R. S.E.W. Leuven, and L. M.C. Buydens. 2001. "Possibilities of Visible-near-Infrared Spectroscopy for the Assessment of Soil Contamination in River Floodplains." In *Analytica Chimica Acta* 446: 97–105. Elsevier. doi:10.1016/S0003-2670(01)01265-X.

Kovacs, Z., G. Bázár, M. Oshima, S. Shigeoka, M. Tanaka, A. Furukawa, A. Nagai, M. Osawa, Y. Itakura, and R. Tsenkova. 2015. "Water Spectral Pattern as Holistic Marker for Water Quality Monitoring." *Talanta* 147 (January): 598–608. Elsevier. doi:10.1016/j.talanta.2015.10.024.

Kovacs, Z., A. Slavchev, G. Bazar, B. Pollner, and R. Tsenkova. 2019. "Rapid Bacteria Selection Using Aquaphotomics and near Infrared Spectroscopy." In *18th International Conference on Near Infrared Spectroscopy*, 65–69. IM Publications Open LLP. doi:10.1255/nir2017.065.

Van de Kraats, E. B., J. Munćan, and R. N. Tsenkova. 2019. "Aquaphotomics – Origin, Concept, Applications and Future Perspectives." *Substantia*, November, 13–28. doi:10.13128/substantia–702.

Krishnan, P., J. D. Alexander, B. J. Butler, and J. W. Hummel. 1980. "Reflectance Technique for Predicting Soil Organic Matter." *Soil Science Society of America Journal* 44 (6): 1282–85. Wiley. doi:10.2136/sss aj1980.03615995004400060030x.

Kuroki, S., R. Tsenkova, D. P. Moyankova, J. Muncan, H. Morita, S. Atanassova, and D. Djilianov. 2019. "Water Molecular Structure Underpins Extreme Desiccation Tolerance of the Resurrection Plant Haberlea Rhodopensis." *Scientific Reports* 9 (1): 3049. doi:10.1038/s41598-019-39443-4.

Lagacherie, P., A. R. Sneep, C. Gomez, S. Bacha, G. Coulouma, M. H. Hamrouni, and I. Mekki. 2013. "Combining Vis-NIR Hyperspectral Imagery and Legacy Measured Soil Profiles to Map Subsurface Soil Properties in a Mediterranean Area (Cap-Bon, Tunisia)." *Geoderma* 209–210 (November): 168–76. Elsevier B.V. doi:10.1016/j.geoderma.2013.06.005.

Li, J., W. Huang, X. Tian, C. Wang, S. Fan, and C. Zhao. 2016. "Fast Detection and Visualization of Early Decay in Citrus Using Vis-NIR Hyperspectral Imaging." *Computers and Electronics in Agriculture* 127 (September): 582–92. Elsevier B.V. doi:10.1016/j.compag.2016.07.016.

St. Luce, M., N. Ziadi, B. Gagnon, and A. Karam. 2017. "Visible near Infrared Reflectance Spectroscopy Prediction of Soil Heavy Metal Concentrations in Paper Mill Biosolid- and Liming by-Product-Amended Agricultural Soils." *Geoderma* 288 (February): 23–36. Elsevier B.V. doi:10.1016/j.geoderma.2016.10.037.

Luck, W.A.P. 1974. "Structure of Water and Aqueous Solutions." In *International Symposium Marburg*, edited by W. A. P. Luck, 248–84. Weinheim, Germany: Verlag Chemie. doi:10.1002/bbpc.19760800719.

Mahlein, A-K. 2016. "Plant Disease Detection by Imaging Sensors–Parallels and Specific Demands for Precision Agriculture and Plant Phenotyping." *Plant Disease* 100 (2): 241–51. American Phytopathological Society.

McBratney, A.B., B. Minasny, and R. Viscarra Rossel. 2006. "Spectral Soil Analysis and Inference Systems: A Powerful Combination for Solving the Soil Data Crisis." *Geoderma* 136 (1–2): 272–78. Elsevier. doi:10.1016/j.geoderma.2006.03.051.

Meilina, H., S. Kuroki, B.M. Jinendra, K. Ikuta, and R. Tsenkova. 2009. "Double Threshold Method for Mastitis Diagnosis Based on NIR Spectra of Raw Milk and Chemometrics." *Biosystems Engineering* 104 (2): 243–49. Academic Press. doi:10.1016/J.BIOSYSTEMSENG.2009.04.006.

Meyer, J. H. 1989. "Rapid Simultaneous Rating of Soil Texture, Organic Matter, Total Nitrogen and Nitrogen Mineralization Potential by near Infra-Red Reflectance." *South African Journal of Plant and Soil* 6 (1): 59–63. Taylor & Francis Group. doi:10.1080/02571862.1989.10634481.

Morita, H., T. Hasunuma, M. Vassileva, A. Kondo, and R. Tsenkova. 2014. "A New Screening Method for Recombinant Saccharomyces Cerevisiae Strains Based on Their Xylose Fermentation Ability Measured by near Infrared Spectroscopy." *Analytical Methods* 6 (17): 6628–34. Royal Society of Chemistry. doi:10.1039/c4ay00785a.

Moron, A., and D. Cozzolino. 2003. "Exploring the Use of near Infrared Reflectance Spectroscopy to Study Physical Properties and Microelements in Soils." *Journal of Near Infrared Spectroscopy* 11 (2): 145–54. N I R Publications. doi:10.1255/jnirs.362.

Mouazen, A. M., M. R. Maleki, J. De Baerdemaeker, and H. Ramon. 2007. "On-Line Measurement of Some Selected Soil Properties Using a VIS-NIR Sensor." *Soil and Tillage Research* 93 (1): 13–27. Elsevier. doi:10.1016/j.still.2006.03.009.

Muncan, J., S. Kuroki, D. Moyankova, H. Morita, S. Atanassova, D. Djilianov, and R. Tsenkova. 2019. "Recent Advancements in Plant Aquaphotomics – Towards Understanding of 'Drying without Dying' Phenomenon and Its Implications." *NIR News* 30 (5–6): 22–25. SAGE Publications Sage UK: London, England. doi:10.1177/0960336019855168.

Muncan, J., V. Matovic, S. Nikolic, J. Askovic, and R. Tsenkova. 2020. "Aquaphotomics Approach for Monitoring Different Steps of Purification Process in Water Treatment Systems." *Talanta* 206 (January): 120253. Elsevier B.V. doi:10.1016/j.talanta.2019.120253.

Muncan, J., I. Mileusnic, V. Matovic, J. Sakota Rosic, and L. Matija. 2016. "The Prospects of Aquaphotomics in Biomedical Science and Engineering." In *The 2nd International Aquaphotomics Symposium*. Kobe, Japan. https://aquaphotomics.com/Past/2016/wp-content/uploads/2016/11/23-MUNCAN-Jelena_abst.pdf.

Muncan, J., and R. Tsenkova. 2019. "Aquaphotomics-From Innovative Knowledge to Integrative Platform in Science and Technology." *Molecules* 24 (15): 2742. doi:10.3390/molecules24152742.

Mura, S., C. Cappai, G. F. Greppi, S. Barzaghi, A. Stellari, and T.M.P. Cattaneo. 2019. "Vibrational Spectroscopy and Aquaphotomics Holistic Approach to Determine Chemical Compounds Related to Sustainability in Soil Profiles." *Computers and Electronics in Agriculture* 159 (February): 92–96. Elsevier. doi:10.1016/j.compag.2019.03.002.

Naidu, R. A., E. M. Perry, F. J. Pierce, and T. Mekuria. 2009. "The Potential of Spectral Reflectance Technique for the Detection of Grapevine Leafroll-Associated Virus-3 in Two Red-Berried Wine Grape Cultivars." *Computers and Electronics in Agriculture* 66 (1): 38–45. doi:10.1016/j.compag.2008.11.007.

Nakakimura, Y., M. Vassileva, T. Stoyanchev, K. Nakai, R. Osawa, J. Kawano, and R. Tsenkova. 2012. "Extracellular Metabolites Play a Dominant Role in Near-Infrared Spectroscopic Quantification of Bacteria at Food-Safety Level Concentrations." *Analytical Methods* 4 (5): 1389. The Royal Society of Chemistry. doi:10.1039/c2ay05771a.

Nocita, M., A. Stevens, C. Noon, and B. Van Wesemael. 2013. "Prediction of Soil Organic Carbon for Different Levels of Soil Moisture Using Vis-NIR Spectroscopy." *Geoderma* 199 (May): 37–42. Elsevier B.V. doi:10.1016/j.geoderma.2012.07.020.

Ozaki, Y. 2002. "Applications in Chemistry." In *Near-Infrared Spectroscopy. Principles, Instruments, Applications*, edited by H. W. Siesler, Y. Ozaki, S. Kawata, and H. Heise, 179–213. Weinheim, Germany: Wiley-VCH Verlag GmbH. doi:10.1002/cem.762.

Palacios-Orueta, A., and S. L. Ustin. 1998. "Remote Sensing of Soil Properties in the Santa Monica Mountains I. Spectral Analysis." *Remote Sensing of Environment* 65 (2): 170–83. Elsevier Science Inc. doi:10.1016/S0034-4257(98)00024-8.

Penuelas, J., I. Filella, C. Biel, L. Serrano, and R. Save. 1993. "The Reflectance at the 950–970 Nm Region as an Indicator of Plant Water Status." *International Journal of Remote Sensing* 14 (10): 1887–905. Taylor & Francis Group. doi:10.1080/01431169308954010.

Penuelas, J., J. Pinol, R. Ogaya, and I. Filella. 1997. "Estimation of Plant Water Concentration by the Reflectance Water Index WI (R900/R970)." *International Journal of Remote Sensing* 18 (13): 2869–75. doi:10.1080/014311697217396.

Pepper, I. L. 2013. "The Soil Health-Human Health Nexus." *Critical Reviews in Environmental Science and Technology* 43 (24): 2617–52. Taylor & Francis.

Polischuk, V P, T M Shadchina, T I Kompanetz, I G Budzanivskaya, A L Boyko, and A A Sozinov. 1997. "Changes in Reflectance Spectrum Characteristic of Nicotiana Debneyi Plant under the Influence of Viral Infection." *Archives of Phytopathology and Plant Protection* 31: 115–9. Taylor & Francis.

Purcell, D.E., M.G. O'Shea, R.A. Johnson, and S. Kokot. 2009, "Near-Infrared Spectroscopy for the Prediction of Disease Ratings for Fiji Leaf Gall in Sugarcane Clones." *Applied Spectroscopy* 63 (4): 450–57. SAGE PublicationsSage UK: London, England. doi:10.1366/000370209787944370.

Putra, A., F. Faridah, E. Inokuma, and R. Santo. 2010. "Robust Spectral Model for Low Metal Concentration Measurement in Aqueous Solution Reveals the Importance of Water Absorbance Bands." *Jurnal Sains Dan Teknologi Reaksi* 8 (1). http://e-jurnal.pnl.ac.id/index.php/JSTR/article/view/105.

Putra, A., M. Vassileva, R. Santo, and R. Tsenkova. 2017. "An Efficient near Infrared Spectroscopy Based on Aquaphotomics Technique for Rapid Determining the Level of Cadmium in Aqueous Solution." *IOP Conference Series: Materials Science and Engineering* 210 (1): 012014. IOP Publishing. doi:10.1088/1757-899X/210/1/012014.

Remagni, M.C. C, H. Morita, H. Koshiba, T.M.P. Cattaneo, and R. Tsenkova. 2013. "Near Infrared Spectroscopy and Aquaphotomics as Tools for Bacteria Classification." In *NIR2013: Picking up Good Vibrations – ST1 – Spectroscopy, Scattering, Absorption, Aquaphotomics*, edited by V. Bellon-Maurel, P. Williams, and G. Downey, 602–608. La Grande-Motte: IRSTEA.

Rinot, O., G. J. Levy, Y. Steinberger, T. Svoray, and G. Eshel. 2019. "Soil Health Assessment: A Critical Review of Current Methodologies and a Proposed New Approach." *Science of the Total Environment* 648 (January): 1484–91. Elsevier B.V. doi:10.1016/j.scitotenv.2018.08.259.

Sakudo, A., R. Tsenkova, T. Onozuka, K. Morita, S. Li, J. Warachit, Y. Iwabu, G. Li, T. Onodera, and K. Ikuta. 2005. "A Novel Diagnostic Method for Human Immunodeficiency Virus Type-1 in Plasma by Near-Infrared Spectroscopy." *Microbiology and Immunology* 49 (7): 695–701. Wiley/Blackwell (10.1111). doi:10.1111/j.1348-0421.2005.tb03648.x.

Sakudo, A., R. Tsenkova, K. Tei, T. Onozuka, K. Ikuta, E. Yoshimura, and T. Onodera. 2006. "Comparison of the Vibration Mode of Metals in HNO3 by a Partial Least-Squares Regression Analysis of near-Infrared Spectra." *Bioscience, Biotechnology and Biochemistry* 70 (7): 1578–83. Japan Society for Bioscience, Biotechnology, and Agrochemistry. doi:10.1271/bbb.50619.

Schwertmann, U. 1993. "Relations between Iron Oxides, Soil Color, and Soil Formation." *Soil Color* 31: 51–69. *Proc. Symposium, San Antonio, 1990* sssaspecialpubl (soilcolor). Soil Science Society of America Inc.; Special Publication. doi:10.2136/sssaspecpub31.c4.

Segtnan, V H, T Isaksson, and Y Ozaki. 2001. "Studies on the Structure of Water Using Spectroscopy and Principal Component Analysis." *Analytical Chemistry* 73 (13): 3153–61. doi:10.1021/ac010102n.

Shepherd, K. D., and M. G. Walsh. 2002. "Development of Reflectance Spectral Libraries for Characterization of Soil Properties." *Soil Science Society of America Journal* 66 (3): 988–98. Wiley. doi:10.2136/sssaj2002.9880.

Shonk, J. L., L. D. Gaultney, D. G. Schulze, and G. E. Van Scoyoc. 1991. "Spectroscopic Sensing of Soil Organic Matter Content." *Transactions of the American Society of Agricultural Engineers* 34 (5): 1978–84. American Society of Agricultural and Biological Engineers. doi:10.13031/2013.31826.

Slavchev, A., Z. Kovacs, H. Koshiba, G. Bazar, B. Pollner, A. Krastanov, and R. Tsenkova. 2017. "Monitoring of Water Spectral Patterns of Lactobacilli Development as a Tool for Rapid Selection of Probiotic Candidates." *Journal of Near Infrared Spectroscopy* 25 (6): 423–31. SAGE Publications Sage UK: London, England. doi:10.1177/0967033517741133.

Slavchev, A., Z. Kovacs, H. Koshiba, A. Nagai, G. Bázár, A. Krastanov, Y. Kubota, and R. Tsenkova. 2015. "Monitoring of Water Spectral Pattern Reveals Differences in Probiotics Growth When Used for Rapid Bacteria Selection." Edited by George-John Nychas. *PLoS One* 10 (7): e0130698. Public Library of Science. doi:10.1371/journal.pone.0130698.

Smith, J. D., C. D. Cappa, K. R. Wilson, P. L. Geissler, R. C. Cohen, and R. J. Saykally. 2005. "Unified Description of Temperature-Dependent Hydrogen-Bond Rearrangements in Liquid Water." *Proceedings of the National Academy of Sciences (PNAS)* 102 (40): 14171.

Stenberg, B. 2010. "Effects of Soil Sample Pretreatments and Standardised Rewetting as Interacted with Sand Classes on Vis-NIR Predictions of Clay and Soil Organic Carbon." *Geoderma* 158 (1–2): 15–22. Elsevier B.V. doi:10.1016/j.geoderma.2010.04.008.

Stenberg, B., A.R. Viscarra Rossel, A.M. Mouazen, and J. Wetterlind. 2010. *Visible and Near Infrared Spectroscopy in Soil Science. Advances in Agronomy.* Vol. 107. Academic Press. doi:10.1016/S0065-2113(10)07005-7.

Sudduth, K. A. and J. W. Hummel. 1993. "Soil Organic Matter, CEC, and Moisture Sensing with a Portable NIR Spectrophotometer." *Transactions of the ASAE* 36 (6): 1571–82. American Society of Agricultural and Biological Engineers. doi:10.13031/2013.28498.

Tanaka, M., A. Shibata, N. Hayashi, T. Kojima, H. Maeda, and Y. Ozaki. 1995. "Discrimination of Commercial Natural Mineral Waters Using near Infrared Spectroscopy and Principal Component Analysis." *Journal of Near Infrared Spectroscopy* 3 (4): 203–10. SAGE Publications Sage UK: London, England. doi:10.1255/jnirs.70.

Todorova, M., M. Mihalache, L. Ilie, and S. Atanassova. 2015. "Visible-NIR Reflectance for Evaluation of Luvisols and Phaeozems." *Agricultural Science and Technology* 7 (3): 339–43.

Tsenkova, R. 2006. "AquaPhotomics: Water Absorbance Pattern as a Biological Marker." *NIR News* 17 (7): 13–20.

Tsenkova, R. 2007a. "Aquaphotomics: Extended Water Mirror Approach Reveals Peculiarities of Prion Protein Alloforms." *NIR News* 18 (6): 14–17.

Tsenkova, R. 2007b. "AquaPhotomics: Water Absorbance Pattern as a Biological Marker for Disease Diagnosis and Disease Understanding." *NIR News* 18 (2): 14–16. doi:10.1255/nirn.1014.

Tsenkova, R. 2008. "Aquaphotomics: The Extended Water Mirror Effect Explains Why Small Concentrations of Protein in Solution Can Be Measured with near Infrared Light." *NIR News* 19 (4): 13–14. SAGE Publications Sage UK: London, England. doi:10.1255/nirn.1079.

Tsenkova, R. 2009. "Aquaphotomics: Dynamic Spectroscopy of Aqueous and Biological Systems Describes Peculiarities of Water." *Journal of Near Infrared Spectroscopy* 17 (6): 303–13. SAGE Publications Sage UK: London, England. doi:10.1255/jnirs.869.

Tsenkova, R., and S. Atanassova. 2002. "Mastitis Diagnostics by near Infrared Spectra of Cow's Milk, Blood and Urine Using Soft Independent Modelling of Class Analogy Classification." In *Near Infrared Spectroscopy: Proceedings of the 10th International Conference*, 123–28. Chichester, UK: IM Publications Open LLP.

Tsenkova, R., Z. Kovacs, and Y. Kubota. 2015. "Aquaphotomics: Near Infrared Spectroscopy and Water States in Biological Systems." In *Membrane Hydration*, edited by E. A. DiSalvo, 189–211. Cham, Switzerland: Springer. doi:10.1007/978-3-319-19060-0_8.

Tsenkova, R., J. Munćan, B. Pollner, and Z. Kovacs. 2018. "Essentials of Aquaphotomics and Its Chemometrics Approaches." *Frontiers in Chemistry* 6 (August): 363. Frontiers. doi:10.3389/fchem.2018.00363.

Tsenkova, R. 2010. "Aquaphotomics: Water in the Biological and Aqueous World Scrutinised with Invisible Light." *Spectroscopy Europe* 22 (6): 6–10.

Tsenkova, R, S Atanassova, S Kawano, and K Toyoda. 2009. "Somatic Cell Count Determination in Cow's Milk by near-Infrared Spectroscopy : A New Diagnostic Tool The Online Version of This Article, along with Updated Information and Services, Is Located on the World Wide Web at : Somatic Cell Count Determinatio." *Journal of Animal Science* 79: 2550–57.

Tsenkova, R. N., I. K. Iordanova, K. Toyoda, and D. R. Brown. 2004. "Prion Protein Fate Governed by Metal Binding." *Biochemical and Biophysical Research Communications* 325 (3): 1005–12. doi:10.1016/j.bbrc.2004.10.135.

Udelhoven, T., C. Emmerling, and T. Jarmer. 2003. "Quantitative Analysis of Soil Chemical Properties with Diffuse Reflectance Spectrometry and Partial Least-Square Regression: A Feasibility Study." *Plant and Soil* 251 (2): 319–29. Springer. doi:10.1023/A:1023008322682.

Vågen, T. G., K. D. Shepherd, and M. G. Walsh. 2006. "Sensing Landscape Level Change in Soil Fertility Following Deforestation and Conversion in the Highlands of Madagascar Using Vis-NIR Spectroscopy." *Geoderma* 133 (3–4): 281–94. Elsevier. doi:10.1016/j.geoderma.2005.07.014.

Vanoli, M., F. Lovati, M. Grassi, M. Buccheri, A. Zanella, T.M.P Cattaneo, and A. Rizzolo. 2018. "Water Spectral Pattern as a Marker for Studying Apple Sensory Texture." *Advances in Horticultural Science* 32 (3): 343–51. doi:10.13128/ahs-22380.

Vasques, G. M., J. A. M. Demattê, R. A. Viscarra Rossel, L. Ramírez-López, and F. S. Terra. 2014. "Soil Classification Using Visible/near-Infrared Diffuse Reflectance Spectra from Multiple Depths." *Geoderma* 223–225 (1): 73–78. Elsevier. doi:10.1016/j.geoderma.2014.01.019.

Viscarra Rossel, R. A., D. J. J. Walvoort, A. B. McBratney, L. J. Janik, and J. O. Skjemstad. 2006. "Visible, near Infrared, Mid Infrared or Combined Diffuse Reflectance Spectroscopy for Simultaneous Assessment of Various Soil Properties." *Geoderma* 131 (1–2): 59–75. Elsevier. doi:10.1016/j.geoderma.2005.03.007.

Vohland, M., J. Besold, J. Hill, and H.C. Fründ. 2011. "Comparing Different Multivariate Calibration Methods for the Determination of Soil Organic Carbon Pools with Visible to near Infrared Spectroscopy." *Geoderma* 166 (1): 198–205. Elsevier B.V. doi:10.1016/j.geoderma.2011.08.001.

Wang, J., J. Ding, A. Abulimiti, and L. Cai. 2018. "Quantitative Estimation of Soil Salinity by Means of Different Modeling Methods and Visible-near Infrared (VIS-NIR) Spectroscopy, Ebinur Lake Wetland, Northwest China." *PeerJ* 2018 (5): e4703. Peer J Inc. doi:10.7717/peerj.4703.

Wang, W., C. Li, R. Gitaitis, E. W. Tollner, G.C. Rains, and S. C. Yoon. 2010. "Near-Infrared Hyperspectral Reflectance Imaging for Early Detection of Sour Skin Disease in Vidalia Sweet Onions." *American Society of Agricultural and Biological Engineers Annual International Meeting 2010, ASABE 2010* 4 (10): 3415–36. St. Joseph, MI: American Society of Agricultural and Biological Engineers. doi:10.13031/2013.29968.

Wcindorf, D. C. 2018. "Advanced Modeling of Soil Biological Properties Using Visible Near Infrared Diffuse Reflectance Spectroscopy." *International Journal of Bioresource Science* 5 (1): 1–20. doi:10.30954/2347-9655.01.2018.1.

Williams, P. 2009. "Influence of Water on Prediction of Composition and Quality Factors: The Aquaphotomics of Low Moisture Agricultural Materials." *Journal of Near Infrared Spectroscopy* 17 (6): 315–28. doi:10.1255/jnirs.862.

Zhang, Y., X. Sun, S. G. Bajwa, S. Sivarajan, J. Nowatzki, and M. Khan. 2018. *Plant Disease Monitoring With Vibrational Spectroscopy. Comprehensive Analytical Chemistry.* 1st ed. Vol. 80. Elsevier B.V. doi:10.1016/bs.coac.2018.03.006.

11 Healthy Soils—Healthy People
Soil and Human Health— The Reality of the Balkan Region

Ratko Ristić
University of Belgrade

Marijana Kapović Solomun
University of Banja Luka

Ivan Malušević
University of Belgrade

Slavko Ždrale
University of East Sarajevo

Boris Radić, Siniša Polovina, and Vukašin Milćanović
University of Belgrade

CONTENTS

11.1 INTRODUCTION

Status of good soil health comprises synergy between key soil properties of high quality and optimal dynamics of relevant soil processes. Favorable soil texture and structure, good infiltration and water retention capacity, content of nutrients, and soil reaction are key soil properties. Relevant soil processes include good aeration, resistance to wind and water erosion, and strong nutrient cycling (Lal 2011). Soil organic matter is a very important aspect of healthy soil both for its agricultural use and environmental functions. Reaching of the optimum level of key soil properties and processes and synergistic interactions among them provide complexity of soil functions which acts as a living system (Kibblewhite et al. 2008), whose health is dependent on anthropogenic and natural impacts. In this way, the health of a soil "organism" can be considered similar to the health of a human organism (Magdoff 2001). Human health depends on soil health due to the amount of food production and the quality of food (the concentrations of micronutrients, protein, and essential amino-acids). Any deficiency, excess, or imbalance of some of these elements in soils can affect human health (Lal 2011; Marlow et al. 2009). Degraded surfaces produce soil dust which has an important role in air quality (Prospero et al. 1983; Smith and Lee 2003). The misuse of soil contributes to the formation of particles smaller than 10 μm in size (PM10), with the following consequences on human health: respiratory problems, lung tissue damage, and even lung cancer (UNEP 2019; Wall et al. 2015). Soils are frequently polluted with heavy metals, harbor antibiotic-resistant organisms, and contain pathogens, which have negative effects on human health. The ongoing climate change is partly influenced by human misusage of soil resources and the increase in the emission of carbonaceous (CO_2, CH_4, soot), nitrogenous (N_2O, NO_x), and other organic and inorganic compounds (Lal 2011). Climate change and global warming endanger food production and water distribution. They intensify the frequency of floods and forest fires with strong effects on both soil and human health.

Serbia (SRB) and Bosnia and Herzegovina (BIH) are characterized by a pedological diversity as a consequence of a long and complex processes in a variety of geological, climate, water, and vegetation conditions (Dragović et al. 2017). The land is a principal natural resource and a medium for the survival and development of living organisms on the planet. Given that land formation processes can take from several decades to thousands of years, protection and responsible management are activities that should be priority. Rational management over land resources and establishment of Sustainable Development Goals (SDGs) may contribute with numerous activities focusing on mitigation and adaptation to the effects of climate change, as well as on the conservation and enhancement of biodiversity (Ristić et al. 2020).

SRB and BIH are South East European (SEE) countries in the central part of the Balkan Peninsula. Both countries are still burdened with the consequences of the Civil War (1991–1995), which led to the disintegration of the former Yugoslavia and NATO bombing (1995, 1999). In addition to the political and economic ones, the ecological effects concerning the land resource are highly emphasized. BIH and SRB ratified the United Nations Convention to Combat Desertification (UNCCD) in 2002 and 2007, recognizing the imperative to stop land degradation and preserve soil as a medium for food production and a component of ecosystem stability. In 2014, BIH voluntarily committed to the Land Degradation Neutrality (LDN) target setting process, SRB in 2019. Between 2016 and 2018, LDN was implemented separately by each administrative unit across BIH (Kapović-Solomun 2018; Čustović and Ljuša 2018), contrary to SRB which conducted the LDN target setting process at the national level (Ristić et al. 2020). Considering the current situation, this paper will discuss soils and relation with human health in SRB and part of BIH, i.e., the Republic of Srpska (RS) entity, on the basis of the available data. One of the most significant impacts of the Civil War in BIH was the destruction of historical land and soil data, including land records. Even 28 years after the conflict, land data availability is still scarce and represents one of the major challenges to assess land degradation status and implement environmental policy frameworks (Kapović-Solomun 2018; Tošić et al. 2019). Generally, the most frequent direct drivers of land degradation across this region were soil erosion, floods, land abandonment, drought, pollution, and urbanization, which are underpinned by the lack of responsible land use planning by local communities where the majority of them do not have local strategic documents for land use planning, despite legal obligations in this regard (Kapović-Solomun 2018). Moreover, soil contamination by depleted uranium (DU) as a consequence of NATO aggression during 1995 (in the RS territory) and 1999 [in the southern Serbian province of Kosovo and Metohija (KM)] directly affects human health. This is a highly sensitive political question and the topic of disagreement between the two entities in BIH. Some local communities in the RS (e.g., Han Pijesak) highlighted this problem but again in a non-official way. Additionally, in the post-conflict period, land mines and associated contamination cover approximately 1.97% of the BIH territory, as a direct consequence of war (MAC, MCABH 2018).

The Republic of Serbia is a continental country located in South-Eastern Europe (Figure 11.1), with a total area of 88,361 km². The territory of the Republic of Serbia is divided into two autonomous provinces (AP), Vojvodina and KM, the territory of the City of Belgrade, and 193 municipalities. Serbia has a population of more than 9 million (2011 census), with an average density of 92.6 inhabitants per km². During the last decade (2010–2019), there is an evident decrease in the population rate, a decline in the birth rate, and the increased concentration of population in urban areas (Statistical Office of the Republic of Serbia 2019).

Bosnia and Herzegovina is also a SEE country that geographically belongs to the Adriatic and the Black Sea basins, with a total area of 51,197 km² of land and 12.2 km² of the sea (Figure 11.1). The RS entity covers 24,641 km² (Republika Srpska Institute of Statistics 2019). BIH has a population of 3.53 million (2013 census), with an average density of 68.9 inhabitants per km².

11.2 NATURAL CHARACTERISTICS OF THE REGION

11.2.1 GEOMORPHOLOGICAL CHARACTERISTICS

The territory of Serbia is situated in the Alpine realm in southern Europe (Figure 11.2), consisting of two different regions: the large Vojvodina plain to the north and the hilly and mountainous area to the south, with the Danube and Sava rivers representing the border between them. It comprises various geological units of composite structures and complex tectonic interrelations (Pavlović et al. 2017). Terrains with a slope of less than 5% and slightly sloped terrains (5%–10%) make about one-third of the total area, while terrains on slopes of 10%–30% occupy 24% of the total area. Steep

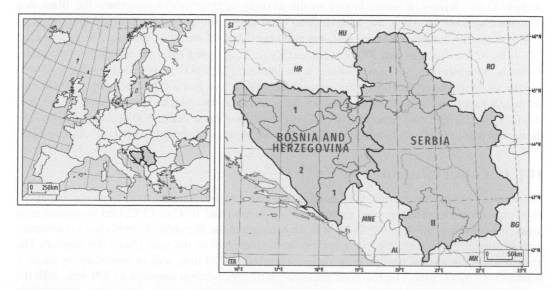

FIGURE 11.1 The spatial location of the Republic of Serbia (I – AP of Vojvodina; II – AP of KM) and Bosnia and Herzegovina (1 – Republic of Srpska; 2 – Federation Bosnia and Herzegovina: 3 – Brčko District).

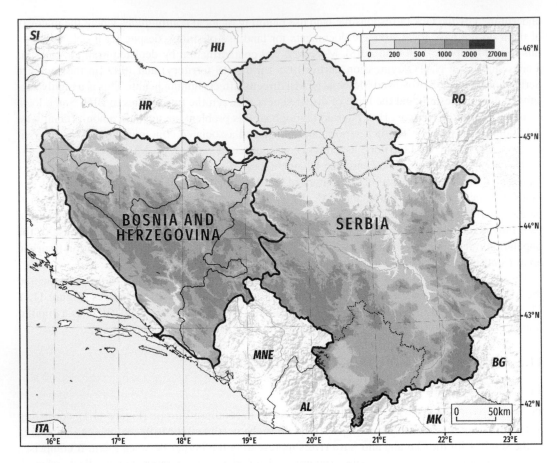

FIGURE 11.2   Geomorphology of Bosnia and Herzegovina and Serbia.

and very steep areas (>30%), on which shallow, erodible soils unsuitable for cultivation are formed, occupy 42.6%. Rivers in Serbia belong to the territory of three major recipients: the Black Sea (Danube Basin, 82,707 km²), the Adriatic Sea (5,004 km²), and the Aegean Sea (650 km²).

BIH is mainly mountainous with lowlands along big rivers. Of the total land area, 5% are lowlands, 24% hills, 42% mountains, and 29% karst areas (Figure 11.2). Going from the north toward the south, lowlands gradually turn into broad hillsides at an altitude of 200–600 m above sea level (m a.s.l.) and into the mountainous region from 600 to more than 2,000 m a.s.l. Areas at heights of up to 500 m a.s.l. are mostly located in the northern and southern parts of BIH and along the valleys of Una, Sava, Vrbas, Bosna, Drina, Spreča, and Neretva Rivers. The elevations of more than 1,800 m a.s.l. is common in the central part of BIH with mountains and plateaus which descend abruptly toward the Adriatic Sea.

11.2.2   Climate Characteristics

The territory of Serbia is dominated by a temperate continental climate, with more or less pronounced local characteristics. Mean annual air temperature varies from 10°C to 12°C, and in mountainous areas, it varies from 3.7°C to 6.7°C. Most of the territory of the Republic of Serbia has a continental precipitation regime, with larger amounts in the warmer half of the year (April–September). The amount of mean annual precipitation varies from 540 to 820 mm, and in mountainous areas, it reaches over 1,000 mm. The recorded maximal daily precipitation amounts to 220 mm, with the strongest rain intensity of 5.5 mm min⁻¹. In the warmer part of the year, north-west and west winds are predominant, while east and south-east winds dominate in the colder part of the year.

BIH has diverse climatic regimes characterized by a Mediterranean climate in the south-west, continental-mountain climate in the central and northern regions and moderate continental climate in the remaining area (Trbić et al. 2010). In the south-west region, due to the vicinity of the Adriatic Sea, the average January temperatures are from 3°C to 5°C, while summers are dry and hot (Popov 2017). The central region of RS has a continental-mountain climate with severe winters (absolute minimum temperatures range from −24°C to −34°C) and warm summers (absolute maximum temperatures range from 30°C to 36°C). Average yearly precipitation ranges from 1,000 to 1,200 mm (Trbić et al. 2010).

11.2.3 Forests

According to the Serbian National Forest Inventory (MAFWM-FD 2009), forests in Serbia cover an area of 2,252,400 ha (29.1% of the total area of the country). A very unfavorable state of the afforested area (equal to 6.37%) was found in the AP Vojvodina. The total (standing) volume reaches 362,487,418 m^3, and the annual volume increment amounts to 9,079,773 m^3. The average values of wood volume and volume increment are 160.9 and 4.0 m^3 ha^{-1}, respectively. Oak and beech forests cover 61.3% of the total forest area, conifer forests cover 10.8%, while poplar plantations and other forests cover 27.9%. In the total forest area, high natural forests occupy 27.6% of the area, while coppice forests account for 64.7% and forest plantations occupy 7.8% of the area.

Forests cover about 53% of the BIH territory (about 53%). High and coppice coniferous and deciduous forests prevail, while the most common type of deciduous trees is beech (*Fagus* spp.), which makes almost 40% compared to the distribution of all types, while oak (*Quercus* spp.) covers around 20%. Spruce and fir, which may be found at higher altitudes growing on steep grounds, account for an additional 20% of the forest cover in BIH. An 80% share of these forests are state-owned (UNEP 2017).

11.2.4 Water Resources

The territory of Serbia has an average flow of 508.8 m^3 s^{-1}, i.e., about 16×10^9 m^3 per year, which represents an average specific runoff of about 5.7 L s^{-1}·km^2 (MAEP, IWMJČ 2015). According to this and based on the annual distribution of surface water of 1,500 m^3 per inhabitant, Serbia is one of the poorest areas of Europe from the aspect of the availability of water resources. The lowland areas that are the most populated and where soil resources are the richest (northern and central Serbia) are the poorest in water yields (specific discharges 2–4 L s^{-1}·km^2). The publicly owned hydro system for irrigation has been built on an area of about 105,500 ha, which is less than 6% of the land that has favorable characteristics for irrigation. BIH has an average flow of 1,200 m^3 s^{-1}, with a specific runoff of about 23.4 L s^{-1}·km^2. Only 3,312 ha are covered by irrigation systems, which is 0.33% of the total arable land (Dragović et al. 2017).

11.2.5 Biodiversity Conservation and Management

Balkan Peninsula, with parts of SRB and BIH, represents one of the Earth's 25 centers of biodiversity. SRB and BIH cover just 1.37% of the European area but contain 39% of its vascular flora, 51% of its fish fauna, 49% of its reptile fauna, 74% of its bird fauna, and 67% of its mammal fauna. An area of 575,310 ha (6.5% of the total land area) has been classified as a protected area in Serbia. In addition to that, 53,804 ha (0.61%) were classified as protected areas in terms of UNESCO MAB and 63,919 ha (0.72%) are under protection in accordance with the Ramsar Convention. The size of protected areas in BIH amounts to 104,880 ha, which is 2% of the country. Protected areas comprise national parks, nature parks, the landscapes of exceptional features, nature reserves and special nature reserves, natural monuments, and protected habitats.

11.3 SOILS OF THE REGION

Integrated soil research in Serbia began in the mid-19th century. In 1919, the Faculty of Agriculture and Forestry was founded, which developed a soil research team that produced the Serbian Soil classification system (Table 11.1). The current Soil Map of Serbia consists of 20 soil maps, with a 1:50,000 scale (Pavlović et al. 2017). The general distribution of soil classes in Serbia is based on the character of its natural humidification. The water-physical properties of the soil are used to distinguish between three large soil groups: the automorphic soil (80%), the hydromorphic soil (19%), and the halomorphic soil (1%).

The Republic of Serbia has 4,690,812 ha of agricultural land, which makes 53.02% of the national territory (Figure 11.3). The largest percentage (25.93%) of that area is occupied by arable land and gardens, while orchards occupy 0.37%, vineyards 0.17%, and pastures 1.9% of that territory. The complex of agricultural areas occupies 12.58%, while agroforestry areas occupy 12.07% of the total agricultural land area (EEA 2018).

Automorph and hydromorph soils are dominant in the territory of the BIH (Table 11.2). The most widespread soils are Mollic Leptosol, Haplic Cambisol, and Leptic Cambisol according to WRB (2014), while for the hydromorph class, the most common are Haplic Planosols and Fluvisols (Kapović-Solomun 2018). The majority of soils have a low plant and nutritional soil potential (UNEP 2017).

According to Corine Land Cover 2018 (EEA 2018), most of the territory of BIH is covered by forests (53%), while agriculture occupies 26.8% of the country (Figure 11.3). Urban and traffic areas together occupy 1.7% of the land. Within the forest land, broadleaved forests cover 32.62% of the area, while coniferous and mixed forests cover 13.72% of the area. Transitional woodland-shrub is identified on 5.53% of the territory, and land principally occupied by agriculture with significant

TABLE 11.1
Soil Types in Serbia, Their Area and Use Restrictions

Soil Type Serbian Class/WRB	Area (ha)	Restrictions
Lithosol/Leptosol	107,000	Unproductive soil
Arenosol	86,000	Severe restrictions due to excessive filtration; Poor to medium productive soil
Rendzinas/Leptosol, calcaric	≈527,000	Severe to medium restrictions
Kalkomelanosol/Leptosol	≈155,000	Severe restrictions
Ranker/Leptosol, eutric, dystric	572,000	Severe restrictions
Chernozem	1,200,000	Fertile soil; No restrictions
Smonitza/Vertisol	780,000	Moderate restrictions
Kalkocambisol/No adequate WRB name	≈350,000	Severe to medium restrictions
Eutric cambisol/Cambisol, eutric	560,000	Moderate restrictions
Dystric cambisol/Cambisol, dystric	≈2,280,000	Severe to very severe restrictions
Luvisol	≈510,000	Moderate to medium restrictions
Pseudogley/Stagnosol	538,000	Moderate to severe restrictions—conditionally productive soil
Podzol	≈17,000	Severe to very severe restrictions
Fluvisol, Humofluvisol/Fluvisol, humic Humogley/Gleysol, humic; Eugley/Gleysol	≈760,000	No restrictions to serious restrictions—conditionally highly productive soil
Solontchak and Solonetz	233,000	Severe restrictions
Histosol	≈3,000	Moderate to severe restrictions
Deposol/Anthrosol	≈50,000	Unproductive soil; Moderate to severe restrictions

Source: Protić et al. (2005), in Pavlović et al. (2017).

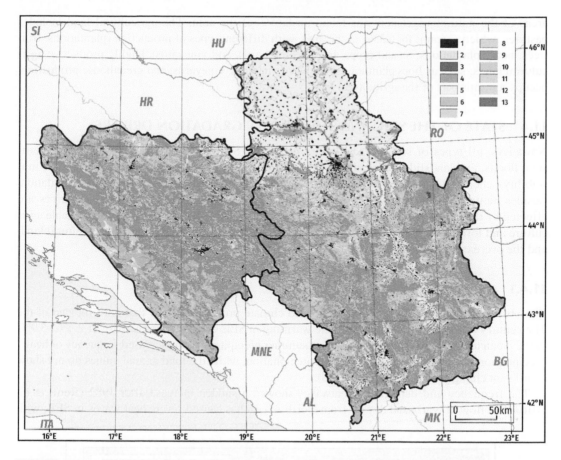

FIGURE 11.3 Land use in Serbia and Bosnia and Herzegovina (EEA 2018) (1 – Urban fabric; 2 – Industrial, commercial and transport units; 3 – Mine, dump and construction sites; 4 – Artificial, non-agricultural vegetated areas; 5 – Arable land; 6 – Permanent crops; 7 – Pastures; 8 – Heterogeneous agricultural areas; 9 – Forests; 10 – Scrub and/or herbaceous vegetation associations; 11 – Open spaces with little or no vegetation; 12 – Inland wetlands; 13 – Inland waters).

TABLE 11.2
Spatial Area of Dominant Soil Types Soils in BIH (Resulović 1999)

Soil Type/WRB	Area (ha)	Area (%)
Lithosols, regosols, stone soils/Lithosols	442,300	8.65
Rendzina/ Leptosol, calcaric and Ranker/Leptosol, eutric, distric	217,200	4.25
Smonitza/Vertisol	98,700	1.96
Kalkocambisol/no adequate WRB name	816,200	15.97
Terra Rossa and Kalkocambisol/ no adequate WRB name	797,700	15.50
Eutric Cambisol/Cambisol, eutric	250,000	4.89
Dystric Cambisol/Cambisol, dystric and Ranker/ Leptosol, eutric, distric	1,469,100	28.73
Luvisol/Luvisol	65,500	1.34
Pseudogley/Stagnosol	317,300	2.90
Fluvisol/Fluvisol	188,300	3.68
Eugley/Gleysol; Humogley/Gleysol, humic	81,600	1.60
Peat soils/Histosol	13,200	0.26

areas of the natural vegetation was found on 8.65% of the area. The agricultural land includes pastures, which account for 6.33% of the area, with different types of production (plantations, fruit trees, berry plantations, annual crops, intensive orchards, and vineyards). Highly productive agricultural lands, such as hops plantations, intensive orchards, vineyards, and greenhouse production, occupy a minor part of the agricultural land.

11.4   STATE OF THE SOIL AND THE MAIN DEGRADATION DRIVERS

Nowadays, all types of soil degradation are present in SEE: soil loss (due to intensive soil erosion, floods, and landslides), degradation of soil physical properties (compaction, loss of structural stability), chemical (salinization, acidification, and nutrient depletion), and biological degradation. Besides natural conditions and processes, primarily ongoing climate change, soil resources are affected by the constant and increasing pressures of human activities, including unsustainable agriculture and forestry practices, urbanization, pollution, the abandonment of fertile agricultural land, and the overexploitation of minerals.

11.4.1   Climate Change

The observed data processing and model-based climate projections show that until the end of the 21st century (Đurđević et al. 2018), climate warming in SEE will cause an increase of 2.5°C–5.0°C in the mean temperature, with a reduction of summer precipitation, an increased frequency of heavy precipitation, and significantly lower snow precipitation, while the total annual values do not show significant changes (Figures 11.4 and 11.5).

The frequency and duration of heatwaves showed a sudden increase after 1982 (Ruml et al. 2016). The flow of warm and dry air from North Africa has caused severe heat waves in SEE, with

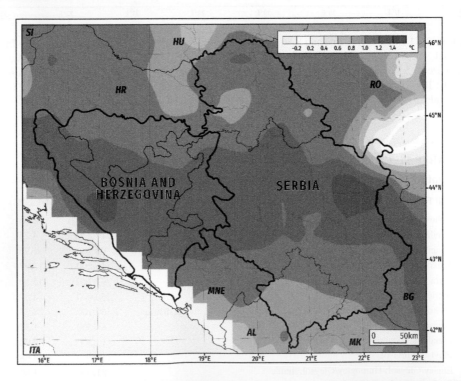

FIGURE 11.4   Changes in the mean air temperature (°C) in summer (VI–VIII) in the period 1985–2014 (reference period 1961–1990) (Đurđević et al. 2018).

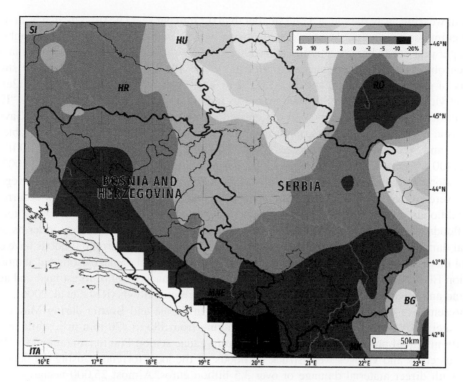

FIGURE 11.5 Changes in the mean precipitation (%) in summer (VI–VIII) in the period 1985–2014 (reference period 1961–1990) (Đurđević et al. 2018).

absolute maximum air temperatures of 44.9°C (2007) recorded in SRB and 43°C in BIH (Popov 2017). During the summer of 2012, the worst drought since the beginning of measurement was recorded at several stations in SEE (Unkašević and Tošić 2014). The number of days with an absolute maximum temperature >30°C (tropical days) is expected to increase, while the total number of days with an absolute minimum temperature <0°C (frost days) will decrease (Kržić et al. 2011). The vegetation period will start earlier and end later, as has already been registered to be happening in the Northern Hemisphere (Frich et al. 2002). The duration of the dry period (maximum number of consecutive days with daily precipitation <1 mm) is expected to be extended. For the period 2079–2100, the reduction of rainfall in the Balkan region might be considerably large, i.e., about 20% of the mean precipitation found during the period 1961–1990 (Gualdi et al. 2008).

The paradigm for altered climate conditions was the giant cyclone formed over the Balkans in May 2014, with the following characteristics: immobility, duration of about 7 days, and huge spatial coverage of more than 50,000 km^2 (UBFF 2014). The cyclone produced intense precipitation which was the dominant input for fast surface runoff generation and torrential flood forming, with recurrence intervals from once in 1,000 years to once in 5,000 years (UBFF 2014). During this event, historical records for 24 and 72 h at rain-gauge stations were overcome at many measuring points in SRB and BIH.

11.4.2 Soil Erosion and Torrential Floods

Erosion processes of different categories of destruction are present on 76,355 km^2 (86.4% territory of Serbia), due to dominant natural characteristics and strong anthropogenic influence. The average annual production of erosive material amounts to 37.25·10^6 m^3, in other words, 487.85 m^3·km^{-2}, which is 4.88 times more than natural (geological) erosion. The average rate of soil formation on the slopes in Serbia is 0.1 mm per year. Geological (natural) erosion is the action of wind, water, ice, and gravity in wearing away the soil at a rate smaller than 0.1 mm per year (100 $m^3 \cdot km^{-2}$). It is a relatively slow, continuous process unlike accelerated erosion, which produces a rate of soil loss higher than 0.1 mm per year due to human activities (Kostadinov 2008). The paradigm of negative human impact has been observed during the construction of the "Stara Planina" ski resort, when the strongest intensity of gully erosion in the world (133,023.2 $m^3 \cdot km^{-2}$) was recorded (Ristić et al. 2012a). Area sediment yields and intensity of erosion processes in Serbia were estimated based on the "Erosion Potential Method" (EPM). This method was created, developed, and calibrated in Serbia (Gavrilović 1972, Ristić 2017). The shares of certain categories of erosion intensity are presented in Table 11.3.

Strong and excessive erosion processes cover 35% of the territory of Serbia (IWMJČ 2001). More than 12,000 torrents of different categories have been recorded in Serbia (Kostadinov 2007). The climate, along with the specific characteristics of the relief, the distinctions of the soil and vegetation cover, severe erosion processes, and social-economic conditions result in the frequent occurrence of torrential floods. Among natural hazards with serious risks for people and their activities, torrential (flash) floods are the most common hazard in Serbia and the most significant one in terms of huge material damage and loss of human lives (Ristić et al. 2012b, 2012c). Torrential floods have caused death of more than 130 people in the period 1950–2014 and material damage estimated at more than 10 billion EUR (Ristić et al. 2015). Overexploitation or mismanagement of forest land and agricultural land and urbanization provoke severe erosion and torrential floods (Ristić et al. 2006).

Representative examples are the torrential floods in Serbia and Bosnia during May of 2014. Local watersheds received 3-day precipitation ranging from 180 to 420 mm m^{-2}, while the absolute daily maximal precipitation recorded at the rain-gauge station Planina amounted to 218 mm. Numerous cities and villages were struck by floods on the local torrents causing the death of 73 people with direct material damage of over 3.5 billion euros. Almost 25,000 hectares of arable land were flooded and covered with deposited sediment or damaged by more than 6,000 activated landslides. The spill threatened numerous municipal waste landfills and mine exploitation products. Only in one locality ("Stolice" dam in Western Serbia), more than 100,000 m^3 of antimony sludge was spilled toward the downstream river valley. The formed cover was 50–75 m wide, 5–10 cm (in some places up to 70 cm) thick, with high concentrations of arsenic, antimony, barium, zinc, and lead, which requires complex remediation works.

Soil erosion is also a very important land degradation driver all over BIH. During the Civil War, significant amounts of soil data were destroyed, including for example, the most complete overview of soil erosion provided by the soil erosion map, developed between 1979 and 1985 (Lazarević 1986). The soil erosion map for the entire territory of BIH has not been updated, but in 2012, the erosion map was reconstructed for the RS (Tošić et al. 2012, 2013). By comparing soil erosion maps from 1985 to 2012 for the entire RS, it was observed that the extent of the affected area remained the same (21,851,04 km^2, covering 86.96% of the RS), but the intensity of soil erosion decreased over this period (Table 11.4).

TABLE 11.3

Erosion Processes in Serbia

Category	Erosion Processes Intensity	$m^3 \cdot km^{-2} \cdot year^{-1}$	km^2	%
I	Excessive erosion	>3,000	2,888	3.27
II	Intensive erosion	1,200–3,000	9,138	10.34
III	Medium erosion	800–1,200	19,386	21.94
IV	Weak erosion	400–800	43,914	49.78
V	Very weak erosion	100–400	13,035	14.75
	Total		88,361	100

Source: Basic Water Resources Management Plan of Serbia, 2001.

TABLE 11.4

State of Erosion in the RS (Tošić et al. 2012, 2013)

Erosion Category[a]	State in 1985		State in 2012		Difference	
	km²	%	km²	%	km²	%
I Excessive	266.7	1.2	191.9	0.9	74.8	0.3
II Strong	392.7	1.8	7.3	0.0	385.4	1.8
III Medium	2,414.8	11.1	1,141.6	5.2	1,273.1	5.8
IV Low	2,000.8	9.2	3,733.6	17.1	−1,732.8	−7.9
V Very low	16,776.2	76.8	16,776.7	76.8	−0.5	0.0
Surface under erosion	21,851.0	100	21,851.0	100	0.0	0.00

[a] Erosion category (in $m^3 \cdot km^{-2} \cdot year^{-1}$) I excessive (>3,000), II Strong (1,200–3,000), III Medium (800–1,200), Low (400–800) and Very low (100–400).

11.4.3 Agriculture and Land Degradation

11.4.3.1 Loss of SOC

The main causes of the organic matter and plant nutrients impoverishment of soils are: the method of agricultural production, the irregular intake of organic fertilizers, the intake of exclusively mineral (nitrogen) fertilizers, the burning of crop residues, the growing of crops that bind large amounts of organic matter from the soil and water and aeolian erosion.

The territory of Serbia stored $705.84 \cdot 10^{12}$ g (Tg) of organic carbon, in the soil layers at 0–30 cm depths. In 50% of the territory of Serbia, the share of soil organic carbon stored in the 0–30 cm soil layer is up to 2%. Organic carbon stock in the topsoil (0–30 cm) is by 40.71% higher in forests and semi-natural areas compared to agricultural land. The analysis of samples obtained from agricultural land, during a systematic control of agricultural soil fertility, indicates that the majority of samples (54.21%) had a low organic carbon content (1.1%–2%). Medium organic carbon content (2.1%–6%) was found in 32.96% of the samples, while a very low content (<1%) was observed in 12.83% of the samples (SEPA 2018a, Dragović et al. 2017). In the plains of the northern part of Serbia, which are very important for food production, the concentrations of organic carbon were determined, at a 100 cm depth, of the most common soil types, expressed in tons per hectare (t·ha⁻¹): Regosols-204; Vertisols-171; Gleysols-165; Chernozems-165; Solonchaks-87; Fluvisols-93; Arenosols-99 (Belić et al. 2013).

BIH and its entities still do not have soil monitoring, i.e., reliable data related to SOC, which is a consequence of the bad socio-economic situation and soils being treated as an issue of secondary importance. Unfortunately, there are no state or entity data that can be used to show the SOC content in BIH. On the other hand, through the LDN process, global data (ISRIC 2016) are used for the SOC estimation despite many uncertainties particularly in shallow soils on limestone that covers the south of the country (Kapović-Solomun 2018; Čustović and Ljuša 2018).

Global data on SOC revealed a low content of SOC in the north of the country with intensive agricultural production, while other regions particularly under forests had sufficient amounts of SOC.

11.4.3.2 Acidification and Salinization

Soil acidification is a consequence of natural pedogenic processes, especially in humid conditions, but also of the impact of anthropogenic factors through the intensive use and application of inadequate agrotechnical measures, as well as pollution due to dry and wet deposition of air-pollutants. Acidic soils cover most of the agricultural land of central Serbia, where 43% of the total surveyed area has increased potential acidity and belongs to the group of strongly acidic to acidic soils (1,197,000 ha), 20% belongs to the group of acidic to slightly acidic soils, while only 35% belongs to the group of slightly acidic to neutral soils. Acidic soils cover around 1/3 of the total soil resources in BIH (UNEP 2017). The reduced intake of organic matter and the use of exclusively mineral (nitrogen) fertilizers have also contributed to the more intensive process of soil acidification over the past decades (Čakmak et al. 2014, Mrvić et al. 2010).

Soil salinization represents an excessive accumulation of salt in the soil profile. It can be caused by rising levels of saline groundwater or the result of flood waves with a high concentration of dissolved salts, and due to the use of mineral fertilizers with a higher content of sodium. The largest area of saline lands is located in the territory of AP Vojvodina (233,000 ha), while only 2% of the land in central Serbia is in the group of alkaline soils (MEMSP 2011). Salinization is recognized as a land degradation driver typical of the agricultural regions of BIH, and the general conclusion is that high-quality soils (classes I, II, and III of soil capability) account for only 15%–16% of all soils (UNEP 2017).

11.4.4 FORESTRY

Fires are a significant factor in deforestation and soil degradation in Serbia. Limestone soils are also particularly sensitive to the effects of fire, due to shallowness, high levels of organic matter and long paedogenesis (Kapović et al. 2011, 2013; Kapović and Knežević 2010). During the period from 2010 to 2018, more than 400 forest fires were recorded, damaging about 15,000 ha of forest, with 150,000 m^3 of wood mass. There was an increased frequency of forest fires (2002, 2007, 2012, and 2017), that corresponds to the appearance of heat waves and drought periods. In the period 2012–2016, in forests managed by the Public Enterprise "Serbian Forests", 316 fires were recorded, on a total area of 8,074.55 ha (Milanović 2017). In 15 cases, the cause was thunder; in 158 cases, the fires were man-induced, and in 143 cases, the causes were unknown (likely man-induced). In BIH, forest fires tend to occur twice over the year, primary in March and then in August, during the larger water deficit periods. Drought periods in summer and human impacts have increased the number of wildfires in the last decades in BIH (Kapović 2011; Kapović-Solomun 2020). In the 2008–2017 period, BIH was highly affected by wildfires, with a total burned area of 203,948 ha, which represents 3.98% of the country (Republika Srpska Institute of Statistics 2019).

Two types of illegal logging can be clearly distinguished in the Western Balkans: poverty-driven illegal logging and commercial illegal logging. The estimated volumes of illegally cut wood in Serbian public forests are between 10,000 and 32,000 m^3 per year, while the total amount of illegally cut wood in private forests is estimated at 500,000 m^3 (Ristić et al. 2019). Illegal logging in Serbia is strongly linked to numerous factors including unfavorable social and economic conditions, low awareness of the importance of forest protection, institutional inefficiency, bad law enforcement, and inefficient judicial and sanctioning systems.

Damage caused by insects (*Ips typographus, Pityogenes chalcographus, Lymantria dispar*) and diseases (*Heterobasidion parviporum*) vary annually, but in the period from 2102 to 2018, it affected about 340,000 ha and more than 140,000 m^3 of wood. Additionally, a strong natural disaster in 2014 (ice breakdown) hit forests in the eastern part of Serbia causing damage on 1.6 million m^3 of wood on an area of 43,000 ha (Ristić et al. 2019). The fragmentation of forests is initiated when forest roads, electric power, water-supply installations, ski trails, and ski lift corridors penetrate the old-growth or mature forest, dividing large surfaces into small elements, changing their habitat conditions (Ristić et al. 2011). Fragmentation is followed by habitat loss that seriously endangers forest wildlife (Kapos 1989; Institute for Nature Protection of Serbia 1999; Ristić et al. 2008). Altered radiation, wind, water, and nutrient regimes create new habitat conditions, inducing tree mortality in fragments and strongly influence forest dynamics and structure.

11.4.5 URBANIZATION

Loss of fertile soil due to urbanization in Serbia is an intensive process (Table 11.5). The average annual occupation of land, in the period from 2000 to 2018 was: for residential, business, sports,

TABLE 11.5
Structure of Surfaces Occupied by Urbanization (2000–2018)

Categories	Occupied by Urbanization (ha)			
	2000–2006	2006–2012	2012–2018	Total
Pastures and mixed farmland	2,280	1,148	2,930	6,358
Arable land and permanent crops	939	1,777	0	2,716
Water bodies	0	14	91	105
Open spaces with little or no vegetation	0	0	0	0
Natural grassland	3	8	0	11
Forest and transitional woodland shrub	1,066	1,264	1,768	4,098
Wetlands	36	30	0	66

Source: The Report on Soil Quality in the Republic of Serbia, SEPA, Belgrade, 2018.

and recreational facilities −243.69 ha; for industrial and commercial facilities −184.55 ha; for road network and supporting infrastructure −88.64 hectares; and for mines, landfills, and construction sites −611.53 ha (SEPA 2018b).

Permanent loss of the most fertile soils through urbanization is common practice in BIH, where in the process of developing spatial and regulation plans, irresponsible decision-making does not consider the fertility rate of soils in this process. Many municipalities do not have spatial development plans resulting from bad socio-economic conditions. The problem of illegal construction is also very pronounced due to inadequate and unsynchronized planning and implementation mechanisms, certain socio-economic factors and war-time migrations of people. Such a situation is quite detrimental for the land and the entire environment in BIH (Kapović-Solomun et al. 2018).

11.4.6 MASS MOVEMENTS ON SLOPES

According to a recent research, 30% of the territory of Serbia is threatened by mass movements (landslides, rockfalls), which affect more than 40,000 locations (more than 1,100 landslides have been recorded only in the territory of the capital city of Belgrade). The occurrence of landslides was especially intensified after a series of floods in 2012 and 2014 (http://geoliss.mre.gov.rs/). A total of 1,883 active landslides have been registered, which cover more than 3,700 ha. Most of them are from 5 to 10 m deep, while in the Neogene rock formations they have greater distribution and depth (often over 10 m). There are around 1,800 active landslides in BIH (UNEP 2017). Depending on the climate conditions, BIH has more than 1,000 landslides annually. Some of them are old landslides that have reactivated, and some are new.

11.4.7 WASTE MANAGEMENT

Inadequate disposal of municipal and other types of waste are a serious cause of land degradation in SEE. In BIH, there are no systemically organized separate collecting, sorting, and recycling of waste. A total of 91 waste disposal sites were registered, and 1,100 illegal dumpsites are still in use (Dragović et al. 2017). In the territory of SRB, 164 landfills and 4,481 illegal landfills were registered in 2018. About 20% of the generated municipal waste in Serbia ends up on illegal landfills, often near the road infrastructure, on arable land, and in riparian areas. The costs of the waste sector to reduce GHG emissions and soil reclamation require an investment of € 297,22·10⁶ (UNDP 2017). The age of these landfills varies from 4 to 63 years, about 70% of them are not foreseen in the spatial planning documents, and no environmental impact assessment studies have been produced for them.

During 2018, the total amount of waste generated in SRB was about 11.6 million tons or about 1.7 tons per capita per year. Of the total amount of waste produced, about 70% was waste from thermal processes.

Leachate from landfills is not collected or treated, which endangers groundwater, surface water, and soil, due to the high content of organic matter and heavy metals. Spontaneous combustions of landfill gases are common. Most of the landfills in the local communities do not meet even the minimum technical requirements.

11.4.8 ENERGETICS, MINING, AND INDUSTRY

The energy sector is the biggest environmental polluter in SEE. During 2016, 16 thermal power plants in the Western Balkans with a total of 8 GW of installed capacity had higher emissions of sulfur dioxide, PM-2.5, and PM-10 particles and nitrogen oxides than 250 European coal power plants (with a total of 156 GW installed power). The main reasons for this unfavorable situation are outdated technologies and the use of lignite as the fuel (UNEP 2019). Surface coal mines of the Electric Power Industry of Serbia in the zone of Kolubara and Kostolac currently cover about 12,000 ha, with a tendency to excavate a new 200 ha every year (Official Gazette of RS, No. 33/2012). In SRB, only 2.7% of ash is used, primarily in the cement industry, while most of it is deposited in the open, so that to date, 250–300 million tons of ash have been deposited on landfills, occupying the area of more than 1,500 ha of arable land (Životić et al. 2012, Pavlović and Mitrović 2013, Popović et al. 2013). The amount of 7.45 million tons of coal dust was produced during 2018. In the four "Kolubara" coal mines, with an area of about 120 km^2, around 1,200 ha of re-cultivated areas are currently in use (about 850 ha under forest and about 300 ha under corn, wheat, and clover). In BIH, coal is exploited on an area of 18,000 ha, whereas the area for disposal of waste materials covers almost 6,000 ha (Dragović et al. 2017).

About 40,000 ha of land at surface mines and tailings dumps have been degraded, and less than 20% have been re-cultivated through the application of landscaping and afforestation measures (Official Gazette of RS, No. 29/2010). Heavy metal pollution, most often lead and zinc, is present near larger industrial plants: cement factories (nickel-contaminated land), battery factory in Sombor (lead-contaminated soil), copper rolling mill in Sevojno (copper and zinc-contaminated soil), mining and smelting in the Bor basin (soil contaminated with copper and arsenic). The annual national costs for managing contaminated sites in the EU amounted to €10 per capita, and in the SEE, it is €40 per capita (SEPA 2015). The exploitation of clay deposits for industrial needs leads to a loss of about 100 ha per year (Official Gazette of RS, No. 33/2012), and due to the lack of reclamation, more than 1,000 ha of agricultural land have been destroyed so far. Activities on the exploitation of gravel and sand along rivers (125 legal and an unknown number of illegal locations) lead to the loss of about 60 ha of agricultural land per year.

In accordance with the National Renewable Energy Action Plan of the Republic of Serbia, (https://energetskiportal.rs) around 120 Small Hydro-power Plants of Derivative Type (SHPPDT) have been built in Serbia, with pipelines in the length of 2–5 km. Additionally, the construction of 856 SHPPDT is planned (http://www.elektrosrbija.rs) in the mountainous regions of Serbia, mainly in protected natural areas (National Parks, Nature Parks, Special Nature Reserves). A total of 100 SHPPDT have been built in BIH and additional 300 are planned for construction. The implementation of this has led to the endangerment or even the disappearance of endemic and protected fish species (*Salmo trutta*, *Austropotamobius torrentium*), the fragmentation of the most valuable aquatic habitats, the fragmentation of forests (due to the construction of access roads and derivative pipelines), the endangered water supply of local communities and intensive erosion along access roads (Ristić et al. 2018). All the planned SHPPs would provide only 2%–3.5% of the annual energy needs of Serbia, but that would mean total devastation of most of the quality watercourses in the mountainous regions of Serbia, piping installations along 2,200 km of stream beds, the removal of 21,400 ha forests, and the destruction of 128,4·10^6 m^3 of high-quality forest soil. Due to the small energy contribution and detrimental environmental consequences, the authorities in the US have removed more than 1,000 (SHPP-DT) in the period 1993–2017 (Eichelmann and Scharl 2017). Similar processes are taking place in France, Spain, Germany, and Sweden. Other ways of producing energy from renewable sources have far less negative effects on the environment, and if the current losses of the Public Enterprise "Electric Power Industry of Serbia" during the transmission of electricity, were reduced by only 2% that would eliminate the need for derivative SHPPs (Ristić et al. 2018). Nevertheless, the construction of as many as 2,800 objects (Neslen 2017) has been planned in the Balkans, although this region is already one of the most endangered by current climate anomalies, which, among other things, leads to significantly reduced flow rates in low flow periods.

The current state of the soil in SRB and BIH has numerous impacts on human health and stability of communities at both the local and regional levels. The ongoing climate change, the constant and increasing anthropogenic pressures, including unsustainable agriculture and forestry practices, urbanization, pollution, the abandonment of fertile agricultural land, and the overexploitation of minerals, disturb soil health. Climate warming in SEE, with prolonged drought periods and extreme heat waves, produced negative effects on crop production in terms of quantity and quality, with an estimated material damage higher than 5 billion euros in the period from 2010 to 2015. (www.rtv. rs). At the same time, the thermal-related mortality cases as well as respiratory problems are significantly increased. The increased frequency of heavy precipitation (higher than 50 mm per day), associated with land degradation, provoked severe erosion and torrential floods, causing human victims, and deposition of sterile or polluted sediment on highly productive soil. The organic matter and plant nutrients impoverishment of soils reduce the quantity and quality of produced food. Leachate from landfills endangers groundwater, surface water and soil, due to the high content of organic matter and heavy metals. Deposited coal dust, after burning of lignite in thermal power plants, is a significant source of air polluters (UNEP 2019). When soil is suspended in the atmosphere as dust, it can be transported and absorbed by the human body (Smith and Lee 2003). Various chemicals and soil additives such as pesticides and fertilizers are used in agriculture, and their toxic residues in the form of soil dust can endanger children and the elderly (Smith and Lee 2003). Exposure to soil dust can result in health problems such as eye irritation, respiratory disorders, pulmonary disease, and an increased risk of lung and skin cancer (Rylander 1986; Clausnitzer and Singer 2000). In addition, inhaling soil dust may result in asthma, an asthma-like syndrome, or chronic obstructive pulmonary disease, and in chronic respiratory symptoms (Schenker 1998). The consumption of food produced in the contaminated soil can provoke digestive diseases. Heavy metals are present in soil as constituents of the rocks and due to human activities (Morgan 2013). They can initiate nervous system disorders, liver and kidney failure, intestine tract distress, anemia, skin cancer (Arsenic); cardiomyopathy, liver and kidney damage, gastroenteritis, pneumonitis, osteomalacia, cancer (Cadmium); lesions in skin, intestinal mucosa, pulmonary edema, lung cancer (Chromium); nervous system disorders, brain damage, hematologic effects, kidney disease, intestine tract distress, hypertension (Lead); nervous and gastric system disorders, kidney and pulmonary damage, a potent teratogen (Mercury); gastric, liver, and kidney defects, neurological effects, emphysema, lung cancer (Nickel); anemia, tissue lesions (Zinc) (Kabata-Pendias and Mukherjee 2007). In addition to that, the deficiency or toxicity of trace elements can affect reproduction (Oliver 1997).

11.5 SOIL POLLUTION AND HUMAN HEALTH

The NATO bombing of Yugoslavia (from 24 March to 10 June 1999) led to releases of many hazardous chemical substances: oil or petroleum products (hydrocarbon components), vinyl chloride monomer (VCM), ethylene dichloride (EDC), metallic mercury (Hg), lead (Pb), polycyclic aromatic hydrocarbons (PAHs), polychlorinated biphenyls (PCBs), dioxins and furans, pyralene, nitrogen oxides (NOXs), sulfur dioxide (SO_2), and so forth (Labus 2000). DU munitions were used in 112 locations, dominantly in the southern Serbian province of KM (Betti 2003). DU contained in projectiles (radioactive isotope[238]U, the half-life of 4.5 billion years) is spread out in the air over several hundred miles after striking the target and subsequently falls out, producing environmental and food chain contamination as a both radioactive and toxic substance (Joksimovich 2000, Giannardia and Dominici 2003). About 31,000 API (armor-piercing incendiary) rounds were allegedly fired in the southern Serbian province of KM (Durante et al. 2003).

The environmental consequences of DU residue will be felt for thousands of years. It is inhaled and passed through the skin and eyes, transferred through the placenta into the fetus, distributed into tissues, and eliminated in urine. The illness and death of NATO veterans in European countries from leukemia are numerous, after serving in the Kosovo War. *In vitro* incubation of human osteoblast cells with DU transformed the cells to tumorigenic phenotype 9.6 times compared with untreated cells (Abu-Qare et al. 2005). Uranium 238 in the soil, plants, and sporadically even in the urine of certain individuals from the southeast part of Serbia has been evidenced (Milačić et al. 2004). More munitions were used in the attack on Yugoslavia than in the whole of the Gulf War, when almost 1 million 30 mm uranium tipped bullets and over 14,000 large-caliber shells were fired, resulting in between 300 and 800 tonnes of DU dust scattered over soil and watercourses. NATO has refused to release any data about the type, quantity, and location of use of DU weapons, claiming that no scientific study has ever yet proved a link with cancer (Clarke 2002).

"Zastava" factory in Kragujevac (central part of Serbia) was bombed several times, and the bombing caused heavy material destruction and endangered the human environment. In the fires caused by the bombing, 2,200 kg of a sealant and 800 kg of pyralene burned. About 3,000 kg of pyralene oil poured out of two damaged transformers and partly spread over the industrial zone and partly flowed into rivers and lakes. The average micronuclei (MN) frequency in peripheral blood lymphocytes of newborns from Kragujevac increased by a factor of 1.4 after the bombing. Changes in the genetic material manifested in the increased MN frequency, compared to those in the control sample of newborns, are a direct consequence of environmental pollution (Milojević-Djordjević et al. 2004). The consequences of industrial complex bombing in the city of Pančevo (25 km from the capital of Serbia, Belgrade) have been serious: over 2,000 tonnes of 1.2 dichloroethane (DCE) and 8 tonnes of metallic mercury were released producing contamination of a wastewater canal which is leaking into the Danube. This was in addition to the release of 460 tonnes of VCM 80,000 tonnes of oil and oil products, and 250 tonnes of liquid ammonia. Atmospheric dioxins recorded in Skopje (Macedonia), downwind of Pančevo immediately following the NATO bombing, are said to have reached seven times the internationally agreed 'safe' levels. Dioxins and other airborne toxins resulting from the NATO bombing have also been recorded in Poland, Hungary, and Greece. Again, further systematic monitoring is required to see whether these emissions, severe but localized in time and space, will lead to significant long-term effects (Clarke 2002). Laboratory analyses of samples taken from the Danube sediment and biota revealed significant chronic pollution, both upstream and downstream of the sites directly affected by the conflict.

The mortality rate in Belgrade in winter 1999 (6 months after the bombing) increased by 40% compared to the year before the NATO bombing (Labus 2000). Maximum allowable concentrations of SO_2, NOXs, and ammonia were exceeded 5–10 times between 18 April and 26 April 1999. In many areas in the south-western region, crops and forests were damaged and leaves fell off from trees. Vineyards and crops in the southern region were also damaged. Bulgarian farmers near the towns of Kula and Belogradčik reported that flowers fell off from fruit trees, and vegetables began to rot on their land. In Macedonia, radiation levels have risen eight times compared to the levels from previous years. Tests in southern Serbia show soil samples containing concentrations of uranium over a thousand times the natural level used as a basis for cleanup considerations (Joksimovich 2000).

In BIH, NATO bombed only the positions inside the RS (Hadžići, Han Pijesak, Sokolac) which left destructive consequences on human and soil health. The number of people with cancer in the bombed regions increased four times in the period 1996–2002, and mortality 2.2 times compared to the period before the bombing. In addition, people who migrated from Hadžići to Bratunac after the bombing, had a 10 times higher mortality rate than the inhabitants of Bratunac (SIF 2019).

Military veterans who participated in the war in BIH, where DU missiles were used, have 14 times more chromosomal abnormalities than normal in their genomes. Until 2009 more than 2,600 Italian soldiers who were in the Serbian province of KM were ill from lymphoma (a type of cancer) and leukemia. In Serbia, the number of cancer patients grows by 2% each year and the number of deaths by 2.5%, which is a direct consequence of NATO bombing with DU: in 1999, there were 19,625 cases of cancer and 12,312 people died of cancer, and in 2012, there were as many as 36,408 cases of cancer and 21,269 people died, which is almost twice than before the NATO aggression. In 2013 and 2014, an increase in the number of malignant tumors in the province of KM increased by 57% (Krivokapić 2019). The cancer mortality rate in Serbia is the highest in Europe, which is partly the consequence of the NATO bombing, with 5,500 new cancer patients per 1 million inhabitants. After 2006, the number of cancer cases has increased by 59%, whereas the number of leukemia and lymphoma caused deaths increased by 118% (Dimitrijević et al. 2019).

The level of landmine and other residual explosive materials contamination represents a special problem for BIH. Unexploded mines placed in the range of 2–5 km on both sides of the demarcation line deserve special attention in consideration of this issue. However, data on the number of mines and minefields in BIH are neither reliable nor complete. In the database of the Mine Action Centre (MAC) BIH there are 19,000 registered reports on minefields. It is estimated that they represent only around 50%–60% of their real number. According to MAC's data, the current size of mine suspected area is 1,262,82 km^2 or 2.5% of the total area of the country. Given the regular daily demining activities, the landmine contaminated area is decreasing. However, considering the large-scale flooding in May 2014, it is estimated that landmine fields were shifted, but there are no official data on that, yet (UNEP 2017).

The consequences of military activities in the region still have strong influence on soil and human health due to the presence of DU, other toxic and carcinogenic matters. DU produced environmental and food chain contamination as a both radioactive and toxic substance. DU residue in the soil was inhaled and passed through the skin and eyes, transferred through the placenta into the fetus and distributed into tissues, causing leukemia, deformation of fetus and newborns, changes in the genetic material, while in adults, there was a multiple increase of chromosomal abnormalities. Soil pollution with radioactive matters, during military actions, influenced catastrophic consequences to human health with increased mortality rates, numerous cancer cases, leukemia, and lymphoma. It is estimated that about 79,000 landmines in BIH endanger 96,700 ha of land in 8,525 micro-locations populated by 545,603 inhabitants (www.bhmac.org). As a result of the NATO bombing in 1999, there are still unexploded cluster munitions in the territory of the Republic of Serbia on an area of about 2,500,000 m^2, as well as dozens of rockets and air bombs at 150 locations (www.czrs.gov.rs).

11.6 LAND USE CHANGES AND LAND DEGRADATION NEUTRALITY

At the 13th session of the UNCCD Conference of the Parties, the Strategic Framework of the Convention was adopted for the period from 2018 to 2030, and one of the decisions is to achieve the status of LDN by 2030 and improve the reporting system about the progress made through sub-indicator 15.3.1. "The percentage of degraded land and soil of the total area of land resources." Land-use changes in the Republic of Serbia were analyzed for the period from 2000 to 2015, and in Bosnia for the period from 2000 to 2012. Table 11.6 contains the basic categories of land cover and observed changes, starting from the "zero" period (the year 2000) until the end period (2015). The areas under forests increased by 1,951.11 km^2 (Table 11.6) in SRB i.e., by 6.36%, and 756.76 km^2 or 1,48% of the territory in BIH (Table 11.7). The negative trend was observed with the areas under grasslands, which decreased by 1,764,88 km^2, i.e., by 30.09% (Table 11.6), but in BIH sparsely vegetated land, grasslands and shrubs area increased by 146.80 km^2, while cropland decreased by 1,057,93 km^2 due to intensive land abandonment during and after the war conflict. For the observed period, the biggest increase in SRB is observed in the artificial areas,

TABLE 11.6

Changes in the Spatial Distribution of Land Cover in SRB for the Period from 2000 to 2015 (Ristić et al. 2020)

Category	Area (km²)		Change	
	2000	2015	(km²)	(%)
Forest	30,664.61	32,615.72	1,951.11	6.36
Grasslands	5,864.74	4,099.85	−1,764.88	−30.09
Croplands	50,131.14	49,282.76	−848.38	−1.69
Wetlands	117.94	119.18	1.24	1.05
Artificial areas	892.62	1,539.48	646.86	72.47
Bare land and other	60.68	72.82	12.14	20.00
Water bodies	756.70	758.62	1.92	0.25
Total	88,488.44	88,488.44		

TABLE 11.7

Changes in the Spatial Distribution of Land Cover for the Period from 2000 to 2012 in BIH (EEA 2012)

Category	Area (km²)		Change	
	2000	2010	(km²)	(%)
Forest	23,708.53	24,465.29	756.76	1.48
Sparsely vegetated land scrubs and grassland	7,516.6	7,663.40	146.80	0.29
Croplands	18,847.67	17,789.74	−1,057.93	2.07
Wetlands	247.26	243.35	−3.91	0.01
Artificial areas	688.58	801.58	112.99	0.22
Bare land and other	50.53	65.59	15.06	0.03
Total	51,059.18	51,028.94	−30.24	4.10

which increased by 1,539,48 km², i.e., by 72.47%. At the same time 112.99 km² of land in BIH was urbanized. The areas of wetlands and water bodies have slightly increased (by 1.30%), while the areas under bare rocky grounds and other areas increased by 20%, i.e., 12.14 km² in Serbia (Ristić et al. 2020) and 15.6 km² in BIH (EEA 2012).

The most important measures in Serbia to achieve LDN are: to increase the area of national territory covered by forests to 41.4% by 2050, to increase the area under forests in the AP of Vojvodina to 14.3% (of the total area of the territory of the AP), primarily by applying the system of forest protection belts, to increase the level of forest cover in areas under bare and degraded soil in mountainous areas south of the Sava and Danube Rivers, to control erosion and torrential processes on an area of 100,000 ha by 2030, to maintain the determined positive trend of LDN, applying appropriate measures and activities, through spatial and planning documentation.

Unlike in other countries, LDN targets in BIH are planned separately by each entity. Therefore, for the entity FBIH the most important measures are (Čustović and Ljuša 2018): the remediation of degraded land on 3,500 ha, the revitalization of land in vulnerable and abandoned areas on 100,000 ha, the demining of land on 83,200 ha, the afforestation of bare and damaged land on 10,000 ha, the revitalization and preservation of pastures on 10,000 ha. LDN targets for the entity RS (Kapović-Solomun 2018) are: to keep 53% of the territory under forests, afforestation, and other measures that will preserve and improve forests and forest land on 40,000 ha, measures for the improvement of sparsely vegetated areas, coppice forests, shrubs and grasslands on 110,000 ha, the revitalization of cropland on 40,000 ha, the rehabilitation of abandoned land and pastures on 115,000 ha, the improvement of the quality of coppice forests and shrubs on 80,000 ha, and the demining of forests and agricultural land on 40,000 ha.

Afforestation measures have a major impact on a range of problems, for which appropriate solutions need to be found: the mitigation of the effects of climate change in rural and urban areas (increase in O_2 emissions, decrease in CO_2 emissions), soil erosion control, the protection of watershed areas of water reservoirs that are part of the water supply systems (located in mountainous areas, with a significant share of eroded areas and barren lands, which leads to the filling of reservoir spaces), the prevention of torrential floods, which are associated with erosion processes and represent the most common natural catastrophe in the territories of SRB and BIH, the realization of the concept of development of mountainous regions through the agroforestry system, and the conservation and restoration of biodiversity. It is of particular importance to increase the forest cover of the AP of Vojvodina, which is the lowland granary of Serbia, primarily through the forming of forest belt systems, in order to protect the soils from aeolian erosion in the lowland parts, protect the railway, road, and water management infrastructure (water system "Danube-Tisa-Danube"), creating a corridor for biodiversity restoration and the ecological network structure. Therefore, afforestation measures are important not only for forestry, but for a whole range of vital activities aimed at restoring ecosystem services, preventing natural hazards, and protecting the economic potentials of SRB and BIH.

The general LDN targets in SEE are: to improve the quality of land and soil ecosystem services; to improve productivity to ensure uncompromised food production; to increase resilience of landscape and its soil features, but also of the population that is dependent on it; to attain synergy-wise relations of LDN with the other environmental protection goals; to achieve land-resource responsible management (Ristić et al. 2020). The implementation of LDN helps landscape restoration, biodiversity recovery, carbon sequestration, the afforestation of bare land, and the identification of protected areas and ecological corridors. In addition, the LDN concept supports agroforestry measures, flood prevention, water supply, erosion control, and the reclamation of sterile or polluted soil, as well as the attenuation of the "heat island" effect in urban areas. Each of aforementioned LDN goals has a direct positive effect on human well-being and health.

SDGs and the Agenda 2030 consider issues related to climate change and global ecosystem stability, accelerated urbanization, rising economic and social inequalities, and human health (UNDP 2019). SDG 3 (good health and well-being) is related to the progress against deadly diseases, the increase of life expectancy, the decline of maternal mortality rates and suppression HIV, malaria, and for sure in close future COVID-19. More than 400 million people do not have basic healthcare, and 1.6 billion live without a favorable diet and sanitary conditions (www.undp.org). SDG 15 (life on land) is related to the protection and sustainable use of terrestrial ecosystems, sustainable forest management, ways to combat desertification and droughts, and the restoration of degraded land and biodiversity (www.globalgoals.org). Target 15.3 is dedicated to "combating desertification, restoring degraded land and soil, including land affected by desertification, drought and floods, and striving to achieve a land degradation-neutral world by 2030." (www.unccd.int).

The concept of LDN has been introduced in the global dialogue to create a more efficient approach within the framework of activities aimed at combating land degradation. Rational management over land resources, in addition to the increase in quantity and quality of food, may contribute with numerous activities focusing on the mitigation and adaptation to the climate change effects, the conservation and enhancement of biodiversity, ending poverty and balancing social and economic issues at the global level. The conservation and protection of soil, as a principal natural resource and a medium for the survival and development of living beings on the planet, has essential importance for human health and well-being (www.undp.org).

11.7 CONCLUDING REMARKS

The current state of the soil in SRB and BIH coupled with the ongoing climate change, floods, droughts, forest fires, and the constant and increasing anthropogenic pressures has a strong impact on human health. An unfavorable status of soil health leads to the reduction of quantity and quality of produced food and disturb soil ecosystem services. Heavy metals from soil with toxic and carcinogenic impacts endanger human health on huge scale of diseases. SRB and BIH are still exposed to the influence of military activities, from 1992 to 2000, when huge soil surfaces were polluted with DU, petroleum products, VCM, EDC, metallic mercury, lead, PAHs, PCBs, dioxins and furans, pyralene, nitrogen oxides, sulfur dioxide, and other toxic and carcinogenic matters. Environmental and food chain contamination with radioactive and toxic substances caused numerous cancer cases, leukemia, lymphoma, chromosomal abnormalities, and increased mortality rates. The achievement of LDN targets helps the improvement and restoration of ecosystems, halts land degradation and pollution, increases the quantity and quality of produced food, supports soil health and the stability of local communities and living conditions, improves the well-being of people, and contributes to the realization of SDGs 3 and 15.

The growing anthropogenic pressure and the observed and predicted climate change in the Balkan region endanger precious, fragile, and finite soil resource. Soil is the "backbone" of life and the foundation of land ecosystem stability. However, the ongoing human activity leads to a permanent decrease of soil area and degradation of its quality. The rigid anthropocentric approach, which is "free" of responsibility toward nature, is an expression of imminent self-destructiveness of *Homo sapiens*. Regardless of the Freud's determination of two primal instincts, Eros and Thanatos (based on the Empedocles' theory of basic elements earth, air, fire, and water, combined or separated by Love and Strife), it is clear that mankind still has the same dilemma: to choose Eros (Love) and unity into being (life) or Thanatos (Strife) and fall into destruction (death) (Russel 1945, Askay and Farquhar 2006).

Soil degradation significantly contributed to the collapse and disappearance of certain civilizations. Soil erosion, deforestation, salinization, overgrazing are the factors, which among other factors influenced the decline and final fall of prosperous ancient societies. The examples of Mesopotamia, Crete, Greenland, Island, Easter Island, Ancient Greece and Rome, Carthage, Phoenicia, Mayans, and Aztecs are not just exotic historical stories but a warning evidence of possible scenarios for our civilization in a next few decades or centuries (Dale and Carter 1955, Montgomery 2007, Bardgett 2015). Unfortunately, historical lessons are not sufficient to convince Balkan societies to stop intensifying the pressures on ecosystems, neglecting public interest and promoting the material interest of individuals and interest groups under the cloak of law.

One of the priorities for achieving sustainable development in SEE refers to the protection and enhancement of the environment and the wise use of natural resources. This implies integration and harmonization of the policy objectives and measures of all sectorial policies, the harmonization of national legislation with principles of the "RIO" Conventions and IPBES, and their full implementation. The impacts of climate change associated with negative anthropogenic influence lead to the degradation of enormous soil surfaces, endanger forests, biodiversity, economic activities, and health in SEE countries. This is followed by the adverse effects of the increasingly frequent occurrence of torrential floods, erosion processes, droughts, forest fires, landslides, and plant diseases.

The observed data processing and model-based climate projections show that until the end of the 21st century, climate warming in SEE will cause a 2.5°C–5.0°C increase in the mean temperature, with a reduction of summer precipitation, increased frequency of heavy precipitation, and significantly less snow precipitation, while the total annual values do not show significant changes. Climate change and natural hazards cannot be prevented, but it is possible to influence the change of human behavior through consistent nature protection, an affirmation of ecosystem services in accordance with the principles of the "RIO" Conventions, IPBES and SDGs, as well as through the activism of all participants in the social life, such as ministries, government agencies, citizens' associations, academies, individuals, and media.

The current state of soil resources in SEE has been marked by a reduction in the anthropogenic pressure in rural areas, primarily due to population migrations, toward urban centers and abroad, and the declining birth rate. For the last 30 years, transitional woodland-shrub areas and areas principally occupied by agriculture, with a significant share of the natural vegetation have been exposed to intensive abandonment as a consequence of the conflict and post-conflict environment. However, areas exposed to deforestation, the mismanagement of agricultural and forest surfaces, man-induced forest fires, uncontrolled urbanization, the lack of erosion control, and flood protection activities are vulnerable to the effects of climate change and a range of disasters.

In order to protect the soil from degradation, SOC must be preserved and satisfactorily maintained. SOC has great importance for providing ecosystem services and its loss contributes to the decline in soil health. Besides the parent material and soil formation processes (Don et al. 2009), SOC reserves are determined by diverse cropping systems (Puget and Lal 2005), land-use change (Szilassi et al. 2012), and the intensity of wind and aeolian erosion (Lal 2018, 2020). Serbia does not systematically perform the monitoring of SOC in the soil for the entire territory. In the coming period (2020–2030), it is necessary to provide the basis for systematic collection of data and information on SOC. One of the drawbacks in the field of soil management is the lack of updated soil and soil erosion maps of Serbia. Some parts of the existing maps were created more than 30 years ago. The restoration of eroded soil and afforestation of bare land is one of the key activities in the process of mitigation and adaptation to the effects of climate change, as well as the reduction of disaster risk. It involves biotechnical works on slopes and technical works in the channel network, coordinated within a precisely defined administrative and spatial framework.

Sustainable spatial management is not possible through a fragmented approach in which different services deal with certain segments of nature (water, soil, forests, nature protection) and often implement legal solutions in a bureaucratic way. That way enables investors to obtain building permits for the construction of facilities in strictly protected zones, which completely violates the nature protection policy. Urban environments are becoming completely opposite of the close-to-nature ambiance, due to the intensive construction of housing and communal infrastructure, with increasing population density, pollution, green spaces reduction, and the destruction of watercourses.

The continuation of SHPPDT (Small Hydro-power Plants of Derivative Type) construction would lead to major ecosystem disorders, environmental degradation and would be an indicator of the inability of the system and the wider community to consider the auto destructiveness of this form of behavior in public life. Among other things, that would also mean denying to every citizen of the SRB and BIH the basic human right to use the unique natural values during their lifetime and preserve them for future generations. It is absurd that this is financially supported by the entire society, through the imposed obligation for all households to pay compensation for the production of energy from renewable sources, and the citizens were not given the opportunity for its approval. Therefore, the construction of SHPPDTs is not a national interest, both in SRB and BIH. On the contrary, it fundamentally violates the concept of environmental protection and endangers the traditional way of living of the local population. So far, the activities on the construction of SHPPDT in Serbia resulted in removing 320 ha of the highest quality forests in mountainous areas, destroying about 280 ha of stream beds with ichthyofauna and almost 1.92 million m^3 of soil in the 60 cm layer. The natural process of soil renewal in a 60 cm layer requires about 6,000 years (Kostadinov 2008).

The cooperation and overcoming of conflicts between the sectors of agriculture, water management, forestry, energy, nature protection and biodiversity preservation, and local economic development are indispensable at the following levels: policy, spatial planning, practice, investments, and education. It is very important to connect these measures with the process of mitigation and adaptation to climate change following the platforms of UNFCC, UNCCD (particularly LDN), CBD, and IPBES, inform and educate all stakeholders about the planned activities and provide subsidies for their implementation as well as media support.

The rights of the Soil (RoS) concept treats soil as a living entity (Lal 2019), which must be protected from degradation, pollution, and depletion. The dominant liberal order is no longer able to respond to the economic and environmental challenges, which will, by all means, be intensified in the coming decades. The omnipotence of the market, the loss of the influence of traditional forms of organizing human communities ("desacralization") and the emphasis on individuality through the weakening of family ties ("fragmentation") (Molnar 1994) produce consumerism that knows no boundaries (Stiglitz 2012). The disappearance of huge forest areas, the complete degradation of land, the endangerment of all ecosystems, and the danger of extinction of almost a million of plant and animal species raises the question of further civilization development. We need to adopt a new viewpoint and raise awareness of the fact that humans are not above nature but just a single particle constituting a much bigger whole.

REFERENCES

Abu-Qare, A.W. and M.B. Abou-Donia. 2002. Depleted uranium–the growing concern. *Journal of Applied Toxicology*, 22: 149–152.

Askay, R. and J. Farquhar. 2006. *Apprehending the Inaccessible: Freudian Psychoanalysis and Existential Phenomenology*. Northwestern University Press, Evanston, IL.

Bardgett, R.D. 2015. *Earth Matters – How Soil Underlies Civilization*. Oxford University Press, Oxford.

Belić, M., M. Manojlović, LJ. Nešić, et al. 2013. Pedo-ecological significance of soil organic carbon stock in Southe-Eastern Panonnian basin. *Carpathian Journal of Earth and Environmental Sciences*, 8(1): 171–178.

Betti, M. 2003. Civil use of depleted uranium. *Journal of Environmental Radioactivity*, 64(2–3): 113–119.

Čakmak, D., B. Sikirić, J. Beloica, et al. 2014. Acidifikacija zemljišta kao limitirajući faktor poljoprivredne proizvodnje opštine Ljubovija. *Glasnik Šumarskog Fakulteta*, 109: 49–62.

Clarke, R. 2002. Yugoslavia. In *Environmental Problems in East Central Europe*, eds. F.W. Carter and D. Turnock, 396–416. Routledge, London, UK.

Clausnitzer, H. and M.J. Singer. 2000. Environmental influences on respirable dust production from agricultural operations in California. *Atmospheric Environment*, V34: 1739–1745.

Čustović, H. and M. Ljuša. 2018. Final report on land degradation neutrality target setting program in the Federation of Bosnia and Herzegovina, Sarajevo. https://knowledge.unccd.int/sites/default/files/ldn_targets/2019-01/Bosnia%20and%20Herzegovina%20LDN%20TSP%20Country%20Report%20and%20Commitments.pdf (accessed May 27th, 2020).

Dale, T. and V.G. Carter. 1995. *Topsoil and Civilization*. University of Oklahoma Press, Norman.

Dimitrijević, B. and M. Dželetović. 2019. Economic, environmental and health effects of the NATO bombing-survey. In *David vs. Goliath: NATO War Against Yugoslavia and Its Implications*, ed. N. Vukovic, 440–460. Institute of International Politics and Economics, Belgrade, Serbia; Faculty of Security Studies at the University of Belgrade.

Don, A., T. Scholten, and E.D. Schulze. 2009. Conversion of cropland into grassland: Implications for soil organic-carbon stocks in two soils with different texture. *Journal of Plant Nutrition and Soil Science*, 172(1): 53–62.

Dragović, N., R. Ristić, B. Stajić, et al. 2017. Overview of the natural resources management in the Republic of Serbia. In *Natural Resources Management in Southeast Europe: Forest, Soil and Water*, eds. N. Dragović, R. Ristić, H. Pülzl, and B. Wolfslehner, 231–269. Deutsche Gesellschaft für Internationale Zusammenarbeit, Skopje, Macedonia.

Durante, M. and M. Pugliese. 2003. Depleted uranium residual radiological risk assessment for Kosovo sites. *Journal of Environmental Radioactivity*, 64: 237–245.

Đurđević, V., A. Vuković, and M. Vujadinović Mandić. 2018. *Osmotrene promene klime u Srbiji i projekcije buduće klime na osnovu različitih scenarija budućih emisija*. Global Environment Facility, Ministarstvo zaštite životne sredine, UNDP, Belgrade, Srbija.

EEA. 2012. Corine land cover datasets (2000–2012). European Environmental Agency. https://land.copernicus.eu/pan-european/corine-land-cover (accessed May 27th, 2020).

EEA. 2018. Corine land cover datasets (2018). European Environmental Agency. https://land.copernicus.eu/pan-european/corine-land-cover (accessed May 27th, 2020).

Eichelmann, U. and A. Scharl. 2017. *Remove the Dams-Free our Rivers (Concept Paper)*, RiverWatch, Manfred HermsenStiftung, Vienna, Austria.

Frich, P., L.V. Alexander, P. Della-Marta, et al. 2002. Observed coherent changes in climatic extremes during the second half of the twentieth century. *Climate Research*. 19: 193–212.

Gavrilović, S. 1972. *Engineering of Torrents and Erosion.* Journal of Construction Special Issue, Belgrade, Yugoslavia.

Giannardia, C. and D. Dominici. 2003. Military use of depleted uranium: assessment of prolonged population exposure. *Journal of Environmental Radioactivity*, 64: 227–236.

Gualdi, S., B. Rajkovic, V. Djurdjevic, et al. 2008. *SINTA – Simulations of Climate Change in the MediTerranean Area, Final Scientific Report.* Istituto Nazionale di Geofisica e Vulcanologia (INGV), Bologna, Italy.

https://energetskiportal.rs/dokumenta/Strategije/Nacionali%20akcioni%20plan%20za%20obnovljive%20izvore%20energije.pdf (accessed May 27th, 2020).

http://geoliss.mre.gov.rs/beware/?page_id=8,Landslidesdatabase (accessed May 27th, 2020).

http://www.czrs.gov.rs/lat/minska-situacija.php (accessed May 27th, 2020). Mine situation, Mine Action Centre of the Republic of Serbia.

http://www.elektrosrbija.rs/me/images/dokumenti/Katastar%20MHE%20u%20Srbiji.pdf (accessed May 27th, 2020).

https://www.globalgoals.org/15-life-on-land (accessed May 27th, 2020).

https://www.rtv.rs/sr_lat/ekonomija/aktuelno/boskovic-stete-vece-od-pet-milijardi-evra_591078.html (accessed May 27th, 2020).

https://www.unccd.int/sites/default/files/sessions/documents/2017-08/ICCD_COP%2813%29_19-1711042E.pdf (accessed May 27th, 2020).

https://www.undp.org/content/undp/en/home/sustainable-development-goals/goal-3-good-health-and-well-being.html (accessed May 27th, 2020).

Institute for Nature Protection of Serbia. 1999. *Red Book of Threatened Species.* Ministry of Environment Protection and University of Belgrade, Faculy of Biology, Belgrade, Serbia.

ISRIC. 2016. International Soil Reference and Information Centre, datasets for 2000. http://www.isric.org/content/soilgrids.

IWMJČ. 2001. *Water Resources Management Basic Plan of Serbia.* Institute for Water Management 'Jaroslav Černi', Ministry of Agriculture, Forestry and Water Resources Management, Belgrade, Serbia.

Joksimovich, V. 2000. Militarism and Ecology: NATO Ecocide in Serbia, *Mediterranean Quarterly.* 11(4): 140–160.

Kabata-Pendias, A. and A. Mukherjee. 2007. *Trace Elements from Soil to Human.* Springer-Verlag, Berlin, Germany.

Kapos, V. 1989. Effects of isolation on the water status of forest patches in the Brazilian Amazon. *Journal of Tropical Ecology*, 5: 173–185.

Kapović, M. 2011. Climate characteristics of the Javor mountain in the Republic of Srpska, *Bulletin of the Faculty of Forestry, University of Banja Luka*, 14: 29–41.

Kapović, M. and M. Knežević. 2010. Characteristics of black soil on Javor mountain limestones in the Republic of Srpska. *First Serbian Forestry Congress, Faculty of Forestry*, November 11th–13th 2010, Belgrade, Serbia, 257–263.

Kapović, M., R. Tošić, M. Knežević, and N. Lovrić. 2013. Assesment of soil properties under degraded forests – case study: Javor mountain – Republic of Srpska. *Archives of Biological Sciences*, 65(2): 631–638, Belgrade.

Kapović-Solomun, M. 2018. *Final Report on Land Degradation Neutrality Target Setting Program in the Republic of Srpska.* Ministry of Agriculture, Forestry and Water management of the Republic of Srpska, Banja Luka.

Kapović-Solomun, M. 2020. *Drought Management Plan for the Republic of Srpska.* Ministry of Agriculture, Forestry and Water Management of the Republic of Srpska, Report, Banja Luka, Bosnia and Herzegovina.

Kibblewhite, M.G., K. Ritz, and M.J. Swift. 2008. Soil health in agricultural systems. *Philosophical Transactions of the Royal Society A*, 363: 685–701.

Kostadinov, S. 2007. Erosion and torrent control in Serbia: Hundred years of experiences. Presented at the Erosion and Torrent Control as a Factor in Sustainable River Basin Management, Belgrade.

Kostadinov, S. 2008. *Torrents and Erosion.* University of Belgrade, Faculty of Forestry, Belgrade, Serbia.

Krivokapić, B. 2019. The NATO bombing of Yugoslavia (1999) 20 years later – the problems of legality, legitimacy and consequences. In *David vs. Goliath: NATO War Against Yugoslavia and Its Implications*, ed. N. Vukovic, 19–56. Institute of International Politics and Economics, Belgrade, Serbia; Faculty of Security Studies at the University of Belgrade.

Kržić, A., I. Tošić, V. Đurđević, et al. 2011. Changes in climate indices for Serbia according to the SRES-A1B and SRES-A2 scenarios. *Climate Research*, 49: 73–86.

Labus, M. 2000. *Policy of Change and Economic Stabilization in Yugoslavia.* Paper delivered at the conference "Yugoslavia: Prospects for Change, Strategies and Roadmaps," Kennedy School of Government, Harvard University, 25 April 2000.

Lal, R. 2011. Soil health and climate change: an overview. In *Soil Health and Climate Change*, eds. B.P. Singh, A.L. Cowie, and K.Y. Chan, 3–24. Springer-Verlag, Berlin Heidelberg.

Lal, R. 2018. Accelerated soil erosion as a source of atmospheric CO_2. *Soil & Tillage Research*, 188: 35–40.

Lal, R. 2019. Rights of soil, *Journal of Soil and Water Conservation*, 74(4) 81–86.

Lal, R. 2020. Soil erosion and gaseous emissions. *Applied Science*, 10: 2784.

Lazarević, R. 1986. SR Bosnia and Herzegovina's erosion map. *Erosion Professional Newsletter*, 14: 87–97.

MAC, MCABH. 2018. Mine Action Report in BiH. Mine Action Center, Ministry of Civil Affairs of Bosnia and Herzegovina. http://www.bhmac.org/wp-content/uploads/2019/05/izvjestaj-PMA-za-2018.-godinu-NACRT.pdf (accessed May 27th, 2020).

MAEP, IWMJČ. 2015. *Strategy for Water Management in the Territory of the Republic of Serbia*. Ministry of Agriculture and Environmental Protection, Institute for Water Management 'Jaroslav Černi', Belgrade, Serbia.

MAFWM-FD. 2009. *The National Forest Inventory of the Republic of Serbia, The growing stock of the Republic of Serbia*, Ministry of Agriculture, Forestry and Water Management of the Republic of Serbia - Forest Directorate, Belgrade, Serbia.

Magdoff, F. 2001. Concept, components and strategies of soil health in agroecosystems. *Journal of Nematology*, 33: 169–172.

Marlow, H.J., W.K. Hayes, S. Soret, R.L. Carter, E.R. Schwab, and J. Sabate. 2009. Diet and the environment: does what you eat matter? *American Journal of Clinical Nutrition*, 89: S1699–S1703.

MEMSP. 2011. *Report on Environment Conditions in Republic of Serbia for 2010*. Ministry of Environment, Minining and Spatial Planning. http://www.sepa.gov.rs/download/Izvestaj2010.pdf (accessed May 27th, 2020).

Milačić, S. and J. Simić. 2004. *The Consequences of NATO Bombing on the Environment in Serbia*. International Congress of the International Radiation Protection Association, Madrid, May 2004. https://www.ipen.br/biblioteca/cd/irpa/2004/files/1d11.pdf (accessed May 27th, 2020).

Milanović, S. 2017. *Unapređenje sistema za zaštitu šuma od požara u Republici Srbiji*. Ministarstvo poljoprivrede, šumarstva i vodoprivrede - Uprava za šume, Beograd. https://upravazasume.gov.rs/wp-content/uploads/2018/03/KONACNI-IZVESTAJ-POZARI-MILANOVIC.pdf (accessed May 27th, 2020).

Milojević-Djordjević, O., D. Grujić, S. Arsenijević, and D. Marinković. 2004. The frequency of micronuclei among newborns from Kragujevac, Central Serbia, after NATO bombing in the spring of 1999. *Russian Journal of Ecology*, 35(6): 426–430.

UNDP, 2017. *Second National Communication of the Republic of Serbia under the United Nations Framework Convention on Climate Change*. Ministry of Environment Protection , Belgrade, Serbia.

Molnar, T. 1996. *Liberal Hegemony*. SKC, Belgrade, Serbia.

Montgomery, D.R. 2007. *Dirt – The Erosion of Civilizations*. University of California Press, CA.

Morgan, R.K. 2013. Soil, heavy metals, and human health. In *Soils and Human Health*, eds. E.C. Brevik and L.C. Burgess, 59–82. CRC Press, Florida.

Mrvić, V., L.M. Kostić-Kravljanac, M.S. Zdravković, et al. 2010. Background limit of Zn and Hg in soils of Eastern Serbia. *Journal of Agricultural Sciences*, 55(2): 157–163.

Neslen, A. 2017. Balkan hydropower projects soar by 300% putting wildlife at risk, research shows. *The Guardian*. https://www.theguardian.com/environment/2017/nov/27/balkan-hydropower-projects-soar-by-300-putting-wildlife-at-risk-research-shows (accessed May 27th, 2020).

Official Gazette of RS, No. 29/2010. The national waste management strategy for the period 2010–2019. https://www.pravno-informacioni-sistem.rs/SlGlasnikPortal/reg/viewAct/011043b3-7cee-4488-ba2c-e95f95271713

Official Gazette of RS, No. 33/2012. National strategy for sustainable use of natural resources and properties. https://www.pravno-informacioni-sistem.rs/SlGlasnikPortal/eli/rep/sgrs/vlada/strategija/2012/33/1

Oliver, M.A. 1997. Soil and human health: a review. *European Journal of Soil Science*, 48: 573–592.

Pavlović, P. and M. Mitrović. 2013. Thermal power plants in Serbia – the impact of ash on soil and plants. In *Energy and Environment, Scientific Conferences, Book 4*, ed. M. Anđelković, 429–433. Serbian Academy of Science and Art (SANU), Belgrade, Serbia.

Pavlović, P., N. Kostić, B. Karadžić, et al. 2017. *The Soils of Serbia*. World Soil Book Series, Springer, Dordrecht, The Netherlands.

Popov, T. 2017. The impact of recent climate variability and potential climate change on phytogeographic characteristics of the Republic of Srpska. Ph.D. diss., University of Belgrade, Faculty of Geography.

Popović, A.R., D.S. Djordjević, D.J. Relić, and J.M. Djinović-Stojanović. 2013. Thermal power plants in Serbia as possible sources of surface and groundwater pollution by macro- and microelements. In *Energy and Environment, Scientific Conferences, Book 4*, ed. M. Anđelković, 373–401. Serbian Academy of Science and Art (SANU), Belgrade, Serbia.

Prospero, J.M., R.J. Charlson, V. Mohnen, et al. 1983. The atmospheric aerosol system: an overview. *Reviews of Geophysics and Space Physics*, 21: 1607–1629.

Protić, N., L. Martinović, B. Miličić, D. Stevanović, and M. Mojasevic. 2005. The status of soil surveys in Serbia and Montenegro. In *Soil resources of Europe*, 2nd edn, eds. R.J.A. Jones, B. Houšková, P. Bullock, and L. Montanarella, 297–315, Research Report No. 9. European Soil Bureau, Luxembourg.

Puget, P. and Lal, R., 2005. Soil organic carbon and nitrogen in a Mollisol in central Ohio as affected by tillage and land uses. *Soil & Tillage Research*, 80(1–2): 201–213.

Republika Srpska Institute of Statistics. 2019. Statistical Yearbook of Republika Srpska, 2019. https://www.rzs. rs.ba/static/uploads/bilteni/godisnjak/2019/StatistickiGodisnjak_2019_WEB.pdf (accessed May 27th, 2020).

Resulović, H. 1999. *Zemljišni resursi u BIH – korištenje u funkciji održivog razvoja (Land Resources in BIH – Use in the Function of Sustainable Development)*. Korištenje tla i vode u funkciji održivog razvoja i zaštite okoliša. Akademija nauka i umjetnosti Bosne i Hercegovine, Sarajevo. Posebna izdanja, knjiga CIX, 33–44.

Ristić, R., A. Marković, B. Radić, et al. 2011. Environmental Impacts in Serbian Ski Resorts. *Carpathian Journal of Earth and Environmental Sciences*, 6(2): 125–134.

Ristić, R., B. Radić, and S. Polovina. 2020. *Land Degradation Neutrality Target Setting Programme*. Report on the applied methodology and identification of targets to achieve Land Degradation Neutrality in the Republic of Serbia. Ministry of Environmental Protection, Republic of Serbia.

Ristić, R., B. Radić, Z. Nikić, and N. Vasiljević. 2008. *Environmental Impact Assessment – Ski run Sunčana dolina – Stara planina*, Faculty of Forestry, Belgrade, Serbia.

Ristić, R., I. Malušević, B. Radić, S. Milanović, V. Milčanović, and S. Polovina. 2019. The role of forest ecosystems in the process of mitigation and adaptation to the effects of climate change. *Contributions, Section of Natural, Mathematical and Biotechnical Sciences, MASA*, 40(1): 25–32.

Ristić, R., I. Malušević, S. Polovina, V. Milčanović, and B. Radić. 2018. Small hydropower plants – derivation type: insignificant energy benefit and immeasurable environmental damage. *Vodoprivreda*, 50(294–296): 311–317.

Ristić, R., M. Kašanin-Grubin, B. Radić, et al. 2012a. Land degradation in ski-resort 'Stara planina'. *Environmental Management*, 49: 580–592.

Ristić, R., S. Kostadinov, B. Abolmasov, et al. 2012b. Torrential floods and town and country planning in Serbia. *Natural Hazards and Earth System Sciences*, 1(12): 23–35.

Ristić, R., S. Kostadinov, B. Radić, G. Trivan, and Z. Nikić. 2012c. Torrential floods in Serbia – man made and natural hazards. *12th Congress INTERPRAEVENT 2012 – Grenoble / France Conference Proceedings*. http://www.interpraevent.at/palm-cms/upload_files/Publikationen/Tagungsbeitraege/2012_2_771.pdf.

Ristić, R., S. Kostadinov, V. Milčanovič, et al. 2015. Torrential floods, spatial and urban planning in Serbia. In *Proceedings of 8th International Symposium "Planned and Normative Protection of Space and Environment", Palić, Serbia*, 16–18 April 2015, eds. Subotica, Serbia.

D. Filipović, V. Šećerov, and Z. Radosavljević, 507–513. Serbian Spatial Planners Association and University of Belgrade – Faculty of Geography.

Ristić, R., S. Polovina, I. Malušević, et al. 2017. Disaster risk reduction based on a GIS case study of the Čađavica River. *SEEFOR (Southeast European Forestry)*, 8(2): 1–8.

Ristić, R., Z. Gavrilović, M. Stefanović, I. Malušević, and I. Milovanović. 2006. *Effects of Urbanization on Appearance of Flood*. BALWOIS, Topic: Droughts and Floods, Ohrid, Macedonia.

Ruml, M., E. Gregorić, A. Vujadinović, et al. 2016. Observed changes of temperature extremes in Serbia over the period 1961–2010. *Atmospheric Research*, 183: 26–41.

Russel, B. 1945. *A History of Western Philosophy*. Simon & Schuster, New York.

Rylander, R. 1986. Lung diseases caused by organic dusts in the farm environment. *American Journal of Industrial Medicine*, 10: 221–227.

Schenker, M.B. 1998. Respiratory health hazards in agriculture. *American Journal of Respiratory and Critical Care Medicine*, 158: S1–S76.

SEPA. 2015. *Report on the State of the Environment in the Republic of Serbia-Short Review*. Serbian Environmental Protection Agency, Belgrade, Serbian.

SEPA. 2018a. Reports on land condition in the Republic of Serbia for the year of 2012–2017. Serbian Environmental Protection Agency. http://www.sepa.gov.rs/download/zemljiste/Zemljiste2016_2017.pdf (accessed May 27th, 2020).

SEPA. 2018b. *Report on Soil Quality in the Republic of Serbia 2016–2017*. Serbian Environmental Protection Agency, Belgrade, Serbian.

SIF. 2019. *Conclusions of the Round Table*. Serbian Intelectual Forum, East Sarajevo, Republika Srpska.

Smith, J.L. and K. Lee. 2003. Soil as a source of dust and implications for human health. *Advances in Agronomy*, 80: 1–32.

Statistical office of the Republic of Serbia. 2019. *Statistical Yearbook of the Republic of Serbia, 2019.* https://publikacije.stat.gov.rs/G2019/PdfE/G20192052.pdf (accessed May 27th, 2020).

Stiglitz, J.E. 2012. *The Prize of Inequality.* W.W. Norton & Company, New York.

Szilassi, P., G., Jordan, F., Kovacs, A. van Rompaey, and W.V., Dessel. 2012. Investigating the link between soil quality and agricultural land use change. A case study in the lake Balaton catchment, Hungray. *Carpathian Journal of Earth and Environment Sciences*, 5(2): 61–70.

Tošić, R., M. Kapović-Solomun, N. Lovrić, and S. Dragićević. 2013. Assessment of Soil erosion potential using RUSLE and GIS: a case study of Bosnia and Herzegovina. *Fresenius Environmental Bulletin*, 22: 3415–3423.

Tošić, R., N. Lovrić, and S. Dragičević. 2019. Assessment of the impact of depopulation on soil erosion: case study – Republika Srpska (Bosnia and Herzegovina). *Carpathian Journal of Earth and Environmental Sciences*, 14: 505–518.

Tošić, R., S., Dragićević, and N. Lovrić. 2012. Assessment of Soil erosion and sediment yield changes using erosion potential method – case study: Republic of Srpska – BIH. *Carpathian Journal of Earth and Environmental Sciences*, 7(4), 147–154. http://www.ubm.ro/sites/CJEES/viewIssue.php?issueId=19 (accessed May 27th, 2020).

Trbić, G., V. Ducic, N. Rudan, et al. 2010. Regional changes of precipitation amount in bosnia and Herzegovina. *6-th International Scientific Conference Dedicated to the International Earth Day*, April 2010, Sofia, Bulgaria, 62–64.

UBFF. 2014. *Hydrological-Hydraulics Study on Causes of Floods in Municipality Krupanj in May 2014.* University of Belgrade Faculty of Forestry, Belgrade, Serbia.

UNDP. 2019. Human Development Report 2019. Beyond income, beyond averages, beyond today: inequalities in human development in the 21st century. http://hdr.undp.org/sites/default/files/hdr2019.pdf (accessed May 27th, 2020).

UNEP. 2017. Action program for combat land degradation and mitigate the effects of drought in Bosnia and Herzegovina. GEF. United Nations Environmental Programme Sarajevo, Bosnia and Herzegovina. https://knowledge.unccd.int/action-programmes/action-programme-combat-land-degradation-and-mit-igate-effects-drought-bosnia-and (accessed May 27th, 2020).

UNEP. 2019. *Air Pollution and Human Health: The Case of the Western Balkans.* United Nations Environmental Programme. https://www.developmentaid.org/api/frontend/cms/uploadedImages/2019/06/Air-Quality-and-Human-Health-Report_Case-of-Western-Balkans_preliminary_results.pdf (accessed May 27th, 2020).

Unkašević, M. and I. Tošić. 2014. Seasonal analysis of cold and heat waves in Serbia during the period 1949–2012. *Theoretical and Applied Climatology*, 120(1–2): 29–40.

Wall, D.H., U.N. Nielsen, and J. Six. 2015. Soil biodiversity and human health. *Nature*, 528(7580): 69–76.

WRB. 2014. *World Soil Resources Reports, World Reference Base for Soil Resources. No 106.* 2nd edition. Food and Agriculture Organization of the United Nations, Rome, Italy.

Životić, M.M., D.D. Stojiljković, M.A. Jovović, and V.V. Čudić. 2012. Possibility of using ash and slag from the landfill of the thermal power plant "Nikola Tesla" as waste with useable value. *Chemical Industry*, 66(3): 403–412.

12 Heavy Metals Bioavailability in Soils and Impact on Human Health

Bal Ram Singh
Norwegian University of Life Sciences

Hamisi Tindwa
Sokoine University of Agriculture

Abul M. Kashem
University of Chittagong

Dheeraj Panghaal
CCS Haryana Agricultural University

Ernest Semu
Sokoine University of Agriculture

CONTENTS

12.1   INTRODUCTION

Heavy metal contamination of soils is an issue of greater concern worldwide because they cause toxicity to plant, animal, and human beings and persist for long period in the soil system due to their unbiodegradable nature. Soil is the primary reservoir of heavy metals in the atmosphere, hydrosphere, and biota, and thus plays a fundamental role in the overall metal cycle in nature (Cao et al., 2010). Metals are added to soils through anthropogenic activities, such as mining, smelting, industry, agriculture, and burning fossil fuels. In recent years, disposal of electronic waste materials containing heavy metals has also contributed to the burden of heavy metal contamination in soils. Heavy metal contamination of soil caused by geogenic materials is also a problem at many places worldwide. Some soils have naturally high levels of heavy metals and, in some cases, plant species, which can take up and store large amounts of heavy metals, have evolved in these locations. A good example is soils developed on alum shale parent materials in southeastern part of Norway, where these soils contain many fold higher concentrations of heavy metals than the other agricultural soils (Mellum et al., 1998), and food crops (cereals, potato, and vegetables) showed concentrations of cadmium many fold higher than the maximum allowable concentration by World Health Organization (WHO) for human consumption (Singh et al., 1995; Mellum et al., 1998).

Heavy metals in soil do not only pose environmental threats but can also create hazards to human health. Several of these elements are necessary for human health, and are beneficial when taken into the body in the form of foods or as supplements at appropriate, low levels because they are quintessential to maintain many physiological and biochemical functions in living organisms at very low concentrations. However, some of these heavy metals in high doses can be harmful to the body while others such as cadmium, mercury, lead, chromium, and arsenic in minute quantities have delirious effects in the body causing acute and chronic toxicities in humans.

Heavy metals are significant environmental pollutants and their toxicity is a problem of increasing significance for ecological, evolutionary, nutritional, and environmental reasons (Jaishankar et al., 2014). The common heavy metals which cause the risks to human health and the environment include arsenic, cadmium, chromium, copper, lead, nickel, and zinc, however, some of them, e.g., Cr, Cu, Ni, and Zn are also essential elements for living organisms including plants, animals, and humans.

These heavy metals do bioaccumulate in living organisms and the human body through various processes and pathways causing adverse effects. The common carriers of metals into human body are heavy metal contaminated wastes, waters, air, plants, and animal products (Li et al., 2017). Heavy metal toxicity, as such, can have several consequences in the human body. It can affect the central nervous function leading to mental disorder, damage the blood constituents, and may damage the lungs, liver, kidneys, and other vital organs promoting several disease conditions (Jaishankar et al., 2014). Also, long-term accumulation of heavy metals in the body may result in slowing the progression of physical, muscular, and neurological degenerative processes that mimic certain diseases such as Parkinson's disease and Alzheimer's disease (Jaishankar et al., 2014). More so, repeated long-term contact with some heavy metals or their compounds may even damage nucleic acids, cause mutation, mimic hormones, thereby disrupting the endocrine and reproductive system and eventually lead to cancer (Jarup, 2003).

Although the study of soils in relation to human health is a complicated endeavor as traditional scientific approaches that isolate a single variable, such as a specific contaminant, and then investigate that variable are not effective in this case because many of the issues that affect human health involve complicated and synergistic relationships (Brevik and Burgess, 2013). However, in this paper, we provide an overview of the current knowledge on heavy metal entry to soils, their mobility and bioavailability to plants in relation to various factors and properties of soils and finally on transfer pathways of heavy metals to humans causing health problems.

12.2 SOURCES OF HEAVY METALS

Heavy metals in soils are derived either from geogenic parent materials or anthropogenic activities, such as agricultural chemicals, industrial wastes, sewage, and transport vehicles. The total concentration of heavy metals in a soil is thus the sum of these two sources. Sources of heavy metals to soils are presented below.

12.2.1 PARENT MATERIAL

The geological parent material, called as lithogenic source, is the key factor deciding the concentration of heavy metals in unpolluted soils. The parent material is a rock or unconsolidated material which on weathering and pedogenesis forms the soil over long period of time. Some important rocks which are rich in heavy metals and contribute to their build up in soils are described below.

12.2.1.1 Black Shales

Black shales are rich source of heavy metals and other trace elements. The soils developed on these rocks contain higher amount of Ag, As, Au, Ba, Cd, Cr, Cu, Hg, Mo, Ni, Pb, Sb, Se, Th, Tl, U, V, W, Zn, rare earths and platinum (Pt) group elements. Black shales have attracted wide attention as sources of elevated Cd concentrations in soils and examples include: up to 4 mg Cd kg^{-1} in soils developed on black shales in Norway (Mellum et al., 1998) and up to 60 mg Cd kg^{-1} in soils in an area of black shales and carbonaceous siltstone in Wushan County in south western China (Tang et al., 2009). These soils in south western China also contain anomalously high concentrations (in mg kg^{-1}) of Ni up to 388, Zn 962, Mo 99, and Sb 15. Control soils nearby contained concentrations of up to (in mg kg^{-1}): Cd 0.76, Ni 39, Zn 108, Mo 1.1, and Sb 3.4. Crops grown in soils on black shales in this area had elevated concentrations of Cd (up to 76.5 mg kg^{-1} DM in maize) and Mo (up to 5.4 mg kg^{-1} DM in beans) which are a major cause of concern for the health of consumers (Tang et al., 2009).

12.2.1.2 Limestones

Soils developed on Jurassic Oolitic limestone in adjacent areas of the Jura Mountains in Switzerland and France have been found to contain anomalously high concentrations of Cd (up to 22 mg kg^{-1}) and Zn (up to 864 mg kg^{-1}) probably due to vertical pedogenic processes involving the weathering of underlying rock and debris from upslope limestones and the accumulation of resistant Zn-containing minerals (Quezada et al., 2009).

12.2.1.3 Phosphorites

Phosphorites are sedimentary rocks containing high concentrations of phosphate minerals, mainly apatite [$Ca_5(F, Cl, OH) (PO_4)_3$] and are used as raw materials for the manufacture of P fertilizers. Mar and Okazaki (2012) reported a large variation in the concentration of Cd in phosphate rocks from different countries (Table 12.1). da Silva et al. (2010) reported concentrations of heavy metals

TABLE 12.1

Cadmium Concentration in Phosphorites of Different Origins

Phosphorite Origin	Cd Concentration (mg kg^{-1})
USA	3–186
Morocco	3–165
Peru	2–186
Russia	0.1–<13
North Africa	60
South Africa	2.0–<13
Brazil	4
Australia (Christmas Island)	7.0–43
Mexico	8
Israel (Arad)	12.0–32

Source: Modified and adopted from Mar and Okazaki (2012).

in phosphorites from a mine in Tunisia and other phosphate rocks from different parts of the world, and the maximum concentrations found (in mg kg^{-1}) were: Cd 62.5, Cr 490, Cu 110, Ni 180, Pb 500, U 150, and Zn 1,850. The Hope phosphorite (rock) from Manchester Parish in Jamaica contained (in mg kg^{1}): Cd up to 6,200, Zn 12,300, and U 166 (Garrett et al., 2008). In Galicia, northern Spain, serpentine soils contain up to 1,162 mg kg^{-1} Cr, 940, mg kg^{-1} Ni, and 150 mg kg^{-1} Cu, compared with up to 50–100 mg kg^{-1} Cr and Ni and 5–25 mg kg^{-1} Cu for local control soils on other parent materials (Miranda et al., 2009).

12.2.1.4   Sedimentary Ironstones

Sedimentary ironstones are found at various places in the world. For example, in Lincolnshire, UK, soils developed on ironstone deposits of Lower Jurassic and Lower Cretaceous ages contained up to 342 mg kg^{-1} total As. Although, due to extremely low inaccessibility of As, it does not cause any significant health hazard (Breward, 2007).

12.2.2   Anthropogenic Sources of Heavy Metals

12.2.2.1   Fertilizers

Macronutrient fertilizers nitrogen (N), phosphorus (P), and potassium (K) are applied to the soil either individually as required, or more frequently together in various combinations as 'compound fertilizers,' such as NPK or NP and often containing the 'secondary' macronutrients calcium (Ca), magnesium (Mg), and sulfur (S). Calcium and Mg are applied as limestone (usually as $CaCO_3$) and dolomite ($MgCO_3.CaCO_3$), respectively, to raise the pH of acid soils. In addition, micronutrients, such as B, Cu, Co, Fe, Mn, Mo, Ni, and Zn, are also applied to either the soil or plant foliage in specialized micronutrient fertilizers and a range of compounds. Most of the inorganic compounds used in macronutrient fertilizers contain significant concentrations of heavy metal contaminants. Phosphatic fertilizers generally contain the highest concentrations of most heavy metals including Cd, Zn, and As.

Nicholson et al. (2003) estimated the amount of Cd added to soils through P fertilizers and found a wide variation in the inputs of Cd to soils through P fertilizers (Table 12.2). Similarly. Eckel et al. (2005) reported that the mean input of heavy metals from fertilizers on a total of 37 farms (both livestock and crops) was: Cd 2.2, Cr 24, Cu 4.3, Ni 6.7, Pb 5.2, and Zn 49 g ha^{-1} year^{-1}. Industrial by-products used for production of Zn fertilizers can give rise to unacceptably high concentrations of various potentially toxic heavy metals including Pb (<5.2%) and Cd (<0.22%). If these

TABLE 12.2

Estimated Input of Heavy Metals in Agricultural Soils through Phosphate Fertilizers Based on the Average Fertilizers Use

Country	P_2O_5 Inputs (kg ha^{-1}year^{-1})	P-fertilizers (g ha^{-1} y^{-1}) Cd	Ni	Pb	Zn	As
Austria	37	1.3	2.8	0.7	62.7	1.9
Denmark	20	0.5	3.4	0.5	11.1	1.2
Finland	29	0.1	0.8	0.8	134.7	2.4
France	49	2.5	5.7	2.3	83.5	1.3
Germany	33	1.3	3.8	0.7	29.8	1.6
Italy	50	2.0	2.8	0.7	38.6	1.0
Netherlands	45	1.9	4.9	0.6	45.0	1.8
Portugal	50	3.1	6.2	0.7	56.4	1.5
Spain	48	1.9	4.8	1.6	82.9	1.9
Sweden	20	0.1	0.7	0.4	16.1	1.1
European Average	43	1.6	3.6	1.0	43.1	2.3

Source: Adopted and modified from Nicholson et al. (2003).

materials are regularly used, in addition to rectifying a deficiency of Zn, the soil could end up by being significantly contaminated with both Pb and Cd.

12.2.2.2 Sewage Effluents and City Wastes

Sewage sludge and waste waters contain many heavy metals depending on their source and can result in contamination of soils if applied in an unregulated manner. In many developed countries, e.g., European Union (EU), USA, Australia, New Zealand, sludge use on agricultural lands is regulated with both the concentration of heavy metals and the amount of sewage sludge applied. These regulations are intended to avoid the accumulation of excessive concentrations of elements which could cause either ecotoxicity in soils and/or phytotoxic effects in crops. However, sewage sludges have been applied to land in some countries for more than 100 years and during this time some of these sludges and the soils to which they were applied would almost certainly have had higher concentrations of some heavy metal(loid)s than are permitted nowadays. However, individual EU Member States can have more conservative limits than those shown above. Moreover, some countries, such as the Netherlands, have banned the application of any sludges to soil. In 2010, total sewage sludge production in the 27 Member States of the EU was around 11.6 Mt (DM) (Milieu Ltd., 2008), and considerable amount of it was used on agricultural lands.

12.2.2.3 Industrial Wastes

Many studies have confirmed a high correlation between industrialization and urbanization activities and heavy metal contamination of soils in China. For example, He et al. (2011) reported that the production of Hg increased by 40% from 2004 to 2007, exceeding 1,500 tons in 2007. Wu et al. (2006) estimated that total Hg emissions from all anthropogenic sources increased at an average annual rate of 2.9% from 1995 to 2003. Ying et al. (2016) found a correlation between Cu, Cr, Cd, and Pb and industrialization and urbanization in Huainan city, Anhui, East China. Zhang et al. (2015) confirmed that human activities such as mining and smelting, industry, sewage, urban development, and fertilizer application released heavy metals into the soil, which resulted in pollution of farmland soil. On-ferrous metal mining and smelting activities are the biggest contributors of heavy metal pollution as they result in large discharges of wastewater, waste gas, and solid waste into the environment.

12.2.2.4 Vehicle Transport

Heavy metals from vehicular emissions can be significant threats to humans and the environment because they have adverse effects on ecosystems inducing contamination of air, water, and soil. Gunawardana et al. (2012) reported that road dust primarily consisted of soil-derived minerals (60%), where 40%–50% of the soil-derived minerals were quartz. Duong and Lee (2011) compared dust from roads, where the average speed ranged from 80 to 90 km h^{-1} with roads where the average speed ranged from 70 to 80 km h^{-1} and found that higher concentrations of heavy metals occurred in dust from roads that had higher average driving speeds. According to Duong and Lee (2011), the concentrations of heavy metals in road dust vary significantly depending on traffic and road features such as roundabouts, motorway roads, and traffic lights. The concentrations of metals in road dust from motorways are approximately twice those found near roundabouts and downtown areas.

12.3 SOIL FACTORS AFFECTING HEAVY METAL MOBILITY AND BIOAVAILABILITY IN SOILS

Technological advancements of human activities have led to the introduction of heavy metals into the soil. Heavy metals are attracting concern all over the world due to their non-biodegradable nature in the environment and their ability to interfere with normal metabolic activities of living organisms. Soil is the main sink for heavy metals. After reaching the soil, heavy metals go through various reactions with different components of soil, which affect their solubility, mobility, and availability in the environment. Heavy metals are in two categories: essential (macro and micro) and non-essential (toxic) elements. Some essential elements include Cr, Co, Cu, Fe, Mn, Ni, Zn, while some non-essential elements include Pb, As, and Cd. The essential elements contribute their significant role to the nutrient requirements of the human organism but possess the potential to be toxic in the case of excessive uptake. In soil, some of these metals are persistent because of their immobile nature. The mobility of metals must be considered when the food chain is concerned. The plant uptake of metals parallels to the bioavailable fractions of the metals existing in soil. The total concentration of metals is not a reliable indicator of the mobility and bioavailability of metals. Metals and their compounds present in different soil fractions and thus their mobility in soils and bioavailability to living organisms must be considered. Water-soluble and exchangeable fractions are generally readily mobile and bioavailable, while metals incorporated into the crystal lattice of clays appear to be relatively immobile and inactive. The most important factors which affect their mobility and availability are soil pH, soil texture, soil organic matter and its form, oxidation–reduction potential, ionic strength, chemical speciation and nature of contamination, and iron and manganese oxides in soils. How these factors affect mobility and bioavailability are described below.

12.3.1 SOIL pH

Soil pH is considered to be one of the most important factors that determine the concentration of heavy metals in the soil solution, their mobility, and availability to plants. Soil pH has the greatest effect of any single factor in the solubility or retention of heavy metal in soils (Ghosh and Singh, 2005). At a low soil pH level, the mobility and concentration in soil solution increase because of the increased hydrogen ion concentration. The mobilization tendency of heavy metals is positively related with the hydrogen ion concentration. The pH of the soil solution maintained at neutral to slightly alkaline condition showed low mobility of all heavy metals. Application of the liming materials in acidic soils reduced the uptake of heavy metals by plants (Subash et al., 2012; Kashem et al., 2010) because an increase in the soil pH value decreased the solubility of heavy metals in the soil (Kashem and Singh, 2001a; Lee et al., 2004). The mobility of heavy metals in soils with low pH increases in the order: Cd < Ni < Zn < Mn < Cu < Pb. However, the effect of pH on the mobility of heavy metals in soils is highly variable, depending on metal types, metal concentration and contents, and the type of organic matter. Lee et al. (2004) reported that the increase in soil pH after application of Ca carbonate decreased the Diethylene triamine penta acetic acid (DTPA) extractable Cd in a sandy soil significantly. Pardo and Guadalix (1996) mentioned that the rising of soil pH increased the pH-dependent charges that increased the heavy metal adsorption on the oxide-bound fraction of soil. This pH-dependent adsorption of Cd and Zn in the Fe–Mn oxide-bound fraction had also been reported by other workers (Kashem and Singh, 2004). Kashem and Singh (2004) found lower proportion of Cd and Zn in the exchangeable fraction while higher in the oxide-bound fraction under flooding condition than under non-flooding condition. They mentioned that flooding raised soil pH of 1.2 units compared to non-flooding soil that deceased the exchangeable Cd, Ni, and Zn by 97%, 47%, and 87%, respectively, in the flooded soil. In another study, where lherzolite was applied as an amendment in a contaminated soil, adsorption mechanism and reduction of metal bioavailability (64%–89%) could mainly be associated with pH rise, that resulted in significantly higher plant growth (133%–282%) and lower Cd (−68% to −89%) and Zn (>−90%) uptake in the plant tissues (Kashem el al., 2010) (Table 12.3).

12.3.2 Soil Texture

Soil texture plays a vital role in the mobility of metals in soil. Soil texture reflects the particle size distribution of the soil and thus the content of fine particles like oxides and clay. The oxides and clay compounds are important adsorption media for heavy metals in soils. Clay fraction is mainly composed of clay minerals and it has high potential capability to bind heavy metals. Soils having granulometric composition characteristic for clay, silt, and dust, and a high content of organic matter have a high sorption capacity and a strong ability to bind heavy metals (Moreira et al., 2013). In general, the clay soil retains high amount of heavy metals compared to sandy soil. However, sandy soils, distinguished by a low sorption capacity and acidity, weakly absorb heavy metals, which lead to relatively high mobility and bioavailability to plants than clayey soil (Sheoran et al., 2009). The land containing a large amount of clay minerals has the ability to accumulate large amounts of heavy metals (Kavamura and Esposito, 2010). In a sorption study for Cd, Zn, Cu, and Pb with different soils, it was shown that the adsorption of these metals was higher in the alum shale (sandy clay loam) soil than in the sandy soil, whereas the opposite was true for desorption of the same metals (Narwal and Singh, 1995). Kashem and Singh (2001b) found a higher accumulation index (AI= metal conc.in

TABLE 12.3
Effect of 5% Lherzolite Application on Soil pH, Exchangeable Cd and Zn, Their Uptake and Dry Matter Yield of Radish at the Third Harvest of Radish (*Raphanus sativus* L.) Grown in Contaminated Soil

Treatment	pH	Cd and Zn conc. (mg kg) in Exchangeable Fraction		Cd Uptake (mg kg) in Radish Plant Parts		Zn Uptake (mg kg) in Radish Plant Parts		Dry Matter Yield (g pot⁻¹)	
		Cd	Zn	Leaves	Roots	Leaves	Roots	Radish Leaves	Radish Bulbs
Without Lherzolite	4.7	2.53a	69.4a	29.9a	24.0a	1,364a	1,707a	1.59b	4.64b
With Lherzolite	5.9	1.12b	7.7b	9.7b	2.7b	138b	65b	3.68a	15.59a
Changes	+1.2	−64%	−89%	−68%	−89%	−90%	−96%	+133%	+282%

Source: Extracted and modified from Kashem et al. (2010).

Means followed by same letter within the column of different parameters are not significantly different at p < 0.05.

+ sign denotes increase and − sign denotes decrease.

rice/ metal conc. in soil) for Cd, Zn, and Ni in rice when plants grown in tannery waste contaminated sandy loam soil than in naturally contaminated alum shale (sandy clay loam) soil indicating higher bioavailability and mobility of metals in light texture soil than in heavy texture soil.

12.3.3 Soil Organic Matter Content and Its Form

Soil organic matter plays an important role on heavy metal mobility and availability depending on soil organic matter fractions and soil pH. Chemically, it increases the cation exchange capacity and water holding capacity of soils. Increasing the amount of organic matter in the soil helps to minimize the uptake of heavy metals by plants. The binding of heavy metals by organic matter is a complex process due to the diversity of its connections with the mineral phase (Fijałkowski et al., 2012). Sorption capacity of organic matter is much higher than the mineral sorption capacity of the soil (Ociepa et al., 2010). Harmful heavy metals and compounds are detoxified by interaction with soil organic matter.

An organic matter application to soil may reduce potential risk of heavy metals in the environment by reducing their mobility and as well as bioavailability. A reduction in the absorption of Cd, Cr, Zn, Ni, and Cu was observed with the addition of sewage sludge, which confirmed that the organic matter supplied to the soil as sewage sludge decreased the availability of heavy metals (Moreira et al., 2013). Organic matter is important for the retention of metals by soil solid particles, thus decreasing mobility and bioavailability. However, because of the complexation of metals by soluble organic matter, the addition of organic matter can result in release of metals from solids to the soil solution. Higher solubility of heavy metals in soil solution at alkaline pH was attributed to enhance the formation of organic matter metal complexes after ionization of weak acid groups. Organic acids, acting as ligands for many metal ions, increase their transport of water along the soil profile (Gang et al., 2010). The mobility of certain metals, e.g., Mn, Zn, Cu, and Fe, decreased with the increasing amount of soil organic matter, while the increasing concentration of humic acid caused an increase in their mobility, and the mobility order was Mn > Zn > Cu > Fe (Khan et al., 1997). Heavy metal mobility increases with the addition of soluble organic matter by formation of organo-mineral complexes, but the addition of solid organic matter can increase soil surface charges and decrease metal mobility (Li et al., 2014). Angelova et al. (2010) investigated peat, compost, and vermicompost resulting in an increase in the starch yield, absolute dry matter yield of potatoes. These amendments of application in soil also led to the effective immobilization of phytoaccessible forms of Pb, Cu, Zn, and Cd in soil. The quantity of the mobile forms of Pb, Zn, Cu, and Cd significantly correlated with the uptake of Pb, Zn, Cu, and Cd by the potato. The amendments decreased heavy metal contents in potato peel and tubers.

Recently, biochar as an organic amendment has been used to reduce the mobility and bioavailability of heavy metals due to its numerous binding sites for heavy metals, large specific surface area, porous structure, high pH, low cost, and availability in large quantities because they are derived from bio-products or industrial by-products (Tan et al., 2015). The effect of grape-pruning-residue (GPR) biochar on Cd, Pb, Cu, and Zn immobilization in a contaminated soil was investigated by a laboratory incubation study up to 8 weeks (Taghlidabad and Sepehr, 2017). The application of GPR biochar in Zn contaminated soil decreased the fractions of Cd, Pb, and Cu by 23%–72% in the exchangeable and by 51%–67% in the carbonate fractions but increased their concentration in oxide and organic bound fractions and consequently decreased the mobility and bioavailability of these metals in the soil investigated. The values of mobility factor of Cd, Pb, Cu, and Zn decreased by 47%, 62%, 70%, and 49%, respectively, with addition 10% of the biochar (Table 12.4).

12.3.4 Oxidation-Reduction Potential

Oxidation–reduction potential (Eh) is an important factor affecting the solubility and availability of heavy metals in soils (Kazemi Poshtmasari et al., 2012). After soil submergence due to aerobic respiration, redox potential (Eh) will reduce, and alternate reductive reactions occurring in

TABLE 12.4

Concentration of Chemical Fractions of Cd, Pb, and Zn (mg kg⁻¹) in Soil with Different Rate of Grape-Pruning-Residue Biochar Addition after 8 Weeks of Incubation

Biochar Rate (w/w %)	Exchangeable	Carbonate	Oxide	Organic	Residual	Total
			Cd			
0	3.8a (13)	8.8a (30)	8.6d (29)	5.5b (19)	2.8a (9)	29.5
2	2.5b (10)	5.6b (21)	9.2c (35)	6.2a (24)	2.8a (11)	26.3
5	2.2bc (9)	4.4c (17)	10.0b (39)	6.2a (24)	2.9a (11)	25.3
10	2.1c (8)	3.7d (15)	10.4a (41)	6.3a (25)	2.9a (11)	25.4
			Pb			
0	16a (3)	136a (26)	150d (29)	61b (12)	157a (30)	520
2	13a (3)	87b (17)	168c (34)	75a (15)	157a (31)	500
5	12b (2)	74c (15)	175b (35)	80a (16)	160a (32)	501
10	11b (2)	41d (9)	181a (38)	81a (17)	161a (34)	475
			Zn			
0	132a (9)	345a (23)	571d (38)	182c (12)	288a (19)	1518
2	96b (7)	201b (14)	616c (43)	229b (16)	290a (20)	1432
5	81c (6)	181c (13)	625b (44)	235a (17)	292a (21)	1414
10	71d (5)	151d (11)	638a (46)	238a (17)	293a (21)	1391

Means followed by the same letter in column within a metal are not significantly different at $P<0.05$ (percent of total in parenthesis) (extracted and modified from Taghlidabad and Sepehr, 2017).

sequence lead to the conversion of oxide forms of $NO3^{-1}$, $SO4^{-2}$, Mn^{+4}, and Fe^{+3} to their reduced forms. These reduced forms of Mn and Fe are more soluble. However, continuous flooding of soil may lead to lower solubility of heavy metals in soil due to enhance adsorption of metal (hydr) oxides and precipitation with sulfide. The oxidation–reduction reaction is a process that involves in the flow of electrons from reducing agents (reducer) to an oxidizing agent (oxidant). The oxidized soils have values ranging from +400 to +700 mV, while the reduced soils may have values from −250 to −300 mV. Soil redox potential can also influence the solubility of heavy metals in the soil. The redox potential of soil significantly determines participation in the form of a mobile element, which can enter the biological cycle, in relation to their total content. Lack of oxygen in the soil causes start-up and increases the mobility of the large portion of heavy metals. Redox reactions can mobilize or immobilize metals, depending on the metal species and micro-environments. When oxidation reactions are involved, the solubility of heavy metals increases with the decreasing pH, whereas in reducing conditions, the solubility of heavy metals (Zn, Cu, Cd, Pb) is higher in alkaline pH as a result of the formation of stable soluble organo-mineral complexes. The differences in the oxygen supply between submerged and non-submerged soils are the causes of changing redox potential and pH. When oxidized soils are submerged, they become anaerobic, reduced, and the pH tends to converge to neutrality irrespective of the initial pH (Kashem and Singh, 2001a). Submergence of soils is reported to decrease the availability of heavy metals in soils attributed to the increase of adsorption of metals on hydrous Mn and Fe oxides. Kashem and Singh (2001a) reported that redox potential (Eh) in soil solution decreased and pH increased after submergence. These changes of Eh and pH caused concentrations of Cd, Ni, and Zn in soil solution decreased with flooding time. They also mentioned that changes of pH, Eh, and heavy metal solubility depend on the soil type, organic matter content, and duration of submergence (Table 12.5).

TABLE 12.5

Changes of pH, Eh, Cd, Ni, and Zn Solubility with Flooding Duration up to 50 Days without Organic Matter Addition

DF	pH			Eh (mvolt)			Cd (µg L^{-1}) in Log			Zn (µg L^{-1}) in Log		
	TC	CC	AC	TC	CC	AC	TC	CC	AC	TC	CC	AC
0	3.9	5.0	5.7	140	168	188	1.50	1.60	0.35	3.58	4.75	2.20
10	4.7	5.1	5.6	−285	100	165	−0.60	0.87	0.10	2.55	4.38	1.81
30	5.9	5.3	5.6	−355	86	−380	−1.25	0.78	0.20	1.45	4.10	1.65
50	6.2	5.6	6.2	−365	70	−395	−1.48	0.60	0.20	1.44	3.75	1.45

Source: Extracted and modified from Kashem and Singh (2001a).

DF, duration of flooding; TC, tannery waste contaminated soil; CC, city sewage contaminated soil; and AC-alum shale naturally contaminated soil.

12.3.5 IONIC STRENGTH

Heavy metals adsorption increases with the decrease of ionic strength for minerals with permanent surface charge density. The adsorption reaction takes place on the negative surface of soil colloids. For these colloidal surfaces, the reduction of ionic strength enhances the surface electric potential more negative and hence ion adsorption is greater. In general, metal sorption decreases with the increase of ionic strengths because of the competition by other cations for the same adsorption sites. This indicates that soil with higher ionic strength may have more risk of heavy metal release from contaminated soils than soil with lower ionic strength.

12.3.6 CHEMICAL SPECIATION AND NATURE OF CONTAMINATION

Chemical speciation is the process of identifying and quantifying different species, forms of phases of elements that are present in soil. The total metal concentration is not a reliable indicator of the bioavailability and mobility metals in soils. Heavy metals and their compounds present in different soil fractions indicate both their mobility in the soils as well as their bioavailability. Metal bioavailability in soils is largely dependent on the partition of the metals between the solid and solution phases. The importance of the metal content of soil solution as a major controlling factor in the bioavailability of metals is acknowledged (Lee et al., 1996). The nature of soil at which heavy metals are associated in various chemical forms depends on soil pH, organic matter content, redox conditions, and soil texture (Kashem and Singh, 2004; Rieuwerts et al., 2006). The nature of this association is referred to as speciation. Thus, the chemical form is of great significance in determining the potential bioavailability and remobilization of the soil metals to other compartments (such as water, plants, and biota) when physicochemical conditions are favorable.

Heavy metals can be added in the soil by both human and natural factors and thus it is necessary to identify properly the geochemical phases in which metals may be bound as heavy metal mobility within soil differs from native to anthropogenic (Awokunmi et al., 2010; Kashem and Singh, 2004). These metals may be distributed in soil components as exchangeable, adsorbed on soil organic matter, precipitated, or complexed (Yuan et al., 2004). Kashem and Singh (2004) observed that the proportion of mobile fractions in the anthropogenically contaminated tannery and city sewage soils of Bangladesh was higher compared to the naturally metal-rich Norwegian alum shale soil. The mobile fractions of metals are correlated with the accumulation index AI of metals. The AI is a good parameter to estimate the relative availability of different metals in different soils. For Cd, Zn, and Ni, the (AI) index decreased in the order: tannery>city sewage> alum shale soil. It implies higher availability of heavy metals in anthropogenically contaminated soils than that of Alum shale naturally metal-rich soil.

12.3.7 Iron and Manganese Oxides

Hydroxides of Fe and Mn occur in clays as coatings on phyllosilicates and as free gels and crystals. They play an important role in the binding of heavy metals by the mineral phase. Mostly their oxides contribute to the formation of the coating on crystalline particles in the soil solid phase. Heavy metals together with the Fe and Mn hydroxides are easily displaced in suitable conditions. Hydrous Fe and Mn oxides may reduce concentrations of heavy metals in soil solution by both precipitation and specific adsorption reactions. Oxides of metals such as Al, Fe, and Mn and oxyhydroxides are also important for immobilization of heavy metals in soils. Because of strong specific adsorption, special affinity for metal oxides and formation and precipitation of specific minerals may account for reducing potential toxic metals mobility in soils (Zeng et al., 2017). Rinklebe et al. (2016) mentioned the role of redox potential as a crucial factor that determines metal oxide concentrations in soils. Generally, under reduced conditions, oxide concentrations (especially for Fe and Mn oxides) are often low, whereas the opposite occurs under oxidized conditions. Thus, heavy metal immobilization by oxides of metal is more effective under oxidized conditions.

Kashem and Singh (2004) used a sequential extraction procedure to assess the transformations in the solid phase species of heavy metals in contaminated soils under flooding and organic matter applications. Flooding over a period of 24 weeks significantly decreased the soluble plus exchangeable Cd, Ni, and Zn concentrations, and there was an increase in Cd and Zn predominantly in the oxide-bound fraction. In the case of Cd and Zn, the oxide fraction was more important. It seems that the breakdown of Fe and Mn oxides caused by flooding provided surfaces with high adsorbing capacity for heavy metals.

12.4 PATHWAYS OF METAL TRANSFER TO HUMANS

Humans come into exposure to heavy metals through a myriad of pathways which could be described by looking at (i) media serving as carriers of heavy metals and (ii) avenues of transfer of the heavy metals from the carriers to humans. The most common media serving as carriers of metals includes heavy metal contaminated wastes, waters, air, plants, and animal products (Kloke et al., 1984; Li et al., 2017).

12.4.1 Media Serving as Carriers of Heavy Metals

Heavy metal polluted air can be a result either of anthropogenic activities or natural environmental disturbances such as earthquakes. In most cases, amounts of heavy metals at the surface of the earth or in the atmosphere due to natural geochemical emissions are many times smaller than depositions and emissions from the anthropogenic activities like mining, power generation, smelting, and internal combustion engines (Nriagu, 1989; Nriagu and Pacyna, 1988; Gharaibeh et al., 2009).

Anthropogenic activities that have resulted in detectable levels of heavy metals in the atmospheric air have been reported. Municipal solid wastes incinerators, for example, have been reported as important sources of unintended emissions into the atmosphere, and the emitted heavy metals can reach humans via inhalation or direct deposition on skin-dermal contact (Li et al., 2017).

Other pathways by which atmospheric air gets contaminated with heavy metals thus making it a carrier of heavy metals as it interacts with human life include the entire chain of activities relating to mining activities (quarrying, beneficiation, smelting). Open-cast quarrying and beneficiation mineral-containing earth and rocks emit dusts into the atmosphere; these dusts contain the sought-after heavy metals (Monjezi et al., 2009; Lin et al., 2016). Smelting also passes the heavy metals into the air; 1st smokes enrich the work-room air with heavy metals and eventually, via chimneys, to the open atmosphere (Žibret et al., 2018) from where vast areas can be affected. Air and soils are reported to be the most affected receptors of a contamination dispersion process from smelters and ore-processing facilities. Following a dominant wind direction and orography, heavy metal effects have been detected in air and soils up to 14km away from the source in a town of Celje in Slovania (Žibret et al., 2018), while Zn smelter impacts in soils are recorded up to 10km away (Yang et al., 2009)

Water is another medium that acts as a potential carrier of heavy metals that would eventually be transmitted to humans. Contamination of important rivers, dams, lakes, and other inland water bodies used by humans for many domestic activities have been reported all over the world (Jackson, 1970; Bhuyan et al., 2017; Patel et al., 2018; Duncan et al., 2018). Rivers and other inland water bodies can receive elevated levels of heavy metals from natural weathering processes of granitic and other rocks lining the banks of such rivers, dams, and lakes. However, levels above the expected natural thresholds from various anthropogenic sources such as residential wastes, sewer outfall, fertilizers, and pesticides have been reported (Patel et al., 2018). Rivers, dams, and other surface and sub-surface water bodies can also receive elevated levels of toxic heavy metals through leachate seepage from a contaminated facility such as landfills to surface and underground waters. Reported exemplary cases include leachate leakage from a landfill in Bangalore, India to other water bodies including bore holes and surrounding open wells (Naveen et al., 2018). Similarly, cases of water bodies being contaminated by heavy metals originating from industrial facilities through underground leaching have been reported (Rehman et al., 2008).

Plant foods are another medium saving as carrier of heavy metals that eventually interact with human life. Plants can accumulate high amounts of toxic heavy metals either through absorption of the metals from a contaminated growth medium such as the soil or through absorption and depositions from foliar application of heavy metal-laden substances such as pesticides and fertilizers. Some plants, both edible and non-edible, have been reported to be good accumulators of various toxins including heavy metals. Such plants can be so good accumulators that scientists have exploited their potential by deploying them as cheap yet sustainable means of decontaminating media such as soils and sediments, an approach collectively referred to as phytoremediation (Malawska and Wiłkomirski, 2000). However, when edible plants have an inherent ability to accumulate such toxins, they present a big challenge by acting as carriers of the toxins, heavy metals, in this case posing a direct risk of transferring these to humans through their ingestion. Cases on elevated concentrations of heavy metals in edible plants have been reported (Giliba et al., 2017).

Solid heavy metal wastes can include scrap metals, metallic devices, batteries, and electronic wastes (Teta and Hikwa, 2017). A big share of the solid heavy metal wastes can come from municipal wastes (UNEP, 2018) although other forms of such wastes can include industrial or mining wastes (Vongdala et al., 2019).

12.4.2 AVENUES OF TRANSFER OF THE HEAVY METALS FROM THE CARRIERS TO HUMANS

From the various media/carriers described above, heavy metals find their way into the human body mainly through two types of avenues, the occupational and environmental avenues. The occupational avenue is used to refer to a scenario in which heavy metals get in contact with humans as they engage with their daily undertakings while the environmental avenue is when humans get directly exposed to heavy metals present in their immediate surroundings. In any of the two scenarios, humans are exposed to the toxic metals mainly through any one of the three channels, namely topical (skin contact), respiratory (inhalation), and gastrointestinal (ingestion).

The occupational avenue is related to incidences of exposure due to the occupation of a person and examples include persons working in large pesticide manufacturing factories (Damalas and Eleftherohorinos, 2011), or people working on the cottage industries such as welding and metal smelting (Sethi et al., 2006), or in the primary mining sites (Qu et al., 2012). Where the workplace involves a confined space like an industrial workshop in an enclosed building, exposure via inhalable agents such as gases, vapors, dusts, and fumes is usually of more concern, especially where adequate ventilation cannot be guaranteed. In an iron/steel production facility, for example, metals such as lead, manganese, zinc, cadmium, chromium, and nickel can be emitted from a furnace as dust, fumes, or vapor. As they get emitted, these metals can be absorbed by particulate matter and get suspended in the surrounding air. Either way, the metals can find their way into or on surfaces of humans through inhalation or dermal contact (Sadovska, 2012; Mousavian et al., 2017).

The environmental avenue of exposure includes incidences in which humans are exposed to the heavy metals directly from contaminated surroundings including soils, atmospheric air, foodstuffs, and/or water. Exposure from soils can be through geophagy, the direct ingestion of soilborne heavy metals by soil-eating persons, especially pregnant women and children, the latter being a group of people with greater hand-to-mouth activities and gastrointestinal absorption (Calabrese et al., 1997). Similarly, transfer of heavy metals from foodstuffs can take various forms. In one form, transfer can be through ingestion of heavy metal contaminated edible plants (Robinson et al., 2009). Contaminated soil particulates suspended in the surrounding air can get into human's body either through inhalation or through dermal contact as they get deposited on the surface of a human body. In one study, elevated levels of heavy metals including arsenic, lead, and cadmium in the blood system of children were observed to be inversely proportional to distance of residence from the copper, lead, zinc smelting plants, respectively (Landrigan and Baker, 1981). In their study, the observed principal routes of exposure were a combination of inhalation and ingestion of heavy metal particulates emitted by the smelters into air, soil, and dust.

Humans may ingest soilborne heavy metals by direct consumption of soil or consumption of plants that either take up heavy metals into the edible portions or have the metals attached to the surfaces of the edible portions. Exposure can also occur through inhalation of suspended soil particulates and dermal contact (De Miguel et al., 1999, Kachenko and Singh, 2006).

12.4.3 TRANSFER OF HEAVY METALS FROM SOIL TO PLANTS TO HUMANS

From a vast list of mineral and non-mineral elements found in soils, only 17 are essential to plant life. Many of the non-essential elements contained in natural soils are usually toxic to plant life at very low concentrations and so plants have developed natural mechanisms to prevent them from being taken up through root absorption (Mishra and Dubey, 2006). Some of such selective mechanisms include plant-controlled avoidance strategies, such as extracellular production of root exudates, organic acids, chelation, and sequestration agents leading to a coordinated immobilization of toxic heavy metals into complexes that would not pass through sites of active or passive entry into the root system (Leita et al., 1996; Salt et al., 2000).

Despite the presence of avoidance mechanisms, there is evidence on the existence of a positive correlation between the concentration of heavy metals in the surrounding, especially the soil and the atmosphere, and the concentration of such metals in tissues of plants growing in that environment. Studies have shown, for example, a positive relationship between atmospheric heavy metal deposition and corresponding concentration of such heavy metals in plants actively grown in that environment (Ugulu et al., 2012) although evidence to the contrary has also been reported (Sağlam, 2013). Both the non-essential heavy metals such as Cd, Pb, As, and Hg and the micronutrient heavy metals ones that are required by plants in minute quantities including Cu^{2+}, Zn^{2+}, Mn^{2+}, Fe^{2+}, Ni^{2+}, and Co^{2+} can overwhelm the natural selectivity mechanisms of the root cell membranes to result into their elevated concentrations in the plant tissues where they become toxic (Krauss et al., 2002; Chojnacka et al., 2005). Depending on the ionic potential of the metal in question, uptake at the root surface can either be through an active transport or passive mechanism (Mishra and Dubey, 2006). During active transport of metals, root cell walls initially bind metal ions from the soil through high affinity binding sites. At the plasma membrane, metals are taken from the exterior to the interior through secondary transporters such as channel proteins and/or H^+-coupled carrier proteins (Mishra and Dubey, 2006).

From the root hairs, water, the heavy metals, and other solutes move down the absorption gradient toward the xylem of the root cortex via one of the three pathways—apoplastic, symplastic, and vacuolar pathways (Lux et al., 2011). Using the apoplastic pathway, water, suspended particles, and solutes can move from one location to another within a root or other organs through the continuum of cell walls before ever entering a cell. The apoplast involves solute transfer through extracellular fluid and gas spaces between and within cell walls of the root. The apoplast is, therefore, a fully permeable route in which the contents move by passive diffusion. In the symplastic pathway, on the other hand, water, suspended particles, and solutes are intracellularly transferred as they pass from cell to cell through specialized tubular channels known as plasmodesmata that connect the cytoplasm of adjacent cells. The symplastic pathway is, therefore, a selectively permeable pathway in which the movement of its contents occurs by osmosis. In addition to the apoplastic and symplastic pathway, water, heavy metals, and other solutes can be transported over long distances such as from the root to the higher parts of the shoot and leaves via the vacuolar pathway. Solutes including heavy metals move from one vacuole to another up the plant through specialized vacuolar membrane protein, the aquaporins (Martinoia et al., 2000).

In the course of uptake and transport within the plant, heavy metals usually get deposited in various plant parts. Depending on both the nature and type of plant and the heavy metal in question, deposition has been reported in various parts of the plant including the cell walls, in tissues of roots, stems, and leaves (Berthelsen et al., 1995; Tupan and Azrianingsih, 2016; Zhang et al., 2017). The research has also shown remarkable differences in amounts of metals plants can accumulate, transport, and distribute among their edible and non-edible parts (Singh et al., 2012; Sağlam, 2013; Sulaiman and Hamzah, 2016). Regardless of their point of deposition, heavy metals present in plants find their ways to humans either directly as humans consume the contaminated edible parts of the plant or indirectly as man consumes meat and other animal-based products from animals that were fed with heavy metal contaminated plants (Zhuang et al., 2008; Lwuanyanwu and Chioma, 2017) (Table 12.6).

12.4.4 Impact of Heavy Metals on Human Health

Toxicity of most heavy metals to humans and other mammalian systems is largely due to chemical reactivity of the metal ions with cellular structural proteins, enzymes, and membrane systems. Consequently, organs that accumulate the highest concentration of the metal in question *in vivo* are the primary target organs of specific metal toxicities. The magnitude of the toxicity and damage is usually dependent on both the route of exposure and the chemical nature of the metal toxin such as its valency state, volatility, and lipid solubility (Mahurpawar, 2015). Table 12.7 summarizes the target organs or systems and clinical manifestations of chronic exposures to selected heavy metals.

TABLE 12.6

Representative Quantities of Heavy Metals Contained in Various Media as Carriers of Heavy Metals Reported by Other Investigators

S/N	Medium/Carrier of Heavy Metals	Quantity of Heavy Metals Reported (ppm)	References
1	Industrial wastes	Mercury - 0.02–7.75 Lead - 98–13,700	
	a. Slag	Cadmium - 0.3–70.5 Chromium - 23–3170	Sankpal and Naikwade (2012) and Ramesh and Damodhram
	b. Effluent waters	Chromium - 0.65–1.72 Copper - 0.072–2.30 Iron - 6.62–19.38 Cadmium - 0.04–1.20 Zinc - 7.04–7.11 Lead - 2.1	(2016)

(Continued)

TABLE 12.6 (*Continued*)

Representative Quantities of Heavy Metals Contained in Various Media as Carriers of Heavy Metals Reported by Other Investigators

S/N	Medium/Carrier of Heavy Metals	Quantity of Heavy Metals Reported (ppm)	References
2	Water (receiving waters)		
	a. Lakes (e.g.)	Copper - 0.031–0.084	Bahnasawy et al. (2011)
	Lake Manzala-Egypt	Zinc - 0.185–0.433	
		Cadmium - 0.016–0.217	
		Lead - 0.009–0.042	
	b. Rivers		
	Bone river-Indonesia	Arsenic - 0.066–82.5	
		Mercury - 0.016–2.08	
	Ganga river-India	Lead - 0.018–1.67	
		Copper - 1.35–4.58	
		Zinc - 4.74–8.4	
		Lead - 0.24–0.85	
		Cromium - 0.32–0.85	
		Cadmium - 0.54–0.85	
3	Crops (grown on contaminated soil)	Iron - 7.23	
	Cabbage	Arsenic - 0.0211	
		Cadmium - 0.036	
		Copper - 0.314	Zwolak et al. (2019) and
		Manganese - 0.99	Tasrina et al. (2014)
		Zinc - 2.65–9.926	
		Lead - 0.12–0.671	
		Arsenic - 0.188	
	Carrot	Cadmium - 0.023–2.521	
		Copper - 0.227–5.570	
		Zinc - 1.591–52.590	
		Lead - 0.233–7.23	
		Arsenic - 0.014	
	Tomato	Cadmium - 0.028	
		Copper - 0.468	
		Zinc - 1.419	
		Lead - 0.078	
4	Phosphatic Fertilizers		
	a. Urea phosphates	Cadmium - 2.76	
		Lead - 0.4	
		Arsenic - 13.74	Chibueze et al. (2012)
	b. Diammonium phosphate (DAP)	Cadmium - 7.9	and AlKhader (2015)
		Lead - 2.1	
	c. Monoammonium phosphate (MAP)	Arsenic - 2.8	
		Cadmium - 0.5	
		Lead - 1.8	
	d. Single super phosphate (SSP)	Arsenic - 43.0	
		Cadmium - 6.1	
		Lead - 2.2	
		Arsenic - 5.5	
	e. NPK (30:10:10)	Cadmium - 0.42	
		Lead - 3.4	
		Arsenic - 7.85	

TABLE 12.7

Health Impacts of Exposure to Selected Heavy Metals

Heavy Metal	Target Organs in Human Body	Clinical Effect of Exposure	References
Arsenic	Pulmonary Nervous System, Skin, other internal organs (bladder, kidney, lungs and liver)	*Acute exposure*: Destruction of blood vessels, brain damage, gastrointestinal tissue damage *Chronic exposure*: Skin disorders such as altered pigmentation and keratosis, arsenicosis, development of various types of cancers (bladder, lung, liver and kidney cancers).	Martin and Griswold (2009) and Huy et al. (2014)
Cadmium	Lungs, Blood system, Pulmonary system, Renal system	*Acute exposure*: Severe pulmonary edema, development of chemical pneumonitis, respiratory failure, death (at high doses). *Chronic exposure*: Osteomyelitis; lung tumors, proximal tubule cell damage; proteinuria glycosuria, amino aciduria, polyuria and decreased absorption of phosphates; chronic obstructive lung disease; emphysema; Itai-Itai (ouch-ouch) disease.	Chakraborty et al. (2013) and Bernard (2008)
Zinc	Brain; Respiratory system; Gastrointestinal tract; Prostate	*Acute exposure*: Chemical pneumonitis; respiratory distress syndrome; emesis; fatigue, chills, fever, myalgias, cough, dyspnea, leukocytosis, thirst, metallic taste and salivation. corrosion of the gut, acute renal tubular necrosis and interstitial nephritis *Chronic exposure*: Interferes with the uptake of copper leading to symptoms of copper deficiency	Plum et al. (2010) and Wang et al. (2017)
Copper	Central nervous system (CNS); Gastrointestinal (GI) system; hepatic system	*Acute exposure*: GI mucosal ulcerations and bleeding; acute hemolysis and hemoglobinuria; hepatic necrosis with jaundice; nephropathy with azotemia; oliguria; cardiotoxicity with hypotension; tachycardia and tachypnea, dizziness, headache, convulsions, lethargy, stupor, and coma. *Chronic exposure* Liver damage	National Research Council (2000) and Taylor et al. (2020)
Mercury	Nervous system; Digestive system; Immune systems, lungs and kidneys	*Acute exposure* Tremors, insomnia, memory loss, neuromuscular effects, headaches and cognitive and motor dysfunction *Chronic exposure* Neuropsychiatric effects (i.e. tremor, anxiety, emotional lability, forgetfulness, insomnia, anorexia, erethism (abnormal irritation, sensitivity, or excitement), fatigue, and cognitive and motor dysfunction), renal impairment, and oropharyngeal inflammation; polyneuropathy and acrodynia (pink disease) in children	Mo et al. (2016) and Vahabzadeh and Balali-Mood (2016)

(Continued)

TABLE 12.7 (*Continued*)
Health Impacts of Exposure to Selected Heavy Metals

Heavy Metal	Target Organs in Human Body	Clinical Effect of Exposure	References
Nickel		*Acute exposure*	ATSDR (2005) and
		Respiratory tract irritation, chemical pneumonia,	Das et al. (2008)
		emphysema and varying degrees of hyperplasia of	
		pulmonary cells, and fibrosis (pneumoconiosis);	
		Severe lung damage; allergic dermatitis	
		Chronic exposure	
		Laryngeal cancer; kidney cancer; and cancer of the	
		prostate or bone; chronic bronchitis; lung cancer	
Lead	Blood system;	*Acute exposure*	Wani et al. (2015),
	Cardiovascular system,	Encephalopathy (with associated symptoms such as	ATSDR (2019), and
	Nervous system,	Ataxia; Coma; Convulsions; Death; Hyperirritability,	Bulka et al. (2020)
		and Stupor).	
		Chronic exposure	
		Severe cramping abdominal pain colic-like pain);	
		Ischemic coronary heart disease; Cerebrovascular	
		accidents, and Peripheral vascular disease	

12.5 MEASURES TO REDUCE BIOAVAILABILITY OF HEAVY METALS

12.5.1 REGULATION OF pH OF THE HOLDING MEDIUM SUCH AS SOIL

Reactivity of most heavy metals is strongly influenced by pH of the surrounding. In the soil, for example, metals tend to form insoluble metal mineral phosphates and carbonates at high pH conditions. However, at low pH, most metals exist as free ionic species or as soluble organometals making them more bioavailable (Sandrin and Hoffman, 2007; Rensing and Maier, 2003). This is because more protons (H^+) are available under acidic conditions enough to saturate metal-binding sites (on the phosphates) making the metals less likely to form insoluble precipitates with phosphates when the pH of the system is lowered (Huges and Poole, 1991). Under basic conditions, metal ions can replace protons to form other species, such as hydroxo-metal complexes.

Overall, regulating the pH to less acidic conditions reduces the bioavailability of heavy metals and hence prevents them from entering the food chain.

12.5.2 AGRICULTURAL MEASURES TO REDUCE METAL TRANSFER TO THE FOOD CHAIN

Another way of limiting the bioavailability of toxic heavy metals is through curtailing their transfer to the food chain. This can be carried out by 1st exploiting the natural differences that crops and varieties have in their heavy metal uptake. Choosing plants with low transfer factors (e.g., legumes, cereals) has been shown to reduce metal concentrations in edible parts of the crops significantly (Puschenreiter et al., 2005). This means that crops with heavy metal uptake capacities such as lettuce or spinach should be avoided especially on heavy metal contaminated hotspots.

The behavior of heavy metals in the soil is controlled by a myriad of complex interacting processes such as adsorption and desorption, complexation and dissociation, precipitation and dissolution, or very slow diffusion into the interior of clay minerals and oxides (Kim et al., 2015). All these processes are in turn influenced by physiochemical parameters such as soil pH, soil redox conditions, total metal content, and contents of organic matter, clay minerals, and oxides (Kim et al., 2015). Although each of these processes is metal specific, the ability to form stable complexes with organic and inorganic ligands is what makes the use of organic wastes in curtailing bioavailability of these toxic metals particularly effective. As a consequence, treating the heavy metal contaminated soils with organic and/or inorganic amendments has been reported to reduce the transfer of such metals to edible parts of the crop plants (Gul et al., 2015). The various types of amendments that have been previously used for this purpose include municipal solid waste compost, biosolid compost, cow manure, sheep manure, sewage sludge, bark chips, woodchips, vegetable waste, vermicompost, red mud, lime, beringite, zeolites, charcoal, fly ash; and biochar (Puschenreiter et al., 2005; Gul et al., 2015). In the study by Alam et al. (2020), organic fertilizers amendment such as vermicompost were shown to reduce uptake of Cd, Cr, Pb, and Mn heavy metals by radish and thus increase the crops growth and productivity. Similar results were also obtained with the application of fresh manure (Walker et al., 2003).

12.5.3   Extraction of the Metals from the Soil

Extraction of the toxic heavy metals from contaminated soils is another approach by which such metals could be prevented from entering the food chain and thus become bio-unavailable. There are several sophisticated and expensive methods of extracting heavy metals from contaminated soil, but use of phytoremediators has proven both relatively cost-effective and reliable. Phytoremediation of heavy metal contaminated soils using non-edible plant species such as trees, ornamentals, and grasses has been proposed as a safer, environmentally friendly, and cost-effective technique (Ji et al., 2011). Researchers have also shown the feasibility of using crop plants that accumulate less of the heavy metals in their edible parts but increasingly high levels of the metals in their non-edible parts as phytoremediation agents (Ciura et al., 2005; Murakami et al., 2009).

12.5.4   Use of Chemical Stabilizers to Prohibit Bioavailability

Several chemical stabilizers have been reported that work to reduce the bioavailability of heavy metals in soils and other media such as sewage sludge and sediments through in situ immobilization of heavy metals. This approach has gained a great deal of attention mainly because of usefulness as a less expensive technique that effectively reduces the risk of groundwater contamination, plant uptake, and exposure to other living organisms (Nejad et al., 2018). Chemical stabilizers that have been used to immobilize heavy metals in the soil include fulvic acid and phosphogypsum which work by significantly lowering the total and available concentration of Zn, Pb, Cd, and Ni in sewage sludge biochar (Huang et al., 2017); acid mine drainage sludge, lime stone, and steel slag all of which have been shown to reduce the bioavailable fraction of heavy metals in contaminated soil and their subsequent accumulation in earthworm (Yoon et al., 2019) or calcined cockle shell powder which was effective at immobilizing Cd, Pb, and Zn in mine tailing soil (Islam et al., 2017).

12.6   SUMMARY AND CONCLUSIONS

Soil is the major reservoir of heavy metals and thus plays an important role in cycling of metals in nature and their availability to food plants and transfer to food chain. Soils provide all food and fodder consumed by humans and animals, but they are also a *source of heavy meals, caused either by geogenic* process in soils or anthropogenic activities which lead to elevated levels of these metals in plant- or animal-based foods implicating the human health either positively or negatively. Metals added to soils through geogenic processes are generally less mobile and bioavailable than those added through anthropogenic activities. However, metals present in soils or added through anthropogenic activities undergo different chemical and physical reactions and hence their mobility and bioavailability are dependent on soil properties, such as soil pH, soil texture, soil organic matter and its form, oxidation–reduction potential, chemical speciation, and nature of contamination, iron and manganese oxides in soils, and thus, metal bioavailability in soils is largely dependent on the partition of the metals between the solid and solution phases and soil solution concentrations. Soil pH has the greatest effect of any single factor in the solubility or retention of heavy metal in soils but its effect is highly variable, depending on metal types, metal concentrations and contents, and the type of organic matter. Because of strong specific adsorption, special affinity for metal oxides and formation and precipitation of specific minerals may account for reducing potential toxic metals mobility in soils and thereby reducing their availability to food crops.

The most common media serving as carriers of metals include heavy metal contaminated wastes, waters, air, plants, and animal products, and their way into the human body occurs through occupational and environmental avenues. Water is a potential carrier of heavy metals and can easily be transmitted to humans. Contamination of important rivers, dams, lakes, and other inland water bodies used by humans for many domestic activities has been reported all over the world.

Plant foods are another medium saving as carrier of heavy metals as plants can accumulate high amounts of toxic heavy metals, heavy metal-laden substances, such as pesticides and fertilizers. Transfer of heavy metals can be through ingestion of heavy metal contaminated edible plants, or contaminated soil particulates suspended in the surrounding air can get into human's body either through inhalation or through dermal contact as they get deposited on the surface of a human body. Thus, metals present in plant foods enter humans either directly through consumption of contaminated edible plant or indirectly through animal-based products from animals that were fed with heavy metal contaminated plants. This shows that metals present in soils end up in plant and animal food and thus contributing negatively to human health.

Metals impact human health through their high accumulation in organs that are the primary target organs of specific metal toxicities such as cardiovascular and nervous systems (lungs, liver, and kidney, etc.). Metals entry to the food chain can be reduced by making them unavailable in soils by addition of organic and/or inorganic amendments and raising of soil pH by liming materials. At the same time, choosing plants with low transfer factors (e.g., legumes, cereals) has been shown to reduce metal concentrations in edible parts of the crops significantly. In addition, removal of metals from soils either by leaching with chemical agents has also been suggested. Phytoremediation using non-edible plant species such as trees, ornamentals, and grasses has been proposed as a safer, environmentally friendly, and cost-effective technique.

The complicated and intricate interactions involved between soil, plant, animal, and human health must be understood and further investigated through multidisciplinary approach by soil and plant scientists, biologists, public health authorities and medical doctors. This opens new avenues for soil and plant scientists to interact with public health authorities and medical professionals to find ways to enhance nutrient densities in edible parts for staple food crops and to make them more bio-accessible to animal and human body. Furthermore, linking soil to human health and making public aware of it is an important area of research, and this will be an essential component of creating funds for such types of research.

REFERENCES

Agency for Toxic Substances and Disease Registry (ATSDR). 2005. *Toxicological Profile for Nickel*. Atlanta, GA: U.S. Department of Health and Human Services, Public Health Service.

Alam, M., Hussain, Z., Khan, A., Khan, M. A., Rab, A., Asif, M., Shah, M.A., and Muhammad, A. 2020. The effects of organic amendments on heavy metals bioavailability in mine impacted soil and associated human health risk. *Scientia Horticulturae* 262:109067.

AlKhader, A.M.F. (2015). The impact of phosphorus fertilizers on heavy metals content of soils and vegetables grown on selected farms in Jordan. *Agrotechnology* 5:137. doi:10.4172/2168-9881.1000137

Angelova, V., Ivanov, R., Pevicharova, G., and Ivanov, K. 2010. *Effect of organic amendments on heavy metals uptake by potato plants*. 19th World Congress of Soil Science, Soil Solutions for a Changing World Brisbane, Australia. Published on DVD, 2010.

ATSDR. 2019. *Lead Toxicity. What Are Possible Health Effects from Lead Exposure?* Atlanta, GA: U.S. Department of Health and Human Services, Public Health Service.

Awokunmi, E. E. Asaolu, S. S., and Ipinmoroti, K. O. 2010. Effect of leaching on heavy metals concentration of soil in some dumpsites. *African Journal of Environmental Science and Technology* 4(8):495–499.

Bahnasawy, M., Khidr, A.A., and Dheina, N. (2011). Assessment of heavy metal concentrations in water, plankton, and fish of Lake Manzala, Egypt. *Turkish Journal of Zoology* 35(2):271–280.

Bernard, A. 2008. Cadmium & its adverse effects on human health. *Indian Journal of Medical Research* 128(4):557–564.

Berthelsen, B.O., Steinnes, E., Solberg, W. and Jingsen, L. 1995. Heavy metal concentrations in plants in relation to atmospheric heavy metal deposition. *Journal of Environmental Quality* 24(5):1018

Bhuyan, S., AbuBakar, M., Akhtar A., Hossain, M.B., Ali, M.M., Islam, S. 2017. Heavy metal contamination in surface water and sediment of the Meghna River, Bangladesh. *Environmental Nanotechnology, Monitoring & Management* 8:273–279.

Brevik, E.C. and Burgess, L.C. (Eds). 2013. *Soils and Human Health*. Boca Raton, FL: CRC Press.

Breward, N. 2007. Arsenic and presumed resistate trace element geochemistry of the Lincolnshire (UK) sedimentary ironstone, as revealed by regional geochemical survey using soil, water and stream sediment sampling. *Applied Geochemistry* 22:1970–1993.

Bulka, C.M., Bryan, M.S., Persky, V.W., Daviglus, M.L., Durazo-Arvizu, R.A., Parvez, F., Slavkovich, V., Graziano, J.H., Islam, JT., Baron, J.A., Ahsan, H., and Argos, M. 2020. Changes in blood pressure associated with lead, manganese, and selenium in a Bangladeshi cohort. *Environmental Pollution* 248:28–35.

Calabrese, E.J., Stanek, E.J., James, R.C., and Roberts S.M. 1997. Soil ingestion: a concern for acute toxicity in children. *Environmental Health Perspect* 105:1354–1358.

Cao, H.B., Chen J.J., Zhang J., Zhang H., Qiao L., Men, Y. 2010. Heavy metals in rice and garden vegetables and their potential health risks to inhabitants in the vicinity of an industrial zone in Jiangsu, China. *Journal of Environmental Sciences* 22(11):1792–1799.

Chakraborty, S., Dutta, A.R., Sural, S., Gupta, D., and Sen, S. 2013. Ailing bones and failing kidneys: a case of chronic cadmium toxicity. *Annals of Clinical Biochemistry* 50(5):492–495.

Chibueze, U., Akubugwo, E., Kingsley, N.A., Nnanna, A.L., Nwokocha, J.N., and Ekekwe, D.N. (2012). Appraisal of heavy metal contents in commercial inorganic fertilizers blended and marketed in Nigeria. *American Journal of Chemistry* 2(4):228–233.

Chojnacka, K., Chojnacki, A., Gorecka, H., and Gorecki, H. 2005. Bioavailability of heavy metals from polluted soils to plants. *Science of the Total Environment* 337:175–182.

Ciura, J., Poniedzialek, M., Sękara, A., and Jędrszczyk, E. 2005. The possibility of using crops as metal phytoremediants. *Polish Journal of Environmental Studies* 14(1):17–22.

da Silva, E. F., Mlayah, A., Gomes, C., Noronha, F., Charef, A., Sequeira, C., Esteves, V., and Marques, A. R. F. 2010. Heavy elements in the phosphorites from Kalaat Khasba mine (North-western Tunisia): Potential implications on the environment and human health. *Journal of Hazardous Materials*. doi:10,10,1016/j.hazmat.2010.06.020.

Damalas, C.A. and Eleftherohorinos, I.G. 2011. Pesticide exposure, safety issues, and risk assessment indicators. *International Journal of Environmental Research and Public Health* 8(5):1402–1419.

Das, K.K, Das, S.N and Dhundasi, S.A. 2008. Nickel, its adverse health effects & oxidative stress. *Indian Journal of Medical Research* 128:412–425.

De Miguel, E., Llamas, J.F., Chacón, E., Mazadiego, L.F. 1999. Sources and pathways of trace elements in urban environments: a multi-elemental qualitative approach. *Science of the Total Environment* 235:355–357.

Duncan, A.E., de Vries, N., and Nyarko, K.B. 2018. Assessment of heavy metal pollution in the sediments of the river pra and its tributaries. *Water Air Soil Pollut* 229(8): 272.

Duong, T. and Lee, B.K. 2011. Determining contamination level of heavy metals in road dust from busy traffic areas with different characteristics. *Journal of Environmental Management* 92(3):554–562.

Eckel, H., Roth, U., D€ohler, H., Nicholson, F. A., and Unwin, R. (Eds). 2005. *Assessment and Reduction of Heavy Metal Input into Agro-Ecosystems (AROMIS)*. Darmstadt, Germany: KTBL.

Fijałkowski, K., Kacprzak, M.,. Grobelak, A., and Placek, A. 2012. The influence of selected soil parameters on the mobility of heavy metals in soils. *Inżynieria i Ochrona Środowiska* 15:81–92.

Gang, W., Hubiao, K., Xiaoyang, Z., Hongbo, S., Liye, C., and Chengjiang, R. 2010. A critical review on the bio-removal of hazardous heavy metals from contaminated soils: Issues, progress, eco-environmental concerns and opportunities. *Journal of Hazardous Materials* 174:1–8.

Garrett, R. G., Porter, A. B., Hunt, P. A., and Lawlor, G. C. 2008. The presence of anomalous trace element levels in present day Jamaican soils and the geochemistry of Late-Miocene or Pliocene phosphorites. *Applied Geochemistry* 23:822–834.

Gharaibeh, A. A., El-Rjoob, A-W.O., and Harb, M.K. 2010. Determination of selected heavy metals in air samples from the northern part of Jordan. *Environmental Monitoring and Assessment* 160:425–429.

Ghosh, M. and Singh, S.P. 2005. A review on phytoremediation of heavy metals and utilization of its by-products. *Asian Journal on Energy and Environment* 6:214–231.

Giliba, R.A., Boon, E.K., Kayombo, C.J., Chirenje, L.I., Musamba, E.B., Kashindye, A.M., and Mushi, J.R. 2017. Assessment of heavy metals in some edible and fodder plants from Mazimbu Village, Morogoro, Tanzania. *Journal of Life Sciences* 3(2):93–96.

Gul, S., Naz, A., Fareed, I., and Irshad, M. 2015. Reducing heavy metals extraction from contaminated soils using organic and inorganic amendments – a review. *Polish Journal of Environmental Studies* 24(3):1423–1426.

Gunawardana, C., Goonetilleke, A., Egodawatta, P., Dawes, L., and Kokot, S. 2012. Source characterization of road dust based on chemical and mineralogical composition. *Chemosphere* 87:163–170.

He, C., Pan, F., and Yan, Y. 2011. Is economic transition harmful to China's urban environment? Evidence from industrial air pollution in Chinese cities. *Urban Studies* 49:1767–1790.

Huang, Z., Lu, Q., Wang, J., Chen, X., Mao, X., and He, Z. 2017. Inhibition of the bioavailability of heavy metals in sewage sludge biochar by adding two stabilizers. *PLoS One* 12(8):e0183617.

Huges, M.N. and Poole, R.K. 1991. Metal speciation and microbial growth-the hard (and soft) facts. *Journal of General Microbiology* 137:725–734.

Huy, T.B., Tuyet-Hanh, T.T., Johnston, R., and Nguyen-Viet, H. 2014. Assessing health risk due to exposure to arsenic in drinking water in Hanam Province, Vietnam. *International Journal of Environmental Research and Public Health* 11:7575–7591.

Islam, M.N., Taki, G., Nguyen, X.N., Jo, Y.T., Kim, Y., and Park, J.H. 2017. Heavy metal stabilization in contaminated soil by treatment with calcined cockle shell. *Environmental Science and Pollution Research* 24(8):7177–7183.

Jackson, T.A. 1970. Sources of heavy metal contamination in a river-lake system. *Environmental Pollution* 18(2):131–138.

Jaishankar, M., Mathew, B.B., Shah, M.S., and Gowda, K.R.S. 2014. Biosorption of few heavy metal ions using agricultural wastes. *Journal of Environment Pollution and Human Health* 2(1):1–6.

Jarup, L. 2003. Hazards of heavy metal contamination. *British Medical Bulletin* 68(1):167–182.

Ji, P., Sun, T., Song, Y., Ackland, M.L., and Liu, Y. 2011. Strategies for enhancing the phytoremediation of cadmium-contaminated agricultural soils by Solanum nigrum L. *Environmental Pollution* 159(3):762–768.

Kachenko, A.G. and Singh, B. 2006. Heavy metals contamination in vegetables grown in urban and metal smelter contaminated sites in Australia. *Water, Air and Soil Pollution* 169:101–123.

Kashem, M. A. and Singh, B. R. 2001a. Metal availability in contaminated soils: I. Effect of flooding and organic matter in changes in Eh, pH and solubility of Cd, Ni and Zn. *Nutrient Cycling and Agroecosystems* 61:247–255.

Kashem, M.A. and Singh, B.R. 2001b. Metal availability in contaminated soils: II. Uptake of Cd, Ni and Zn in rice plants grown under flooded culture with organic matter addition. *Nutrient Cycling in Agroecosystems* 61:257–266.

Kashem, M.A. and Singh, B.R. 2004. Transformations in solid phase species of metals as affected by flooding and organic matter. *Communications in Soil Science and Plant Analysis* 35:1435–1456.

Kashem, M.A., Singh, B.R., Kubota, H., Nagashima, R.S., Kitajima, N., Kondo, T., and Kawai, S. 2010. Effect of lherzolite on chemical fractionation of Cd and Zn in contaminated soils and on the growth in plants. *Water Air and Soil Pollution* 207:241–251.

Kavamura, N. V. and Esposito, E. 2010. Biotechnological strategies applied to the decontamination of soils polluted with heavy metals. *Biotechnology Advances* 28:61–69.

Kazemi Poshtmasari, H., Tahmasebi Sarvestani, Z., Kamkar, B., Shataei, S., and Sadeghi, S. 2012. Comparison of interpolation methods for estimating pH and EC in agricultural fields of Golestan province North of Iran. *International Journal of Agriculture and Crop Sciences* 44:157–167.

Khan, S., Quresho, M.A., and Singh, J. 1997. Influence of Heavy metal composition on the mobility of some micronutrients through soil. *Indian Journal of Environmental Health* 39(3):217–221.

Kim, H.C., Jang, T.W., Chae, H.J., Choi, W.J., Ha, M.H., Ye, B.J., Kim, B.W., Jeon, M.J., Kim, S.Y., and Hong, Y.S. 2015. Evaluation and management of lead poisoning. *Annals of Occupational and Environmental Medicine* 27:30.

Kloke, A., Sauerbeck, D.R., Vetter, H. (1984) The Contamination of Plants and Soils with Heavy Metals and the Transport of Metals in Terrestrial Food Chains. In: Nriagu, J.O. (ed.), *Changing Metal Cycles and Human Health. Dahlem Workshop Reports, Life Sciences Research Report*, vol 28. Berlin, Heidelberg: Springer, 113–141.

Krauss, M., Wilcke, W., Kobza, L., and Zech, W. 2002. Predicting heavy metal transfer from soil to plant: potential use of Freundlich-type functions. *Journal of Plant Nutrition and Soil Science* 165(1):3–8.

Landrigan, P.J. and Baker, E.J. 1981. Exposure of children to heavy metals from smelters: epidemiology and toxic consequences. *Environmental Research* 25(1):204–224.

Lee, S. Z., Allen, H. E., Huang, C.P., Sparks, D.L., Sanders, P.F., and Peijnenburg, W.J.G.M. 1996. Predicting soil water partition coefficients for cadmium. *Environmental Science and Technology* 30:3418–3424.

Lee, T. M., Lai, H. Y., and Chen, Z. S. 2004. Effect of chemical amendments on the concentration of cadmium and lead in long-term contaminated soils. *Chemosphere* 57:1459–1471.

Leita, L., Dc Nobili, M., Cesco, S., and Mondini, C. 1996. Analysis of intercellular cadmium forms in roots and leaves of bush bean. *Journal of Plant Nutrition* 19:527–533.

Li, J., Pu, L., Zhu, M., Zhang, J., Li, P., Dai, X., and Liu, L. 2014. Evolution of soil properties following reclamation in coastal areas: a review. *Geoderma* 226–227:130–139.

Li, T., Wan, Y., Ben, Y., Fan, S., and Hu, J. 2017. Relative importance of different exposure routes of heavy metals for humans living near a municipal solid waste incinerator. *Environmental Pollution* 226:385–393.

Lin, W., Lin, Y., and Wang, Y. 2016. A decision-making approach for delineating sites which are potentially contaminated by heavy metals via joint simulation. *Environmental Pollution* 211:98–110.

Lux, A., Martinka, M., Vaculík, M., and White, P.J. 2011. Root responses to cadmium in the rhizosphere: a review. *Journal of Experimental Botany* 62(1):21–37.

Lwuanyanwu, K.P. and Chioma, N.C. 2017. Evaluation of heavy metals content and human health risk assessment via consumption of vegetables from selected markets in Bayelsa State, Nigeria. *Biochem Anal Biochem* 6:3. doi:10.4172/2161-1009.1000332.

Mahurpawar, M. 2015. Effects of heavy metals on human health. *International Journal of Research – GRANTHAALAYAH* 530:1–7.

Malawska, M. and Wiłkomirski, B. 2000. Soil and plant contamination with heavy metals in the area of the old railway junction Tarnowskie Góry and near two main railway routes. *Rocz Panstw Zakl Hig* 51(3):259–267.

Mar, S. S. and Okazaki, M. 2012. Investigation of Cd contents in several phosphate rocks used for the production of fertilizer. *Microchemical Journal* 104:17–21.

Martin, S. and Griswold, W. 2009. Human health effects of heavy metals. *Environmental Science and Technology Briefs for Citizens* 15:1–6.

Martinoia, E., Massonneau, A., and Frangne, N. 2000. Transport processes of solutes across the vacuolar membrane of higher plants. *Plant and Cell Physiology* 41(11):1175–1186. doi:10.1093/pcp/pcd059.

Mellum, H.K., Arnesen, A.K., and Singh, B.R. 1998. Extractbility and plant uptake of heavy metals in alum shale soils. *Communications in Soil Science and Plant Analysis* 29:1183–1198.

Milieu Ltd., WRc, and RPA. 2008. Environmental, economic and social impacts of the use of sewage sludge on land (Final Report for the European Commission, DG Environment under Study Contract DG ENV.G.4/ETU/2008/0076r).

Miranda, M., Benedito, J. L., Blanco-Penedo, I., Lopez-Lamas, C., Merino, A., and Lopez- Alonso, M. 2009. Metal accumulation in cattle raised in a serpentine-soil area: Relationship between metal concentrations in soil, forage and animal tissues. *Journal of Trace Elements in Medicine and Biology* 23:231–238.

Mishra, S. and Dubey, R. S. 2006. Heavy metal uptake and detoxification mechanisms in plants. *International Journal of Agricultural Research* 1:122–141.

Mo, T., Sun, S., Wang, Y., Luo, D., Peng, B., and Xia, Y. 2016. Mercury poisoning caused by Chinese folk prescription (CFP): a case report and analysis of both CFP and quackery. *Medicine* 95(44):e5162.

Monjezi, M., Shahriar, K., Dehghani, H., and Samimi Namin, F. S. 2009. Environmental impact assessment of open pit mining in Iran. *Environmental Geology* 58:205–216.

Moreira, R.S., Ronaldo, L.M., and Santos, B.R. 2013. Heavy metals availability and soil fertility after land application of sewage sludge on dystroferric Red Latosol. *Agricultural Sciences* 37:61413–7054.

Mousavian, N. A., Mansouri, N., and Nezhadkurki, F. 2017. Estimation of heavy metal exposure in workplace and health risk exposure assessment in steel industries in Iran. *Measurement* 102:286–290.

Murakami, M., Nakagawa, F., Ae, N., Ito, M., and Arao, T. 2009. Phytoextraction by rice capable of accumulating Cd at high levels: reduction of Cd content of rice grain. *Environmental Science & Technology* 43(15):5878–5883.

Narwal, R.P. and Singh, B.R. 1995. Sorption of cadmium, zinc, copper and lead by soils developed on alum shales and other materials. *Norwegian Journal of Agricultural Sciences* 9:177–188.

National Research Council. 2000. *Copper in Drinking Water*. Washington, DC: The National Academies Press. doi:10.17226/9782/

Naveen, B.P., Sumalatha, J., and Malik, R.K. 2018. A study on contamination of ground and surface water bodies by leachate leakage from a landfill in Bangalore, India. *International Journal of Geo-Engineering* 9:27.

Nejad, Z.D., Jung, M. C., and Kim, K.H. 2018. Remediation of soils contaminated with heavy metals with an emphasis on immobilization technology. *Environmental Geochemistry and Health* 40(3):927–953.

Nicholson, F.A., Smith, S.R., Alloway, B.J., Carlton-Smith, C., and Chambers, B.J. 2003. An inventory of heavy metaltrace metal inputs to agricultural soils in England and Wales. *Science of the Total Environment* 311:205–219.

Nriagu, J.O. 1989. A global assessment of natural sources of atmospheric trace metals. *Nature* 338:47–49.

Nriagu, J.O. and Pacyna, J.F. 1988. Quantitative assessment of worldwide contamination of air, water, and soils by trace metals. *Nature* 333:134–139.

Ociepa, E., Kisiel, A., and Lach, J. 2010. Effect of fertilization with sewage sludge and composts on the change of cadmium and zinc solubility in soils. *Journal of Environmental Studies* 2:171–175.

Pardo, M. T. and Guadalix, M. E. 1996. Zinc sorption-desorption by two andept: effect of pH and support medium. *European Journal of Soil Science* 47:257–263.

Patel, P., Raju, N.J., Reddy, B.C.S.R., Suresh, U., Sankar, D.B., and Reddy, T.V.K. 2018. Heavy metal contamination in river water and sediments of the Swarnamukhi River Basin, India: risk assessment and environmental implications. *Environmental Geochemistry and Health* 40(2):609–623.

Plum, L.M., Rink, L., and Haase, H. 2010. The essential toxin: impact of zinc on human health. *International Journal of Environmental Research and Public Health* 7(4):1342–1365.

Puschenreiter, M., Horak, O., Friesl-Hanl, W., and Hartl, W. 2005. Low-cost agricultural measures to reduce heavy metal transfer into the food chain – a review. *Plant Soil and Environment* 51(1):1–11.

Qu, C-S., Ma, Z-W., Yang, J., Liu, Y., Bi, J., and Huang, L. 2012. Human exposure pathways of heavy metals in a lead-zinc mining area, Jiangsu Province, China. *PLoS One* 7(11):e46793.

Quezada-Hinojosa, R. P., Matera, V., Adatte, T., Rambeau, C., and F€ollmi, K. B. 2009. Cadmium distribution in soils covering Jurassic oolitic limestone with high Cd contents in the Swiss Jura. *Geoderma* 150:287–301.

Ramesh, P. and Damodhram, T. (2016). Determination of heavy metals in industrial waste waters of Tirupati Region, Andhra Pradesh. *International Journal of Science and Research* 5(5):2452–2455.

Rehman, W., Zeb, A., Noor, N., and Nawaz, M. 2008. Heavy metal pollution assessment in various industries of Pakistan. *Environmental Geology* 55(2):353–358.

Rensing, C. and Maier, R.M. 2003. Issues underlying use of biosensors to measure metal bioavailability. *Ecotoxicology and Environmental Safety* 56:140–147.

Rieuwerts, J. S., Ashnore, M. R., Farago, M. E., and Thornton, I. 2006. The influence of soil characteristics on the extractability of Cd, Pb and Zn in upland and moorland soils. *Science of the Total Environment* 366:864–875.

Rinklebe, J., Shaheen, S.M., and Frohne, T. 2016. Amendment of biochar reduces the release of toxic elements under dynamic redox conditions in a contaminated floodplain soil. *Chemosphere* 142:41–47.

Robinson, G. R., Larkins, P., Boughton, C. J., Bradley, W. R., and Sibrell, P. L. 2007. Assessment of contamination from arsenic pesticide use on orchards in the Great Valley Region, Virginia and West Virginia. *Journal of Environmental Quality* 36:654–663.

Sadovska, V. 2012. Health risk assessment of heavy metals adsorbed in particulates. *World Academy of Science, Engineering and Technology* 6:211–214.

Sağlam, C. 2013. Heavy metal accumulation in the edible parts of some cultivated plants and media samples from a volcanic region in Southern Turkey. *Ekoloji* 22(86):1–8. doi:10.5053/ekoloji.2013.861.

Salt, D.E., N. Kato, U. Kramer, R.D. Smith and I. Raskin, (2000). The role of root exudates in nickel hyperaccumulation and tolerance in accumulator and non accumulator species of Thlaspi. In: Terry, E. and Banuelos, G. (eds.), *Phytoremediation of Contaminated Soil and Water*. Boca Raton: Lewis Publishers Inc, 189–200.

Sandrin T.R. and Hoffman D.R. (2007) Bioremediation of organic and metal co-contaminated environments: Effects of metal toxicity, speciation, and bioavailability on biodegradation. In: Singh, S.N. and Tripathi, R.D. (eds.), *Environmental Bioremediation Technologies*. Berlin, Heidelberg: Springer, 1–37.

Sankpal, S.T. and Naikwade, P.V. (2012). Heavy metal concentration in effluent discharge of pharmaceutical industries. *Science Research Reporter* 2(1):88–90.

Sethi, P.K., Khandelwal, D., and Sethi, N. 2006. Cadmium exposure: health hazards of silver cottage industry in developing countries. *Journal of Medical Toxicology* 2(i):14–15.

Sheoran, V., Sheoran, A. S., and Poonia, P. 2009. Phytomining: a review. *Minerals Engineering* 22:1007–1019.

Singh, B.R., Narwal, R.P., Jeng, A.S., and Almås, A. 1995. Crop uptake and extractability of cadmium in soils naturally high in metals at different pH levels. *Communications in Soil Science and Plant Analysis* 26:2123–2142.

Singh, S., Zacharias, M., Kalpana S., and Mishra, S. 2012. Heavy metals accumulation and distribution pattern in different vegetable crops. *Journal of Environmental Chemistry and Ecotoxicology* 4(10):170–177.

Subash, C.S., Kashem, M.A., and Osman, K.T. 2012. Effect of lime and farmyard manure on the concentration of cadmium in water spinach (Ipomoea aquatic). *International Scholarly Research Network ISRN Agronomy* 2012:1–6.

Sulaiman, F.R. and Hamzah, H.A. 2016. Heavy metals accumulation in suburban roadside plants of a tropical area (Jengka, Malaysia). *Ecological Processes* 7:28. doi:10.1186/s13717-018-0139-3.

Taghlidabad, H. R. and Sepehr, E. 2018. Heavy metals immobilization in contaminated soil by grape-pruning-residue biochar. *Archives of Agronomy and Soil Science* 64:1041–1052.

Tan, X., Liu, Y., Gu, Y., Zeng, G., Wang, X., Hu, X., Sun, Z., and Yang, Z. 2015. Immobilization of Cd (II) in acid soil amended with different biochars with a long term of incubation. *Environmental Science and Pollution Research* 22:12597–12604.

Tang, G., Tangfu, X., Wang, S., Lei, J., Zhand, M., Yuanyuan, G., Li, H., Ning, Z., and He, L. 2009. High cadmium concentrations in areas with endemic fluorosis: a serious hidden toxin? *Chemosphere* 76:300–305.

Tasrina, R.C., Rowshon, A., Mustafizur, A.M.R., Rafiqul, I., Ali, M.P. (2014). Heavy metals contamination in vegetables and its growing soil. *International Journal of Environmental Analytical Chemistry* 2:142. doi:10.4172jreac.1000142

Taylor, A.A., Tsuji, J.S., Garry, M.R. McArdle, M.E., Goodfellow Jr., W.L., William, J. Adams, W.J., and Menzie, C.A. 2020. Critical review of exposure and effects: implications for setting regulatory health criteria for ingested copper. *Environmental Management* 65:131–159.

Teta, C. and Hikwa, T. 2017. Heavy metal contamination of ground water from an unlined landfill in Bulawayo, Zimbabwe. *Journal of Health and Pollution* 7(15):18–27.

Tupan, C. I. and Azrianingsih, R. 2016. Accumulation and deposition of lead heavy metal in the tissues of roots, rhizomes and leaves of seagrass Thalassia hemprichii (Monocotyledoneae, Hydrocharitaceae). *AACL Bioflux* 9(3):580–589.

Ugulu, I., Dogan, Y., Baslar, S., and Varol, O. 2012. Biomonitoring of trace element accumulation in plants growing at Murat Mountain. *International Journal of Environmental Science and Technology* 9:527–534.

United Nations Environment Programme (UNEP) Waste Management in ASEAN Countries. [(accessed on 03 November 2019)] Available online: https://wedocs.unep.org/bitstream/handle/20.500.11822/21134/waste_mgt_asean_summary.pdf?sequence=1&isAllowed=y.

Vahabzadeh, M. and Balali-Mood, M. 2016. Occupational metallic mercury poisoning in gilders. *The International Journal of Occupational and Environmental Medicine* 7(2):116–122.

Vongdala, N., Tran, H.-D., Xuan, T.-D., Teschke, R., and Khanh, T.D. 2019. Heavy metal accumulation in water, soil, and plants of municipal solid waste landfill in Vientiane, Laos. *International Journal of environmental Research and Public Health* 16(1):22.

Walker, D.J., Clemente, R., Roig, A., and Berna, M.P. 2003. The effects of soil amendments on heavy metal bioavailability in two contaminated mediterranean soils. *Environmental Pollution* 122(2):303–312.

Wang, C., Cheng, K., Zhou, L., He, J., Zheng, X., Zhang, L., Zhong, X., and Wang, T. 2017. Evaluation of long-term toxicity of oral zinc oxide nanoparticles and zinc sulfate in mice. *Biological Trace Element Research* 178(2):276–282.

Wani, A.L., Ara, A., and Usmani, J.A. 2015. Lead toxicity: a review. *Interdisciplinary Toxicology* 8(2):55–64.

Wu, Y., Wang, S., Streets, D.G., Hao, J., Chan, M. and Jiang, J. 2006. Trends in anthropogenic mercury emissions in China from 1995 to 2003. *Environmental Science & Technology* 40:5312–5318.

Yang, Y. G., Jin, Z. S., Bi, X. Y., Li, F. L., Sun, L., Liu, J., and Fu, Z. Y. 2009. Atmospheric deposition carried Pb, Zn, and Cd from a zinc smelter and their effect on soil microorganisms. *Pedosphere* 19:422–433.

Ying, L., Shaogang, L., and Xiaoyang, C. 2016. Assessment of heavy metal pollution and human health risk in urban soils of a coal mining city in East China. *Human and Ecological Risk Assessment: An International Journal* 22:1359–1374.

Yoon, D., Choi, W.S., Hong, Y.K., Lee, Y.B., and Kim, S.C. 2019. Effect of chemical amendments on reduction of bioavailable heavy metals and ecotoxicity in soil. *Applied Biological Chemistry* 62:53

Yuan, C., Shi, J., He, B., Liu, J., Lang, L., and Jiang, G. 2004. Speciation of heavy metals in marine sediments from the East China Sea by ICP-MS with sequential extraction. *Elsevier* 30:769–783.

Zhang, T., Bai, Y., Hong, X., Sun, L., and Liu, Y. 2017. Particulate matter and heavy metal deposition on the leaves of Euonymus japonicus during the East Asian monsoon in Beijing, China. *PLoS One* 12(6):e0179840.

Zhang, X., Zhong, T., Liu, L., and Ouyang, X. 2015. Impact of soil heavy metal pollution on food safety in China. *PLoS One* 10:e0135182.

Zeng, G., Wan, J., Huang, D. L., Hu, L., Huang, C., Cheng, M., Xue, W., Gong, X., Wang, R., and Jiang, D. 2017. Precipitation, adsorption and rhizosphere effect: the mechanisms for phosphate-induced Pb immobilization in soils – a review. *Journal of Hazardous Materials* 339:354–367.

Zhuang, P., McBride, M.B., Xia, H., Li, N., and, Li, Z. 2008. Health risk from heavy metals via consumption of food crops inthe vicinity of Dabaoshan mine, South China. *Science of The Total Environment* 407(5):1551–1561.

Žibret, G., Gosar, M., Miler, M., and Alijagić, J. 2018. Impacts of mining and smelting activities on environment and landscape degradation – slovenian case studies. *Land Degradation & Development* 29:4457–4470.

Zwolak, A., Sarzyńska, M., Szpyrka, E., et al. (2019). Sources of soil pollution by heavy metals and their accumulation in vegetables: a review. *Water Air Soil Pollution* 230:164.

13 Managing Soil Biology for Multiple Human Benefits

André L.C. Franco, Steven J. Fonte, and Diana H. Wall
Colorado State University

CONTENTS

13.1 INTRODUCTION: THE MANY BENEFITS THAT HUMANS OBTAIN FROM SOIL BIODIVERSITY

Globally there is increasing recognition that all life aboveground, including humans, is dependent on life in soils and the benefits soil biodiversity provides (Brondizio et al. 2019). Biodiversity, the variability among living organisms and the ecological complexes they are part of, underpins all goods and ecosystem services required for life above and belowground to survive (CBD 1994). Research on soil biology at all scales from DNA to ecosystem ecology has revealed an enormous wealth and variation of organisms interacting underground, including plant roots and vertebrates; however, in this chapter, we narrow our focus to discuss the complexity of processes mediated by microbes and invertebrate animals and adaptations to sustainably manage soil biodiversity (Coyle et al. 2017). While attention to soil biodiversity in intensively managed agricultural systems has primarily focused on those organisms providing either symbiotic benefits to plants such as bacteria, Rhizobia, fungal mycorrhizae, or pathogens and parasites of roots (some bacteria, fungi, plant parasitic nematodes, and insects), or the role of earthworms in compost and soil aeration, recent research has broadened to include new information on the wealth of soil biodiversity and the contribution to multiple ecosystem services (Brondizio et al. 2019).

An estimated 25% of earth's biodiversity inhabits soil (Decaëns et al. 2006; Bardgett and van der Putten 2014) and supports life above and belowground by providing ecosystem services like nutrient cycling, climate regulation, erosion control, water purification, disease regulation in plants and animals, including humans, and pollution remediation (Table 13.1). Additionally, the organisms themselves are a reservoir of food for wildlife, potential medicines, and novel discoveries for biotechnology (Bach et al. 2020). The explosion of global scientific research in the past 20 years is revealing who the organisms are, where they live, what they do as dynamic populations and in complex food webs and their response to climate change and misuse of land. This greater interest in soil biodiversity and interdisciplinary collaborations of scientists with indigenous and primary users

TABLE 13.1
Evidence Supporting the Mechanisms by Which Soil Invertebrates and Microbes Provide Essential Benefits to Society

Benefit	Mechanism	Reference
Nutrient cycling	Conversion of atmospheric N into readily assimilable compounds by microorganisms.	Postgate (1982) and Santos, Nogueira, and Hungria (2019)
	Microbial consumption by soil animals increases mineralization rates and enhanced litter decomposition.	Ingham et al. (1985) and García-Palacios et al. (2013a, 2013b)
Climate regulation	Soil biodiversity regulates greenhouse gas emissions and storage of C in soils through organic matter decomposition.	Tsiafouli et al. (2015), de Vries et al. (2013), Birgé et al. (2016), Creamer et al. (2016), Jackson et al. (2017), and Franco et al. (2020a)
Erosion control	Formation of stable soil aggregates and macropores by fungi and soil invertebrates (ants, termites, earthworms, coleopterans).	Jouquet et al. (2012) and Lehmann et al. (2017)
Water capture/storage	Creation of tunnels and air spaces by the movement of soil invertebrates that extend vertically and horizontally in the soil profile, thus facilitating infiltration of water to deeper soil layers.	Evans et al. (2011)
Disease regulation	Suppressive soils controlling plant pathogens.	Kerry (2000), Latz et al. (2016), Mendes et al. (2011), Postma et al. (2008), and Weller et al. (2002)
	Entomopathogenic fungi and nematodes to control insect and phytophagous mite populations.	Sinha, Choudhary, and Kumari (2016), Litwin, Nowak, and Różalska (2020), and Rasmann et al. (2005)
	Control of phytophagous mites, thrips and whiteflies by predatory soil- and leaf-inhabiting mites.	Gerson, Smiley, and Ochoa (2003) and Tixier (2018)
	Source of novel antibiotics such as penicillin.	Wall et al. (2015)
Pollution remediation	Earthworm vermicomposting alleviates contamination by antibiotic-resistance genes and human-pathogenic bacteria in wastewater sludge transformation though changes in the microbial community. It also accelerates the bio-degradation of chemical toxicants and bio-stabilization of heavy metals.	Huang et al. (2020), Hénault-Ethier, Martin, and Gélinas (2016), Soobhany (2018), Lv, Xing, and Yang (2018), and Bhat et al. (2018)
	Use of soil organisms as indicators of varying levels of contamination.	Park et al. (2016), Sharma, Cheng, and Grewal (2015), and Pankhurst et al. (1997)

forms a constructive basis for managing land sustainably for above and belowground biodiversity and potentially for multiple ecosystem services: information that is frequently ignored when scientific disciplines remain insular and conservation, restoration, and management plans focus primarily aboveground (Geisen, Wall, and van der Putten 2019; Adhikari and Hartemink 2016; Birgé et al. 2016; Bach et al. 2020).

Nutrient cycling and retention are governed by soil organisms through the movement and transformation of complex organic matter into inorganic nutrients (N, P) and soil biotic control of the rate of release and use by plants (Ingham et al. 1985; Robertson and Groffman 2007). A diversity of bacteria, fungi, protists, and invertebrates of various sizes alter the quality, size, and quantity of soil organic matter. Microbes are consumed by some soil animals thus increasing mineralization rates (Ingham et al. 1985). Research studies at molecular and taxonomic levels of microbes and invertebrates show the tight connections between nutrient cycling and above and belowground biology. A study of an agricultural field where soils were enriched with bacteria, fungi, protozoa, and nematodes showed increased nutrient uptake and plant yield and reduced nutrient losses (Bender and van der Heijden 2015). At larger biome and global scales, a metanalysis showed soil fauna enhanced litter decomposition by ~37%, although climate and litter quality had different effects between biomes (García-Palacios et al. 2013a, 2013b). These and other studies indicate that microbes and invertebrates together influence nutrient availability and retention in soil and provide benefits to plant growth, including food for people, in many ecosystems. In this chapter, we discuss how practical applications of soil biota can enhance the sustainable use of nutrient resources for food production and ecological restoration.

Climate regulation in soil is mediated by the abundance and biomass of soil microbes and invertebrates, their respiration and ability to facilitate C storage on mineral surfaces and in soil aggregates (Jackson et al. 2017). For example, globally, living soil nematode biomass is ~0.03 Gt C (van den Hoogen et al. 2019), which with other microbial and invertebrate taxa contributes a considerable portion of C to soils. de Vries et al. (2013) examined ecosystem services such as N retention, C storage, and water filtration across Europe and showed that they were consistently explained by soil food web composition and not just soil properties or land-use systems. A network analysis was used as a basis for determining how soil biodiversity, as related to ecosystem functions such as C and N cycling and soil quality, could be incorporated into land-use policy in a pan-European study (Creamer et al. 2016). Disruptions to soil physiochemical habitats reduce soil biodiversity and biomass and increase CO_2 release (Birgé et al. 2016). One of the largest losses of soil CO_2 across Europe was shown with intensive agriculture which reduced soil C stored and lowered the complexity of soil taxa and functional groups in food webs causing a decline in multiple ecosystem services such as climate and nutrient regulation and retention (Tsiafouli et al. 2015; de Vries et al. 2013). However, when restoration of soils occurs naturally, the number of taxa and functional groups in the food web become more connected and they take up more C (Morriën et al. 2017). In the subsequent sections of this chapter, we discuss how management strategies impact soil biota and their role on C cycling and climate regulation, as well as climate change feedbacks on processes governed by soil biota.

Erosion control and water regulation are ecosystem services that are generally attributed to bioturbators or ecosystem engineers, the visible soil invertebrates such as termites, earthworms, ants, and vertebrates. Their activities are all also critical for water capture and storage and contribute to the resilience of diverse ecosystems, especially rainfed agriculture (Evans et al. 2011). These animals are examples of taxa that perform multiple functions such as decomposition, soil structure formation, carbon sequestration, and water transport (Soliveres et al. 2016; Wagg et al. 2014; Thurman, Northfield, and Snyder 2019; Bach et al. 2020; Franco et al. 2020a). They transform soil and organic matter through the formation of stable soil aggregates and by their movement, creating tunnels and air spaces that extend vertically and horizontally in the soil profile, thus facilitating infiltration of water to deeper soil layers. The economic benefit of bioturbators such as earthworms in agricultural soils is being recognized (Plaas et al. 2019). Other organisms such as bacteria and fungi are often transported in or on earthworms and also participate in soil aggregation and the creation of associated soil pores, creation of smaller soil aggregates, while plant roots and water move through aerated channels. The bioturbators expand patches of available nutrients for plants across the landscape increasing availability of food for herbivores. The protection of soil through biotic structural stabilization is a vulnerable ecosystem service as climate change and land-use change continue, but it can be adapted for management (Birgé et al. 2016; Orgiazzi et al. 2016). Globally, projections indicate that erosion will increase in regions with high soil biodiversity, affecting availability of soils for plant growth (Guerra et al. 2020).

Regulation of disease is an ecosystem service supplied by biodiversity in natural systems, and soil biodiversity is no exception. With the advent of the 2020 pandemic, attention is drawn to the evidence for biological controls of soil pathogens and pests among people, other animals, and plants, and the need for fields specializing in soil organisms to share findings. Belowground food webs include beneficial microbes and invertebrates, and the predators and pathogens of other organisms. About 40 human diseases from soilborne pathogens occur globally (Jeffery and van der Putten 2011). Land and soil management resulting in loss of soil biodiversity has been linked to increase in human infectious diseases such as Lyme disease, hookworm, and anthrax (Patz et al. 2004; Wall, Nielsen, and Six 2015).

The search for beneficial bacteria, fungi, protists, and invertebrates for use as controls of pathogens, parasites, and pests continues as a major research area to regulate diseases (Ciancio, Pieterse, and Mercado-Blanco 2019; Meisner and De Boer 2018; Keesstra et al. 2016). Examples directly affecting human and animal health include soil-transmitted helminths (Bethony et al. 2006; Keiser and Utzinger 2019) and indirectly include suppressive soils controlling plant pathogens (Latz et al. 2016), entomopathogenic nematodes (EPNs) controlling insect pests (Rasmann et al. 2005), and an increasing number of studies showing differences in urban and farm soil biota and relationship to human disease and health (Steffan et al. 2018). Many biocontrols and other ways of managing soil biodiversity, discussed in this chapter, develop from ecological knowledge of soil biodiversity (Wall, Nielsen, and Six 2015; Bender, Wagg, and van der Heijden 2016; Geisen, Wall, and van der Putten 2019).

Soil pollution and contamination have implications for human, plant, and animal health because of naturally occurring elements that may be transferred through soil–plant food web interactions or metabolites produced by soil organisms or plants. Besides naturally occurring elements in soils that may cause deficiencies such as iodine in human diets or elements like lead that can be toxic, introduced pollutants from mining, industries, urban, and agriculture settings are also increasing (Bech 2020; Steffan et al. 2018). Less is known on whether soil organisms transform the multitude of contaminants but as taxa and food webs are affected, the use of soil organisms as indicators of varying levels of contamination continues (Park et al. 2016; Sharma, Cheng, and Grewal 2015). Disease regulation also includes soil organisms as a source of novel antibiotics, such as penicillin from the soil fungus *Penicillium* that naturally produces anti-microbial compounds. On the other hand, wide use of antibiotics has increased antibiotic resistance to pathogens in humans and livestock and through inputs of antibiotic manures have been shown to negatively affected soil microbial communities, nutrient cycling, making this a growing concern for humans, livestock, and wildlife (Wepking et al. 2019; Cycoń, Mrozik, and Piotrowska-Seget 2019; Bardgett and van der Putten 2014). Also, wide use of veterinary anthelmintics on livestock against parasitic infections, most of which are excreted in manure, can have negative effects on the survival and reproduction of soil organisms (González-Tokman et al. 2017), including non-target effects on populations of biocontrol agents such as EPNs (Barrón-Bravo et al. 2020).

Soil biota provide a source of food for wildlife and humans, and for soil invertebrates, this occurs whether they are transient occupants of soil during their life cycle or permanent residents of soils. Many invertebrates are a source of food and protein for wildlife and humans (Decaëns et al. 2006; Wall, Nielsen, and Six 2015; Govorushko 2019). The dependence of animals such as large vertebrates, birds, lizards, and other invertebrates on soil organisms as food in all ecosystems represent another benefit that soil biota provide, but this can be rapidly lost when soil is disturbed (Orgiazzi et al. 2016). In fact, soil food webs, based on organic matter and detritus, can help support and stabilize aboveground food webs by providing a relatively constant source of food and energy throughout the year, thus supplementing energy inputs from actively growing plants (and herbivores) which can vary markedly across seasons (Moore et al. 2004).

This chapter presents current and emerging evidence of practical ways in which soil biota is being used to improve human benefits from agricultural production, ecological restoration, organic waste treatment, and for informing ecosystem responses to anthropogenic global environmental changes. Challenges and opportunities are discussed for improved management and conservation, the further development of soil biota and their potential contributions and applications for solving multiple agricultural and environmental challenges. Given the continuing nature of these areas of research, this chapter is not meant to be a comprehensive review, rather special attention is given to the rapidly increasing knowledge on the benefits of soil invertebrates.

13.2 PROMOTING SPECIFIC ECOSYSTEM FUNCTIONS WITH TARGETED INTRODUCTIONS OF SOIL BIOTA

The goal of targeted introductions of soil organisms is to promote specific ecosystem services that enhance the environmental and economic sustainability of agriculture production, ecological restoration, and other processes such as organic waste treatment (Bender, Wagg, and van der Heijden 2016; Birgé et al., 2016; Smith and Collins, 2007). There exist numerous management approaches in which soil species or their products (e.g., vermicomposts) have been directly used to enhance nutrient availability and plant growth (e.g., N-fixing and P-mobilizing microbes), to provide biological control of pests and plant disease suppression (e.g., EPNs and fungi), and to detoxify and suppress human pathogens from environmental pollutants (e.g., vermicomposting of sewage sludge). Here, we present examples of targeted soil biota additions, some of which have become standard agriculture practices for many years, while others are only more recently being explored. We have attempted to demonstrate how soil organisms can benefit the sustainability of the systems where they or their direct products are applied and discuss some of the most recent developments.

13.2.1 INTRODUCTION OF SOIL ORGANISMS TO IMPROVE THE SUSTAINABLE USE OF NUTRIENT RESOURCES

13.2.1.1 Nitrogen

The benefits of soil biodiversity to nutrient management in agriculture are best illustrated by the long history of use of N-fixing microorganisms in legume crops (Santos, Nogueira, and Hungria 2019; Hungria et al. 2006). Nitrogen is a critical limiting element for plant growth and development, and the conversion of atmospheric N into readily assimilable compounds is carried out by specialized prokaryotes, including free-living soil bacteria of the genus *Azotobacter*, plant-associative bacteria such as *Azospirillum*, and bacteria that form symbioses with legumes and other plants, such as *Rhizobium* and *Bradyrhizobium* (Postgate 1982). Soybean is still the most inoculant-consuming crop worldwide (Santos, Nogueira, and Hungria 2019). In Brazil alone, 36.5 million hectares of soybean cropping area (approximately 78% of the total crop area) are inoculated every year (ANPII 2018; Santos, Nogueira, and Hungria 2019). The variety of inocula doses commercialized in the last decade has rapidly grown due to increased inoculation of other crops such as maize, cowpea, common beans, and co-inoculation of soybean and common bean with rhizobia and *Azospirillum* (Hungria et al. 2010, 2015).

While the ability to form N-fixing symbiosis with rhizobia has promoted the inclusion of legume plants in most cropping rotation systems worldwide, there is also growing interest to develop biological N fixation strategies for the world's major cereal crops—rice, wheat, and maize—which do not associate with rhizobia. These three grass crops combined account for approximately half of all synthetic N fertilizer produced (Ladha et al. 2016). Hence, there is much to be gained from developing soil organisms and inocula to support N fixation in non-legume crops (Ryu et al. 2020; Pankievicz et al. 2019). Recent studies clearly indicate that grass species can obtain enough N via biological fixation to promote robust plant growth (Pankievicz et al. 2015), showing that these plants hold promise for the expansion of associative N fixation and biological plant growth promotion. Significant progress has been made in understanding the multitude of physiological, biochemical, and ecological processes governing cereal crops associations with diazotroph (Van Deynze et al. 2018). Besides promoting plant growth through enhancing N availability, diazotrophs can increase of the acquisition of other nutrients, including phosphate solubilization, and stimulate plant hormones (Pérez-Montaño et al. 2014; Fukami, Cerezini, and Hungria 2018). The combined effects of all these mechanisms increase root growth and reduce plant nutrient deficiencies, which in turn can further increase plant access to N.

Larger soil fauna also enhance N mineralization (Postma-Blaauw et al. 2006) (Table 13.2). Earthworms can excrete large amounts of mineralized N (Blair et al. 1997), but their gut- and cast-associated processes also promote N mineralization indirectly by stimulating bacterial turnover and activity (Brown et al. 2004). Ants and termites increase the supply of soil mineral N that in turn support crop growth (Evans et al. 2011). In particular, termites have shown N-fixing capacity through microbial associations in field conditions (Ohkuma, Noda, and Kudo 1999; Breznak et al. 1973), suggesting that these macro-invertebrates may also increase N supply for plants. Other soil fauna

TABLE 13.2
Beyond Microbial Inoculants – Contribution of Soil Invertebrates to Nitrogen and Phosphorus Availability in Soils

Taxa	Mechanism	References
Nitrogen		
Earthworms	Excretion of large amounts of mineralized N, and indirect promotion of N mineralization by stimulating bacterial turnover and activity in the gut- and cast-associated processes.	Blair et al. (1997) and Brown et al. (2004)
Termites	N-fixing capacity through microbial associations.	Ohkuma, Noda, and Kudo (1999) and Breznak et al. (1973)
Nematodes	Microbivorous nematodes increase N mineralization rates through stimulating microbial turnover.	Gebremikael et al. (2016), Bouwman et al. (1994), Freckman (1988), Ekschmitt et al. (1999), Djigal et al. (2004), Ferris et al. (1998), and Hunt et al. (1987)
Phosphorus		
Earthworms	The gut alters soil mineral sorption complexes to remobilize mineral-associated P and enhance its availability.	Chapuis-Lardy et al. (2011) and Van Groenigen et al. (2019)
Termites	Termites transform organic P through the enzymatic activity in the fresh biostructures that forms its nests and gallery networks.	Chapuis-Lardy et al. (2011)
Coleopterans	The gut passage of some coleopterans causes significant solubilization and enzymatic hydrolysis of organic P, and desorption of mineral-associated inorganic P.	Li et al. (2006)
Nematodes	Inoculating phosphobacteria along with their invertebrate grazers (microbivorous nematodes) increases P availability.	Irshad et al. (2012) and Mezeli et al. (2020)

like microbivorous nematodes are known to increase N mineralization rates and plant growth when added in the laboratory and field (Gebremikael et al. 2016; Bouwman et al. 1994; Freckman 1988; Ekschmitt et al. 1999; Djigal et al. 2004; Ferris et al. 1998) and particularly increase N availability when added with other soil fauna (Hunt et al. 1987). Therefore, adding multiple taxa to improve the connections within the soil food web is a promising research area.

13.2.1.2 Phosphorus

Soil biota have been increasingly part of the solution to address unsustainable phosphorus (P) management through biotechnologies based on the capacity of certain organisms to remobilize mineral-associated soil P (Menezes-Blackburn et al. 2018; Richardson 2001; Mezeli et al. 2020; Coulis et al. 2014). Most crops rely heavily on P inputs to sustain production; however, low plant P use efficiency is pervasive in some soils due to natural chemical sorption properties of clay minerals and transformations that follow P fertilizer application. In some cases, meeting crop P requirements means saturating systems with P fertilizer, which is often cost-prohibitive and results in extremely low use efficiency of this diminishing mineral resource (Sharpley 1995; Stutter et al. 2015; Shen et al. 2011; Sharpley et al. 2018).

Soil organisms from microbes to micro- and macro-invertebrates are instrumental in facilitating plant access to mineral-associated P (Richardson 2001; Ros et al. 2017; Mezeli et al. 2020). While the benefits from soil fauna have yet to be scaled up for use in agriculture systems, microbes have been intensively screened for traits that promote the efficient mobilization of P, and some phosphobacteria and mycorrhizal inoculants have been available commercially for several years (Owen et al. 2015; Baas et al. 2016). These products typically rely on microorganisms that release metabolites to solubilize the mineral-associated P into plant-available forms and phosphatase enzymes that enhance organic P mineralization (Malboobi et al. 2009; Osorio and Habte 2014; Tawaraya, Naito, and Wagatsuma 2006; Richardson et al. 2009; Shropshire and Bordenstein 2016), exude plant hormones such as auxin (Spaepen 2015) and increase plant P uptake by stimulating plant root growth (Bal et al. 2013; Rashid, Charles, and Glick 2012). Microbial consortia (multiple species), as opposed to single species isolates have been developed to promote these multiple pathways by which microorganisms can increase available P and plant uptake (Baas et al. 2016).

One common limitation of microbial inoculants is that even when sufficient mineral-associated P is mobilized and made available by the microorganisms, the microbial biomass can fix a significant amount of P and thus impede P uptake and benefits to plant growth (Menezes-Blackburn et al. 2014). Although this can be beneficial in low pH soils with high P fixation capacity (Oberson et al. 2006), recent studies have shown that increasing trophic complexity by inoculating phosphobacteria along with their invertebrate grazers (microbivorous nematodes) can increase P availability and plant uptake, as well as shoot biomass and P concentration (Irshad et al. 2012). This evidence highlights the importance of trophic interactions to re-mineralize P from the microbial pool, thereby reducing competition between plants and inoculated microbes and increasing the cycling of mobilized P (Mezeli et al. 2020).

In addition, several studies have demonstrated a significant increase of plant-available P forms caused by the activity of soil macro-invertebrates such as earthworms (Ouedraogo et al. 2005), coleoptera (Li et al. 2006), and termites (Ruckamp et al. 2010) (Table 13.2). Findings from Li et al. (2006) indicated significant solubilization and enzymatic hydrolysis of organic P, and desorption of mineral-associated inorganic P during gut passage of scarabaeid beetles (Coleoptera), causing profound P transformations and increases in bio-available P in the soil. The gut content of earthworms has a higher pH compared to the soil they ingest; this alters mineral sorption complexes due to the competition for sorbing sites between phosphate and carboxyl groups of a mucus glycoprotein produced in its gut, finally enhancing P availability (Chapuis-Lardy et al. 2011; Van Groenigen et al. 2019). Termites transform organic P through the enzymatic activity in the fresh biostructures that forms its nests and gallery networks, producing high levels of labile P in the soils they inhabit (Chapuis-Lardy et al. 2011).

Unlocking part of the mineral-associated P already stocked in soils is fundamental to enhance the sustainability of the use of nutrient resources in agriculture (Menezes-Blackburn et al. 2018). Recent literature indicates that considering the ecological complexity of soil communities (that is the number and nature of interactions) as opposed to the diversity or abundance levels within specific groups may better advance our ability to utilize soil biology for sustainable P management in agricultural soils (Mezeli et al. 2020).

13.2.2 Introduction of Natural Enemies for Pest Control Services

EPNs, or soil nematodes that parasitize insects, have been utilized in biological control of insect pests for decades (Lacey and Georgis 2012), with EPN species from the genera *Steinernema* and *Heterorhabditis* being the most used in biocontrol programs (Leite et al. 2019). Some of these EPN species are generalist parasites that can infect hundreds of different insect species from more than ten different orders (e.g., *S. carpocapsae*; Hodson et al. 2011, 2012), while other EPN species are specialist parasites with a much narrower range of host insect species (e.g., *S. scapterisci* and *S. scarabaei*; Nguyen and Smart 1990; Stock and Koppenhöfer 2003). These soil nematodes have evolved a mutualistic association with bacteria, which is a deadly combination for insects, with the nematode as the mobile vector finding an insect host to invade and the bacteria multiplying as a food source for the nematode and also killing the insect. This causes septicemia in the hosts and provides an adequate environment for their reproduction (Leite et al. 2019). EPNs use chemical cues from roots of herbivore-damaged plants to locate potential hosts, often while insects are feeding either below- or above-ground (Ali, Alborn, and Stelinski 2010; Rasmann et al. 2005). Their high reproduction rate and short generation time allows for mass-production in culture media, and that associated with their ability to rapidly kill their hosts and cause limited harm to non-target organisms in most cases make them effective biological control agents (Sandhi et al. 2020; Dillman et al. 2012; Adams et al. 2006).

Use and development of EPNs have been increasing dramatically in recent years, driven by advances in production technology and mass-rearing, the continuous discovery of new efficacious species of EPNs, and the urgent need to reduce pesticide usage (Lacey and Georgis 2012). Some examples of commercial application of EPNs are the control of Diaprepes root weevil in citrus, fungus gnats in mushroom fields, mole crickets and scarab larvae in lawn and turf, and black vine weevil in nursery plants (Lacey and Georgis 2012). More recent studies have indicated the effectiveness of EPNs on the control of many other crop and public health pests, including control of disease-vector insects associated with major public health concerns such as *Aedes Aegypti*, the vector of dengue, zika, and chikungunya viruses (Suwannaroj et al. 2020). However, EPNs still represent only a small share of the pest control market (Subramanian and Muthulakshmi 2016), limited primarily by their higher cost relative to other chemical or biological options. With production costs dropping rapidly, EPNs have considerable potential to expand as a soil-biota based solution for biocontrol in the near future.

Other examples of soil organisms that have been tested and used for their insecticidal ability include entomopathogenic fungi (EPF) and predatory mites, among others. Organic farmers commonly use several species of EPF of the order Hypocreales, including species of the genera *Beauveria*, *Metarhizium*, *Isaria*, *Aschersonia*, *Hirsutella*, and *Lecanicillium*, to control insect and phytophagous mite populations in their fields (Sinha, Choudhary, and Kumari 2016; Litwin, Nowak, and Różalska 2020). The use of predatory soil- and leaf-inhabiting mites in open-field systems is still rare, but more common use for commercial biological control occurs in protected vegetable and ornamental cultivation systems to control phytophagous mites, thrips and whiteflies (Gerson, Smiley, and Ochoa 2003; Tixier 2018). Most of the commercially available predatory mites are leaf-inhabiting species from the Phytoseiidae family (Tixier 2018), while less research has been done on the application of soil predatory mites (e.g., Laelapidae and Macrochelidae families) as biocontrol agents (Knapp et al. 2018).

13.2.3 VERMICOMPOST AND OTHER ORGANIC MATERIAL APPLICATION

It is well established that earthworms improve soil properties and structure (Blouin et al. 2013), increase nutrient availability to plants (Van Groenigen et al. 2019), and stimulate populations of microbes and metabolites that promote plant growth (Tomati, Grappelli, and Galli 1988). Many studies have demonstrated that these beneficial effects of earthworms on soil physical, biological, and chemical attributes can have significant positive effects on crop yields in agriculture fields (as reviewed in van Groenigen et al. 2014). More recently, earthworms have been increasingly applied to produce vermicomposts by the breakdown of organic residues such as industrial and animal wastes, crop residues, and sewage sludge (Huang et al. 2020; Sharma and Garg 2019; Cen et al. 2020; Biruntha et al. 2020). In Europe, *Eisenia fetida*, *Eisenia andrei*, and *Dendrobaena veneta* are most commonly used earthworm species in vermicomposting (Boruszko 2020). These and other earthworms produce vermicompost by fragmenting the organic waste substrates, efficiently converting the organic residues into fine-structured humic substances with a more diverse and metabolically active microbial community and higher rates of nutrient mineralization (Zhao et al. 2020; Kolbe et al. 2019; Atiyeh et al. 2001).

The beneficial effects of vermicomposted organic wastes on plant growth of legumes and cereal crops, vegetables, ornamental and flowering plants have been consistently demonstrated (Huang et al. 2020; Atiyeh et al. 2001, 2002). They are commonly used as either a component of horticultural soil-less container media or soil additives to improve seed germination, seedling growth, and plant productivity through increases in nutrient availability and plant growth regulators. Vermicomposts can suppress plant pathogens (Szczech et al. 1993) and populations of plant parasitic nematodes (Singh et al. 2020), as well as enhance the activity of mycorrhizae fungi (Atiyeh et al. 2002). Reduced Al toxicity and improved cation exchange capacity via vermicompost application in urban horticulture soil have also been reported (Senapati et al. 1999).

The environmental benefits of vermicomposts go beyond the reduction of chemical inputs into soils. Recent studies have shown that vermicomposting alleviates contamination by antibiotic-resistance genes and human-pathogenic bacteria in wastewater sludge transformation though dramatic changes in the microbial community (Huang et al. 2020). Most studies indicated reductions of sludge human-pathogenic bacteria that could exceed 85% for total coliforms, 93% for fecal coliforms, and 99% for fecal enterococci during sludge vermicomposting (Huang et al. 2020; Hénault-Ethier, Martin, and Gélinas 2016; Soobhany 2018; Lv, Xing, and Yang 2018). Earthworms processing of sludge also accelerates the bio-degradation of chemical toxicants and bio-stabilization of heavy metals (Bhat et al. 2018). Therefore, vermicomposting technology is a demonstrated solution to detoxify wastewater sludge and other materials (e.g., coal ash generated in brick kilns from the construction industry in developing countries (Paul et al. 2020; Mondal et al. 2020)) into nutrient-rich soil fertility amendments for agricultural application.

13.2.4 SOIL TRANSPLANTATION TO RESTORE DEGRADED FIELDS

Soil biota are important drivers of plant community development (Wardle et al. 2004; Yang et al. 2018). Soil organisms and the plants they associate with interact continuously causing net positive and negative effects on plant survival, growth, and reproduction. These plant–soil interactions lay the basis of an ecological restoration technique called soil transplantation, where soil from a donor area (usually a nature reserve) is distributed (transplanted) thinly over the surface of a degraded area subjected to ecological restoration. By doing this, both the target soil organisms and plant seeds are transferred to the area under recovery. However, evidence shows that, even when differences in the seeds present in donor soils are excluded, donor-soil biota tend to drive plant community development into the direction of the vegetation composition found in the donor sites (Wubs et al. 2016).

Soil transplantation accelerates plant community development on former arable lands, using soil inocula from nature reserves to kick-start ecological restoration in the new location. In soil transplantation, the origin of the donor soil has great influence on plant community development. For example, introduction of late-successional soil communities stimulates the establishment of late-successional plant species (Vécrin and Muller 2003; Pywell et al. 2011; Buisson et al. 2018). Soils from arable lands, on the other hand, have a greater abundance of root-feeding invertebrates, and their use in mixing soil inocula suppresses early successional ruderal plants which are especially sensitive to antagonists (Wubs, Melchers, and Bezemer 2018). The effects of single introductions of soil biota via soil transplantation on plant communities last decades (Wubs et al. 2019), as soil biota networks become more connected as nature restoration progresses (Morriën et al. 2017).

Many examples have emerged showing that inoculation with soil biota from a donor area can restore ecosystems in different situations, including petroleum-contaminated soils (Yergeau et al. 2015), post-mining restoration in arid lands (Kneller et al. 2018), abandoned arable land (Wubs et al. 2016), and degraded Mediterranean grasslands (Buisson et al. 2018). Therefore, besides being increasingly used for steering the soil biota in crop lands, soil inoculations also offer a promising method for the restoration of degraded ecosystems (Mariotte et al. 2018). A deeper understanding of the specific mechanisms leading to these beneficial effects of soil inoculations may provide a general framework for local land managers selecting appropriate donor soils for ecological restoration projects.

13.3 MANAGEMENT CONSIDERATIONS FOR SUPPORTING WHOLE SOIL COMMUNITIES

Given the range of potential benefits from diverse soil organisms for human and ecosystem health, it is important to identify management levers for enhancing existing soil biodiversity and associated soil functions. While the direct additions of targeted soil organisms (discussed above) can contribute substantially towards meeting particular management goals, modification of soil habitats can result in large-scale transformation of whole soil communities (Brussaard, de Ruiter, and Brown 2007). Here we discuss management levers that are most common within food production systems.

13.3.1 TILLAGE

Of the common management practices in agricultural systems, tillage is perhaps the most disruptive for soil communities. Tillage takes many forms, but complete inversion of the soil (e.g., via mouldboard plow) or implements designed to thoroughly break apart soil clods (e.g., rototiller) are the most detrimental for soil biological communities due to the extensive loss of soil structure and associated microhabitats. Tillage can help aerate compacted surface soils and incorporate organic residues, thus stimulating decomposition by putting organic residues in greater contact with soil and associated microbial communities (Roger-Estrade et al. 2010). However, tillage also disrupts soil aggregates and macropore continuity, thus homogenizing surface layers and potentially reducing the diversity of habitats in the soil profile (Hendrix et al. 1986). In general, tillage is thought to have greater impacts on larger soil fauna (e.g., earthworms, arthropods, enchytraeids), that are physically damaged by mechanical disturbance or rely upon larger soil pores for movement and habitat (Kladivko 2001; Postma-Blaauw et al. 2010; Tsiafouli et al. 2015). However, intensive tillage also has long-term negative impacts for soil microbial biomass and activity (Zuber and Villamil 2016), suggesting that the impacts extend across entire soil food webs and have important consequences for both species and functional diversity in agricultural soils (Tsiafouli et al. 2015). Looking specifically at effects on soil biodiversity across a range of studies, de Graaff et al. (2019) found tillage to reduce overall soil fauna (micro-, meso-, and macrofauna) and bacterial diversity, but observed no significant effects on soil fungal diversity.

In light of the well-documented negative effects of tillage on soil biodiversity and biological activity and adverse impacts on soil water dynamics, erosion control and soil organic matter (Holland 2004; Bronick and Lal 2005), there has been considerable effort to develop no-till or reduced tillage systems, particularly in the USA. These systems rely more on herbicides and/or special farm equipment and the maintenance of a mulch layer to help control weeds. Using a meta-analysis approach, Briones and Schmidt (2017) found reduced tillage systems to significantly increase earthworm abundance and biomass, with no-till practices and those that maintain a surface residue layer showing the greatest benefit. Earthworms, as ecosystem engineers, can in turn have considerable impacts on soil structure and a range of other soil organisms through creation of new resources and habitats (Eisenhauer 2010). A number of studies have shown positive effects of reduced tillage practices (including no-till) on smaller soil organisms, including microarthropods and fungal feeding nematodes (Brennan, Fortune, and Bolger 2006; Tabaglio, Gavazzi, and Menta 2009; Hendrix et al. 1986), but in some cases, lower trophic levels (i.e., bacterial and fungal feeding organisms) can benefit from tillage (Wardle 1995; Okada and Harada 2007). While reducing tillage largely appears to benefit soil biological communities, the overall effect of tillage likely depends on the timing and intensity of disturbance, as well as the soil organisms in question (Kladivko 2001; Giller et al. 1997).

13.3.2 Nutrient Management

Nutrient inputs are a key driver of agroecosystem productivity and soil biological activity. Synthetic fertilizers, in particular, are widely used and contribute considerably to global agricultural production, but they are often thought to have negative impacts on a range of soil organisms, especially when used in higher quantities than recommended or necessary (Bach et al. 2018). Synthetic fertilizers can have indirect negative effects on soil biota, especially ammonium-based fertilizers that are known to acidify soil habitats (Bünemann, Schwenke, and Van Zwieten 2006). However, adding fertilizer to a nutrient limited system, may actually support some soil organisms indirectly via enhanced root and residue return associated with more productive, reduced-tillage crops. In fact, a meta-analysis by Geisseler and Scow (2014) suggested that synthetic fertilizers increase soil microbial biomass by an average of 15% in various agricultural systems and that this is associated with an increase in soil C. At the same time, de Graaff et al. (2019) found low rates of synthetic N inputs (<150kg ha^{-1}) to enhance bacterial diversity, but also to significantly reduce the diversity of arbuscular mycorrhizal fungi and soil fauna. This suggests that while moderate use of fertilizers can support crop production and biological activity, the effects on soil communities likely depend on the specific organism and trophic level being considered (Bender and van der Heijden 2015). The negative effects of synthetic fertilizers are perhaps most pronounced when they replace or are compared against organic fertilizers. Organic nutrient inputs (e.g., manure, compost, crop residues) not only provide essential nutrients for plant growth, but also organic C that serves as a fundamental energy source for soil food webs. A meta-analysis by Liu et al. (2016) suggested that nematode abundance, species richness and community structure all tend to increase with increasing C inputs. Others have suggested a similar trend for soil macrofauna communities, where abundance and diversity tend to increase with organic matter inputs (Mutema et al. 2013; Melman et al. 2019). These findings suggest that increased reliance on organic nutrient inputs, possibly with some judicious use of synthetic fertilizers, represents a much more important management lever for enhancing soil biological communities and the various soil functions they provide.

13.3.3 Pesticides

In contrast to fertilizer inputs, many pesticides are toxic to a variety of soil organisms and considerable improvements to soil biodiversity management could be achieved via reduced and more strategic use of pesticides and or use of biocontrol agents. Research on pesticides, however, is complicated by the fact that new pesticides are constantly being developed and much of the literature on pesticide impacts considers products that are now banned in many parts of the world (Pelosi et al. 2014). Regardless, we can make some generalizations, as many pesticides, including insecticides, fungicides, and nematicides were developed specifically to control aboveground or belowground pests that are broadly related to non-target organisms in the soil, can greatly impact soil communities. At the same time, effects of herbicides on soil fauna and microorganisms are less clear (Bünemann, Schwenke, and Van Zwieten 2006; Rose et al. 2016; Pisa et al. 2014), and more research is needed. While negative effects are reported, no-till agriculture offers one example where herbicides may not be so detrimental, since the benefits of reduced soil disturbance on soil communities often appear to outweigh the negative impacts of the herbicides that are used to avoid tillage (Hagner et al. 2019; Hendrix et al. 1986), but this likely depends on the type of herbicide in question, as well as the intensity of tillage it replaces (Rose et al. 2016).

13.3.4 IRRIGATION

Farmland under irrigated agriculture is expected to undergo large-scale shifts around the globe as demand for food increases, thus leading to the expansion of irrigated land in some areas, while at the same time, water availability is declining in other areas, resulting in a shift to rain fed systems (Elliott et al. 2014). Conversion to irrigated agriculture can increase microbial activity and the abundance and diversity of soil communities in dryland areas (Calderón et al. 2016; Li et al. 2018). Such improvements are likely related to greater soil organic matter inputs associated with increased productivity in irrigated systems. However, depending on the location, the maintenance of higher moisture levels during the growing season could increase pathogens and colonization of non-drought adapted biota. In addition to considering the complete conversion of irrigated to rain fed systems (or vice versa), water delivery timing and strategies in fields that are currently irrigated are like to change due to shifting priorities in regional water allocation, with important implications for soil communities. For example, Holland et al. (2013) observed that deficit irrigation in vineyards (a strategy to reduce overall crop water use) led to an increase in fungal activity, but a decrease in the abundance of protozoa, collembola, nematodes and mites. Similarly, a shift from furrow irrigation to more efficient drip irrigation strategies in California were shown to increase spatial heterogeneity in soils, with drier areas between crop rows displaying reduced soil biological activity (Schmidt et al. 2018).

13.3.5 GRAZING

The integration of livestock into cropping systems offers a number of potential benefits for soil communities, especially when perennial forages are included in rotation with annual crops (e.g., Lavelle et al., 2014), since they offer multiple years with minimal disturbance and high amounts of rhizosphere C inputs. However, grazing management (in both rotational systems and permanent pastures) plays a large role in determining the overall impact of grazing. Overgrazing is a concern for soil biodiversity, since it can lead to poor soil cover, erosion, and overall soil degradation (Andriuzzi and Wall 2017; Bardgett and Wardle 2003). This is especially relevant for permanent grazing lands, which are often situated in more marginal areas that are prone to erosion. Several studies have reported that intensive grazing can negatively impact the diversity and abundance of soil microarthropods and nematodes (Hu et al. 2015; Gruss et al. 2018); however, others have reported that intermediate levels of grazing may best support soil biological activity and diversity (Zhou et al. 2010).

13.3.6 DIVERSIFICATION OF CROPPING SYSTEMS

Diversification of agricultural systems offer considerable potential for enhancing soil biodiversity. Increased spatial diversity of plants has been shown to increase the diversity of multiple soil fauna (De Deyn et al. 2004; Sabais, Scheu, and Eisenhauer 2011; Veen et al. 2019) as well as soil microbial communities (LeBlanc, Kinkel, and Kistler 2015). This is likely due to increased heterogeneity of habitats and resource quality, but this may also be related to greater biomass production and soil

C inputs in more diverse plant mixtures (Fornara, Tilman, and Hobbie 2009). Increased temporal diversity of plants can also contribute to more active and diverse soil communities. For example, McDaniel et al. (2014) found that crop rotations increase soil microbial biomass relative to mono-cultures, and that this effect was further amplified by the inclusion of cover crops in crop rotations. Others showed inclusion of cover crops in rotations can significantly increase microbial activity and diversity (Kim et al. 2020). Spatial heterogeneity at farm and landscape scales can also contribute to diversity. For example, hedgerows and weedy field margins can serve as reservoirs of soil biodiversity (Smukler et al. 2010) and potentially contribute to recolonization of production fields after management related disturbance (Pardon et al. 2019). At larger scales, soil biodiversity may be influenced by overall landscape complexity and intensity (Diekötter et al. 2010; Culman et al. 2010).

13.4   THE THREAT OF ANTHROPOGENIC GLOBAL ENVIRONMENTAL CHANGES TO SOIL BIOTA AND ITS ECOSYSTEM FUNCTIONS

Human activities are causing many global environmental changes that threaten species and their functional roles in global ecosystems. Increasing atmospheric CO_2 concentrations, rising global temperatures, changing precipitation patterns, among other elements of global change, force organisms to adapt, disperse or go extinct, thus leading to major shifts in species composition, community structure, and ecosystem functioning (e.g., Chomel et al., 2019). Responses of belowground microbes and invertebrates are increasingly understood based on experimental and observational research across multiple temporal and spatial scales (Bastida et al. 2020). For example, we now have a better understanding of how nematodes, the most abundant animals on Earth, are affected from polar deserts to highly productive grasslands by anthropogenic global changes related to climate (warming (Mueller et al. 2016a; Andriuzzi et al. 2018; Thakur et al. 2017), droughts (Preisser and Strong 2004; Franco et al. 2019a), floods (Landesman, Treonis, and Dighton 2011; Song et al. 2016) and perturbations to biogeochemical cycles (atmospheric N deposition (Ettema, Lowrance, and Coleman 1999; Eisenhauer et al. 2012; Shaw et al. 2019), and elevated CO_2 (Hoeksema et al. 2000; Neher et al. 2004; Mueller et al. 2016a)).

Elevated CO_2 and N deposition tend to increase the domination of microbial feeders (bacterivores and fungivores) over predator and root feeders in nematode communities (Mueller et al. 2016a; Shaw et al. 2019) Elevated CO_2 can also reduce the effectiveness of EPNs as biocontrol agents (Hiltpold, Moore, and Johnson 2020). Temperature warming suppresses top soil predators (Bakonyi and Nagy 2000; Thakur et al. 2017), and this effect is attributed to the drying of soil microhabitats given that nematodes dwell in soil water films, and larger sized predators require thicker water films than smaller nematodes in lower trophic groups. Due to this strong dependence on water availability, changes in rainfall regimes are pointed as a dominant driver of shifts in nematode community composition and functioning (Blankinship, Niklaus, and Hungate 2011; Franco et al. 2019a). Changes in rainfall appear to disrupt the balance between the abundance of root-feeding nematodes, which are major constraints to ecosystem primary production, and their predators in favor of the root feeders (Franco et al. 2019a). These nematode responses increase in magnitude from arid to moist environments, and further aggravate the effects of drought on grassland primary productivity and above-belowground plant biomass partitioning (Franco et al. 2020b). Other examples have emerged of soil biota responses mediating the impacts of global change factors on ecosystems (see Table 13.3).

The effects of various types of land-use changes on soil biota from tropical to temperate ecosystems are well documented (de Vries et al. 2012; Paula et al. 2014; Emmerling 2014; Mendes et al. 2015; Vazquez et al. 2019; Franco et al. 2016). Deforestation and land-use intensification commonly cause the loss of specialist soil species across broad taxonomic groups, which in turn lead to decreased taxonomic and functional diversity and functional homogenization (Franco et al. 2019b; Mueller et al. 2016b; Clavel, Julliard, and Devictor 2011; Nordén et al. 2013; Rodrigues et al. 2013). These impacts together can have important implications for forest ecosystem function

TABLE 13.3

Evidence Supporting the Role of Soil Invertebrates and Microbes in Driving Ecosystem Responses to Global Environmental Changes

Global Change Factor	Influence of Soil Biota on Ecosystem Response	References
Drought	Termites maintain ecosystem functioning during periods of drought in grasslands and tropical rainforests.	Ashton et al. (2019) and Bonachela et al. (2015)
	Drought-induced increases in root-feeding nematode populations further aggravate its effects on grassland primary productivity and above-belowground plant biomass partitioning.	Franco et al. (2019a) and Franco et al. (2020b)
	Dung beetles alleviate the effects of predicted drought events on plants via physical manipulation of the soil matrix.	Johnson et al. (2016)
Warming	Warming reverses the effect of high spider densities on fungal-feeding Collembola and ultimately leads to slower decomposition rates in Artic tundra.	Koltz et al. (2018) and Creamer et al. (2015)
	The growth of distinct microbial communities under warming can alter the turnover and fate of organic matter and its response to temperature.	
N deposition	Responses of soil microbes to N deposition are positively associated with responses of plant biomass, soil C cycle, and soil N cycle globally.	Garcia-Palacios et al. (2015)
Elevated CO_2	Responses of fungal abundance to elevated CO_2 are positively correlated with those of plant biomass but negatively with those of the N cycle.	Garcia-Palacios et al. (2015)
Land-use change	Changes in food web structure of soil invertebrate communities alter biomass decomposition rates in disturbed tropical rainforests.	Cárdenas et al. (2017)
	Agricultural intensification can decrease soil biodiversity and alter food webs, with direct implications for ecosystem functioning.	de Vries et al. (2013), de Graaff et al. (2019), and Franco et al (2020a)

and resistance to climate change (Table 13.3), since soil organisms such as termites are key regulators of decomposition, nutrient heterogeneity, and moisture retention in forests (Ashton et al. 2019; Franco et al. 2020a). It appears that agriculture- and urbanization-related land-use changes are having comparable effects on the extinction risk of soil biota to other global change drivers related to atmospheric pollution and climate change (Veresoglou, Halley, and Rillig 2015; Eisenhauer, Bonn, and Guerra 2019), with pronounced functional consequences for ecosystems (Cárdenas et al. 2017).

Although we have gained important mechanistic insights from studies examining soil responses to a particular global change factor in isolation, soils and soil biota are simultaneously affected by a multitude of anthropogenic pressures and perturbations. It is becoming increasingly recognized that important interactive effects cannot be predicted from studying the independent effect of each global change factor separately (Tylianakis et al. 2008). Are soil biota responses exacerbated by synergistic interactions between anthropogenic drivers that affect their ecosystem functions more strongly than single factors (see theoretial example in Bardgett and Caruso, 2020)? Recent findings from Rillig et al. (2020) using a multifactor experiment, with as many as 10 simultaneous drivers, support the idea that the effect of many drivers on soil communities and the processes that they control is unpredictable from single- or two-factor studies. They found that some soil properties such as soil aggregation and soil water repellency responded unpredictably to the simultaneous exposure to multiple drivers, with stronger responses compared to the effects of individual drivers (Rillig et al.

2020). A previous experiment had manipulated four global change drivers and found both additive and interactive changes to soil biotic community structure and energy channels (Thakur et al. 2019). Such evidence strongly encourages research on the complex multifactor nature of global change and its impacts on soil biota and their ecosystem functions.

13.5 CONCLUDING COMMENTS

In this chapter, we aimed to synthesize standard and developing management practices that influence soil biodiversity and the ecosystem services provided by diverse soil biota. The examples discussed here show that soil biota are increasingly considered and used as a nature-based solution for improving agriculture production and sustainability, to restore degraded ecosystems, and to decontaminate soils and the environment. The science of soil ecology has moved far beyond the simple study of soil species and communities as indicators of environmental changes. Our rapidly increasing ecological understanding of life in soils has laid a solid foundation for the development of sustainable solutions to some pressing issues in our society and for future generations. The benefits of soil biota and their regulation of several ecosystem services are gained from both well-established and also emerging management approaches that either promote soil biodiversity indirectly by improving the soil habitat, or manipulate soil species and their products (e.g., vermicompost) in a targeted manner to promote specific ecosystem functions. In the most successful cases, improved management of soil biota may be able to largely replace the application of synthetic fertilizers and pesticides for plant nutrition and pest control, thus dramatically reducing environmental pollution caused by agriculture. Soil ecology has also provided solutions to detoxifying wastes from human activities, to steer ecological restoration in degraded ecosystems, and to increase our understanding of how ecosystems respond to global change pressures.

REFERENCES

Adams, B. J., A. Fodor, H. S. Koppenhöfer, E. Stackebrandt, S. P. Stock, and M. G. Klein. 2006. Biodiversity and Systematics of Nematode-Bacterium Entomopathogens. *Biological Control* 37 (1): 32–49. Academic Press Inc. doi:10.1016/j.biocontrol.2005.11.008.

Adhikari, K., and A. E. Hartemink. 2016. Linking Soils to Ecosystem Services – A Global Review. *Geoderma* 262: 101–111. Elsevier B.V. doi:10.1016/j.geoderma.2015.08.009.

Ali, J. G., H. T. Alborn, and L. L. Stelinski. 2010. Subterranean Herbivore Induced Volatiles Released by Citrus Roots upon Feeding by Diaprepes Abbreviatus Recruit Entomopathogenic Nematodes. *Journal of Chemical Ecology* 36 (4): 361–368. Springer. doi:10.1007/s10886-010-9773-7.

Andriuzzi, W. S., B. J. Adams, J. E. Barrett, R. A. Virginia, and D. H. Wall. 2018. Observed Trends of Soil Fauna in the Antarctic Dry Valleys: Early Signs of Shifts Predicted under Climate Change. *Ecology* 99: 312–321. doi:10.1002/ecy.2090.

Andriuzzi, W. S., and D. H. Wall. 2017. Responses of Belowground Communities to Large Aboveground Herbivores: Meta-Analysis Reveals Biome-Dependent Patterns and Critical Research Gaps. *Global Change Biology* 23 (9): 3857–3868. doi:10.1111/gcb.13675.

ANPII. 2018. Levantamento Do Uso de Inoculantes No Brasil. Associação Nacional Dos Produtores e Importadores de Inoculantes. In *VIII Congresso Brasileiro de Soja*. Brasilia: Embrapa, https://ainfo.cnptia.embrapa.br/digital/bitstream/item/178745/1/CBSoja-2018.pdf.

Ashton, L. A., H. M. Griffiths, C. L. Parr, T. A. Evans, R. K. Didham, F. Hasan, Y. A. Teh, H. S. Tin, C. S. Vairappan, and P. Eggleton. 2019. Termites Mitigate the Effects of Drought in Tropical Rainforest. *Science* 363: 174–177. http://science.sciencemag.org/.

Atiyeh, R. M., C. A. Edwards, S. Subler, and J. D. Metzger. 2001. Pig Manure Vermicompost as a Component of a Horticultural Bedding Plant Medium: Effects on Physicochemical Properties and Plant Growth. *Bioresource Technology* 78 (1): 11–20. doi:10.1016/S0960-8524(00)00172-3.

Atiyeh, R. M., S. Lee, C. A. Edwards, N. Q. Arancon, and J. D. Metzger. 2002. The Influence of Humic Acids Derived from Earthworm. *Bioresource Technology* 84: 7–14. http://www.prairieswine.com/pdf/3263.pdf.

Baas, P., C. Bell, L. M. Mancini, M. N. Lee, R. T. Conant, and M. D. Wallenstein. 2016. Phosphorus Mobilizing Consortium Mammoth PTM Enhances Plant Growth. *PeerJ* 2016 (6): e2121. PeerJ Inc. doi:10.7717/peerj.2121.

Bach, E. M., K. S. Ramirez, T. D. Fraser, and D. H. Wall. 2020. Soil Biodiversity Integrates Solutions for a Sustainable Future. *Sustainability* 12 (7): 2662. MDPI AG. doi:10.3390/su12072662.

Bach, E. M., R. J. Williams, S. K. Hargreaves, F. Yang, and K. S. Hofmockel. 2018. Greatest Soil Microbial Diversity Found in Micro-Habitats. *Soil Biology and Biochemistry* 118 (March): 217–226. Elsevier Ltd. doi:10.1016/j.soilbio.2017.12.018.

Bakonyi, G., and P. Nagy. 2000. Temperature- and Moisture-Induced Changes in the Structure of the Nematode Fauna of a Semiarid Grassland – Patterns and Mechanisms. *Global Change Biology* 6 (6): 697–707. John Wiley & Sons, Ltd (10.1111). doi:10.1046/j.1365-2486.2000.00354.x.

Bal, H. B., L. Nayak, S. Das, and T. K. Adhya. 2013. Isolation of ACC Deaminase Producing PGPR from Rice Rhizosphere and Evaluating Their Plant Growth Promoting Activity under Salt Stress. *Plant and Soil* 366 (1–2): 93–105. Springer. doi:10.1007/s11104-012-1402-5.

Bardgett, R. D., and T. Caruso. 2020. Soil Microbial Community Responses to Climate Extremes: Resistance, Resilience and Transitions to Alternative States. *Philosophical Transactions of the Royal Society B: Biological Sciences* 375 (1794). Royal Society Publishing. doi:10.1098/rstb.2019.0112.

Bardgett, R. D., and W. H. van der Putten. 2014. Belowground Biodiversity and Ecosystem Functioning. *Nature* 515 (7528): 505–511. Nature Publishing Group. doi:10.1038/nature13855.

Bardgett, R. D., and D. A. Wardle. 2003. Herbivore-Mediated Linkages between Aboveground and Belowground Communities. *Ecology* 84 (9): 2258–2268. Wiley-Blackwell. doi:10.1890/02-0274.

Barrón-Bravo, O. G., J. A. Hernández-Marín, A. J. Gutiérrez-Chávez, E. Franco-Robles, J. Molina-Ochoa, C. R. Cruz-Vázquez, and C. A. Ángel-Sahagún. 2020. Susceptibility of Entomopathogenic Nematodes to Ivermectin and Thiabendazole. *Chemosphere* 253 (August): 126658. Elsevier Ltd. doi:10.1016/j.chemosphere.2020.126658.

Bastida, F., D. J. Eldridge, S. Abades, F. D. Alfaro, A. Gallardo, L. García-Velázquez, C. García, et al. 2020. Climatic Vulnerabilities and Ecological Preferences of Soil Invertebrates across Biomes. *Molecular Ecology* 29 (4): 752–761. doi:10.1111/mec.15299.

Bech, J. 2020. Soil Contamination and Human Health: Part 1—Preface. *Environmental Geochemistry and Health* 42 (1): 1–6. Springer Netherlands. doi:10.1007/s10653-019-00513-1.

Bender, S. F., and M. G.A. van der Heijden. 2015. Soil Biota Enhance Agricultural Sustainability by Improving Crop Yield, Nutrient Uptake and Reducing Nitrogen Leaching Losses. *Journal of Applied Ecology* 52 (1): 228–239. Blackwell Publishing Ltd. doi:10.1111/1365-2664.12351.

Bender, S. F., C. Wagg, and M. G.A. van der Heijden. 2016. An Underground Revolution: Biodiversity and Soil Ecological Engineering for Agricultural Sustainability. *Trends in Ecology and Evolution* 31 (6): 440–452. Elsevier Ltd. doi:10.1016/j.tree.2016.02.016.

Bethony, J., S. Brooker, M. Albonico, S. M. Geiger, A. Loukas, D. Diemert, and P. J. Hotez. 2006. Soil-Transmitted Helminth Infections: Ascariasis, Trichuriasis, and Hookworm. *Lancet* 367 (9521): 1521–1532. doi:10.1016/S0140-6736(06)68653-4.

Bhat, S. A., S. Singh, J. Singh, S. Kumar, Bhawana, and A. P. Vig. 2018. Bioremediation and Detoxification of Industrial Wastes by Earthworms: Vermicompost as Powerful Crop Nutrient in Sustainable Agriculture. *Bioresource Technology* 252 (March): 172–179. Elsevier Ltd. doi:10.1016/j.biortech.2018.01.003.

Birgé, H. E., R. A. Bevans, C. R. Allen, D. G. Angeler, S. G. Baer, and D. H. Wall. 2016. Adaptive Management for Soil Ecosystem Services. *Journal of Environmental Management* 183 (December): 371–378. Academic Press. doi:10.1016/j.jenvman.2016.06.024.

Biruntha, M., N. Karmegam, J. Archana, B. K. Selvi, J. A. J. Paul, B. Balamuralikrishnan, S. W. Chang, and B. Ravindran. 2020. Vermiconversion of Biowastes with Low-to-High C/N Ratio into Value Added Vermicompost. *Bioresource Technology* 297 (February): 122398. Elsevier Ltd. doi:10.1016/j.biortech.2019.122398.

Blair, J. M., R. W. Parmelee, M. F. Allen, D. A. Mccartney, and B. R. Stinner. 1997. Changes in Soil N Pools in Response to Earthworm Population Manipulations in Agroecosystems with Different N Sources. *Soil Biology and Biochemistry* 29 (3–4): 361–367. Pergamon. doi:10.1016/S0038-0717(96)00098-3.

Blankinship, J. C., P. A. Niklaus, and B. A. Hungate. 2011. A Meta-Analysis of Responses of Soil Biota to Global Change. *Oecologia* 165 (3): 553–565. Springer-Verlag. doi:10.1007/s00442-011-1909-0.

Blouin, M., M. E. Hodson, E. A. Delgado, G. Baker, L. Brussaard, K. R. Butt, J. Dai, et al. 2013. A Review of Earthworm Impact on Soil Function and Ecosystem Services. *European Journal of Soil Science* 64 (2): 161–182. John Wiley & Sons, Ltd. doi:10.1111/ejss.12025.

Bonachela, J. A., R. M. Pringle, E. Sheffer, T. C. Coverdale, J. A. Guyton, K. K. Caylor, S. A. Levin, C. E. Tarnita. 2015. Termite mounds can increase the robustness of dryland ecosystems to climatic change. *Science* 347 (6222): 651–655. doi:10.1080/19443994.2014.939496.

Boruszko, D. 2020. Vermicomposting as an Alternative Method of Sludge Treatment. *Journal of Ecological Engineering* 21 (2): 22–28. Polish Society of Ecological Engineering (PTIE). doi:10.12911/22998993/116352.

Bouwman, L. A., J. Bloem, P. H. J. F. van den Boogert, F. Bremer, G. H. J. Hoenderboom, and P. C. de Ruiter. 1994. Short-Term and Long-Term Effects of Bacterivorous Nematodes and Nematophagous Fungi on Carbon and Nitrogen Mineralization in Microcosms. *Biology and Fertility of Soils* 17 (4): 249–256. . Springer-Verlag doi:10.1007/BF00383977.

Brennan, A., Tony F., and T. Bolger. 2006. Collembola Abundances and Assemblage Structures in Conventionally Tilled and Conservation Tillage Arable Systems. *Pedobiologia* 50: 135–145. doi:10.1016/j.pedobi.2005.09.004.

Breznak, J. A., W. J. Brill, J. W. Mertins, and H. C. Coppel. 1973. Nitrogen Fixation in Termites. *Nature* 244 (5418): 577–580. . Nature Publishing Group doi:10.1038/244577a0.

Briones, M. J. I., and O. Schmidt. 2017. Conventional Tillage Decreases the Abundance and Biomass of Earthworms and Alters Their Community Structure in a Global Meta-Analysis. *Global Change Biology* 23 (10): 4396–4419. doi:10.1111/gcb.13744.

Brondizio, E.S., J. Settele, S. Diaz, and H. T. Ngo, eds. 2019. *Global Assesment Report on Biodiversity and Ecosystem Services of the Intergovernmental Science-Policy Platform on Biodiversity and Ecosystem Services*. IPBES Secretariat, Bonn, Germany.

Bronick, C. J., and R. Lal. 2005. Soil Structure and Management: A Review. *Geoderma* 124 (1–2): 3–22. doi:10.1016/J.Geoderma.2004.03.005.

Brown, G. G., A. G. Moreno, I. Barois, C. Fragoso, P. Rojas, B. Hernández, and J. C. Patrón. 2004. Soil Macrofauna in SE Mexican Pastures and the Effect of Conversion from Native to Introduced Pastures. *Agriculture, Ecosystems and Environment* 103 (2): 313–327. Elsevier. doi:10.1016/j.agee.2003.12.006.

Brussaard, L., P. C. de Ruiter, and G. G. Brown. 2007. Soil Biodiversity for Agricultural Sustainability. *Agriculture, Ecosystems and Environment* 121 (3): 233–244. Elsevier. doi:10.1016/j.agee.2006.12.013.

Buisson, E., R. Jaunatre, C. Römermann, A. Bulot, and T. Dutoit. 2018. Species Transfer via Topsoil Translocation: Lessons from Two Large Mediterranean Restoration Projects. *Restoration Ecology* 26 (June): S179–S188. Blackwell Publishing Inc. doi:10.1111/rec.12682.

Bünemann, E. K., G. D. Schwenke, and L. Van Zwieten. 2006. Impact of Agricultural Inputs on Soil Organisms – A Review. *Australian Journal of Soil Research* 44 (4): 379–406. doi:10.1071/SR05125.

Calderón, F. J., D. Nielsen, V. Acosta-Martínez, M. F. Vigil, and D. Lyon. 2016. Cover Crop and Irrigation Effects on Soil Microbial Communities and Enzymes in Semiarid Agroecosystems of the Central Great Plains of North America. *Pedosphere* 26 (2): 192–205. doi:10.1016/S1002–0160(15)60034-0.

Cárdenas, R. E., D. A. Donoso, A. Argoti, and O. Dangles. 2017. Functional Consequences of Realistic Extinction Scenarios in Amazonian Soil Food Webs. *Ecosphere* 8 (2): e01692. doi:10.1002/ecs2.1692.

CBD1994. *Convention on Biological Diversity. Text and Annexes.* UNEP/CBD/94/1, Geneva: UNEP.

Cen, Y., L. Li, L. Guo, C. Li, and G. Jiang. 2020. Organic Management Enhances Both Ecological and Economic Profitability of Apple Orchard: A Case Study in Shandong Peninsula. *Scientia Horticulturae* 265 (April): 109201. Elsevier B.V. doi:10.1016/j.scienta.2020.109201.

Chapuis-Lardy, L., R. Le Bayon, M. Brossard, D. Lopez-Hernandez, and E. Blanchart. 2011. Role of Soil Macrofauna in Phosphorus Cycling. In *Phosphorus in Action: Biological Processes in Soil Phsphorus Cycling*, edited by E. K. Bunemann, A. Oberson, and E. Frossard, 477. Berlin, Germany: Springer.

Chomel, M., J. M. Lavallee, N. Alvarez-Segura, F. de Castro, J. M. Rhymes, T. Caruso, F. T. de Vries, et al. 2019. Drought Decreases Incorporation of Recent Plant Photosynthate into Soil Food Webs Regardless of Their Trophic Complexity. *Global Change Biology* 25 (10): 3549–3561. doi:10.1111/gcb.14754.

Ciancio, A., C. M. J. Pieterse, and J. Mercado-Blanco. 2019. Editorial: Harnessing Useful Rhizosphere Microorganisms for Pathogen and Pest Biocontrol-Second Edition. *Frontiers in Microbiology* 10 (August): 1–5. doi:10.3389/fmicb.2019.01935.

Clavel, J., R. Julliard, and V. Devictor. 2011. Worldwide Decline of Specialist Species: Toward a Global Functional Homogenization? *Frontiers in Ecology and the Environment* 9 (4): 222–228. doi:10.1890/080216.

Coulis, M., L. Bernard, F. Gérard, P. Hinsinger, C. Plassard, M. Villeneuve, and E. Blanchart. 2014. Endogeic Earthworms Modify Soil Phosphorus, Plant Growth and Interactions in a Legume-Cereal Intercrop. *Plant and Soil* 379 (1–2): 149–160. doi:10.1007/s11104-014-2046-4.

Coyle, D. R., U. J. Nagendra, M. K. Taylor, J. H. Campbell, C. E. Cunard, A. H. Joslin, A. Mundepi, C. A. Phillips, and M. A. Callaham. 2017. Soil Fauna Responses to Natural Disturbances, Invasive Species, and Global Climate Change: Current State of the Science and a Call to Action. *Soil Biology and Biochemistry* 110 (July): 116–133. Elsevier Ltd. doi:10.1016/j.soilbio.2017.03.008.

Creamer, C. A., A. B. de Menezes, E. S. Krull, J. Sanderman, R. Newton-Walters, and M. Farrell. 2015. Microbial community structure mediates response of soil C decomposition to litter addition and warming. *Soil Biology and Biochemistry* 80: 175–188. Elsevier Ltd. doi:10.1016/j.soilbio.2014.10.008.

Creamer, R. E., S. E. Hannula, J. P.Van Leeuwen, D. Stone, M. Rutgers, R. M. Schmelz, P. C. de Ruiter, et al. 2016. Ecological Network Analysis Reveals the Inter-Connection between Soil Biodiversity and Ecosystem Function as Affected by Land Use across Europe. *Applied Soil Ecology* 97: 112–124. doi:10.1016/j.apsoil.2015.08.006.

Culman, S. W, A. Young-Mathews, A. D. Hollander, H. Ferris, S. Sánchez-Moreno, A. T. O'geen, and L. E. Jackson. 2010. Biodiversity Is Associated with Indicators of Soil Ecosystem Functions over a Landscape Gradient of Agricultural Intensification. *Landscape Ecology* 25 (9): 1333–1348. doi:10.1007/s10980-010-9511-0.

Cycoń, M., A. Mrozik, and Z. Piotrowska-Seget. 2019. Antibiotics in the Soil Environment–Degradation and Their Impact on Microbial Activity and Diversity. *Frontiers in Microbiology* 10 (March). doi:10.3389/fmicb.2019.00338.

De Deyn, G. B., C. E. Raaijmakers, J. Van Ruijven, F. Berendse, and W. H. Van Der Putten. 2004. Plant Species Identity and Diversity Effects on Different Trophic Levels of Nematodes in the Soil Food Web. *Oikos* 106 (3): 576–586. doi:10.1111/j.0030-1299.2004.13265.x.

de Graaff, M. A., N. Hornslein, H. L. Throop, P. Kardol, and L. T.A. van Diepen. 2019. Effects of Agricultural Intensification on Soil Biodiversity and Implications for Ecosystem Functioning: A Meta-Analysis. *Advances in Agronomy* 155: 1–44. Elsevier Inc. doi:10.1016/bs.agron.2019.01.001.

de Vries, F. T., M. E. Liiri, L. Bjørnlund, M. A. Bowker, S. S. Christensen, H. M. Setälä, and R. D. Bardgett. 2012. Land Use Alters the Resistance and Resilience of Soil Food Webs to Drought. *Nature Climate Change* 2 (4): 276–280. Nature Publishing Group. doi:10.1038/NCLIMATE1368.

de Vries, F. T., E. Thébault, M. Liiri, K. Birkhofer, M. Tsiafouli, L. Bjørnlund, H. B. Jørgensen, et al. 2013. Soil Food Web Properties Explain Ecosystem Services across European Land Use Systems. *Proceedings of the National Academy of Sciences of the United States of America* 110 (35): 14296–14301. doi:10.1073/pnas.1305198110.

Decaëns, T., J. J. Jiménez, C. Gioia, G. J. Measey, and P. Lavelle. 2006. The Values of Soil Animals for Conservation Biology. *European Journal of Soil Biology* 42 (SUPPL. 1): S23–S28. doi:10.1016/j.ejsobi.2006.07.001.

Diekötter, T., S. Wamser, V. Wolters, and K. Birkhofer. 2010. Landscape and Management Effects on Structure and Function of Soil Arthropod Communities in Winter Wheat. *Agriculture, Ecosystems and Environment* 137 (1–2): 108–112. doi:10.1016/j.agee.2010.01.008.

Dillman, A. R., J. M. Chaston, B. J. Adams, T. A. Ciche, H. Goodrich-Blair, S. P. Stock, and P. W. Sternberg. 2012. An Entomopathogenic Nematode by Any Other Name. Edited by Glenn F. Rall. *PLoS Pathogens* 8 (3): e1002527. Public Library of Science. doi:10.1371/journal.ppat.1002527.

Djigal, D., A. Brauman, T.A. Diop, J.L. Chotte, and C. Villenave. 2004. Influence of Bacterial-Feeding Nematodes (Cephalobidae) on Soil Microbial Communities during Maize Growth. *Soil Biology and Biochemistry* 36 (2): 323–331. Pergamon. doi:10.1016/J.SOILBIO.2003.10.007.

Eisenhauer, N. 2010. The Action of an Animal Ecosystem Engineer: Identification of the Main Mechanisms of Earthworm Impacts on Soil Microarthropods. *Pedobiologia* 53 (6): 343–352. Elsevier GmbH. doi:10.1016/j.pedobi.2010.04.003.

Eisenhauer, N., A. Bonn, and C. A. Guerra. 2019. Recognizing the Quiet Extinction of Invertebrates. *Nature Communications* 10: 50. doi:10.1038/s41467-018-07916-1.

Eisenhauer, N., S. Cesarz, R. Koller, K. Worm, and P. B. Reich. 2012. Global Change Belowground: Impacts of Elevated CO2, Nitrogen, and Summer Drought on Soil Food Webs and Biodiversity. *Global Change Biology* 18 (2): 435–447. John Wiley & Sons, Ltd (10.1111). doi:10.1111/j.1365-2486.2011.02555.x.

Ekschmitt, K., G. Bakonyi, M. Bongers, T. Bongers, S. Boström, H. Dogan, A. Harrison, et al. 1999. Effects of the Nematofauna on Microbial Energy and Matter Transformation Rates in European Grassland Soils. *Plant and Soil* 212 (1): 45–61. Kluwer Academic Publishers. doi:10.1023/A:1004682620283.

Elliott, J., D. Deryng, C. Müller, K. Frieler, M. Konzmann, D. Gerten, M. Glotter, et al. 2014. Constraints and Potentials of Future Irrigation Water Availability on Agricultural Production under Climate Change. *Proceedings of the National Academy of Sciences of the United States of America* 111 (9): 3239–3244. doi:10.1073/pnas.1222474110.

Emmerling, C. 2014. Impact of Land-Use Change towards Perennial Energy Crops on Earthworm Population. *Applied Soil Ecology* 84 (December): 12–15. doi:10.1016/j.apsoil.2014.06.006.

Ettema, C. H., R. Lowrance, and D. C. Coleman. 1999. Riparian Soil Response to Surface Nitrogen Input: The Indicator Potential of Free-Living Soil Nematode Populations. *Soil Biology and Biochemistry* 31 (12): 1625–1638. Pergamon. doi:10.1016/S0038-0717(99)00072-3.

Evans, T. A., T. Z. Dawes, P. R. Ward, and N. Lo. 2011. Ants and Termites Increase Crop Yield in a Dry Climate. *Nature Communications* 2 (1): 1–7. Nature Publishing Group. doi:10.1038/ncomms1257.

Ferris, H., R.C. Venette, H.R. van der Meulen, and S.S. Lau. 1998. Nitrogen Mineralization by Bacterial-Feeding Nematodes: Verification and Measurement. *Plant and Soil* 203 (2): 159–171. Kluwer Academic Publishers. doi:10.1023/A:1004318318307.

Fornara, D. A., D. Tilman, and S. E. Hobbie. 2009. Linkages between Plant Functional Composition, Fine Root Processes and Potential Soil N Mineralization Rates. *Journal of Ecology* 97 (1): 48–56. doi:10.1111/j.1365–2745.2008.01453.x.

Franco, A.L.C., L. A. Gherardi, C. M. De Tomasel, W. S Andriuzzi, K. E. Ankrom, E. A. Shaw, E. M. Bach, O. E. Sala, and D. H. Wall. 2019a. Drought Suppresses Soil Predators and Promotes Root Herbivores in Mesic, but Not in Xeric Grasslands. *Proceedings of the National Academy of Sciences* 116: 12883–12888. doi:10.1073/pnas.1900572116.

Franco, A.L.C., B. W. Sobral, A. L.C. Silva, and D. H. Wall. 2019b. Amazonian Deforestation and Soil Biodiversity. *Conservation Biology* 33 (January): 590–600. John Wiley & Sons, Ltd (10.1111). doi:10.1111/cobi.13234.

Franco, A.L.C., M. R. Cherubin, C. E. P. Cerri, J. Six, D. H. Wall, and C. C. Cerri. 2020a. Linking soil engineers, structural stability, and organic matter allocation to unravel soil carbon responses to land-use change. *Soil Biology and Biochemistry*. 150: 107998. doi:10.1016/j.soilbio.2020.107998

Franco, A.L.C, L. A. Gherardi, C. M. De Tomasel, W. S. Andriuzzi, K. E. Ankrom, E. M. Bach, P. Guan, O. E. Sala, and D. H. Wall. 2020b. Root herbivory controls the effects of water availability on the partitioning between above and belowground grass biomass. *Functional Ecology*. doi:10.1111/1365-2435.13661.

Franco, A.L.C, M. L. C. Bartz, M. R. Cherubin, D. Baretta, C. E. P. Cerri, B. J. Feigl, D. H. Wall, C. A. Davies, and C. C. Cerri. 2016. Loss of Soil (Macro)Fauna Due to the Expansion of Brazilian Sugarcane Acreage. *The Science of the Total Environment* 563–564 (April): 160–168. doi:10.1016/j.scitotenv.2016.04.116.

Freckman, D. W. 1988. Bacterivorous Nematodes and Organic-Matter Decomposition. *Agriculture, Ecosystems & Environment* 24 (1–3): 195–217. Elsevier. doi:10.1016/0167–8809(88)90066-7.

Fukami, J., P. Cerezini, and M. Hungria. 2018. Azospirillum: Benefits That Go Far beyond Biological Nitrogen Fixation. *AMB Express* 8 (1): 1–12. Springer Verlag. doi:10.1186/s13568-018-0608-1.

García-Palacios, P., F.T. Maestre, J. Kattge, and D.H. Wall. 2013a. Climate and Litter Quality Differently Modulate the Effects of Soil Fauna on Litter Decomposition across Biomes. *Ecology Letters* 16 (8): 1045–1053. doi:10.1111/ele.12137.

García-Palacios, P., F.T. Maestre, J. Kattge, and D.H. Wall. 2013b. Corrigendum to García-Palacios et Al. [Ecol. Lett., 16, (2013) 1045–1053]. *Ecology Letters* 16 (11). doi:10.1111/ele.12179.

Garcia-Palacios, P., M. L. Vandegehuchte, E. A. Shaw, M. Dam, K. H. Post, K. S. Ramirez, Z. A. Sylvain, C. M. de Tomasel, D. H. Wall. 2015. Are there links between responses of soil microbes and ecosystem functioning to elevated CO_2, N deposition and warming? A global perspective. *Global Change Biology*, 21 (4): 1590–1600. doi:10.1111/gcb.12788.

Gebremikael, M. T., H. Steel, D. Buchan, W. Bert, and S. De Neve. 2016. Nematodes Enhance Plant Growth and Nutrient Uptake under C and N-Rich Conditions. *Scientific Reports* 6: 32862. doi:10.1038/srep32862.

Geisen, S., D. H. Wall, and W. H. van der Putten. 2019. Challenges and Opportunities for Soil Biodiversity in the Anthropocene. *Current Biology* 29 (19): R1036–R1044. Elsevier BV. doi:10.1016/j.cub.2019.08.007.

Geisseler, D., and K. M. Scow. 2014. Long-Term Effects of Mineral Fertilizers on Soil Microorganisms – A Review. *Soil Biology and Biochemistry* 75 (August): 54–63. Elsevier Ltd. doi:10.1016/j.soilbio.2014.03.023.

Gerson, U., R. L. Smiley, and T. Ochoa. 2003. *Mites (Acari) for Pest Control*. Oxford, UK: Blackwell Science.

Giller, K. E., M. H. Beare, P. Lavelle, A. M.N. Izac, and M. J. Swift. 1997. Agricultural Intensification, Soil Biodiversity and Agroecosystem Function. *Applied Soil Ecology* 6 (1): 3–16. Elsevier. doi:10.1016/S0929-1393(96)00149-7.

González-Tokman, D., I. Martínez, I. Villalobos-Ávalos, R. Munguía-Steyer, M. del Rosario Ortiz-Zayas, M. Cruz-Rosales, and J. P. Lumaret. 2017. Ivermectin Alters Reproductive Success, Body Condition and Sexual Trait Expression in Dung Beetles. *Chemosphere* 178 (July): 129–135. Elsevier Ltd. doi:10.1016/j.chemosphere.2017.03.013.

Govorushko, S. 2019. Economic and Ecological Importance of Termites: A Global Review. *Entomological Science* 22 (1): 21–35. Blackwell Publishing. doi:10.1111/ens.12328.

Gruss, I., K. Pastuszko, J. Twardowski, and M. Hurej. 2018. Effects of Different Management Practices of Organic Uphill Grasslands on the Abundance and Diversity of Soil Mesofauna. *Journal of Plant Protection Research* 58 (4): 376–380. doi:10.24425/jppr.2018.124652.

Guerra, C. A., I. M. D. Rosa, E. Valentini, F. Wolf, F. Filipponi, D. N. Karger, A. N. Xuan, J. Mathieu, P. Lavelle, and N. Eisenhauer. 2020. Global Vulnerability of Soil Ecosystems to Erosion. *Landscape Ecology* 35: 823–842. Springer. doi:10.1007/s10980-020-00984-z.

Hagner, M., J. Mikola, I. Saloniemi, K. Saikkonen, and M. Helander. 2019. Effects of a Glyphosate-Based Herbicide on Soil Animal Trophic Groups and Associated Ecosystem Functioning in a Northern Agricultural Field. *Scientific Reports* 9 (1): 1–13. Nature Publishing Group. doi:10.1038/s41598-019-44988-5.

Hénault-Ethier, L., V. J. J. Martin, and Y. Gélinas. 2016. Persistence of Escherichia Coli in Batch and Continuous Vermicomposting Systems. *Waste Management* 56 (October): 88–99. Elsevier Ltd. doi:10.1016/j.wasman.2016.07.033.

Hendrix, P. F., R. W. Parmelee, D. A. Crossley, D. C. Coleman, E. P. Odum, and P. M. Groffman. 1986. Detritus Food Webs in Conventional and No-Tillage Agroecosystems. *BioScience* 36 (6): 374–380. doi:10.2307/1310259.

Hiltpold, I., B. D. Moore, and S. N. Johnson. 2020. Elevated Atmospheric Carbon Dioxide Concentrations Alter Root Morphology and Reduce the Effectiveness of Entomopathogenic Nematodes. *Plant and Soil* 447 (1–2): 29–38. Springer. doi:10.1007/s11104-019-04075-0.

Hodson, A. K., M. L. Friedman, L. N. Wu, and E. E. Lewis. 2011. European Earwig (Forficula Auricularia) as a Novel Host for the Entomopathogenic Nematode Steinernema Carpocapsae. *Journal of Invertebrate Pathology* 107 (1): 60–64. Academic Press. doi:10.1016/j.jip.2011.02.004.

Hodson, A. K., J. P. Siegel, and E. E. Lewis. 2012. Ecological Influence of the Entomopathogenic Nematode, Steinernema Carpocapsae, on Pistachio Orchard Soil Arthropods. *Pedobiologia* 55 (1): 51–58. Urban & Fischer. doi:10.1016/j.pedobi.2011.10.005.

Hoeksema, J. D., J. Lussenhop, and J. A. Teeri. 2000. Soil Nematodes Indicate Food Web Responses to Elevated Atmospheric CO2. *Pedobiologia* 44 (6): 725–735. Urban & Fischer. doi:10.1078/S0031-4056(04)70085-2.

Holland, J. M. 2004. The Environmental Consequences of Adopting Conservation Tillage in Europe: Reviewing the Evidence. *Agriculture, Ecosystems and Environment* 103 (1): 1–25. doi:10.1016/j.agee.2003.12.018.

Holland, T. C., A. G. Reynolds, P. A. Bowen, C. P. Bogdanoff, M. Marciniak, R. B. Brown, and M. M. Hart. 2013. The Response of Soil Biota to Water Availability in Vineyards. *Pedobiologia* 56 (1): 9–14. Elsevier GmbH. doi:10.1016/j.pedobi.2012.08.004.

Hu, J., J. Wu, M. Ma, U. N. Nielsen, J. Wang, and G. Du. 2015. Nematode Communities Response to Long-Term Grazing Disturbance on Tibetan Plateau. *European Journal of Soil Biology* 69: 24–32. Elsevier Masson SAS. doi:10.1016/j.ejsobi.2015.04.003.

Huang, K., H. Xia, Y. Zhang, J. Li, G. Cui, F. Li, W. Bai, Y. Jiang, and N. Wu. 2020. Elimination of Antibiotic Resistance Genes and Human Pathogenic Bacteria by Earthworms during Vermicomposting of Dewatered Sludge by Metagenomic Analysis. *Bioresource Technology* 297 (February): 122451. Elsevier Ltd. doi:10.1016/j.biortech.2019.122451.

Huang, L., M. Gu, P. Yu, C. Zhou, and X. Liu. 2020. Biochar and Vermicompost Amendments Affect Substrate Properties and Plant Growth of Basil and Tomato. *Agronomy* 10 (2): 224. MDPI AG. doi:10.3390/agronomy10020224.

Hungria, M., J. C. Franchini, R. J. Campo, and P. H. Graham. 2006. The Importance of Nitrogen Fixation to Soybean Cropping in South America. In *Nitrogen Fixation in Agriculture, Forestry, Ecology, and the Environment*, 25–42. Springer-Verlag. doi:10.1007/1-4020-3544-6_3.

Hungria, M., R. J. Campo, E. M. Souza, and F. O. Pedrosa. 2010. Inoculation with Selected Strains of Azospirillum Brasilense and A. Lipoferum Improves Yields of Maize and Wheat in Brazil. *Plant and Soil* 331 (1): 413–425. Springer. doi:10.1007/s11104-009-0262-0.

Hungria, M., I. C. Mendes. 2015. Nitrogen Fixation with Soybean: The Perfect Symbiosis? Bioindicators of Soil Quality in Brazilina Agroecossytems View Project Additional Inoculations in Soybean Crop View Project. *Biological Nitrogen Fixation* 2 (July): 1009–1024. Hoboken, NJ, USA: John Wiley & Sons, Inc. doi:10.1002/9781119053095.ch99.

Hunt, H. W., D. C. Coleman, E. R. Ingham, R. E. Ingham, E. T. Elliott, J. C. Moore, S. L. Rose, C. P. P. Reid, and C. R. Morley. 1987. *The Detritai Food Web in a Shortgrass Prairie. Biology and Fertility of Soils.* Vol. 3. https://link.springer.com/content/pdf/10.1007%2FBF00260580.pdf.

Ingham, R. E., J. A. Trofymow, E. R. Ingham, and D. C. Coleman. 1985. Interactions of Bacteria, Fungi, and Their Nematode Grazers: Effects on Nutrient Cycling and Plant Growth. *Ecological Monographs* 55 (1): 119–140. doi:10.2307/1942528.

Irshad, U., A. Brauman, C. Villenave, and C. Plassard. 2012. Phosphorus Acquisition from Phytate Depends on Efficient Bacterial Grazing, Irrespective of the Mycorrhizal Status of Pinus Pinaster. *Plant and Soil* 358 (1–2): 155–168. doi:10.1007/s11104-012-1161-3.

Jackson, R. B., K. Lajtha, S. E. Crow, G. Hugelius, M. G. Kramer, and G. Piñeiro. 2017. The Ecology of Soil Carbon: Pools, Vulnerabilities, and Biotic and Abiotic Controls. *Annual Review of Ecology, Evolution, and Systematics* 48 (1): 419–445. doi:10.1146/annurev-ecolsys-112414-054234.

Jeffery, S., and W. van der Putten. 2011. *Soil Borne Human Diseases.* Luxembourg: European Commission, Joint Research Centre Scientific and Technical Report.

Johnson, S. N., G. Lopaticki, K. Barnett, S. L. Facey, J. R. Powell, and S. E. Hartley. 2016. An insect eco-system engineer alleviates drought stress in plants without increasing plant susceptibility to an above-ground herbivore. *Functional Ecology* 30 (6): 894–902. doi:10.1111/1365-2435.12582.

Jouquet, P., J. L. Janeau, A. Pisano, H. T. Sy, D. Orange, L. T. N. Minh, and C. Valentin. 2012. Influence of Earthworms and Termites on Runoff and Erosion in a Tropical Steep Slope Fallow in Vietnam: A Rainfall Simulation Experiment. *Applied Soil Ecology* 61: 161–168.

Keesstra, S. D., J. Bouma, J. Wallinga, P. Tittonell, P. Smith, A. Cerdà, L. Montanarella, et al. 2016. The Significance of Soils and Soil Science towards Realization of the United Nations Sustainable Development Goals. *Soil* 2: 111–128. doi:10.5194/soil-2-111-2016.

Keiser, J., and J. Utzinger. 2019. Community-Wide Soil-Transmitted Helminth Treatment Is Equity-Effective. *The Lancet* 393 (10185): 2011–2012. doi:10.1016/S0140-6736(18)32981-7.

Kerry, B. R. 2000. Rhizosphere Interactions and the Exploitation of Microbial Agents for the Biological Control of Plant-Parasitic Nematodes. *Annual Review of Phytopathology* 38: 423–441.

Kim, N., M. C. Zabaloy, K. Guan, and M. B. Villamil. 2020. Do Cover Crops Benefit Soil Microbiome? A Meta-Analysis of Current Research. *Soil Biology and Biochemistry* 142: 107701. Elsevier Ltd. doi:10.1016/j.soilbio.2019.107701.

Kladivko, E. J. 2001. Tillage Systems and Soil Ecology. *Soil and Tillage Research* 61 (1–2): 61–76. Elsevier. doi:10.1016/S0167-1987(01)00179-9.

Knapp, M., Y. van Houten, E. van Baal, and T. Groot. 2018. Use of Predatory Mites in Commercial Biocontrol: Current Status and Future Prospects. *Acarologia* 58 (Suppl): 72–82. Les Amis d'Acarologia. doi:10.24349/ACAROLOGIA/20184275.

Kneller, T., R. J. Harris, A. Bateman, and M. Muñoz-Rojas. 2018. Native-Plant Amendments and Topsoil Addition Enhance Soil Function in Post-Mining Arid Grasslands. *Science of the Total Environment* 621 (April): 744–752. Elsevier B.V. doi:10.1016/j.scitotenv.2017.11.219.

Kolbe, A. R., M. Aira, M. Gómez-Brandón, M. Pérez-Losada, and J. Domínguez. 2019. Bacterial Succession and Functional Diversity during Vermicomposting of the White Grape Marc Vitis Vinifera v. Albariño. *Scientific Reports* 9 (1): 1–9. Nature Publishing Group. doi:10.1038/s41598-019-43907-y.

Koltz, A. M., A. T. Classen, J. P. Wright. 2018. Warming reverses top-down effects of predators on below-ground ecosystem function in Arctic tundra. *Proceedings of the National Academy of Sciences of the United States of America*, 115 (32): E7541–E7549. doi:10.1073/pnas.1808754115.

Lacey, L. A., and R. Georgis. 2012. Entomopathogenic Nematodes for Control of Insect Pests above and below Ground with Comments on Commercial Production. *Journal of Nematology* 44 (2): 218–225. Society of Nematologists.

Ladha, J. K., A. Tirol-Padre, C. K. Reddy, K. G. Cassman, S. Verma, D. S. Powlson, C. Van Kessel, D. B. De Richter, D. Chakraborty, and H. Pathak. 2016. Global Nitrogen Budgets in Cereals: A 50-Year Assessment for Maize, Rice, and Wheat Production Systems. *Scientific Reports* 6 (1): 1–9. Nature Publishing Group. doi:10.1038/srep19355.

Landesman, W. J., A. M. Treonis, and J. Dighton. 2011. Effects of a One-Year Rainfall Manipulation on Soil Nematode Abundances and Community Composition. *Pedobiologia* 54 (2): 87–91. doi:10.1016/j.pedobi.2010.10.002.

Latz, E., N. Eisenhauer, B. Christian Rall, S. Scheu, and A. Jousset. 2016. Unravelling Linkages between Plant Community Composition and the Pathogen-Suppressive Potential of Soils. *Scientific Reports* 6 (March): 1–10. Nature Publishing Group. doi:10.1038/srep23584.

Lavelle, P., N. Rodriguez, O. Arguello, J. Bernal, C. Botero, P. Chaparro, Y. Gomez, et al. 2014. Soil Ecosystem Services and Land Use in the Rapidly Changing Orinoco River Basin of Colombia. *Agriculture Ecosystems & Environment* 185: 106–117. doi:Doi 10.1016/J.Agee.2013.12.020.

LeBlanc, N., L. L. Kinkel, and H. C. Kistler. 2015. Soil Fungal Communities Respond to Grassland Plant Community Richness and Soil Edaphics. *Microbial Ecology* 70 (1): 188–195. doi:10.1007/s00248-014-0531-1.

Lehmann, A., W. Zheng, and M. C. Rillig, (2017). Soil biota contributions to soil aggregation. *Nature Ecology & Evolution* 1: 1828–1835. doi:10.1038/s41559-017-0344-y.

Leite, L. G., J. E. Marcondes de Almeida, J. G. Chacon-Orozco, and C. Y. Delgado. 2019. Entomopathogenic Nematodes. In *Natural Enemies of Insect Pests in Neotropical Agroecosystems*, 213–221. Cham, Switzerland: Springer International Publishing. doi:10.1007/978-3-030-24733-1_18.

Li, F. R., J. L. Liu, W. Ren, and L. L. Liu. 2018. Land-Use Change Alters Patterns of Soil Biodiversity in Arid Lands of Northwestern China. *Plant and Soil* 428 (1–2): 371–388. Plant and Soil. doi:10.1007/s11104-018-3673-y.

Li, X. Z., R. Ji, A. Schaffer, and A. Brune. 2006. Mobilization of Soil Phosphorus during Passage through the Gut of Larvae of Pachnoda Ephippiata (Coleoptera : Searabaeidae). *Plant and Soil* 288 (1–2): 263–270. doi:10.1007/S11104-006-9113-4.

Litwin, A., M. Nowak, and S. Różalska. 2020. Entomopathogenic Fungi: Unconventional Applications. *Reviews in Environmental Science and Biotechnology* 19 (1): 23–42. Springer. doi:10.1007/s11157-020-09525-1.

Liu, T., X. Chen, F. Hu, W. Ran, Q. Shen, H. Li, and J. K. Whalen. 2016. Carbon-Rich Organic Fertilizers to Increase Soil Biodiversity: Evidence from a Meta-Analysis of Nematode Communities. *Agriculture, Ecosystems and Environment* 232: 199–207. doi:10.1016/j.agee.2016.07.015.

Lv, B., M. Xing, and J. Yang. 2018. Exploring the Effects of Earthworms on Bacterial Profiles during Vermicomposting Process of Sewage Sludge and Cattle Dung with High-Throughput Sequencing. *Environmental Science and Pollution Research* 25 (13): 12528–12537. Springer Verlag. doi:10.1007/s11356-018-1520-6.

Malboobi, M. A., P. Owlia, M. Behbahani, E. Sarokhani, S. Moradi, B. Yakhchali, A. Deljou, and K. M. Heravi. 2009. Solubilization of Organic and Inorganic Phosphates by Three Highly Efficient Soil Bacterial Isolates. *World Journal of Microbiology and Biotechnology* 25 (8): 1471–1477. Springer. doi:10.1007/s11274-009-0037-z.

Mariotte, P., Z. Mehrabi, T. Martijn Bezemer, G. B. De Deyn, A. Kulmatiski, B. Drigo, G. F. (Ciska) Veen, M. G. A. van der Heijden, and P. Kardol. 2018. Plant–Soil Feedback: Bridging Natural and Agricultural Sciences. *Trends in Ecology and Evolution* 33 (2): 129–142. Elsevier Ltd. doi:10.1016/j.tree.2017.11.005.

McDaniel, M. D., L. K. Tiemann, and A. S. Grandy. 2014. Does Agricultural Crop Diversity Enhance Soil Microbial Biomass and Organic Matter Dynamics? A Meta-Analysis. *Ecological Applications* 24 (3): 560–570. doi:10.1890/13-0616.1.

Meisner, A., and W. De Boer. 2018. Strategies to Maintain Natural Biocontrol of Soil-Borne Crop Diseases during Severe Drought and Rainfall Events. *Frontiers in Microbiology* 9 (November): 1–8. doi:10.3389/fmicb.2018.02279.

Melman, D. A., C. Kelly, J. Schneekloth, F. Calderón, and S. J. Fonte. 2019. Tillage and Residue Management Drive Rapid Changes in Soil Macrofauna Communities and Soil Properties in a Semiarid Cropping System of Eastern Colorado. *Applied Soil Ecology* 143 (June): 98–106. Elsevier. doi:10.1016/j.apsoil.2019.05.022.

Mendes, L. W., M. J. de Lima Brossi, E. E. Kuramae, and S. M. Tsai. 2015. Land-Use System Shapes Soil Bacterial Communities in Southeastern Amazon Region. *Applied Soil Ecology* 95: 151–160. doi:10.1016/j.apsoil.2015.06.005.

Mendes, R., M. Kruijt, I. De Bruijn, E. Dekkers, M. Van Der Voort, J. H. M. Schneider, Y. M. Piceno, T. Z. DeSantis, G. L. Andersen, P. A. H. M. Bakker, and J. M. Raaijmakers. 2011. Deciphering the rhizosphere microbiome for disease-suppressive bacteria. *Science* 332 (6033): 1097–1100.

Menezes-Blackburn, D., C. Giles, T. Darch, T. S. George, M. Blackwell, M. Stutter, C. Shand, et al. 2018. Opportunities for Mobilizing Recalcitrant Phosphorus from Agricultural Soils: A Review. *Plant and Soil* 427 (1–2): 5–16. Plant and Soil. doi:10.1007/s11104-017-3362-2.

Menezes-Blackburn, D., M. A. Jorquera, L. Gianfreda, R. Greiner, and M. de la Luz Mora. 2014. A Novel Phosphorus Biofertilization Strategy Using Cattle Manure Treated with Phytase-Nanoclay Complexes. *Biology and Fertility of Soils* 50 (4): 583–592. Springer Verlag. doi:10.1007/s00374-013-0872-9.

Mezeli, M. M., S. Page, T. S. George, R. Neilson, A. Mead, M. S.A. Blackwell, and P. M. Haygarth. 2020. Using a Meta-Analysis Approach to Understand Complexity in Soil Biodiversity and Phosphorus Acquisition in Plants. *Soil Biology and Biochemistry* 142 (March): 107695. Elsevier Ltd. doi:10.1016/j.soilbio.2019.107695.

Mondal, A., L. Goswami, N. Hussain, S. Barman, E. Kalita, P. Bhattacharyya, and S. S. Bhattacharya. 2020. Detoxification and Eco-Friendly Recycling of Brick Kiln Coal Ash Using Eisenia Fetida: A Clean Approach through Vermitechnology. *Chemosphere* 244 (April): 125470. Elsevier Ltd. doi:10.1016/j.chemosphere.2019.125470.

Moore, J. C., E. L. Berlow, D. C. Coleman, P. C. De Suiter, Q. Dong, A. Hastings, N. C. Johnson, et al. 2004. Detritus, Trophic Dynamics and Biodiversity. *Ecology Letters* 7 (7): 584–600. doi:10.1111/j.1461-0248.2004.00606.x.

Morriën, E., S. E. Hannula, L. B. Snoek, N. R. Helmsing, H. Zweers, M. de Hollander, R. L. Soto, et al. 2017. Soil Networks Become More Connected and Take up More Carbon as Nature Restoration Progresses. *Nature Communcations* 8: 14349. Anne Winding. doi:10.1038/ncomms14349.

Mueller, K. E., D. M. Blumenthal, Y. Carrillo, S. Cesarz, M. Ciobanu, J. Hines, S. Pabst, et al. 2016a. Elevated CO2 and Warming Shift the Functional Composition of Soil Nematode Communities in a Semiarid Grassland. *Soil Biology and Biochemistry* 103 (December): 46–51. Pergamon. doi:10.1016/J.SOILBIO.2016.08.005.

Mueller, R. C., J. L. M. Rodrigues, K. N. €. Usslein, and B. J. M. Bohannan. 2016b. Land Use Change in the Amazon Rain Forest Favours Generalist Fungi. *Functional Ecology* 30: 1845–1853. doi:10.1111/1365-2435.12651.

Mutema, M., P. Mafongoya, I. Nyagumbo, and L. Chikukura. 2013. Effects of Crop Residues and Reduced Tillage on Macrofauna Abundance. *Journal of Organic Systems* 8 (1): 5–16.

Neher, D. A., T. R. Weicht, D. L. Moorhead, and R. L. Sinsabaugh. 2004. Elevated CO2 Alters Functional Attributes of Nematode Communities in Forest Soils. *Functional Ecology* 18 (4): 584–591. John Wiley & Sons, Ltd (10.1111). doi:10.1111/j.0269–8463.2004.00866.x.

Nguyen, K. B., and G. C. Smart Jr. 1990. Steinernema Scapterisci n. Sp. (Rhabditida: Steinernematidae). *Journal of Nematology* 22 (2): 187–199. Society of Nematologists.

Nordén, J., R. Penttilä, J. Siitonen, E. Tomppo, and O. Ovaskainen. 2013. Specialist Species of Wood-Inhabiting Fungi Struggle While Generalists Thrive in Fragmented Boreal Forests. *Journal of Ecology* 101 (3): 701–712. doi:10.1111/1365-2745.12085.

Oberson, A., D. K Friesen, I. M. Rao, E. K. Bu, C. Smithson, B. L. Turner, and E. Frossard. 2006. Improving Phosphorus Fertility in Tropical Soils through Biological Interventions. In *Biological Approaches to Sustainable Soil Systems*, edited by N. Uphoff, A.S. Ball, E. Fernandes, H. Herren, O. Husson, M. Laing, C. Palm, et al., 531–546. Boca Raton, FL: CRC.

Ohkuma, M., S. Noda, and T. Kudo. 1999. Phylogenetic Diversity of Nitrogen Fixation Genes in the Symbiotic Microbial Community in the Gut of Diverse Termites. *Applied and Environmental Microbiology* 65 (11): 4926–4934. American Society for Microbiology. doi:10.1128/aem.65.11.4926-4934.1999.

Okada, H., and H. Harada. 2007. Effects of Tillage and Fertilizer on Nematode Communities in a Japanese Soybean Field. *Applied Soil Ecology* 35 (3): 582–598. doi:10.1016/j.apsoil.2006.09.008.

Orgiazzi, A., R.D. Bardgett, E. Barrios, V. Behan-Pelletier, M. J. I. Briones, J. L. Chotte, G. B. De Deyn, et al. 2016. *Global Soil Biodiversity Atlas. European Commission.* Luxembourg: Publications Office of the European Union. https://www.cabdirect.org/cabdirect/abstract/20173130918.

Osorio, N. W., and M. Habte. 2014. Soil Phosphate Desorption Induced by a Phosphate-Solubilizing Fungus. *Communications in Soil Science and Plant Analysis* 45 (4): 451–460. Taylor & Francis. doi:10.1080/00103624.2013.870190.

Ouedraogo, E., L. Brussaard, A. Mando, and L. Stroosnijder. 2005. Organic Resources and Earthworms Affect Phosphorus Availability to Sorghum after Phosphate Rock Addition in Semi-Arid West Africa. *Biology and Fertility of Soils* 41 (6): 458–465. doi:10.1007/S00374-005-0840-0.

Owen, D., A. P. Williams, G. W. Griffith, and P. J.A. Withers. 2015. Use of Commercial Bio-Inoculants to Increase Agricultural Production through Improved Phosphrous Acquisition. *Applied Soil Ecology* 86: 41–54. doi:10.1016/j.apsoil.2014.09.012.

Pankhurst, C E., B.M. Doube, and V.V.S.R. Gupta. 1997. Biological indicators of soil health: Synthesis. In *Biological Indicators of Soil Health*, edited by C.E. Pankhurst, B.M. Doube, and V.V.S.R. Gupta, 451. Wallingford: CAB International.

Pankievicz, V. C. S., F. P. do Amaral, K. F. D. N. Santos, B. Agtuca, Y. Xu, M. J. Schueller, A. C. M. Arisi, et al. 2015. Robust Biological Nitrogen Fixation in a Model Grass-Bacterial Association. *The Plant Journal* 81 (6): 907–919. Blackwell Publishing Ltd. doi:10.1111/tpj.12777.

Pankievicz, V. C.S., T. B. Irving, L. G.S. Maia, and J. M. Ané. 2019. Are We There yet? The Long Walk towards the Development of Efficient Symbiotic Associations between Nitrogen-Fixing Bacteria and Non-Leguminous Crops. *BMC Biology* 17 (1): 1–17. BioMed Central Ltd. doi:10.1186/s12915-019-0710-0.

Pardon, P., D. Reheul, J. Mertens, B. Reubens, P. De Frenne, P. De Smedt, W. Proesmans, L. Van Vooren, and K. Verheyen. 2019. Gradients in Abundance and Diversity of Ground Dwelling Arthropods as a Function of Distance to Tree Rows in Temperate Arable Agroforestry Systems. *Agriculture, Ecosystems and Environment* 270–271 (December 2017): 114–128. Elsevier. doi:10.1016/j.agee.2018.10.017.

Park, B., J. Lee, H. Ro, and Y. H. Kim. 2016. Short-Term Effects of Low-Level Heavy Metal Contamination on Soil Health Analyzed by Nematode Community Structure. *The Plant Pathology Journal* 32 (4): 329–339.

Patz, J. A., P. Daszak, G. M. Tabor, A. A. Aguirre, M. Pearl, J. Epstein, N. D. Wolfe, et al. 2004. Unhealthy Landscapes: Policy Recommendations on Land Use Change and Infectious Disease Emergence. *Environmental Health Perspectives* 112 (10): 1092–1098. Public Health Services, US Dept of Health and Human Services. doi:10.1289/ehp.6877.

Paul, S., H. Kauser, M. S. Jain, M. Khwairakpam, and A. S. Kalamdhad. 2020. Biogenic Stabilization and Heavy Metal Immobilization during Vermicomposting of Vegetable Waste with Biochar Amendment. *Journal of Hazardous Materials* 390 (May): 121366. Elsevier B.V. doi:10.1016/j.jhazmat.2019.121366.

Paula, F. S., J. L. M. Rodrigues, J. Zhou, L. Wu, R. C. Mueller, B. S. Mirza, B. J. M. Bohannan, et al. 2014. Land Use Change Alters Functional Gene Diversity, Composition and Abundance in Amazon Forest Soil Microbial Communities. *Molecular Ecology* 23 (12): 2988–2999. doi:10.1111/mec.12786.

Pérez-Montaño, F., C. Alías-Villegas, R. A. Bellogín, P. Del Cerro, M. R. Espuny, I. Jiménez-Guerrero, F. J. López-Baena, F. J. Ollero, and T. Cubo. 2014. Plant Growth Promotion in Cereal and Leguminous Agricultural Important Plants: From Microorganism Capacities to Crop Production. *Microbiological Research* 169 (5–6): 325–336. Urban und Fischer Verlag Jena. doi:10.1016/j.micres.2013.09.011.

Pisa, L. W., V. Amaral-Rogers, L. P. Belzunces, J. M. Bonmatin, C. A. Downs, D. Goulson, D. P. Kreutzweiser, et al. 2014. Effects of Neonicotinoids and Fipronil on Non-Target Invertebrates. *Environmental Science and Pollution Research* 22 (1): 68–102. Springer Verlag. doi:10.1007/s11356-014-3471-x.

Plaas, E., F. Meyer-Wolfarth, M. Banse, J. Bengtsson, H. Bergmann, J. Faber, M. Potthoff, T. Runge, S. Schrader, and A. Taylor. 2019. Towards Valuation of Biodiversity in Agricultural Soils: A Case for Earthworms. *Ecological Economics* 159 (May): 291–300. Elsevier B.V. doi:10.1016/j.ecolecon.2019.02.003.

Postgate, J.R. 1982. Biological Nitrogen Fixation: Fundamentals. *Philosophical Transactions of the Royal Society of London. B, Biological Sciences* 296 (1082): 375–385. The Royal Society. doi:10.1098/rstb.1982.0013.

Postma-Blaauw, M. B., J. Bloem, J. H. Faber, J. W. van Groenigen, R. G.M. de Goede, and L. Brussaard. 2006. Earthworm Species Composition Affects the Soil Bacterial Community and Net Nitrogen Mineralization. *Pedobiologia* 50 (3): 243–256. Elsevier GmbH. doi:10.1016/j.pedobi.2006.02.001.

Postma-Blaauw, M. B., R. G. M. De Goede, J. Bloem, J. H. Faber, and L. Brussaard. 2010. Soil Biota Community Structure and Abundance under Agricultural Intensification and Extensification. *Ecology* 91 (2): 460–473. doi:10.1890/09-0666.1.

Postma, J., M. T. Schilder, J. Bloem, and W. K. van Leeuwen-Haagsma. 2008. Soil Suppressiveness and Functional Diversity of the Soil Microflora in Organic Farming Systems. *Soil Biology and Biochemistry* 40 (9):2394–2406.

Preisser, E. L., and D. R. Strong. 2004. Climate Affects Predator Control of an Herbivore Outbreak. *The American Naturalist* 163 (5): 754–762. doi:10.1086/383620.

Pywell, R. F., W. R. Meek, N. R. Webb, P. D. Putwain, and J. M. Bullock. 2011. Long-Term Heathland Restoration on Former Grassland: The Results of a 17-Year Experiment. *Biological Conservation* 144 (5): 1602–1609. Elsevier. doi:10.1016/j.biocon.2011.02.010.

Rashid, S., T. C. Charles, and B. R. Glick. 2012. Isolation and Characterization of New Plant Growth-Promoting Bacterial Endophytes. *Applied Soil Ecology* 61 (October): 217–224. Elsevier. doi:10.1016/j.apsoil.2011.09.011.

Rasmann, S., T. G. Köllner, J. Degenhardt, I. Hiltpold, S. Toepfer, U. Kuhlmann, J. Gershenzon, and T. C. J. Turlings. 2005. Recruitment of Entomopathogenic Nematodes by Insect-Damaged Maize Roots. *Nature* 434 (7034): 732–737. Nature Publishing Group. doi:10.1038/nature03451.

Richardson, A. E. 2001. Prospects for Using Soil Microorganisms to Improve the Acquisition of Phosphorus by Plants. *Australian Journal of Plant Physiology* 28 (9): 897–906. doi:10.1071/pp01093.

Richardson, A. E., J. M. Barea, A. M. McNeill, and C. Prigent-Combaret. 2009. Acquisition of Phosphorus and Nitrogen in the Rhizosphere and Plant Growth Promotion by Microorganisms. *Plant and Soil* 321 (1–2): 305–339. Springer. doi:10.1007/s11104-009-9895-2.

Rillig, M. C., M. Ryo, A. Lehmann, C. A. Aguilar-Trigueros, A. Buchert, A. Wulf, A. Iwasaki, J. Roy, and G. Yang. 2020. The Role of Multiple Global Change Factors in Driving Soil Functions and Microbial Biodiversity. *Science* 366: 886–890. http://science.sciencemag.org/.

Robertson, G. P., and P. M. Groffman. 2007. Nitrogen Transformations. In *Soil Microbiology, Chemistry, and Ecology*, edited by E. A. Paul, 341–364. Amsterdam, Netherlands: Academic Press. doi:10.1016/B978-0-08-047514-1.50017-2.

Rodrigues, J. L. M., V. H. Pellizari, R. Mueller, K. Baek, E. D. C. Jesus, F. S. Paula, B. Mirza, et al. 2013. Conversion of the Amazon Rainforest to Agriculture Results in Biotic Homogenization of Soil Bacterial Communities. *Proceedings of the National Academy of Sciences* 110 (3): 988–993. National Academy of Sciences. doi:10.1073/pnas.1220608110.

Roger-Estrade, J., C. Anger, M. Bertrand, and G. Richard. 2010. Tillage and Soil Ecology: Partners for Sustainable Agriculture. *Soil and Tillage Research* 111 (1): 33–40. Elsevier. doi:10.1016/j.still.2010.08.010.

Ros, M. B.H., T. Hiemstra, J. W. van Groenigen, A. Chareesri, and G. F. Koopmans. 2017. Exploring the Pathways of Earthworm-Induced Phosphorus Availability. *Geoderma* 303: 99–109. doi:10.1016/j.geoderma.2017.05.012.

Rose, M. T., T. R. Cavagnaro, C. A. Scanlan, T. J. Rose, T. Vancov, S. Kimber, I. R. Kennedy, R. S. Kookana, and L. Van Zwieten. 2016. *Impact of Herbicides on Soil Biology and Function. Advances in Agronomy.* Vol. 136. Elsevier Inc. doi:10.1016/bs.agron.2015.11.005.

Ruckamp, D., W. Amelung, N. Theisz, A. G. Bandeira, and C. Martius. 2010. Phosphorus Forms in Brazilian Termite Nests and Soils: Relevance of Feeding Guild and Ecosystems. *Geoderma* 155 (3–4): 269–279. doi:10.1016/J.Geoderma.2009.12.010.

Ryu, M. H., J. Zhang, T. Toth, D. Khokhani, B. A. Geddes, F. Mus, A. Garcia-Costas, et al. 2020. Control of Nitrogen Fixation in Bacteria That Associate with Cereals. *Nature Microbiology* 5 (2): 314–330. Nature Research. doi:10.1038/s41564-019-0631-2.

Sabais, A. C. W., S. Scheu, and N. Eisenhauer. 2011. Plant Species Richness Drives the Density and Diversity of Collembola in Temperate Grassland. *Acta Oecologica* 37 (3): 195–202. Elsevier Masson SAS. doi:10.1016/j.actao.2011.02.002.

Sandhi, R. K., D. Shapiro-Ilan, A. Sharma, and G. V. P. Reddy. 2020. Efficacy of Entomopathogenic Nematodes against the Sugarbeet Wireworm, Limonius Californicus (Mannerheim) (Coleoptera: Elateridae). *Biological Control* 143 (April): 104190. Academic Press Inc. doi:10.1016/j.biocontrol.2020.104190.

Santos, M. S., M. A. Nogueira, and M. Hungria. 2019. Microbial Inoculants: Reviewing the Past, Discussing the Present and Previewing an Outstanding Future for the Use of Beneficial Bacteria in Agriculture. *AMB Express* 9 (1): 205. Springer. doi:10.1186/s13568-019-0932-0.

Schmidt, J. E., C. Peterson, D. Wang, K. M. Scow, and A. C. M. Gaudin. 2018. Agroecosystem Tradeoffs Associated with Conversion to Subsurface Drip Irrigation in Organic Systems. *Agricultural Water Management* 202 (April): 1–8. Elsevier B.V. doi:10.1016/j.agwat.2018.02.005.

Senapati, B. K., P. Lavelle, E. Blanchart, M. Mahieux, R. Thomas, K. Pradeep, S. Giri, T. Decaëns, and B. Pashanasi. 1999. In-Soil Earthworm Technologies for Tropical Agroecosystems. In *Earthworms Management in Tropical Agroecosystems*, edited by P. Lavelle, L. Brussaard, and P. Hendrix, 199–237. Wallingford, UK: CAB International.

Sharma, K., and V. K. Garg. 2019. Recycling of Lignocellulosic Waste as Vermicompost Using Earthworm Eisenia Fetida. *Environmental Science and Pollution Research* 26 (14): 14024–14035. Springer Verlag. doi:10.1007/s11356-019-04639-8.

Sharma, K., Z. Cheng, and P. S. Grewal. 2015. Relationship between Soil Heavy Metal Contamination and Soil Food Web Health in Vacant Lots Slated for Urban Agriculture in Two Post-Industrial Cities. *Urban Ecosystems* 18 (3): 835–855. doi:10.1007/s11252-014-0432-6.

Sharpley, A., H. Jarvie, D. Flaten, and P. Kleinman. 2018. Celebrating the 350th Anniversary of Phosphorus Discovery: A Conundrum of Deficiency and Excess. *Journal of Environmental Quality* 47 (4): 774–777. Wiley. doi:10.2134/jeq2018.05.0170.

Sharpley, A. N. 1995. Dependence of Runoff Phosphorus on Extractable Soil Phosphorus. *Journal of Environmental Quality* 24 (5): 920–926. Wiley. doi:10.2134/jeq1995.00472425002400050020x.

Shaw, E. A., C. M. Boot, J. C. Moore, D. H. Wall, and J. S. Baron. 2019. Long-Term Nitrogen Addition Shifts the Soil Nematode Community to Bacterivore-Dominated and Reduces Its Ecological Maturity in a Subalpine Forest. *Soil Biology and Biochemistry* 130 (March): 177–184. doi:10.1016/j.soilbio.2018.12.007.

Shen, J. B., L. X. Yuan, J. L. Zhang, H. G. Li, Z. H. Bai, X. P. Chen, W. F. Zhang, and F. S. Zhang. 2011. Phosphorus Dynamics: From Soil to Plant. *Plant Physiology* 156 (3): 997–1005. doi:10.1104/Pp.111.175232.

Shropshire, J. D., and S. R. Bordenstein. 2016. Speciation by Symbiosis: The Microbiome and Behavior. *MBio* 7 (2): e01785-15. American Society for Microbiology. doi:10.1128/mBio.01785-15.

Singh, B. D., K. K. Hazra, U. Singh, and S. Gupta. 2020. Eco–Friendly Management of Meloidogyne Javanica in Chickpea (Cicer Arietinum L.) Using Organic Amendments and Bio–Control Agent. *Journal of Cleaner Production* 257 (June): 120542. Elsevier Ltd. doi:10.1016/j.jclepro.2020.120542.

Sinha, K. K., A. K. Choudhary, and P. Kumari. 2016. Entomopathogenic Fungi. In *Ecofriendly Pest Management for Food Security*, 475–505. Elsevier Inc. doi:10.1016/B978-0-12-803265-7.00015-4.

Smith, J. L., and H. P. Collins. 2007. Management of Organisms and Their Processes in Soils. In *Soil Microbiology, Ecology and Biochemistry*, edited by E. A. Paul, 471–502. Amsterdam, Netherlands: Elsevier. doi:10.1016/B978-0-08-047514-1.50021-4.

Smukler, S. M., S. Sánchez-Moreno, S. J. Fonte, H. Ferris, K. Klonsky, A. T. O'Geen, K. M. Scow, K. L. Steenwerth, and L. E. Jackson. 2010. Biodiversity and Multiple Ecosystem Functions in an Organic Farmscape. *Agriculture, Ecosystems and Environment* 139 (1–2): 80–97. Elsevier B.V. doi:10.1016/j.agee.2010.07.004.

Soliveres, S., F. Van Der Plas, P. Manning, D. Prati, M. M. Gossner, S. C. Renner, F. Alt, et al. 2016. Biodiversity at Multiple Trophic Levels Is Needed for Ecosystem Multifunctionality. *Nature* 536: 456–459. doi:10.1038/nature19092.

Song, M., X. Li, S. Jing, L. Lei, J. Wang, and S. Wan. 2016. Responses of Soil Nematodes to Water and Nitrogen Additions in an Old-Field Grassland. *Applied Soil Ecology* 102 (June): 53–60. Elsevier. doi:10.1016/J. APSOIL.2016.02.011.

Soobhany, N. 2018. Preliminary Evaluation of Pathogenic Bacteria Loading on Organic Municipal Solid Waste Compost and Vermicompost. *Journal of Environmental Management* 206 (January): 763–767. Academic Press. doi:10.1016/j.jenvman.2017.11.029.

Spaepen, S. 2015. Plant Hormones Produced by Microbes. In *Principles of Plant-Microbe Interactions: Microbes for Sustainable Agriculture*, edited by B. Lugtenberg, 247–256. Switzerland: Springer International Publishing. doi:10.1007/978-3-319-08575-3_26.

Steffan, J. J., E. C. Brevik, L. C. Burgess, and A. Cerdà. 2018. The Effect of Soil on Human Health: An Overview. *European Journal of Soil Science* 69 (1): 159–171. Blackwell Publishing Ltd. doi:10.1111/ejss.12451.

Stock, S. P., and A. M. Koppenhöfer. 2003. Steinernema Scarabaei n. Sp. (Rhabditida: Steinernematidae), a Natural Pathogen of Scarab Beetle Larvae (Coleoptera: Scarabaeidae) from New Jersey, USA. *Nematology* 5 (2): 191–204. Brill. doi:10.1163/156854103767139680.

Stutter, M. I., C. A. Shand, T. S. George, M. S. A. Blackwell, L. Dixon, R. Bol, R. L. MacKay, A. E. Richardson, L. M. Condron, and P. M. Haygarth. 2015. Land Use and Soil Factors Affecting Accumulation of Phosphorus Species in Temperate Soils. *Geoderma* 257–258 (November): 29–39. Elsevier. doi:10.1016/j. geoderma.2015.03.020.

Subramanian, S., and M. Muthulakshmi. 2016. *Entomopathogenic Nematodes. Ecofriendly Pest Management for Food Security*. doi:10.1016/B978-0-12-803265-7.00012-9.

Suwannaroj, M., T. Yimthin, C. Fukruksa, P. Muangpat, T. Yooyangket, S. Tandhavanant, A. Thanwisai, and A. Vitta. 2020. Survey of Entomopathogenic Nematodes and Associate Bacteria in Thailand and Their Potential to Control Aedes Aegypti. *Journal of Applied Entomology* 144 (3): 212–223. doi:10.1111/jen.12726.

Szczech, M., W. Rondomański, M. W. Brzeski, U. Smolińska, and J. F. Kotowski. 1993. Suppressive Effect of a Commercial Earthworm Compost on Some Root Infecting Pathogens of Cabbage and Tomato. *Biological Agriculture and Horticulture* 10 (1): 47–52. Taylor & Francis Group. doi:10.1080/01448765 .1993.9754650.

Tabaglio, V., C. Gavazzi, and C. Menta. 2009. Physico-Chemical Indicators and Microarthropod Communities as Influenced by No-till, Conventional Tillage and Nitrogen Fertilisation after Four Years of Continuous Maize. *Soil and Tillage Research* 105 (1): 135–142. doi:10.1016/j.still.2009.06.006.

Tawaraya, K., M. Naito, and T. Wagatsuma. 2006. Solubilization of Insoluble Inorganic Phosphate by Hyphal Exudates of Arbuscular Mycorrhizal Fungi. *Journal of Plant Nutrition* 29 (4): 657–665. Taylor & Francis Group. doi:10.1080/01904160600564428.

Thakur, M. P., I. M. Del Real, S. Cesarz, K. Steinauer, P. B Reich, S. Hobbie, M. Ciobanu, R. Rich, K. Worm, and N. Eisenhauer. 2019. Soil Microbial, Nematode, and Enzymatic Responses to Elevated CO_2, N Fertilization, Warming, and Reduced Precipitation. *Soil Biology & Biochemistry* 135: 184–193. doi:10.1016/j.soilbio.2019.04.020.

Thakur, M. P., D. Tilman, O. Purschke, M. Ciobanu, J. Cowles, F. Isbell, P. D. Wragg, and N. Eisenhauer. 2017. Climate Warming Promotes Species Diversity, but with Greater Taxonomic Redundancy, in Complex Environments. *Science Advances* 3: e1700866. http://advances.sciencemag.org/content/advances/3/7/e1700866.full.pdf.

Thurman, J. H., T. D. Northfield, and W. E. Snyder. 2019. Weaver Ants Provide Ecosystem Services to Tropical Tree Crops. *Frontiers in Ecology and Evolution* 7 (May): 120. Frontiers Media SA. doi:10.3389/fevo.2019.00120.

Tixier, M. S. 2018. Predatory Mites (Acari: Phytoseiidae) in Agro-Ecosystems and Conservation Biological Control: A Review and Explorative Approach for Forecasting Plant-Predatory Mite Interactions and Mite Dispersal. *Frontiers in Ecology and Evolution* 6: 192. (DEC) Frontiers Media S.A. doi:10.3389/fevo.2018.00192.

Tomati, U., A. Grappelli, and E. Galli. 1988. The Hormone-like Effect of Earthworm Casts on Plant Growth. *Biology and Fertility of Soils* 5 (4): 288–294. Springer-Verlag. doi:10.1007/BF00262133.

Tsiafouli, M. A., E. Thébault, S. P. Sgardelis, P. C. de Ruiter, W. H. van der Putten, K. Birkhofer, L. Hemerik, et al. 2015. Intensive Agriculture Reduces Soil Biodiversity across Europe. *Global Change Biology* 21 (2): 973–985. doi:10.1111/gcb.12752.

Tylianakis, J. M., R. K. Didham, J. Bascompte, and D. A. Wardle. 2008. Global Change and Species Interactions in Terrestrial Ecosystems. *Ecology Letters* 11: 1351–1363. doi:10.1111/j.1461-0248.2008.01250.x.

van den Hoogen, J., S. Geisen, D. Routh, H. Ferris, W. Traunspurger, D. A. Wardle, R. G. M. de Goede, et al. 2019. Soil Nematode Abundance and Functional Group Composition at a Global Scale. *Nature* 572 (July): 194–198. Nature Publishing Group. doi:10.1038/s41586-019-1418-6.

Van Deynze, A., P. Zamora, P. Delaux, C. Heitmann, D. Jayaraman, S. Rajasekar, D. Graham, et al. 2018. Nitrogen Fixation in a Landrace of Maize Is Supported by a Mucilage-Associated Diazotrophic Microbiota. Edited by Eric Kemen. *PLoS Biology* 16 (8): e2006352. Public Library of Science. doi:10.1371/journal.pbio.2006352.

van Groenigen, J. W., I. M. Lubbers, H. M. J. Vos, G. G. Brown, G. B. De Deyn, and K. J. van Groenigen. 2014. Earthworms Increase Plant Production: A Meta-Analysis. *Scientific Reports* 4 (2): 6365. doi:10.1038/srep06365.

Van Groenigen, J. W., K. J. Van Groenigen, G. F. Koopmans, L. Stokkermans, H. M. J. Vos, and I. M. Lubbers. 2019. How Fertile Are Earthworm Casts? A Meta-Analysis. *Geoderma* 338 (March): 525–535. Elsevier B.V. doi:10.1016/j.geoderma.2018.11.001.

Vazquez, C., R.G.M. de Goede, G.W. Korthals, M. Rutgers, A.J. Schouten, and R.E. Creamer. 2019. The Effects of Increasing Land Use Intensity on Soil Nematodes: A Turn towards Specialism. *Functional Ecology* 33 (July): 2003–2016. doi:10.1111/1365-2435.13417.

Vécrin, M. P., and S. Muller. 2003. Top-Soil Translocation as a Technique in the Re-Creation of Species-Rich Meadows. *Applied Vegetation Science* 6: 271–278. Wiley. doi:10.2307/1479027.

Veen, G. F., E. R. J. Wubs, R. D. Bardgett, E. Barrios, M. A. Bradford, S. Carvalho, G. B. De Deyn, et al. 2019. Applying the Aboveground-Belowground Interaction Concept in Agriculture: Spatio-Temporal Scales Matter. *Frontiers in Ecology and Evolution* 7 (August): 300. Frontiers Media S.A. doi:10.3389/fevo.2019.00300.

Veresoglou, S. D., J. M. Halley, and M. C. Rillig. 2015. Extinction Risk of Soil Biota. *Nature Communcations* 6: 8862. doi:10.1038/ncomms9862.

Wagg, C., S. F. Bender, F. Widmer, and M. G. A. Van Der Heijden. 2014. Soil Biodiversity and Soil Community Composition Determine Ecosystem Multifunctionality. *Proceedings of the National Academy of Sciences of the United States of America* 111 (14): 5266–5270. National Academy of Sciences. doi:10.1073/pnas.1320054111.

Wall, D. H., U. N. Nielsen, and J. Six. 2015. Soil Biodiversity and Human Health. *Nature* 528 (7580): 69–76. Nature Publishing Group. doi:10.1038/nature15744.

Wardle, D. A. 1995. Impacts of Disturbance on Detritus Food Webs in Agro-Ecosystems of Contrasting Tillage and Weed Management Practices. *Advances in Ecological Research* 26 (C): 105–185. doi:10.1016/S0065-2504(08)60065-3.

Wardle, D. A., R. D. Bardgett, J. N. Klironomos, H. Setälä, W. H. Van Der Putten, and D. H. Wall. 2004. Ecological Linkages between Aboveground and Belowground Biota. *Science* 304 (5677): 1629–1633. American Association for the Advancement of Science. doi:10.1126/science.1094875.

Weller, D. M., J. M. Raaijmakers, B. B. M. Gardener, and L. S. Thomashow. 2002. Microbial Populations Responsible for Specific Soil Suppressiveness to Plant Pathogens. *Annual Review of Phytopathology* 40:309–348.

Wepking, C., B. Badgley, J. E. Barrett, K. F. Knowlton, J. M. Lucas, K. J. Minick, P. P. Ray, S. E. Shawver, and M. S. Strickland. 2019. Prolonged Exposure to Manure from Livestock-Administered Antibiotics Decreases Ecosystem Carbon-Use Efficiency and Alters Nitrogen Cycling. *Ecology Letters* 22 (12): 2067–2076. doi:10.1111/ele.13390.

Wubs, E. R. J., W. van der Putten, S. R. Mortimer, G. W. Korthals, H. Duyts, R. Wagenaar, and T. M. Bezemer. 2019. Single Introductions of Soil Biota and Plants Generate Long-term Legacies in Soil and Plant Community Assembly. Edited by Katharine Suding. *Ecology Letters* 22 (April): 1145–1151. John Wiley & Sons, Ltd (10.1111). doi:10.1111/ele.13271.

Wubs, E. R. J., P. D. Melchers, and T. M. Bezemer. 2018. Potential for Synergy in Soil Inoculation for Nature Restoration by Mixing Inocula from Different Successional Stages. *Plant and Soil* 433 (1–2): 147–156. Springer International Publishing. doi:10.1007/s11104-018-3825-0.

Wubs, E. R. J., W. Van Der Putten, M. Bosch, and T. M. Bezemer. 2016. Soil Inoculation Steers Restoration of Terrestrial Ecosystems. *Nature Plants* 2 (8): 1–5. Palgrave Macmillan Ltd. doi:10.1038/NPLANTS.2016.107.

Yang, G., C. Wagg, S. D. Veresoglou, S. Hempel, and M. C. Rillig. 2018. How Soil Biota Drive Ecosystem Stability. *Trends in Plant Science* 23 (12): 1057–1067. doi:10.1016/j.tplants.2018.09.007.

Yergeau, E., T. H. Bell, J. Champagne, C. Maynard, S. Tardif, J. Tremblay, and C. W. Greer. 2015. Transplanting Soil Microbiomes Leads to Lasting Effects on Willow Growth, but Not on the Rhizosphere Microbiome. *Frontiers in Microbiology* 6: 1436. (DEC) Frontiers Media S.A. doi:10.3389/fmicb.2015.01436.

Zhao, F., Y. Zhang, Z. Li, J. Shi, G. Zhang, H. Zhang, and L. Yang. 2020. Vermicompost Improves Microbial Functions of Soil with Continuous Tomato Cropping in a Greenhouse. *Journal of Soils and Sediments* 20 (1): 380–391. Springer. doi:10.1007/s11368-019-02362-y.

Zhou, X., J. Wang, Y. Hao, and Y. Wang. 2010. Intermediate Grazing Intensities by Sheep Increase Soil Bacterial Diversities in an Inner Mongolian Steppe. *Biology and Fertility of Soils* 46 (8): 817–824. doi:10.1007/s00374-010-0487-3.

Zuber, S. M., and M. B. Villamil. 2016. Meta-Analysis Approach to Assess Effect of Tillage on Microbial Biomass and Enzyme Activities. *Soil Biology and Biochemistry* 97: 176–187. Elsevier Ltd. doi:10.1016/j.soilbio.2016.03.011.

14 Structural Attributes of Disease-Suppressive Soils and Their Impact on Human Health

Rattan Lal
The Ohio State University

CONTENTS

14.1 INTRODUCTION

The need to produce an adequate amount of high-quality food with minimal environmental footprint is more now than ever before. Plant diseases caused by fungi, bacteria, and viruses can reduce crop yield by 10%–20% (Toyota and Shirai 2018) and aggravate the existing serious problem of food insecurity. The problem of the food and nutritional security is also being aggravated by the COVID-19 pandemic and the current and projected climate change (Lal et al. 2020). Global pesticide use from 2010 to 2014 was 2.8 kg ha^{-1} or 0.65 g of pesticide use per kg of crop production (Zhang 2018). More than 2 million Mg of global pesticides used during 2010s may increase to 3.5 million Mg year^{-1} during 2020 (Sharma et al. 2019). In 2011–2012, pesticide use was 2.7 million Mg in the world and 0.5 million Mg in the U.S. (Marquez 2018). Severe environmental problems have been caused by heavy inputs of pesticides (Carson 1962) and chemical fertilizers (Savci 2012; Sharma and Singhvi 2017). About 3 million cases of pesticide poisoning and 0.22 million deaths per annum are reported in developing countries (Mahmood et al. 2016). Whereas the Paracelsus (1493–1541, Swiss Alchemist and Physician) dictum "the right dose differentiates a poison from a remedy" is still valid, there are a range of industrial chemicals whose impacts are complex and long-lived, and thus, they necessitate cautious approach about the validity of the dose concept in a complex world (Grandjean 2016).

Anthropogenic activities (i.e., soil sealing, agricultural land-use intensification, biological invasions resulting from the introduction of non-native species, and heavy use of pesticides and fertilizers since 1960s) have drastically and adversely impacted communities of soil biota (Geisen,

Wall, and van der Putten 2019) with weakening of ecosystem services (ESs) and creation of some serious disservices. Thus, there is an urgency to determine alternative measures to control plant diseases. It has been widely understood since the 1980s that use of organic amendments, which also enhance soil structure and sequester organic carbon in soil, can control plant diseases by increasing disease-suppressive properties (Hoitink and Fahy 1986; Stone, Scheurell, and Darby 2004,). Therefore, a prudent strategy is to reduce pesticide use by enhancing disease-suppressive attributes of cropland soils through restoration of soil structure and adoption of best management practices (BMPs). Soil aggregation, aeration, and other determinants of soil structure include soil organic matter (SOM) content and its management (Figure 14.1). Disease-suppressive soils have the capacity to moderate the environment and reduce incidence of disease, even in the presence of pathogens and susceptible plants (Allard and Micallef 2019).

Soil structure refers to arrangement of soil particles and voids through interactions between clay and fine silt particles on the one side and organic (i.e., byproducts of microbial processes, fungal hyphae) and inorganic cementing agents (i.e., salts of polyvalent cations and sesquioxides) on the other (Greenland 1965; Tisdall and Oades 1982; Six et al. 2004; Bronick and Lal 2005). Soil fauna (macro, meso, and micro) and flora (e.g.,plant roots,fungi,mycorrhizae) play a critical role in the formation of secondary particles or aggregates and their stability. Formation and stabilization of these secondary particles or organo-mineral complexes determine bulk density, total porosity, pore size distribution, continuity, and stability/durability of the pores/voids formed through arrangements of the particles. These structural characteristics are critical to numerous soil functions and ESs. Important among these are composition of soil air, aeration, or the gaseous exchange between soil and the atmosphere, water transmission (infiltration rate, hydraulic conductivity), and retention (field moisture capacity, plant available water capacity) or the so called "green"water supply on which depends the growth net primary productivity of the terrestrial biosphere. It is these structure-induced changes in soil properties and their dynamics, mediated by microbial processes through input of biomass-C, that determine the disease-suppressive capacity of soil. Naturally, therefore, disease-suppressive properties of soil are enhanced by input of compost that has recalcitrant organic compounds (Hoitink, McSpadden, and Miller 2008). Stability of organic compounds in compost

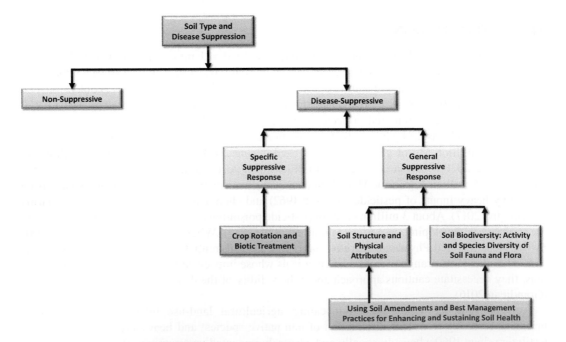

FIGURE 14.1   Common types of disease-suppressive soils.

and other amendments affects the efficacy of disease suppression. Hoitink and colleagues observed that compost with recalcitrant compounds can effectively suppress *Phytophthora*, *Pythium*, and *Thielaviopsis* root rots and other diseases.

Soil health, primarily an ecological characteristic (van Bruggen and Semenov 2000), is determined by a strong interaction between soil biological and physical properties or the interaction among solid, liquid, and gaseous phases of soil under natural and managed ecosystems. Soil structure is a dynamic property and is strongly affected by tillage, or lack of it, and crop rotations (Munkholm, Heck, and Deen 2013). Crop rotations affect soil structure by influencing microbial communities (Ball et al. 2005). In addition to altering porosity (i.e., pore size distribution, continuity, and stability), appropriate crop rotations also affect the habitats of soil biota, especially the micro-organisms. Thus, there is a strong relationship between soil structure, SOM decomposition and turnover, and the activity and species diversity of soil biota (Six et al. 2004). Inclusion of a non-host crop in the rotation can affect the community composition and influence the population of disease-causing organisms (Peters et al. 2003). Therefore, the objective of this chapter is to describe the importance of soil structure to disease-suppressive attributes of a soil, and how to manage structure and soil bio-physical health to enhance disease-suppression capacity of soils of agroecosystems.

14.2 DISEASE-SUPPRESSIVE ATTRIBUTES IN SOILS

One century ago, Sir Albert Howard, working in Indore, India, recognized the influence of soil factors on disease resistance (Howard 1921). Considering the susceptibility to plant diseases, soils can be grouped into two categories: (i) conducive soils, which cause severe diseases in crops and (ii) disease-suppressive soils (Figure 14.1). Disease-conducive soils are those in which plant diseases readily occur because abiotic and biotic conditions are favorable to pathogens and crops are susceptible. Some examples of the diseases found in conducive soils include take-all of wheat, root rot, damping-off, and oat cyst nematode (Agrios 2005). In contrast, disease-suppressive soils are those soils in which the pathogen does not establish or persist, establishes but causes little or no damage, or establishes and causes disease for a while but thereafter the disease is less important, although the pathogen may persist in the soils(Baker and Cook 1974).

In general, disease-suppressive soils have the capacity to limit the infection and its severity by pathogens because of its high biodiversity even in the presence of a susceptible host.

The pathogen-suppressive capacity of soil may be due to pathogens inability to do the following (Cook 2014): (i) establish, (ii) cause disease, or (iii) disease declines over time. Thus, the pathogen-suppressive soils may have one or all of those attributes. Soil's capacity to limit diseases may be due to its capacity to suppress pathogens or suppress diseases (Figure 14.2), and these two characteristics may not be coupled (Hoeper and Alabouvette 1996). Disease suppression is related to microbial characteristics in which rhizospheric organisms disrupt fungal infections through formation of antibiotics, nutrient availability, and creation of host-specific resistance (Ball et al. 2005). Sturz and Christie (2003) highlighted the positive role of the beneficial microbial allelopathy in the root zone which can suppress diseases. These microbial processes are impacted by management-induced changes in soil quality. Disease-suppression abilities may also be caused by predation of fungal hyphae by soil fauna viz arbuscular mycorrhizae fungi commonly abbreviated as AMF (Ball et al. 2005), which are also natural biofertilizers (Berruti et al. 2016). In other words, disease suppression is an attribute of an ecosystem stability, as determined by soil biodiversity. Soil microbes can suppress pathogens by a range of mechanisms: (i) improve plant health, (ii) increase natural plant defense, (iii) excrete antibiotics, (iv) compete against pathogens, and (iv) parasitize the pathogen (Mousa and Raizada 2016). Active myceles in soil have beneficial impact on soil and plant health (Bhatti, Haq, and Bhat 2017). Thus, increase in activity and species diversity of soil biota through inputs of organic amendments may enhance ecosystem stability and increase disease-suppressive capacity (Figure 14.2).

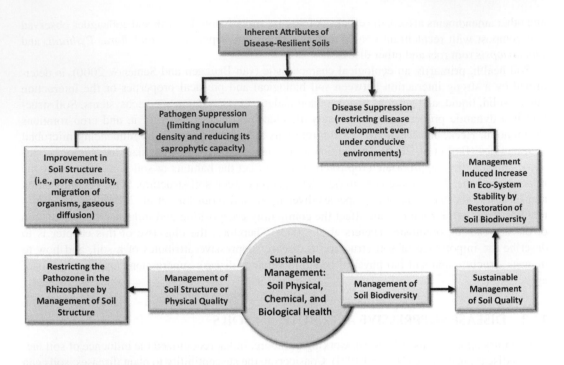

FIGURE 14.2   Disease-resilience of a soil, dependent on the interaction between structural properties and other factors, depends on its capacity to either suppress pathogens by limiting their inoculum density and reducing its saprophytic activity or suppress disease (even in the presence of host and inoculum) by restricting disease development. Pathogen suppression and disease suppression may not be coupled and may be mutually exclusive.

Therefore, disease suppression in soils is a microbiological phenomenon (Gómez Expósito et al. 2017) whereby natural biome-based plant defense; through rhizodeposition and plant root exudates; stimulate and support micro-organisms that suppress the incidence of disease. Such a disease-suppressive capacity of soil may be general or specific (Schlatter et al. 2017) (Figure 14.2). Disease-suppressive behavior of soil because of specific microbial communities is the general disease suppression based on the concentration of soil organic carbon (SOC) and its chemical properties (Mousa and Raizada 2016). General suppressiveness of soils is a function of the activity of the collective microbial community and is controlled by competition for available resources (Mazzola 2002; Weller et al. 2002; Gómez Expósito et al. 2017). It is the activity and composition of microbial communities that govern the general disease-suppressive behavior. In addition to the bacteria, Penton et al. (2014) observed that fungi in the rhizosphere can also influence disease suppression, including a number of endophytic spp. and mycoparasites (i.e., *Xylorsia spp.*). Penton and colleagues also observed that non-suppressive soils may be dominated by *Alternaria*, *Gibberella*, and *Penicillium*.

Soils with general suppression are those with high SOC content, and thus, high activity and species diversity of soil biota. There are several mechanisms (biotic and abiotic) which impart the general disease suppression in soils. Thus, such a general suppression can be enhanced by improving and sustaining the SOC content through adoption of site-specific BMPs (i.e., using conservation of agriculture, cover cropping, green manuring, and using compost) which have a positive impact on soil structure.

In contrast, specific suppression is related to a high concentration of specific microbial communities that suppress a specific pathogen such as root-borne diseases. In comparison, specific suppression implies control of a particular disease-causing pathogen through antibiotics or parasitism (Hoitink, McSpadden, and Miller 2008). Disease-suppressive soil against specific pathogens is decreased by continuous monoculture that enhances a community of *Pseudomonas* florescence (Mousa and Raizada 2016). Specific suppression can also be enhanced by inoculants of biocontrol agents into compost-amended substrates (Hoitink and Boehm 1999). Therefore, biotic-suppressive measures are based on the concept of the transfer of soilborne microbes (e.g., fungi, bacteria) to the rhizosphere for suppression of plant diseases (Poltronieri and Reca 2020). Use of microbiomes can lead to formation of disease-suppressive soil for a specific pathogen (Toyota and Shirai 2018). Beneficial soil micro-organisms are used to strengthen ESs including suppression of soilborne diseases (Stirling et al. 2016). Rather than using chemical biocides, use of the microbiome to control plant disease is an environment-friendly option (Ghorbani et al. 2008), with also positive effects on human health and well-being. Soilborne diseases also adversely affect crop health (Janvier et al. 2007), the productivity and nutritional quality, and human health and well-being.

14.3 MANAGEMENT OF DISEASE SUPPRESSION BY MANAGING SOIL STRUCTURE

Soil biodiversity and its effects on soil structure are key determinants of disease suppression in soils. Tillage and crop rotations can impact disease-suppressive attributes of soil (Conway 1996) though changes in soil structure. Therefore, management of disease suppression in soils is primarily management of soil biodiversity, and of soil structure through adoption of sustainable agricultural practices which optimize these attributes under site-specific conditions. Crop rotation, conservation agriculture (CA) with residue mulch and input of compost are among important practices for enhancing disease-suppressive attributes of agricultural soils through their impact on soil health.

14.3.1 MANAGING SOIL HEALTH THROUGH STRENGTHENING BIODIVERSITY BY RESTORING SOC CONTENT

Decline in soil structure by land misuse and soil mismanagement may aggravate soilborne fungal diseases (through an increase in colony density in compacted soil and by altering the nature and composition of root exudates) (Ball et al. 2005). Assessment of microbial community structure and its management is essential to disease suppression (Mazzola 2004). The pathogenic suppression of soil depends on soil structure, more specifically on the pore size distribution and pore continuity which facilitate aeration, gaseous exchange, etc. (Figure 14.3). Soil structure and pore continuity strongly determine the movement of soil organisms, which govern the predator–prey interaction,

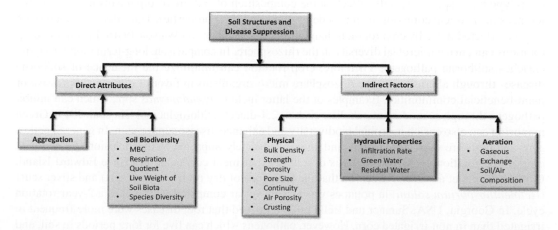

FIGURE 14.3 Parameters related to soil structure in relation to suppression of pathogens and suppression of disease.

with strong implications to the survival and colonization (spread) of the pathogens. This is the basis of pathogens concept proposed by Otten et al. (2001). It refers to the volume of soil in the root–soil interphase (rhizosphere) which contains an adequate fungal propagate to infect the host plant. Otten et al. (2001) observed that the increase in compaction and decrease in aggregate size reduced colonization and thus the infection by *Rhizoctonia solani* because of the decline in volume and continuity of macropores. An increase in soil biodiversity by restoration of the SOC stock is important in improving soil structure and enhancing disease-suppressive attributes of soil.

14.3.2 COMPOST

Foliar and root pathogens may be suppressed by compost. Greenhouse studies have shown that use of compost (upto 20% by volume) can suppress several soilborne diseases such as damping off, roor rots, and wilts (Noble and Coventry 2005). Use of compost from hardwood bark suppressed Rhizoctonia damping-off in the containerized experiment (Nelson, Kuter, and Hoitink 1983). Use of compost in the field plots have shown to suppress Fusarium patch, red thread, damping off, and brown patch in turf grass, and the disease-suppressive effects increase with increase in the rate of compost application (Noble and Coventry 2005). Based on field experiments, Tilston, Pitt, and Groenhof (2002) reported that maturity, formulation, and degree of processing determine compost chemistry and disease suppression. Tilston and colleagues also concluded that disease suppression is pathogenic-specific and may be modified by the activities of micro-organisms. In addition to enhancing biodiversity (Hoitink, McSpadden, and Miller 2008), favorable soil structure, especially through regular inputs of compost, enhances general suppressive attributes of a soil (Hoitink and Fahy 1986). Effectiveness of disease suppression by compost may depend on several factors. Hoitink, Stone, and Grebus (1996) observed that induction of disease suppression may depend on feedstock, the composting environment, as well as conditions during curing and utilization which determine the potential for recolonization of composts by biocontrol agents. Whereas, particle size of compost is also considered important, Lozano, Blok, and Termorshuizen (2008) observed that raw feedstock or substrate quality is relatively more important than particle size in disease suppression by compost. Antoniou et al. (2017) reported that even the rhizosphere microbiome recruited from a suppressive compost could protect plants against the wilt pathogens of tomatoes and improve plant health and productivity.

14.3.3 CROP ROTATION

Crop type has a species-specific effect on the composition of soil micro-organisms because of differences in chemical composition of root depositions and litter properties. Thus, disease suppression may be affected more by crop rotation than by soil type (Huber and Watson 1970). In general, crop rotations can enrich microbial diversity in the rhizosphere. In comparison, long-term monocropping enriches soilborne pathogens. Therefore, crop rotation can minimize the incidence of soilborne diseases through alterations of the rhizosphere micro-organisms in favor of disease-suppressive or plant-beneficial communities. Examples of the latter include *Pseudomonas* spp., which can inhibit pathogens through secondary metabolites such as 2,4-diacetylphloroglucinol (Jin et al. 2019). Green manuring can also enrich microbial biodiversity and enhance disease suppression. In general, pathogens and pests are short-lived in soil, and thus are adversely impacted by the inclusion of a non-host crop in the rotation cycle. On the basis of a field experiment conducted at Prince Edward Island, Canada, Peters et al. (2003) observed that the incidence of dry rot (*Fusarium* spp.) and sliver scurf (*Helmenthosporium solani*) in potatoes was less in 3-year compared with that in a 2-year rotation cycle. In Georgia, USA, Sumner and Bell (1982) observed that root diseases were more frequent in irrigated than in non-irrigated corn. However, pathogens which can live for long periods in soil, and those that can survive in SOM saprophytically (i.e., fungi or bacteria that live on dead organisms or SOM pool), may not be suppressed by change in crop rotation. Examples of species which can survive saprophytically are *Pythium* spp., *Phytophthora* spp., *Fusarium* spp., and *Sclerotium rolfsii* (Sumner 1982; Ball et al. 2005).

Jin et al. (2019) observed that green manuring with Brassiceae crops (i.e., Indian mustard, wild rocket) could reduce incidence of pathogenic *Fusarium* spp. and also decrease Fusarium wilt of cucumbers and tomatoes. The disease-suppressive attributes of *Brassiceae* spp. are attributed to the release of antifungal compounds such as isothiocyanates (Jin et al. 2019). Crop rotations create a layered defense against pathogens through the addition of specific bacteria as a consortium to a conducive soil (Tringe 2019; Carrión et al. 2019).

14.4 MANAGEMENT OF SOIL STRUCTURE TO ENHANCE DISEASE SUPPRESSION BY CONSERVATION AGRICULTURE

Soil functional attributes can be strongly altered by choice of tillage methods and residue management practices (Figure 14.4). CA, already practiced on 180M ha globally (Kassam, Friedrich, and Derpsch 2019), is gaining momentum because of its effectiveness in erosion control and moisture conservation (Lal 2015). Because conversion to CA can influence soil microbial communities, it is

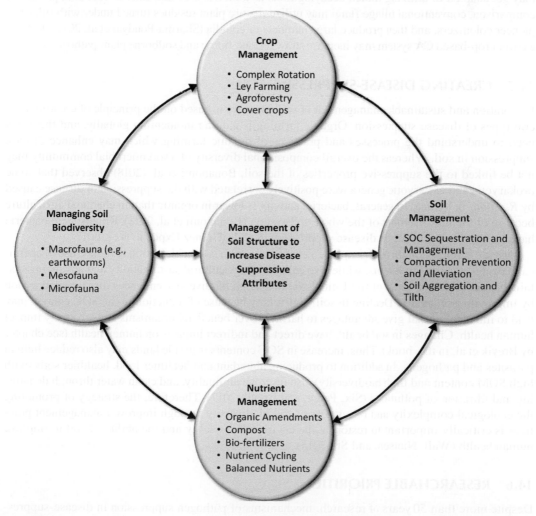

FIGURE 14.4 Strategies of improving soil structure through management of soil, nutrients, biodiversity, and crops.

possible to influence the soil processes with management decisions with regard to choice of tillage practices (Sipilä et al. 2012). Continuous use of CA over a long time leads to the concentration of biomass on the soil surface, decreases losses of water by runoff and evaporation, and moderates soil temperature. These conditions are similar to those under native undisturbed systems. Furthermore, differences in soil physical properties (e.g., compaction, aeration) in CA versus conventional tillage lead to differences in microbial community structures among tillage systems (Sipilä et al. 2012). In Sweden, Löbmann et al. (2016) observed that sites with biological suppression of root rot symptoms had permanent soil cover or a "green bridge" due to a permanent presence of plant roots, and biotic effects were associated with high sand and SOM contents. Summers et al. (2014) concluded that disease-suppressive effects related to cover crop treatments were associated with long-term changes in soil structure. Therefore, improvement of soil aeration through restoration of soil structure (i.e., macropores and continuous pores) may inhibit the development of such pathogens. In accordance with this hypothesis, van Bruggen et al. (2015) also reported that organic systems may have higher disease-suppressive potential because organic fertilizers improve soil health, resistance, and resilience. In the dryland Pacific Northwest (USA) wheat cropping systems, Sharma-Poudyal et al. (2017) observed that a larger proportion of fungal communities were not affected by tillage systems. However, no-till fungi may get adapted to utilizing intact, decaying roots as a food source and exist as root endophytes. In comparison, conventional tillage fungi may utilize mature plant residues turned under with tillage as pioneer colonizers, and then produce large numbers of conidia (Sharma-Poudyal et al. 2017). Indeed, a cover crop-based CA system may increase straw residue-borne and soilborne plant-pathogenic fungi.

14.5 CREATING DISEASE-SUPPRESSIVE SOILS

Restoration and sustainable management of soil structure is based on the principle of creating general types of disease suppression. Organic farming is gaining momentum globally, and there is a need to understand the processes and practices of organic farming which may enhance disease suppression in soil. Whereas the overall compositional diversity of a soil microbial community may not be linked to the suppressive properties of the soil, Bonanomi et al. (2018) observed that some prokaryotic and eukaryotic genera were positively correlated with the suppression of disease caused by *R. solani* in lettuce. In general, bacterial activity is lower in organic than in chemical agriculture because of the improvement of the whole soil system (Bonanomi et al. 2018). Rhizosphere bacteria have an important impact on disease suppression in soils (Gómez Expósito et al. 2017).

Disease suppression through improvement of soil health is a viable option to reduce use of chemical biocides in agroecosystems, while also enhancing agricultural sustainability and advancing sustainable development goals of the United Nations. High biodiversity enhances disease suppression by improving soil health. Decline in soil biodiversity because of reduction in the SOC content may lead to imbalances that give advantages to harmful over beneficial organisms and adversely impact human health. Changes in soil health have direct and indirect impacts on human health (see chapter by Brevik et al. in this book). Thus, increase in SOC contents in arable lands may also reduce human parasites and pathogens. In addition to producing abundant and healthier food, healthier soils (with high SOM content and high biodiversity) also improve air quality, and clean water through denaturing and filtration of pollutants (Six, Pereg, and Brevik 2017). Therefore, the strategy of promoting the ecological complexity and robustness of soil biodiversity through improved management practices is critically important to restoring the environment quality and the ability of soil to improve human health (Wall, Nielsen, and Six 2015).

14.6 RESEARCHABLE PRIORITIES

Despite more than 50 years of research, mechanisms of pathogen suppression in disease-suppressive soils remain poorly understood (Löbmann et al. 2016). Improved understanding of the disease inhibitory mechanisms of suppressive soils is necessary to the development of environmentally friendly tools to control diseases with minimal adverse impacts on the environment by reducing the use of pesticides. It is important to understand how indigenous microbiomes can reduce incidence of plant diseases even when plants are susceptible to the pathogens present in the root zone and environments are conducive to disease? Microbes in soil can suppress diseases. Therefore, researchable priorities include:

1. Identification of soil and crop-specific mechanisms for inducing disease suppression in soils,
2. Development of soil and crop management practices which enhance general and specific disease suppression,
3. Assessment of the properties of compost which strengthen disease suppression,
4. Conduct microbiome research to understand mechanisms of general and specific disease suppression,
5. Establishment of the cause–effect relationship between soil health and human health through nutritional quality of the food produced,
6. Evaluate impact of organic farming in creating disease suppression in soils,
7. Understand determinants of soil biodiversity and how they affect disease suppressiveness and human health,
8. Strengthen understanding of the positive effects of soil structure,aeration and porosity on disease suppression,
9. Relate soil physical attributes to microbial community structure and compostion in soil,and
10. Enhance understanding of the soil–plant–animal–human health nexus.

14.7 CONCLUSIONS

Pesticide use has increased agricultural production, However, indiscriminate and excessive use has adversely impacted biodiversity and quality of water and air. Exposure to pesticides also have severe impacts on human health. Appropriate and discriminate use through a right dose is important to using pesticide as a remedy rather than as a poison. Furthermore, pesticide use in crops can also be reduced by creating disease-suppressive soils. These soils are characterized by a high SOM content, high biodiversity, good soil structure, favorable aeration, and optimal soil moisture and temperature regimes. These characteristics are imparted by regular inputs of compost and biomass-C, and adoption of CA. The latter comprises no-till, retention of crop residue mulch, complex rotations, inclusion of a cover crop in the rotation cycle, integrated nutrient management (judicious combination of organic and inorganic fertilizers along with the use of rhizobial and mycorrhizae), and integration of crops with trees and livestock. Despite more than 50 years of research on disease-suppressive soils, the underlying mechanisms are not clearly understood. Nonetheless, maintenance of a permanent ground cover (i.e., pastures, cover cropping) may change the microbiome community structure that suppress diseases even in the presence of pathogens and host plants. There is a strong need to conduct multi-disciplinary research in understanding of soil and crop-specific mechanisms of disease suppression and in identification of management practices that enhance and strengthen these attributes so that use of pesticides in agricultural ecosystems can be reduced and food safety and environment quality is enhanced. The relation between soil health and human health is supported by the dictum that "the health of soil, plants, animals, people and environment is one and indivisible."

REFERENCES

Agrios, G.N.B.T. 2005. Control of Plant Diseases. In *Plant Pathology*, ed. G.N.B.T. Agrios, 293–353. 5th ed. San Diego, CA: Academic Press. http://www.sciencedirect.com/science/article/pii/B9780080473789500154.

Allard, S.M., and S.A. Micallef. 2019. Chapter 11- The Plant Microbiome: Diversity, Dynamics, and Role in Food Safety. In *Safety and Practice for Organic Food*, ed. D. Biswas and S.A. Micallef, 229–257. Academic Press. http://www.sciencedirect.com/science/article/pii/B9780128120606000118.

Antoniou, A., M.-D. Tsolakidou, I.A. Stringlis, and I.S. Pantelides. 2017. Rhizosphere Microbiome Recruited from a Suppressive Compost Improves Plant Fitness and Increases Protection against Vascular Wilt Pathogens of Tomato. *Frontiers in Plant Science* 8 (November 29): 2022. https://pubmed.ncbi.nlm.nih.gov/29238353.

Baker, K.F., and R.J. Cook. 1974. *Biological Control of Plant Pathogens*. Ed. W.H. Freeman. San Francisco, CA: American Phytopathological Society.

Ball, B.C., I. Bingham, R.M. Rees, C.A. Watson, and A. Litterick. 2005. The Role of Crop Rotations in Determining Soil Structure and Crop Growth Conditions. *Canadian Journal of Soil Science* 85, no. 5 (November 1): 557–577. doi:10.4141/S04-078.

Berruti, A., E. Lumini, R. Balestrini, and V. Bianciotto. 2016. Arbuscular Mycorrhizal Fungi as Natural Biofertilizers: Let's Benefit from Past Successes. *Frontiers in Microbiology* 6 (January 19): 1559. https://pubmed.ncbi.nlm.nih.gov/26834714.

Bhatti, A.A., S. Haq, and R.A. Bhat. 2017. Actinomycetes Benefaction Role in Soil and Plant Health. *Microbial Pathogenesis* 111: 458–467. Academic Press.

Bonanomi, G., G. Cesarano, V. Antignani, C. Di Maio, F. De Filippis, and F. Scala. 2018. Conventional Farming Impairs Rhizoctonia Solani Disease Suppression by Disrupting Soil Food Web. *Journal of Phytopathology* 166, no. 9 (September 1): 663–673. https://doi.org/10.1111/jph.12729.

Bronick, C.J., and R. Lal. 2005. Soil Structure and Management: A Review. *Geoderma* 124, no. 1–2 (January): 3–22.

van Bruggen, A.H.C., and A.M. Semenov. 2000. In Search of Biological Indicators for Soil Health and Disease Suppression. *Applied Soil Ecology* 15, no. 1: 13–24. http://www.sciencedirect.com/science/article/pii/S0929139300000688.

van Bruggen, A.H.C., K. Sharma, E. Kaku, S. Karfopoulos, V. V Zelenev, and W.J. Blok. 2015. Soil Health Indicators and Fusarium Wilt Suppression in Organically and Conventionally Managed Greenhouse Soils. *Applied Soil Ecology* 86: 192–201. http://www.sciencedirect.com/science/article/pii/S0929139314002984.

Carrión, V.J., J. Perez-Jaramillo, V. Cordovez, V. Tracanna, M. de Hollander, D. Ruiz-Buck, L.W. Mendes, et al. 2019. Pathogen-Induced Activation of Disease-Suppressive Functions in the Endophytic Root Microbiome. *Science* 366, no. 6465 (November 1): 606–612. http://science.sciencemag.org/content/366/6465/606.abstract.

Carson, R. 1962. *Silent Spring*. Boston, MA: Houghton Mifflin.

Conway, K.E. 1996. An Overview of the Influence of Sustainable Agricultural Systems on Plant Diseases. *Crop Protection* 15, no. 3: 223–228. http://www.sciencedirect.com/science/article/pii/0261219495001190.

Cook, R.J. 2014. Plant Health Management: Pathogen Suppressive Soils. In *Encyclopedia of Agriculture and Food Systems*, ed. N.K. Van Alfen, 441–455. Cambridge, MA: Academic Press Inc. https://linkinghub.elsevier.com/retrieve/pii/B9780444525123001820.

Geisen, S., D.H. Wall, and W.H. van der Putten. 2019. Challenges and Opportunities for Soil Biodiversity in the Anthropocene. *Current Biology* 29, no. 19: R1036–R1044. http://www.sciencedirect.com/science/article/pii/S0960982219310231.

Ghorbani, R., S. Wilcockson, A. Koocheki, and C. Leifert. 2008. Soil Management for Sustainable Crop Disease Control: A Review. *Environmental Chemistry Letters* 6, no. 3: 149–162. doi:10.1007/s10311-008-0147-0.

Gómez Expósito, R., I. de Bruijn, J. Postma, and J.M. Raaijmakers. 2017. Current Insights into the Role of Rhizosphere Bacteria in Disease Suppressive Soils. *Frontiers in Microbiology* 8 (December 18): 2529. https://pubmed.ncbi.nlm.nih.gov/29326674.

Grandjean, P. 2016. Paracelsus Revisited: The Dose Concept in a Complex World. *Basic & Clinical Pharmacology & Toxicology* 119, no. 2 (August 1): 126–132. doi:10.1111/bcpt.12622.

Greenland, D.J. 1965. Interaction between Clays and Organic Compounds in Soils. II: Adsorption of Organic Compounds and This Effect on Soil Properties. *Soil Fertility* 28: 521–532.

Hoeper, H., and C. Alabouvette. 1996. Importance of Physical and Chemical Soil Properties in the Suppressiveness of Soils to Plant Diseases. Ed. FAO. *European Journal of Soil Biology* 32, no. 1: 41–58.

Hoitink, H.A.J., and M. Boehm. 1999. Biocontrol Within the Context of Soil Microbial Communities: A Substrate-Dependent Phenomenon. *Annual Review of Phytopathology* 37, no. 1 (September 1): 427–446. doi:10.1146/annurev.phyto.37.1.427.

Hoitink, H.A.J., and P.C. Fahy. 1986. Basis for the Control of Soilborne Plant Pathogens with Composts. *Annual Review of Phytopathology* 24, no. 1 (September 1): 93–114. doi:10.1146/annurev.py.24.090186.000521.

Hoitink, H.A.J., B.B. McSpadden, and S.A. Miller. 2008. Current Knowledge on Disease Suppressive Properties of Compost. In *Proceedings of the International Congress CODIS 2008*, ed. J.G. Fuchs, T. Kupper, L. Tamm, and K. Schenk, 19–26. Solothum, Switzerland: 27–29 February 2008.

Hoitink, H.A.J., A.G. Stone, and M.E. Grebus. 1996. Suppression of Plant Diseases by Composts BT - The Science of Composting. In ed. M. de Bertoldi, P. Sequi, B. Lemmes, and T. Papi, 373–381. Dordrecht: Springer Netherlands. doi:10.1007/978-94-009-1569-5_35.

Howard, A. 1921. The Influence of Soil Factors on Diseas Resistance. *Annals of Applied Biology* 7, no. 4 (February 1): 373–389. doi:10.1111/j.1744-7348.1921.tb05525.x.

Huber, D.M., and R.D. Watson. 1970. Effect of Organic Amendment on Soil-Borne Plant Pathogens. *Phytopathology* 60, no. January: 22.

Janvier, C., F. Villeneuve, C. Alabouvette, V. Edel-Hermann, T. Mateille, and C. Steinberg. 2007. Soil Health through Soil Disease Suppression: Which Strategy from Descriptors to Indicators? *Soil Biology and Biochemistry* 39, no. 1: 1–23. http://www.sciencedirect.com/science/article/pii/S0038071706003142.

Jin, X., J. Wang, D. Li, F. Wu, and X. Zhou. 2019. Rotations with Indian Mustard and Wild Rocket Suppressed Cucumber Fusarium Wilt Disease and Changed Rhizosphere Bacterial Communities. *Microorganisms* 7, no. 2 (February): 57.

Kassam, A., T. Friedrich, and R. Derpsch. 2019. Global Spread of Conservation Agriculture. *International Journal of Environmental Studies* 76, no. 1 (January 2): 29–51. doi:10.1080/00207233.2018.1494927.

Lal, R. 2015. A System Approach to Conservation Agriculture. *Journal of Soil and Water Conservation* 70, no. 4 (July 1): 82A–88A.

Lal, R., E.C. Brevik, L. Dawson, D. Field, B. Glaser, A. Hartemink, R. Hatano, et al. 2020. Managing Soils for Recovering from the COVID-19 Pandemic. *Soil Systems* 4, no. 3: 46.

Löbmann, M.T., R.R. Vetukuri, L. de Zinger, B.W. Alsanius, L.J. Grenville-Briggs, and A.J. Walter. 2016. The Occurrence of Pathogen Suppressive Soils in Sweden in Relation to Soil Biota, Soil Properties, and Farming Practices. *Applied Soil Ecology* 107: 57–65. http://www.sciencedirect.com/science/article/pii/S0929139316301639.

Lozano, J., W.J. Blok, and A.J. Termorshuizen. 2008. Effect of Compost Particle Size on Suppression of Plant Diseases. *Environmental Engineering Science* 26, no. 3 (December 27): 601–607. doi:10.1089/ees.2008.0002.

Mahmood, I., S.R. Imadi, K. Shazadi, A. Gul, and K.R. Hakeem. 2016. Effects of Pesticides on Environment BT. In *Plant, Soil and Microbes: Volume 1: Implications in Crop Science*, ed. K.R. Hakeem, M.S. Akhtar, and S.N.A. Abdullah, 253–269. Cham, Switzerland: Springer International Publishing. doi:10.1007/978-3-319-27455-3_13.

Marquez, E. 2018. In the U.S. and the World, Pesticide Use Is Up. *Pesticide Action Network*. http://www.panna.org/blog/us-and-world-pesticide-use.

Mazzola, M. 2002. Mechanisms of Natural Soil Suppressiveness to Soilborne Diseases. *Antonie van Leeuwenhoek* 81, no. 1: 557–564. doi:10.1023/A:1020557523557.

Mazzola, M. 2004. Assessment and Management of Soil Microbial Community Structure for Disease Suppression. *Annual Review of Phytopathology* 42: 35–59.

Mousa, W.K., and M.N. Raizada. 2016. Natural Disease Control in Cereal Grains. In *Encyclopedia of Food Grains*, eds. C. Wrigley, H. Corke, and K. Seetharaman, J. Faubion, , 2nd edn., vol. 4, 257–263. Oxford, UK: Academic Press.

Munkholm, L.J., R.J. Heck, and B. Deen. 2013. Long-Term Rotation and Tillage Effects on Soil Structure and Crop Yield. *Soil and Tillage Research* 127: 85–91. http://www.sciencedirect.com/science/article/pii/S0167198712000554.

Nelson, E.B., A. Kuter, and H.A.J. Hoitink. 1983. Effects of Fungal Antagonists and Compost Age on Suppression of Rhizoctonia Damping-Off in Container Media Amended with Compost Hardwood Bark. *Ecology and Epidemiology* 73: 1458–1462.

Noble, R., and E. Coventry. 2005. Suppression of Soil-Borne Plant Diseases with Composts: A Review. *Biocontrol Science and Technology* 15, no. 1 (February 1): 3–20. doi:10.1080/09583150400015904.

Otten, W., D. Hall, K. Harris, K. Ritz, I.M. Young, and C.A. Gilligan. 2001. Soil Physics, Fungal Epidemiology and the Spread of Rhizoctonia Solani. *New Phytologist* 151, no. 2 (August 1): 459–468. doi:10.1046/j.0028-646x.2001.00190.x.

Penton, C.R., V.V.S.R. Gupta, J.M. Tiedje, S.M. Neate, K. Ophel-Keller, M. Gillings, P. Harvey, A. Pham, and D.K. Roget. 2014. Fungal Community Structure in Disease Suppressive Soils Assessed by 28S LSU Gene Sequencing. *PLoS One* 9, no. 4 (April 3): e93893. doi:10.1371/journal.pone.0093893.

Peters, R.D., A. V Sturz, M.R. Carter, and J.B. Sanderson. 2003. Developing Disease-Suppressive Soils through Crop Rotation and Tillage Management Practices. *Soil and Tillage Research* 72, no. 2: 181–192. http://www.sciencedirect.com/science/article/pii/S0167198703000874.

Poltronieri, P., and I.B. Reca. 2020. Chapter 9- Microbial Products and Secondary Metabolites in Plant Health. In, ed. P. Poltronieri and Y.B.T.-A.P.B. for I.R. to B.S. Hong, 189–202. Academic Press. http://www.sciencedirect.com/science/article/pii/B9780128160305000094.

Savci, S. 2012. Investigation of Effect of Chemical Fertilizers on Environment. *APCBEE Procedia* 1: 287–292. http://www.sciencedirect.com/science/article/pii/S2212670812000486.

Schlatter, D., L. Kinkel, L. Thomashow, D. Weller, and T. Paulitz. 2017. Disease Suppressive Soils: New Insights from the Soil Microbiome. *Phytopathology*^{TM} 107, no. 11 (June 26): 1284–1297. https://doi.org/10.1094/PHYTO-03-17-0111-RVW.

Sharma-Poudyal, D., D. Schlatter, C. Yin, S. Hulbert, and T. Paulitz. 2017. Long-Term No-till: A Major Driver of Fungal Communities in Dryland Wheat Cropping Systems. *PLoS One* 12, no. 9 (September 12): e0184611. doi:10.1371/journal.pone.0184611.

Sharma, A., V. Kumar, B. Shahzad, M. Tanveer, G.P.S. Sidhu, N. Handa, S.K. Kohli, et al. 2019. Worldwide Pesticide Usage and Its Impacts on Ecosystem. *SN Applied Sciences* 1, no. 11: 1446. doi:10.1007/s42452-019-1485-1.

Sharma, N., and R. Singhvi. 2017. Effects of Chemical Fertilizers and Pesticides on Human Health and Environment: A Review. *International Journal of Agriculture, Environment and Biotechnology* 10: 675–680.

Sipilä, T.P., K. Yrjälä, L. Alakukku, and A. Palojärvi. 2012. Cross-Site Soil Microbial Communities under Tillage Regimes: Fungistasis and Microbial Biomarkers. *Applied and Environmental Microbiology* 78, no. 23 (December 1): 8191–8201. http://aem.asm.org/content/78/23/8191.abstract.

Six, J., H. Bossuyt, S. Degryze, and K. Denef. 2004. A History of Research on the Link between (Micro) Aggregates, Soil Biota, and Soil Organic Matter Dynamics. *Soil and Tillage Research* 79, no. 1: 7–31. http://www.sciencedirect.com/science/article/pii/S0167198704000881.

Six, J., L. Pereg, and E. Brevik. 2017. Soil Biodiversity and Human Health. In *EGU General Assembly Conference Abstracts*, 19207. https://ui.adsabs.harvard.edu/abs/2017EGUGA..1919207S.

Stirling, G., H. Hayden, T. Pattison, and M. Stirling. 2016. *Soil Health, Soil Biology, Soilborne Diseases and Sustainable Agriculture: A Guide*. Clayton, Australia: CSIRO Publishing.

Stone, A.G., S. Scheurell, and H.M. Darby. 2004. Supression of Soil-Borne Diseases in Field Agricultural Systems: Organic Matter Management, Cover Cropping and Other Cultural Practices. In *Soil Organic Matter in Sustainable Agriculture*, ed. F. Magdoff and R.R. Weil, 131–177. Boca Raton, FL: CRC Press.

Sturz, A.V, and B.R. Christie. 2003. Beneficial Microbial Allelopathies in the Root Zone: The Management of Soil Quality and Plant Disease with Rhizobacteria. *Soil and Tillage Research* 72, no. 2: 107–123. http://www.sciencedirect.com/science/article/pii/S0167198703000825.

Summers, C.F., S. Park, A.R. Dunn, X. Rong, K.L. Everts, S.L.F. Meyer, S.M. Rupprecht, M.D. Kleinhenz, B. McSpadden Gardener, and C.D. Smart. 2014. Single Season Effects of Mixed-Species Cover Crops on Tomato Health (Cultivar Celebrity) in Multi-State Field Trials. *Applied Soil Ecology* 77: 51–58. http://www.sciencedirect.com/science/article/pii/S0929139314000274.

Sumner, D.R. 1982. Crop Rotation and Plant Productivity. In *Handbook of Agricultural Productivity. Volume I: Plant Productivity*, ed. M. Rechcigl, 273–315. Boca Raton, FL: CRC Press.

Sumner, D.R., and D.K. Bell. 1982. Root Diseases Induced in Corn by Rhizoctonia Solani and Rhizoctonia Zeae. Ed. F A O of the UN. *Phytopathology* 72, no. 1: 86.

Tilston, E.L., D. Pitt, and A.C. Groenhof. 2002. Composted Recycled Organic Matter Suppresses Soil-Borne Diseases of Field Crops. *New Phytologist* 154, no. 3 (June 1): 731–740. doi:10.1046/j.1469-8137.2002.00411.x.

Tisdall, J.M., and J.M. Oades. 1982. Organic Matter and Water-stable Aggregates in Soils. *Journal of Soil Science* 33, no. 2 (June 1): 141–163. doi:10.1111/j.1365-2389.1982.tb01755.x.

Toyota, K., and S. Shirai. 2018. Growing Interest in Microbiome Research Unraveling Disease Suppressive Soils against Plant Pathogens. *Microbes and Environments* 33, no. 4: 345–347. https://pubmed.ncbi.nlm.nih.gov/30606975.

Tringe, S.G. 2019. A Layered Defense against Plant Pathogens. *Science* 366, no. 6465 (November 1): 568–569. http://science.sciencemag.org/content/366/6465/568.abstract.

Wall, D., U. Nielsen, and J. Six. 2015. Soil Biodiversity and Human Health. *Nature* 528, no. 7580 (November 23): 69–76.

Weller, D.M., J.M. Raaijmakers, B.B.M. Gardener, and L.S. Thomashow. 2002. Microbial Populations Responsible for Specific Soil Suppressiveness to Plant Pathogens. *Annual Review of Phytopathology* 40: 309–348.

Zhang, W. 2018. Global Pesticide Use: Profile, Trend, Cost/Benefit and More. *Proceedings of the International Academy of Ecology and Environmental Sciences* 8, no. 1 (March 1): 1–27.

15 Soil Health and Human Nutrition

Rattan Lal
The Ohio State University

CONTENTS

15.1 INTRODUCTION

Malnutrition is a global problem for the 7.8 billion world population in 2020 (UN 2019), of which 1.9 billion or 39% of the world's adults are obese (UNICEF 2018) and 462 million are underweight (Branca 2019). Globally, 52 million children under five are stunted and prone to low weight for height. Lack of adequate nutrition is the major cause of death and disease (Branca 2019). While considerable progress has been made in reducing stunting among children under five from 32.6% in 2000 to 22.2% in 2017 (UNICEF 2018), there is still a long way toward eliminating malnourishment. It is precisely in this context, that restoration and sustainable management of soil health is a high priority because the health of soil, plants, animals, people and the environment is one and indivisible.

There are four types of malnutrition (Stephenson and Latham 2000): (i) protein-energy malnutrition, (ii) iron deficiency and anemia (IDA), (iii) vitamin A deficiency (VAD), and (iv) iodine deficiency disorders (IDD). In addition, zinc deficiency can also lead to decline in and resistance and increased vulnerability in infections (Stephenson and Latham 2000). The present statistics on global hunger and malnourishment, as shown in Table 15.1, indicate that as much as 25% of the world population in 2017 was prone to food insecurity, and 22% of children under five were stunted.

TABLE 15.1
Global Statistics on Hunger and Malnourishment

Parameter	Global Statistics	
	Million	% of Total Population
Number of people globally undernourished	820	11
Number of severely undernourished people	697	9
Number of moderately or severely food-insecure	1,900	25
Stunted children under five	-	22

Source: Adapted from Roser and Ritchie (2020).

315

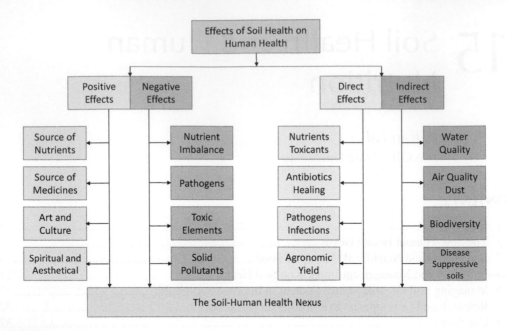

FIGURE 15.1 Human malnourishment is caused by the interaction between a poor diet and a poor health environment. Three components affecting human health are: (i) soil, (ii) water, and (iii) air. (IDA=iron deficiency and anemia, VAD=vitamin A deficiency, IDD=iodine deficiency disorder. Refer to the text for details).

In view of the data listed in Table 15.1, achieving the Sustainable Development Goal (SDG) of the U.N. to eliminate hunger by 2030 (#2 Zero Hunger) may be a daunting challenge. Just as was the case with the Agenda 21 of 1992, and Millennium Development Goals of 2000, SDGs of 2030 may also turn out to be yet another inspirational slogan, and thus, a pie in the sky. The challenge is aggravated by the COVID-19 pandemic.

Malnutrition is caused by the interaction of a poor-quality diet with that of a poor health environment (Figure 15.1). Important among the environmental factors affecting human health both directly and indirectly is soil health (refer to Chapter 1 in this book). Therefore, the objective of this concluding chapter is to describe the impact of soil health on human health and to outline management strategies which can alleviate soil-related constraints and advance SDG #2 (Zero Hunger) and SDG #3 (Human Health and Wellbeing) of the U.N. by 2030.The specific objective is to focus on the "One Health" concept, and adopt a wholistic approach to management of human health through management of the soil–plants–animals–people–environment nexus or the continuum.

15.2   THE SOIL–HUMAN HEALTH NEXUS

Soil affects human health positively and negatively, directly and indirectly (Figure 15.2). In the Anthropocene (Crutzen and Stoermer 2000), human activities can also improve or degrade soil health depending on the specific land use and soil management (Pepper 2013). Soil can positively impact human health by providing nourishment, medicaments, or antibiotics, and through environmental quality in relation to air we breathe and water we drink. Therefore, soil restoration and judicious management, though advancing soil security, can advance/improve human health and well-being (Pepper 2013; Steffan et al. 2018). Brevik (2013a) estimated that 78% of the average per capita calorie consumption worldwide comes from crops grown directly in soil, and another nearly 20% comes from terrestrial food sources that rely indirectly on soil. Thus, the quality of food consumed depends on soil health. Climate change may have adverse impacts on soil health and thus on

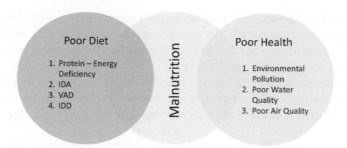

FIGURE 15.2 Interactive effects of soil health on human health.

nutritional quality of the food produced from soil (Brevik 2013b). Therefore, directly, soil affects human growth through quantity and quality of food produced and the nutrition provided through it (Brevik and Burgess 2014). Food may also adversely affect human health through toxic elements (e.g., heavy metals, Pb, Cd, As, Hg) contained in food (Hu 2002; Hubert et al. 2010). This is also the reason that soil degradation is a major cause of human malnutrition (Lal 2009). Soil is also a source of organic and inorganic pollutants that can adversely affect human health (Figure 15.2).

15.3   SOIL HEALTH AND NUTRITIONAL QUALITY OF FOOD

Soil, environmental factors, farming systems, land use and management, and post-harvest handling can all strongly impact the nutritional quality of food (Figure 15.3). Soil quality and functionality; comprising physical, chemical, biological, and ecological parameters; can impact nutrient uptake by plants (crops and pastures) and impact nutrients in plant- and animal-based food products. The impacts of soil factors can be moderated by interactions with climate and weather during the growing season, farming systems, and land use and management. Impacts of land use and management, along with that of farming systems can be both positive and negative

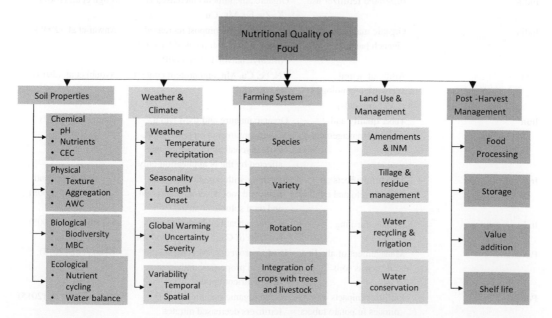

FIGURE 15.3 Factors affecting nutritional quality of food (CEC = Cation exchange capacity, AWC = Available water capacity, MBC = Microbial biomass carbon, INM = Integrated nutrient management).

(Figure 15.2). Therefore, farmers and land managers must be aware of the site-specific recommended management practices which can enhance the beneficial aspects while mitigating the harmful effects. The importance of relevant education about soil properties and processes cannot be overemphasized (Brevik and Burgess 2014).

There are numerous examples of the impacts of soil management on the nutritional quality of food (Table 15.2). The objective of soil management is to increase the micronutrient (Fe, Zn, Cu, I) density in grains consumed by humans to alleviate micronutrient deficiency and improve human health

TABLE 15.2

Some Examples of the Impact of Soil Management on Nutritional Quality of Food Grains

Region/Country	Management	Impact	Reference
Australia	Organic and inorganic fertilizers of wheat	Positive impact on compost on reducing N use to maximize yield of wheat protein and gluten	Abedi, Alemzadeh, and Kazemeini (2010)
China (Henan Province)	Organic and chemical fertilizers on maize and wheat	Deficiency of N or P decreased yield but increased concentration of micronutrients	Li et al. (2007)
Czech	Protein and nitrates in potato tubers	No differences in nitrate and protein contents among ecological and conventional systems	Lachman et al. (2005)
Egypt	Bio-organic fertilizers on wheat	25% bio-organic and 75% NPK produced the best yield, protein, P, K in grains	Hafez and Badawy (2018)
India	Nitrogen application	Fe, Zn, and protein	Chandel et al. (2010) and Mishra et al. (2006)
India	Organic manure in rice–wheat system	Application of organic manures is essential for uptake of Zn and Cu in rice–wheat	
India	Integrated fertilizer use	Organic amendments increased P, K, Zn, Fe, Mn, Cu	Singh et al. (2007)
India	Organic manure and French basil	INM and vermicompost increased production and content of basil oil (Methylchaivol and linalool)	Anwar et al. (2005)
Iran	Artificial neural network technology (ANN models)	ZN, Fe, Cu, Mn, and protein content in wheat	Ayoubi et al. (2014)
Iran	Tuber quality and fertility management	Organic manure and INM had a significant impact on tuber yield. Total glycoalkaloid content was affected by the N application only	Najm et al. (2012)
Iran	Mulching effects on carrot yield and quality	Mulching with organic materials improved root development, yield and total soluble contents	Olfati, Peyvast, and Nosrati-Rad (2008)
Italy	Organic farming	Organic wheat had higher grain contents of Cu, Mg, Mn, P and Zn	Murphy et al. (2008)
Pakistan	Management of alkaline calcareous soils	INM improved SOM content, soil physical properties, and micronutrient uptake	Khan et al. (2018)
Poland	Fertilizer impacts on nitrates in potato tubers	Use of organic manures and fertilizers decreased nitrates in potatoes	Pobereżny et al. (2015)

(Continued)

TABLE 15.2 (*Continued*)

Some Examples of the Impact of Soil Management on Nutritional Quality of Food Grains

Region/Country	Management	Impact	Reference
South Africa	Response of carrot to organic fertilizers	Organic fertilizers are beneficial to carrot yield and quality	Mbatha, Ceronio, and Coetzer (2014)
Turkey (Anatolia)	Application of Zn to wheat	Soil application of Zn was economical with a long-term effect	Yilmaz et al. (1997)
United Kingdom	Biochar	Micronutrients were reduced in wheat grains in biochar-amended soils	Hartley, Riby, and Waterson (2016)
United Kingdom	Protein profile of potato tubers	Organic fertilization leads to an increased stress response and protein composition	Lehesranta et al. (2007)
USA	Rotation impact on potato protein	Tuber protein yield increased when grown in rotation with alfalfa, hairy vetch, white lupin, oats, etc.	Honeycutt (1998)
USA	Effect of humic acid (HA) application on micronutrients in wheat: hydroponic study	HA ameliorated lent interveinal chlorosis	Mackowiak, Grossl, and Bugbee (2001)
USA (Palouse region)	Split N application	Reduced N rate and split N application in produce	Sowers, Miller, and Pan (1994)

(Rengel, Batten, and Crowley 1999; Rengel 2015). Indeed, fertilization with a judicious combination of organic and inorganic micronutrient fertilizers can increase their concentration in soils and in grains (Dhaliwal et al. 2019; Shiwakoti et al. 2019). These nutrients can be applied via soil (ZN, Cu), foliar spray (Fe), or through irrigation water or fertigation (I). However, over-fertilization can lead to toxicity (Rengel, Batten, and Crowley 1999). In the Zhejiang Province of China, Wang et al. (2009) observed that differences in soil properties over short distances because of anthropogenic contamination can also affect the concentrations of micronutrients in rice grains. Based on a study in Bikaner, Rajasthan, India, Kumar and Babel (2011) reported that the micronutrient levels in wheat grains and straw were positively correlated with silt, clay, soil organic matter (SOM) content, and cation exchange capacity (CEC) but negatively with sand and $CaCO_3$ contents and soil pH. Chen (1996) observed that SOM content can have a positive effect on uptake of Fe, Mn, Zn, and Cu by higher plants. Thus, use of micronutrient-enriched organic by-products can be used as soil amendments to improve soil health and nutritional value of food products.

15.4 LAND USE AND MANAGEMENT IMPACTS ON SOIL HEALTH

There are many definitions of soil health. It has been defined in relation to the health of plants (Idowu et al. 2019) and health of plants, animals, and people as a living system (Doran and Zeiss 2000). However, the definition of soil health is complete only if it indicates that the "importance of the continuum" health of soil, plants, animals, people, and environment is one and indivisible (Lal 2016). It is this continuum that is strongly influenced by the SOM content and its dynamic as moderated by land use change. The historic land use change since the dawn of settled agriculture has created the global SOC debt of ~133 Pg C (Sanderman, Hengl, and Fiske 2017; Lal 2018) with strong implications to soil health and functionality. Therefore, urgency to improve soil health to mitigate human malnutrition has increased interest about the impact of land use and management on soil health. Land use (e.g., forest land, grazing land, cropland, urban land, recreational land, mine land) impacts soil by moderating the quality and quantity of the SOM content in the root zone. Land use which increases input of biomass-C in soil and provides a continuous ground cover would enhance soil health through an increase in soil, organic carbon (SOC) content, quality, and dynamics. For example, Thapa et al. (2018) evaluated the impact of land use in semiarid drylands on SOC components as indicators of soil health. Thapa and colleagues observed that the SOC content was 36.9% greater in grasslands than cropland irrespective of the grazing system. They concluded that restoring grasslands can improve soil health and resilience in semi-arid regions. Whereas increasing the SOC content in soil of arid regions is a daunting task, positive impacts on soil health and functionality are also evident by the research data from Iran (Havaee et al. 2013) and from the Birr Watershed of the Upper Blue Nile River Basin in Ethiopia (Amanuel, Yimer, and Karltun 2018). Lal (2020a) observed that integration of livestock with crops and trees have a positive impact on soil health and strengthening of ecosystem services. Land use can be a principal driver of the SOC content and its spatial distribution (Fusaro et al. 2019), especially in an ecologically fragile ecosystem. Changes in SOC quality by change in land use can also affect structures and functions of microbial communities and having strong implications to disease suppressions in plants (Dignam et al. 2018, also see Chapter 14 in this book). Based on a comparative study using[13]C NMR and DRIFT spectroscopy (Leimona et al. 2015), Yeasmin et al. (2020) observed effects of land use on the SOM composition in density fractions of contrasting soils and the attendant impact on soil health.

15.5   MANAGING SOIL ORGANIC MATTER CONTENT TO IMPROVE SOIL HEALTH

Restoring and maintaining the SOM content above the critical threshold, which varies among soils and farming systems (Lal 2020b, 2020c), are important to enhancing soil health and improving productivity (Figure 15.4). Therefore, use of the strategy of integrated nutrient management (INM), based on a judicious combination of organics (recycling of biomass and biofertilizer) and inorganic fertilizers, is important to enhancing productivity and nutritional quality of food (Table 15.2). Several long-term experiments conducted globally have documented the positive impacts of using organics on soil quality and functionality. The effects of different types and amounts of organic materials on productivity have been quantified for different crops from around the world, including

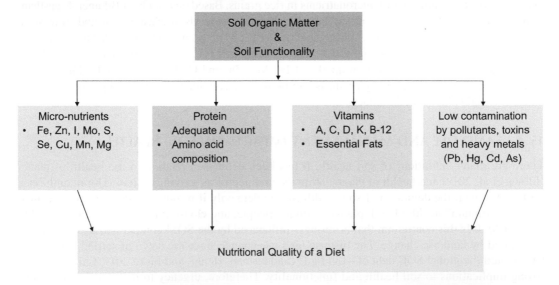

FIGURE 15.4   Effects of soil organic matter content on enhancing nutritional quality of the food in relation to human health.

for cereals (Barzegar, Yousefi, and Daryashenas 2002; Daneshmand, Bakhshandeh, and Rostami 2012; Sarwar et al. 2007; Sary and El-Naggar 2009; Osman, Hassan, and Yassin 2014; Gupta, Narwal, and Antil 2003; Singh et al. 2010; Bayu et al. 2006), winter canola seed (Kazemeini, Hamzehzarghani, and Edalat 2010), sweet potato (Laxminarayana et al. 2015), and roots and tuber crops (Olaleye, Akinbola, and Akintade 2011). These are merely a few examples of the beneficial impacts of organic fertilizers on soil quality and productivity of food and oil seed crops. Yet, the widespread use of organics is limited by the availability, bulk amount, and other logistical issues, including the competing uses of this precious but finite resource.

15.6 RESEARCH AND DEVELOPMENT PRIORITIES

The literature is replete with reports stating the importance of organics to sustaining productivity, improving soil quality, and enhancing the environment. Yet fine-tuning is needed for adaptation of these concepts under site-specific conditions with due consideration of the biophysical (soil, climate, terrain, farming systems), socio-economic (farm size, infrastructure, access to market, credit facilities, inputs), and the human dimensions (land tenure, gender, education, social, and cultural traditions).

Competing uses of organics (i.e., feed, fuel, construction) is another important factor that must be considered. Availability and affordability of clean cooking fuel are essential to sparing agricultural by-products (i.e., crop residues, animal dung) for mulching and for composting to be used as a soil amendment. Use of cover crops in a rotation cycle and during the off-season is a pertinent technological option. In general, cover cropping is more appropriate in humid and sub-humid regions than in semi-arid and arid climates because of the need to store water in the rootzone for food crops (Unger and Vigil 1998). Site-specific research is needed to choose an appropriate cover crop species, growing duration, seed-setting canopy height, growth characteristics, and the ease of suppression by mechanical, rather than chemical, technologies (i.e., mowing). Above all, a cover crop must not compete for water, nutrients, light, or other resources.

Soil- and ecoregion-specific research is also needed for assessing the rate of SOC and soil inorganic carbon (SIC) sequestration in relation to methods of soil fertility management. In addition to assessing the gross rate, the net rate of SOC/SIC sequestration must be determined with due consideration to emissions from farm operations (Lal 2004). In addition to assessing the rate of soil-C sequestration for different type, rates, and mode of application, the amount of biomass-C needed to maintain SOC concentrations must be assessed for soil type, climate, and land use. Furthermore, the critical/threshold level of SOC must be determined below which productivity and nutritional quality is jeopardized (Lal 2020b, 2020c).

The effect of an incremental increase in the SOC stock in the rootzone must be assessed on both the amount of grains/food produced but also on the protein and micro-nutrients (Cu, Zn, Fe, I, Mo) harvested. The focus is on nutritional contents rather than just on the agronomic yield. Thus, nutrition-sensitivity of agriculture is a high priority for the 21st century. It is also important to translate the known and proven science into action by establishing communication with policymakers, so that agriculture is always nutrition-sensitive.

The COVID-19 pandemic has brought the issue of food and nutrition to the forefront. In this context, not only should the food wastage be minimized, but also the role of home gardens and urban farming must be objectively and critically reconsidered toward ensuring food and nutrient security at the household level (Lal 2020d).The goal is to strengthen the local food production system, enhance their resilience against disruptions, and increase their nutritional quality. Another important issue is the role of small landholders and resource-poor farmers who will become even more vulnerable to the isolation caused by the pandemic. These farmers and their dependent may be prone to malnourishment because of the lockdown and reduced access to basic necessities, markets, tools, inputs, and credit facilities.

15.7 CONCLUSIONS

The daunting challenge of global malnutrition may be aggravated by uncertainties related to global warming and the incidents such as the COVID-19 pandemic. Most vulnerable members of society are children, nursing mothers, elderly, and the impoverished who can neither afford nor have the means to grow an adequate amount of nutrient-rich food on degraded and depleted soils managed by extractive farming practices. Soil and crop management, especially that related to the inputs of macro- and micronutrients through integrated use of biofertilizers and chemical fertilizers, can improve the concentrations of protein and micronutrients in food products. In general, use of compost and other organic amendments is important to enhancing and sustaining soil biodiversity and quality and nutritional contents in crop and livestock products. A judicious combination of organic amendments and chemical fertilizers, through the strategy of INM, along with the restoration of soil health of degraded and depleted soils, may be the best strategy to improve nutritional security and make farming a nutrition-sensitive profession.

Anthropogenic climate change and the tragic incidence of global pandemic (i.e., COVID-19) can adversely impact food security and aggravate malnutrition of the vulnerable sectors of society. Translating science into action through identification and promotion of the "One Health" concept is important. It is pertinent to enhance awareness that the "health of soil, plants, animals, people, and the environment is one and indivisible." To reduce waste, it is prudent to produce just enough with fewer inputs by enhancing use efficiency and increasing productivity per unit input of the finite resources.

REFERENCES

Abedi, T., A. Alemzadeh, and S.A. Kazemeini. 2010. Effect of Organic and Inorganic Fertilizers on Grain Yield and Protein Banding Pattern of Wheat. *Australian Journal of Crop Science* 4, no. 6: 384–389. https://www.researchgate.net/publication/259180833.

Amanuel, W., F. Yimer, and E. Karltun. 2018. Soil Organic Carbon Variation in Relation to Land Use Changes: The Case of Birr Watershed, Upper Blue Nile River Basin, Ethiopia. *Journal of Ecology and Environment* 42, no. 1: 16. doi:10.1186/s41610-018-0076-1.

Anwar, M., D.D. Patra, S. Chand, K. Alpesh, A.A. Naqvi, and S.P.S. Khanuja. 2005. Effect of Organic Manures and Inorganic Fertilizer on Growth, Herb and Oil Yield, Nutrient Accumulation, and Oil Quality of French Basil. *Communications in Soil Science and Plant Analysis* 36, no. 13–14: 1737–1746.

Ayoubi, S., A. Mehnatkesh, A. Jalalian, K.L. Sahrawat, and M. Gheysari. 2014. Relationships between Grain Protein, Zn, Cu, Fe and Mn Contents in Wheat and Soil and Topographic Attributes. *Archives of Agronomy and Soil Science* 60, no. 5 (May): 625–638. https://www.researchgate.net/publication/258565398.

Barzegar, A.R., A. Yousefi, and A. Daryashenas. 2002. The Effect of Addition of Different Amounts and Types of Organic Materials on Soil Physical Properties and Yield of Wheat. *Plant and Soil* 247, no. 2 (December): 295–301.

Bayu, W., N.F.G. Rethman, P.S. Hammes, and G. Alemu. 2006. Effects of Farmyard Manure and Inorganic Fertilizers on Sorghum Growth, Yield, and Nitrogen Use in a Semi-Arid Area of Ethiopia. *Journal of Plant Nutrition* 29, no. 2 (February): 391–407.

Branca, F. 2019. *Malnutrition Is a World Health Crisis - World Food Day 2019. WHO.* World Health Organization. https://www.globalcause.co.uk/world-food-day/malnutrition-is-a-world-health-crisis-says-who-expert/.

Brevik, E.C. 2013a. Soils and Human Health: An Overview. In *Soils and Human Health*, ed. E.C. Brevik and L. Burgess, 29–56. Boca Raton, FL: CRC Press.

Brevik, E.C. 2013b. The Potential Impact of Climate Change on Soil Properties and Processes and Corresponding Influence on Food Security. *Agriculture* 3, no. 3: 398–417.

Brevik, E.C., and L.C. Burgess. 2014. The Influence of Soils on Human Health. *Nature Education Knowledge* 5, no. 12: 1. https://www.nature.com/scitable/knowledge/library/the-influence-of-soils-on-human-health-127878980/.

Chandel, G., S. Banerjee, S. See, R. Meena, D.J. Sharma, and S.B. Verulkar. 2010. Effects of Different Nitrogen Fertilizer Levels and Native Soil Properties on Rice Grain Fe, Zn and Protein Contents. *Rice Science* 17, no. 3: 213–227. https://www.sciencedirect.com/science/article/pii/S1672630809600202.

Chen, Y. 1996. Organic Matter Reactions Involving Micronutrients in Soils and Their Effect on Plants. In *Humic Substances in Terrestrial Ecosystems*, ed. A. Piccolo, 507–529. Elsevier Science B.V.

Crutzen, P.J., and E.F. Stoermer. 2000. *The Anthropocene, Global Change.* International Geosphere–Biosphere Programme (IGBP).

Daneshmand, N.G., A. Bakhshandeh, and M.R. Rostami. 2012. Biofertilizer Affects Yield and Yield Components of Wheat. *International Journal of Agriculture* 2, no. 6: 699–704. https://www.cabdirect.org/cabdirect/abstract/20123377781.

Dhaliwal, S.S., R.K. Naresh, A. Mandal, R. Singh, and M.K. Dhaliwal. 2019. Dynamics and Transformations of Micronutrients in Agricultural Soils as Influenced by Organic Matter Build-up: A Review. *Environmental and Sustainability Indicators* 1–2: 100007. http://www.sciencedirect.com/science/article/pii/S2665972719300078.

Dignam, B.E.A., M. O'Callaghan, L.M. Condron, G.A. Kowalchuk, J.D. Van Nostrand, J. Zhou, and S.A. Wakelin. 2018. Effect of Land Use and Soil Organic Matter Quality on the Structure and Function of Microbial Communities in Pastoral Soils: Implications for Disease Suppression. *PLoS One* 13, no. 5: e0196581.

Doran, J.W., and M.R. Zeiss. 2000. Soil Health and Sustainability: Managing the Biotic Component of Soil Quality. *Applied Soil Ecology* 15, no. 1 (August): 3–11. http://www.sciencedirect.com/science/article/pii/S0929139300000676.

Fusaro, C., Y. Sarria-Guzmán, Y.A. Chávez-Romero, M. Luna-Guido, L.C. Muñoz-Arenas, L. Dendooven, A. Estrada-Torres, and Y.E. Navarro-Noya. 2019. Land Use Is the Main Driver of Soil Organic Carbon Spatial Distribution in a High Mountain Ecosystem. *PeerJ* 7 (November 14): e7897–e7897. https://pubmed.ncbi.nlm.nih.gov/31741782.

Gupta, A.P., R.P. Narwal, and R.S. Antil. 2003. Influence of Soil Organic Matter on the Productivity of Pearl Millet - Wheat Cropping System. *Archives of Agronomy and Soil Science* 49, no. 3: 325–332.

Hafez, E.M., and S.A. Badawy. 2018. Effect of Bio Fertilizers and Inorganic Fertilizers on Growth, Productivity and Quality of Bread Wheat Cultivars. *Cercetari Agronomice in Moldova* 51, no. 4: 1–16. https://content.sciendo.com/view/journals/cerce/51/4/article-p1.xml.

Hartley, W., P. Riby, and J. Waterson. 2016. Effects of Three Different Biochars on Aggregate Stability, Organic Carbon Mobility and Micronutrient Bioavailability. *Journal of Environmental Management* 181 (October 1): 770–778.

Havaee, S., S. Ayoubi, M.R. Mosaddeghi, and T. Keller. 2013. Impacts of Land Use on Soil Organic Matter and Degree of Compactness in Calcareous Soils of Central Iran. *Soil Use and Management* 30, no. 1 (March 1): 2–9. doi:10.1111/sum.12092.

Honeycutt, C.W. 1998. Crop Rotation Impacts on Potato Protein. *Plant Foods for Human Nutrition* 52, no. 4: 279–292.

Hu, H. 2002. Human Health and Heavy Metals. In *Life Support: The Environment and Human Health*, ed. M. McCally, 65. Cambridge, MA: MIT Press.

Hubert, B., M. Rosegrant, M.A.J.S. van Boekel, R. Ortiz, and R. Ortiz. 2010. The Future of Food: Scenarios for 2050. *Crop Science* 50 (March 1): 33–50.

Idowu, J., R. Ghimire, R. Flynn, and A. Ganguli. 2019. Soil Health—Importance, Assessment, and Management. In *Circular 694B*, 16. NMSU College of Agricultural, Consumer and Environmental Sciences (ACES). https://aces.nmsu.edu/pubs/_circulars/CR694B/welcome.html.

Kazemeini, S.A., H. Hamzehzarghani, and M. Edalat. 2010. The Impact of Nitrogen and Organic Matter on Winter Canola Seed Yield and Yield Components. *Australian Journal of Crop Science* 4, no. 5: 335–342. https://www.researchgate.net/publication/268298099.

Khan, I., Z. Shah, W. Ahmad, F. Khan, and M. Sharif. 2018. Integrated Nutrient and Tillage Management Improve Organic Matter, Micronutrient Content and Physical Properties of Alkaline Calcareous Soil Cultivated with Wheat. *Sarhad Journal of Agriculture* 34, no. 1: 144–157. doi:10.17582/journal.sja/2018/34.1.144.157.

Kumar, M., and A.L. Babel. 2011. Available Micronutrient Status and Their Relationship with Soil Properties of Jhunjhunu Tehsil, District Jhunjhunu, Rajasthan, India. *Journal of Agricultural Science* 3, no. 2: 97–106.

Lachman, J., K. Hamouz, P. Dvořák, and M. Orsák. 2005. The Effect of Selected Factors on the Content of Protein and Nitrates in Potato Tubers. *Plant, Soil and Environment* 51, no. 10: 431–438. http://citeseerx.ist.psu.edu/viewdoc/download?doi=10.1.1.631.9466&rep=rep1&type=pdf.

Lal, R. 2004. Carbon Emission from Farm Operations. *Environment International* 30, no. 7 (September): 981–990.

Lal, R. 2009. Soil Degradation as a Reason for Inadequate Human Nutrition. *Food Security* 1, no. 1 (February): 45–57.

Lal, R. 2016. Soil Health and Carbon Management. *Food and Energy Security* 5, no. 4 (November 1): 212–222. http://doi.wiley.com/10.1002/fes3.96.

Lal, R. 2018. Digging Deeper: A Holistic Perspective of Factors Affecting Soil Organic Carbon Sequestration in Agroecosystems. *Global Change Biology* 24, no. 8 (August 1): 3285–3301. http://doi.wiley.com/10.1111/gcb.14054.

Lal, R. 2020a. Integrating Livestock with Crops and Trees. *Frontiers in Food Systems*: In press.

Lal, R. 2020b. Food Security Impacts of the "4 per Thousand" Initiative. *Geoderma* 374: 114427. http://www.sciencedirect.com/science/article/pii/S0016706119321585.

Lal, R. 2020c. Soil Organic Matter Content and Crop Yield. *Journal of Soil and Water Conservation* 75, no. 2: 27A–32A.

Lal, R. 2020d. Home Gardening and Urban Agriculture for Advancing Food and Nutritional Security in Response to the COVID-19 Pandemic. *Food Security*: Submitted.

Laxminarayana, K., K.S. John, A. Mukherjee, and C.S. Ravindran. 2015. Long-Term Effect of Lime, Mycorrhiza, and Inorganic and Organic Sources on Soil Fertility, Yield, and Proximate Composition of Sweet Potato in Alfisols of Eastern India. *Communications in Soil Science and Plant Analysis* 46, no. 5 (March 9): 605–618. https://www.tandfonline.com/action/journalInformation?journalCode=lcss20.

Lehesranta, S.J., K.M. Koistinen, N. Massat, H. V. Davies, L.V.T. Shepherd, J.W. McNicol, I. Cakmak, et al. 2007. Effects of Agricultural Production Systems and Their Components on Protein Profiles of Potato Tubers. *Proteomics* 7, no. 4 (February): 597–604.

Leimona, B., S. Amaruzaman, B. Arifin, F. Yasmin, F. Hasan, H. Agusta, P. Sprang, S. Jaffee, and J. Frias. 2015. Indonesia's "Green Agriculture" Strategies and Policies: Closing the Gap between Aspirations and Application. *ICRAF Occasional Paper*. http://www.worldagroforestry.org/sea/Publications/files/occasionalpaper/OP0003-15.pdf.

Li, B.Y., D.M. Zhou, L. Cang, H.L. Zhang, X.H. Fan, and S.W. Qin. 2007. Soil Micronutrient Availability to Crops as Affected by Long-Term Inorganic and Organic Fertilizer Applications. *Soil and Tillage Research* 96, no. 1–2: 166–173.

Mackowiak, C.L., P.R. Grossl, and B.G. Bugbee. 2001. Beneficial Effects of Humic Acid on Micronutrient Availability to Wheat. *Soil Science Society of America Journal* 65, no. 6: 1744–1750.

Mbatha, A.N., G.M. Ceronio, and G.M. Coetzer. 2014. Response of Carrot (Daucus Carota L.) Yield and Quality to Organic Fertiliser. *South African Journal of Plant and Soil* 31, no. 1 (January 2): 1–6. https://www.tandfonline.com/action/journalInformation?journalCode=tjps20.

Mishra, B.N., R. Prasad, B. Gangaiah, and B.G. Shivakumar. 2006. Organic Manures for Increased Productivity and Sustained Supply of Micronutrients Zn and Cu in a Rice-Wheat Cropping System. *Journal of Sustainable Agriculture* 28, no. 1 (May 1): 55–66. https://www.tandfonline.com/action/journalInformation?journalCode=wjsa21.

Murphy, K., L. Hoagland, P. Reeves, and S. Jones. 2008. Effect of Cultivar and Soil Characteristics on Nutritional Value in Organic and Conventional Wheat. In *16th IFOAM Organic World Congress*, 4. Modena, Italy, June 16–20, 2008. http://orgprints.org/view/projects/conference.html.

Najm, A.A., M.R.H.S. Hadi, F. Fazeli, M.T. Darzi, and A. Rahi. 2012. Effect of Integrated Management of Nitrogen Fertilizer and Cattle Manure on the Leaf Chlorophyll, Yield, and Tuber Glycoalkaloids of Agria Potato. *Communications in Soil Science and Plant Analysis* 43, no. 6 (March): 912–923.

Olaleye, A.O., G.E. Akinbola, and B.O. Akintade. 2011. Gravel, Soil Organic Matter, and Texture in Fallowed Alfisols, Entisols and Ultisols: Implications for Root and Tubercrops. *Communications in Soil Science and Plant Analysis* 42, no. 21: 2624–2641. https://www.tandfonline.com/action/journalInformation?journalCode=lcss20.

Olfati, J.A., G. Peyvast, and Z. Nosrati-Rad. 2008. Organic Mulching on Carrot Yield and Quality. *International Journal of Vegetable Science* 14, no. 4 (September 30): 362–368. http://www.haworthpress.com.

Osman, E.E.A.M., M.A. Hassan, and D.A. Yassin. 2014. Mutual Effect of Organic and Inorganic Fertilizers on Productivity and Economic Evaluation of Bread Wheat. *Journal of Soil Sciences and Agricultural Engineering* 5, no. 4: 543–555. https://jssae.journals.ekb.eg/article_49302.html.

Pepper, I.L. 2013. The Soil Health-Human Health Nexus. *Critical Reviews in Environmental Science and Technology* 43, no. 24: 2617–2652.

Pobereżny, J., E. Wszelaczyńska, D. Wichrowska, and D. Jaskulski. 2015. Content of Nitrates in Potato Tubers Depending on the Organic Matter, Soil Fertilizer, Cultivation Simplifications Applied and Storage. *Chilean Journal of Agricultural Research* 75, no. 1: 42–49. https://scielo.conicyt.cl/scielo.php?pid=S0718-58392015000100006&script=sci_arttext&tlng=en.

Rengel, Z. 2015. Availability of Mn, Zn and Fe in the Rhizosphere. *Journal of Soil Science and Plant Nutrition* 15, no. 2 (December 1): 397–409.

Rengel, Z., G.D. Batten, and D.E. Crowley. 1999. Agronomic Approaches for Improving the Micronutrient Density in Edible Portions of Field Crops. *Field Crops Research* 60, no. 1–2 (January 1): 27–40.

Roser, M., and H. Ritchie. 2020. Hunger and Undernourishment. *Ourworldindata.Org*. https://ourworldindata. org/hunger-and-undernourishment?utm_campaign=The Preface&utm_medium=email&utm_source= Revue newsletter.

Sanderman, J., T. Hengl, and G.J. Fiske. 2017. Soil Carbon Debt of 12,000 Years of Human Land Use. *Proceedings of the National Academy of Sciences* 114, no. 36 (September 5): 9575–9580. http://www. pnas.org/content/114/36/9575.abstract.

Sarwar, G., N. Hussain, H. Schmeisky, and S. Muhammad. 2007. Use of Compost an Environment Friendly Technology for Enhancing Rice-Wheat Production in Pakistan. *Pakistan Journal of Botany* 39, no. 5: 1553–1558. https://www.academia.edu/download/40769374/Use_of_compost-_An_environment_friendly.pdf.

Sary, G., and H. El-Naggar. 2009. Effect of Bioorganic Fertilization and Some Weed Control Treatments on Yield and Yield Components of Wheat. *World Journal of Science* 5, no. 1: 55–62. http://citeseerx.ist.psu. edu/viewdoc/download?doi=10.1.1.415.731&rep=rep1&type=pdf.

Shiwakoti, S., V.D. Zheljazkov, H.T. Gollany, M. Kleber, B. Xing, and T. Astatkie. 2019. Micronutrients in the Soil and Wheat: Impact of 84 Years of Organic or Synthetic Fertilization and Crop Residue Management. *Agronomy* 9, no. 8 (August 19): 464.

Singh, S.K., A.K. Singh, B.K. Sharma, and J.C. Tarafdar. 2007. Carbon Stock and Organic Carbon Dynamics in Soils of Rajasthan, India. *Journal of Arid Environments* 68, no. 3 (February 1): 408–421.

Singh, Y., R.K. Gupta, Jagmohan-Singh, Gurpreet-Singh, Gobinder-Singh, and J.K. Ladha. 2010. Placement Effects on Rice Residue Decomposition and Nutrient Dynamics on Two Soil Types during Wheat Cropping in Rice–Wheat System in Northwestern India. *Nutrient Cycling in Agroecosystems* 88, no. 3: 471–480. doi:10.1007/s10705-010-9370-8.

Sowers, K.E., B.C. Miller, and W.L. Pan. 1994. Optimizing Yield and Grain Protein in Soft White Winter Wheat with Split Nitrogen Applications. *Agronomy Journal* 86, no. 6: 1020–1025.

Steffan, J.J., E.C. Brevik, L.C. Burgess, and A. Cerdà. 2018. The Effect of Soil on Human Health: An Overview. *European Journal of Soil Science* 69, no. 1 (January 1): 159–171.

Stephenson, L., and M. Latham. 2000. Global Malnutrition. *Parasitology* 121, no. S1: S5–S22. https://www. cambridge.org/core/journals/parasitology/article/global-malnutrition/306826FC3986D922C69D7B2D 9583DFCF.

Thapa, V.R., R. Ghimire, M.M. Mikha, O.J. Idowu, and M.A. Marsalis. 2018. Land Use Effects on Soil Health in Semiarid Drylands. *Agricultural & Environmental Letters* 3, no. 1 (January 1): 180022. doi:10.2134/ ael2018.05.0022.

UN. 2019. *World Population Prospects 2019: Highlights (ST/ESA/SER.A/423)*. Rome, Italy: United Nations, Department of Economic and Social Affairs, Population Division. https://population.un.org/wpp/ Publications/Files/WPP2019_Highlights.pdf.

Unger, P.W., and M.F. Vigil. 1998. Cover Crop Effects on Soil Water Relationships. *Journal of Soil and Water Conservation* 53, no. 3: 200–207. http://www.jswconline.org/content/53/3/200.short.

UNICEF. 2018. *2018 Global Nutrition Report*. https://globalnutritionreport.org/reports/global-nutrition-report–2018/.

Wang, L., J.P. Wu, Y.X. Liu, H.Q. Huang, and Q.F. Fang. 2009. Spatial Variability of Micronutrients in Rice Grain and Paddy Soil. *Pedosphere* 19, no. 6 (December 1): 748–755.

Yeasmin, S., B. Singh, R.J. Smernik, and C.T. Johnston. 2020. Effect of Land Use on Organic Matter Composition in Density Fractions of Contrasting Soils: A Comparative Study Using 13C NMR and DRIFT Spectroscopy. *Science of The Total Environment* 726: 138395. http://www.sciencedirect.com/ science/article/pii/S0048969720319082.

Yilmaz, A., H. Ekiz, B. Torun, I. Gültekin, S. Karanlik, S.A. Bagci, and I. Cakmak. 1997. Effect of Different Zinc Application Methods on Grain Yield and Zinc Concentration in Wheat Cultivars Grown on Zinc-Deficient Calcareous Soils. *Journal of Plant Nutrition* 20, no. 4–5: 461–471.

Index

Note: **Bold** page numbers refer to tables and *italic* page numbers refer to figures.

Printed and bound by CPI Group (UK) Ltd, Croydon, CR0 4YY

23/10/2024

01778247-0004